Hazards XIV
Cost Effective Safety

Institution of Chemical Engineers, Rugby, UK

Hazards XIV
Cost Effective Safety

Orders for this publication should be directed as follows:

Institution of Chemical Engineers,
Davis Building,
165–189 Railway Terrace, RUGBY,
Warwickshire CV21 3HQ, UK

Tel: +44 1788–578214
Fax: +44 1788–560833

**Copyright © 1998
Institution of Chemical Engineers
A Registered Charity**

All rights reserved. No part of this publication may be reproduced, stored in a retrieval system or transmitted in any forms or by any means: electronic, electrostatic, magnetic tape, mechanical, photocopying or otherwise, without permission in writing from the copyright owner. Opinions expressed in the papers in this volume are those of the individual authors and not necessarily those of the Institution of Chemical Engineers or of the Organising Committee.

Hazards XIV
Cost Effective Safety

A three-day symposium organised by the Institution of Chemical Engineers
(North Western Branch) and held at UMIST, Manchester, UK, 10–12 November 1998.

Organising Committee

N. Gibson (Chairman)	Burgoyne Consultants Ltd
J.A.S. Ashurst	Ashurst Consultancy
S. Beattie	Zeneca plc
D.C. Bull	Shell Research
A.L. Clarke	Huntsman Chemical Company Ltd
P. Cleaver	BG plc
K. Dixon-Jackson	Ciba Speciality Chemicals
P. Doran	GDG Associates
R.F. Evans	HSE
I. Kempsell	British Nuclear Fuels plc
T.A. Kletz	Consultant
G.A. Lunn	Health and Safety Laboratory
R.S. Mason	Consultant
G.S. Melville	Burgoyne Consultants Ltd
N. Morton	HSE
M.F. Pantony	Consultant
M.L. Preston	ICI Engineering Technology
R.C. Santon	HSE
A.I. Thompson	Consultant

INSTITUTION OF CHEMICAL ENGINEERS

SYMPOSIUM SERIES No. 144
ISBN 0 85295 416 6

Printed by Hobbs the Printers Ltd, Brunel Road, Totton, Hampshire, SO40 3WX, UK

Preface

Cost effective safety means prioritising and targeting effort to where it does most good. With chemical and other potentially hazardous installations, major accident experience has demonstrated the value of the preferential hierarchy of risk reduction measures — first articulated in the aftermath of Flixborough — of:

- avoidance — inherently safer processes;
- control — rigorous assessment of risk and implementation of appropriate control measures;
- mitigation — prudent action to deal with residual risk, such as emergency and land use planning.

Papers for this symposium reflect this balance of priorities and the meeting should fulfil the important role of continuing to widen and deepen our understanding of the techniques and tools that can be used to improve the management of risk.

However, twenty-five years after Flixborough we need to ask ourselves — managers, scientists, engineers and regulators — if improving *our* understanding is good enough.

Public opinion suggests otherwise. Despite the very many improvements that have been made, the public appear to be increasingly sceptical about the reassurance they receive about hazardous installations from those working in the industry and from regulators. Much of this scepticism stems from our reluctance to engage with the public, especially local communities, proactively and openly and in a way that involves them in the risk assessment and management process.

Increased emphasis on the rights of the citizen, the developing role of regional and local authorities and proposed legislation such as the Freedom of Information Act and the COMAH Regulations, means that this must change. The challenge for us as scientists and engineers is to facilitate this change by explaining our methodologies in language which the layman can understand and, in particular, by explaining the uncertainties involved in our approach and how we deal with them.

I hope we can start this process at this symposium.

Paul Davies
Head of Chemical and Hazardous Installtions, HSE

Contents

COMAH — An HSE view and application

Paper 1 Page 1		The control of major accident hazards regulations 1999 (COMAH) G. MacDonald (*HSE, UK*) and L. Varney (*DETR, UK*)
Paper 2 Page 11		The regulators approach to assessing COMAH safety reports T.J. Britton (*HSE, UK*)
Paper 3 Page 27		The assessment of technical aspects of COMAH safety reports R.F. Evans (*HSE, UK*)
Paper 4 Page 41		Assessment of COMAH safety reports: Emergency response criteria K.K. McDonald (*HSE, UK*)
Paper 5 Page 53		Safety management system assessment criteria D. Snowball (*HSE, UK*)
Paper 6 Page 65		Assessment of the predictive aspects of COMAH safety reports S. Welsh (*HSE, UK*)
Paper 7 Page 95		Industry experience from the pilot exercise I. Hamilton (*BP, UK*)
Paper 8 Page 101		Meeting the demands of the regulator and litigator: An international approach E. Blackmore (*Eli Lilly and Company Ltd, UK*)

Paper 9   Safety implications of self-managed teams
Page 109   R. Lardner (*The Kiel Centre, UK*)

Paper 10   The impact of the new Seveso II (COMAH)
Page 123   regulations on industry
             J.C. Ansell, J. Mullins, and R. Voke
             (*AEA Technology Consulting, UK*)

Management of safety

Paper 11   The accident database: Capturing corporate
Page 133   memory
             M. Powell-Price (*IChemE, UK*), J. Bond and
             B. Mellin (*Consultants, UK*)

Paper 12   Operational safety reviews
Page 143   J.K. Broadbent (*ICI Eutech, UK*) and
             T. O'Donoghue (*DuPont Melinex, UK*)

Paper 13   CARMAN: A systematic approach to risk
Page 153   reduction by improving compliance to
             procedures
             D. Embrey (*Human Reliability
             Associates Ltd, UK*)

Paper 14   A methodology for assessing and minimising
Page 167   the risks associated with firewater run-off on
             older manufacturing plants.
             C.J. Beale (*Ciba Speciality Chemicals Plc, UK*)

Paper 15   More effective permit-to-work systems
Page 181   R.E. Iliffe, P.W.H. Chung and T.A. Kletz
             (*Loughborough University, UK*)

Paper 16   A qualitative approach to criticality in the
Page 195   allocation of maintenance priorities to
             manufacturing plant.
             C.J. Atkinson (*Burgoyne Consultants Ltd, UK*)

Paper 17 Page 209	European state-of-the-art research: Integrating technical and management/organisational factors in major hazard risk assessment M. Anderson (*Health and Safety Laboratory, UK*)
Paper 18 Page 221	Design for safety applying IEC 6-1508 "from the manufacturer's point of view". S. De Vries, M. Van den Schoor and R. Bours (*Fike Europe, Belgium*)
Paper 19 Page 235	THESIS: The health environment and safety information system — keeping the management system "live" and reaching the workforce A. Lidstone (*EQE International Ltd, UK*)
Paper 20 Page 249	Safety issues and the year 2000 R. Storey (*Allianz Cornhill International, UK*)
Paper 21 Page 257	Information technology and training in safety D. Fernando (*IChemE, UK*)

Fire and explosions

Paper 22 Page 265	The sensitivity of risk assessment of flash fire events to modelling assumptions P.J. Rew, H. Spencer (*WS Atkins Safety and Reliability, UK*) and T. Maddison (*HSE, UK*)
Paper 23 Page 279	Turbulent Reynolds number and turbulent-flame-quenching influences on explosion severity with implications for explosion scaling C.L. Gardner, H. Phylaktou and G.E. Andrews (*University of Leeds, UK*)

Paper 24 Page 293	Evaluation of CFD modelling of gas dispersion near buildings and complex terrain R.C. Hall (*WS Atkins Safety and Reliability, UK*)
Paper 25 Page 305	Explosion venting: The predicted effects of inertia S. Cooper (*Stuvex Safety Systems Ltd, UK*)
Paper 26 Page 321	VOC abatement and vent collection systems P.J. Hunt (*Eutech, UK*)
Paper 27 Page 345	The dangers of grating floors: Dispersion and explosion A.E. Holdo (*University of Hertfordshire, UK*), G. Munday (*ISAA Consultants, UK*) and D.B. Spalding (*Heat and Momentum Ltd, UK*)
Paper 28 Page 359	Suppression of high violence dust explosions using non-pressurised systems S. Cooper and P. Cooke (*Stuvex Safety Systems Ltd, UK*)

Chemical and reaction hazards

Paper 29 Page 373	Managing hazards and risks in fine chemical and peroxygen operations P.G. Lambert (*Laporte Organics Ltd, UK*)
Paper 30 Page 387	Understanding vinyl acetate polymerization accidents J-L. Gustin and F. Laganier (*Rhône-Poulenc Industrialisation, France*)
Paper 31 Page 405	A methodological approach to process intensification M. Wood and A. Green (*BHR Group Ltd, UK*)

Paper 32　Criteria for autoignition of combustible fluids
Page 417　in insulation materials
　　　　　J. Brindley, J.F. Griffiths, N. Hafiz,
　　　　　A.C. McIntosh and J. Zhang
　　　　　(*University of Leeds, UK*)

Paper 33　Assessment of the thermal and toxic effects of
Page 433　chemical and pesticide pool fires based on
　　　　　experimental data obtained using the Tewarson
　　　　　apparatus
　　　　　C. Costa, G. Treand and J-L. Gustin
　　　　　(*Rhône-Poulenc Industrialisation, France*)

Risk assessment and simulation techniques

Paper 34　Top level risk study — A cost effective
Page 447　quantified risk assessment
　　　　　R.I. Facer, J.A.S. Ashurst and K.A. Lee
　　　　　(*EQE International Ltd, UK*)

Paper 35　Application of case-based reasoning to safety
Page 461　evaluation of process configuration
　　　　　A-M. Heikkilä, T. K. Koiranen and M. Hurme
　　　　　(*Helsinki University of Technology, Finland*)

Paper 36　Index method for cost-effective assessment of
Page 475　risk to the environment from accidental releases
　　　　　A.J. Wilday (*HSL, UK*), M.W. Ali (*Universiti Teknologi Malaysia*) and Y. Wu
　　　　　(*University of Sheffield, UK*)

Paper 37　Incorporation of building wake effects into
Page 489　quantified risk assessments
　　　　　I.G. Lines, D.M. Deaves and R.C. Hall
　　　　　(*WS Atkins, Safety and Reliability, UK*)

Paper 38   The use of risk-based assessment techniques to
Page 503   optimise inspection regimes
             G.R. Bennett, M.L. Middleton and P. Topalis
             (*Det Norske Veritas, UK*)

Paper 39   Categorization of risk: Good, bad or
Page 515   indifferent, site risk grading
             P. Clarke (*Allianz Cornhill International, UK*)

Paper 40   Thirty years of quantifying hazards
Page 527   J.T. Illidge, M.L. Preston and A.G. King
             (*ICI Technology, UK*)

Page 535   Index

THE CONTROL OF MAJOR ACCIDENT HAZARDS REGULATIONS 1999 (COMAH)

G MacDonald, L Varney

> The COMAH Regulations are due to come into force on 3 February 1999. They implement the Seveso II Directive and will replace the Control of Industrial Major Accident Hazards Regulations 1984. There are similarities between the two regimes: they apply where certain quantities of dangerous substances are present; there are two levels of duties depending on the quantities held; and there are requirements for safety reports, emergency plans and information to the public. But there are important differences too, such as: broadening the scope of the regime to cover sectors which were previously exempt; greater emphasis on management systems; greater clarity in the purpose and content of safety reports; more explicit requirements for environmental protection; a new duty to test emergency plans; greater public access to information; new requirements on land use planning.
> The paper describes the background to the Seveso regime, and sets out the key features of the new regime.

INTRODUCTION

The Seveso II Directive (96/82/EC published in the Official Jopurnal of the European Communities on 9 December 1996) must be implemented in UK law by 3 February 1999. HSE, the Department of the Environment, Transport and the Regions and the Scottish Office have undertaken a consultation exercise based on a consultative document with proposals for the Control of Major Accident Hazards Regulations (COMAH), which will implement most aspects of the Directive. The consultation period ran from May to September. The speaker will present the results of that consultation exercise at the symposium.

This paper provides:

- background information on the approach to major hazards legislation in the UK and Europe and traces the development of the Seveso Directive to the present day (this information may be well known to some readers and is not essential to understanding of the rest of the paper);

- sets out the main requirements of the Seveso II Directive;

- sets out the key features of the proposals for the COMAH Regulations.

BACKGROUND

The current framework for the control of industrial major hazards in the UK and elsewhere in Europe can be traced back to the work of the Advisory Committee on Major Hazards (ACMH) in the 1970s and early 1980s.

The Committee was set up by the Health and Safety Commission following the disastrous explosion at Flixborough in 1974. ACMH produced three reports, in 1976, 1979 and 1984, the first of which proposed a three part strategy for the control of major hazards consisting of:

- identification;
- prevention/control;
- mitigation.

Identification is carried out by reference to threshold quantities of hazardous substances above which there is thought to be the potential for major accidents giving rise to serious consequences. For practical reasons thresholds do not encompass all installations with such potential, only those with the greatest hazard are included.

Prevention/control requires operators to assess the risks and consequences of major accidents and then apply appropriate precautions to reduce or maintain the risks at tolerable levels.

Mitigation, the third part of the strategy, recognises that accidents cannot be entirely prevented and requires steps to be taken to reduce the effects of such accidents to people and the environment. These steps include land use planning controls to separate vulnerable populations from hazardous installations, emergency planning both on and off-site and information to the public.

Development of UK Legislation

Development of legislation in the UK to implement the ACMH strategy preceded moves within Europe by a number of years. The process started with the introduction of the Notification of Installations Handling Hazardous Substances Regulations (NIHHS) 1982. These required notification to HSE of any site at which hazardous substances listed in the regulations were present above certain thresholds.

Plans for further UK legislation were then overtaken by events in Europe where a series of major accidents persuaded the European Commission that concerted Community wide action in this field was necessary.

The "Seveso" Directive

The efforts of the European Commission and member states resulted in the Seveso Directive. The Directive depended on threshold limits of listed substances to define application and introduced the concept of two level or tiers of control. The Directive differed significantly from the original UK proposals in that it applied to both human health and safety and the environment.

Preparation and subsequent development of the first Directive was driven by a series of accidents in Europe and India. The directive was developed in the aftermath of an incident in Seveso, northern Italy in 1976 where the accidental production and release of a dioxin as an unwanted by-product from a runaway chemical reaction led to widespread environmental contamination. In 1987 the directive was amended to reflect the lessons learnt from the Bhopal disaster. The 1988 amendment broadened the scope to storage premises following the fire in the Sandoz warehouse in Switzerland which led to far reaching pollution of the Rhine.

In the UK the Directive was implemented through the Control of Industrial Major Accident Hazards (CIMAH) Regulations 1984 and its subsequent amendments. CIMAH has an underpinning duty on all sites within scope to identify major accident hazards and take steps to ensure that they are prevented. However there are more specific requirements on top tier sites to produce safety reports and emergency plans and to provide information to members of the public who may be affected by a major accident.

The Review of the Seveso Directive

Though Seveso was a good first attempt at Community wide legislation to control major hazards, it suffered from a number of weaknesses. The directive was complex and difficult to implement, a position not improved by the later amendments. The application annexes were long and inflexible with over 170 named substances. Also, the directive lacked any reference to a major part of the mitigatory package, land use planning. Furthermore, it was felt that the Directive was unevenly implemented in member states.

For a number of years the UK and other member states pressed for a fundamental review of Seveso. In 1989 the Council of Ministers recommended that land use planning should be made part of the Directive precipitating the review which was started in 1990.

The review of Seveso, carried out by the European Commission in conjunction with the Committee of Competent Authorities, made up of representatives of all the governmental bodies enforcing the Seveso Directive in their respective countries, was sufficiently far reaching to require the drafting of a new Directive. The draft, entitled "Proposals for a Council Directive on the Control of Major Accident Hazards involving Dangerous Substances" (COMAH or Seveso II) was published in April 1994. Following intensive negotiations which began in February 1995, the Directive reached common position in June of that year. It was finally adopted on 9 December 1996.

The Seveso II Directive: Requirements

The new Directive, whilst similar in nature to Seveso and following the same two tier format for duties, differs in a number of important ways. It reflects more clearly the relatively recent emphasis on safety management systems, seen as the key to high and sustainable levels of safety. It also contains a mechanism to allow the Directive to be kept up to date with technical progress and includes greater detail to ensure a more uniform implementation by member states.

The chief features of the new Directive are:

(a) Application will depend solely on the presence on site of the threshold quantity of dangerous substance. Previous distinctions between process and storage activities will be removed.

(b) The scope will be extended to include chemical hazards at nuclear installations and explosives. This will leave exemptions for the extractive industries, transport related establishments, pipelines outside of establishments and military installations.

(c) There will be a greater use of generic categories of substances eg "highly flammable" or "toxic" to define application enabling the number of named substances to be reduced to 37. The generic categories are separately defined by the directives dealing with the classification and labelling of dangerous goods for supply. The use of this mechanism should ensure that new substances are covered as soon as they receive a supply classification.

(d) A new "ecotoxic" general category will be introduced to cover substances which present a hazard to the environment without necessarily being dangerous to people.

(e) The increased emphasis on safety management systems is reflected at both top and lower tier. Operators of lower tier establishments are required to have a "major accident prevention policy" with the organisation and arrangements to put the policy into effect. This policy forms part of the safety report that operators must submit for all top tier installations.

(f) In addition to the new management requirements, the contents of safety reports will be set out more precisely, for example making it clear that hazard and risk assessments should cover the whole range of potential accident scenarios.

(g) Land use planning requirements will be introduced but with a need to take into account risks to the environment.

(h) There will be a continuing requirement to prepare on and off-site emergency plans with additional duties to test those plans and put them into effect and to include measures to be taken for remediation and clean up of the environment.

(i) Openness has been extended by making safety reports available to the public.

(j) Criteria for the reporting of major accidents are included to improve the consistency of reporting from member states to the European Commission.

(k) The duties of competent authorities will be extended and there is a new obligation on them to communicate the conclusions of the examination of the safety report or prohibit start up or continued operation where there is evidence that the measure taken for the prevention and mitigation of major accidents are seriously deficient. There will also be a requirement for member states to set up a system for the inspection of installations covered by the Directive.

(l) The European Commission will be empowered to set up a Committee of Member States which will be able to agree amendments to certain annexes to the Directive.

(n) There is a new requirement on competent authorities to designate establishments which might give rise to domino effects.

KEY FEATURES OF THE COMAH REGULATIONS

All Measures Necessary

All operators have to ensure that they have taken "all measures necessary" for the prevention and mitigation of major accidents. We have set out an interpretation of this duty which recognises that, by requiring measures for prevention and mitigation, the duty does not set an objective of zero risk. Therefore a judgement has to be made as to whether the measure is necessary in relation to the major hazard and the associated risks that the measure addresses.

These factors allow us to retain the concept of proportionality in enforcement of the regime and we believe the current approach based on reducing risk to as low a level as reasonably practicable will be sufficient to meet the "all measures necessary" test.

Major Accident Prevention Policy

Operators must set out and implement a policy for major accident prevention - the MAPP. Top-tier operators have to include the MAPP in their safety reports. A stand alone document is required at lower-tier establishments. The Regulations list the elements of a safety management system which must be considered in the MAPP, and all these should be familiar to UK industry:

- roles and responsibilities of personnel
- procedures for hazard identification;
- operational procedures;
- modification procedures;

- procedures for handling emergencies;
- monitoring, auditing and reviewing procedures.

We expect the MAPP usually to be a short and simple document. It should set down what is to be achieved, with an indication of how this is to be done, but not in any great detail. The detail will be contained in other documentation on site eg plant operating procedures, training records, job descriptions, audit reports, to which the MAPP can refer.

It is important to emphasise that an essential element of a safety management system is a procedure for handling emergencies. For lower tier establishments there is no list of requirements which emergency plans must address and there is no set period at which plans have to be tested, as there is for top tier sites, but handling emergencies effectively and efficiently are every bit as important.

Safety Reports

As in Seveso I and CIMAH the safety report must provide all the data and information which is listed in the Regulations. However the new Regulations list, for the first time, the purposes of safety reports. Essentially the report must <u>demonstrate</u> that necessary measures have been taken for major accident prevention and mitigation in terms of both organisational and technical factors. We are therefore moving from CIMAH where the safety report is mainly descriptive, to COMAH where it is, additionally, a justification for an operator's approach.

The regime for submission, review and revision of the safety report by the operator differs in COMAH from CIMAH. So too is the action required from the competent authority. For new establishments, the proposal is that operators submit safety reports in two parts. The first part, submitted prior to construction, should describe substances, accident scenarios, processes and activities on site and the surrounding environment, together with such information as the operator has on the design of plant which might affect safety. The operator cannot start construction without having received the assessment conclusions from the competent authority. The second part of the safety report is submitted before operation and it adds further relevant design detail together with the operational procedures etc which it would have been unreasonable to expect the operator to have before construction. Again the operator cannot start operation before receiving the competent authority's conclusions of the assessment.

Existing establishments will submit their safety reports and continue operating whilst the regulator undertakes its assessment. For both new and existing establishments the competent authority is under a duty to prohibit operations if they have evidence that there is a serious deficiency in the measures taken for major accident prevention or mitigation.

Safety reports should be reviewed and, where necessary, revised:

- at least every 5 years;

- at any time where justified by new facts or to take account of new technical knowledge;
- when modifications are proposed which could have significant repercussions on major accidents.

Emergency Planning

There are similarities between the emergency planning regimes in CIMAH and COMAH: the operator must produce an on-site plan and the local authority must prepare on off-site plan. Both must meet objectives and contain information which are set out in the Regulations. However, there are also significant differences: providing for restoration and clean-up of the environment after a major accident has been added to the objectives of emergency plans; on-site and off-site emergency plans have to be tested at least once every 3 years.

Environmental remediation and clean-up raises some difficult questions. For example, what sort of baseline needs to be established for the quality of the environment against which the adequacy of remediation and clean-up will be judged? And who pays for it? These issues are being considered by DETR, who have the policy lead in these areas, at the moment.

The requirement to test emergency plans every 3 years is an important new duty. Testing is essential to give confidence in the accuracy, completeness and practicability of the plan. The Directive does not define what constitutes an adequate test so we have set out in the consultative document an interpretation. Central to this is the idea that testing is not a single activity. For instance, it does not necessarily mean a full scale live exercise at each installation at an establishment or, for the off site plan, at each establishment every 3 years. Rather testing may consist of a range of activities such as:

- table top exercises, where all the appropriate resources are brought together in one place to work through their roles in the event of an emergency;
- "control post" communication exercises which examine the adequacy of the communication arrangements between the key players in an emergency;
- use of virtual reality systems to support table top or other exercises;
- live exercises which deploy on the ground all the appropriate resources in a simulation of their actual response to an accident.

There may be scope for employing economies of scale in testing regimes. For instance, for an off site plan, it may be possible for one live or table top exercise to test the off site components of two or more sites. This will depend upon similarities of the location and of the risks posed to the adjacent population and environment. Similarly, for on site plans at multi-installation sites, it may be possible to use lessons learned from live exercises on some installations, supported by table top exercises for other installations. This will depend on similarities in the hazards and risks posed, and in the type of emergency response.

Whatever the precise details of the arrangements it will be essential that lessons learned from tests are fed back to all relevant personnel and organisations.

Emergency plans must be reviewed as well as tested. Review is a fundamental process which must take into account:

- all material changes in the activity;

- any changes in the emergency services relevant to the operation of the plant;

- advances in technical knowledge, for example new, more effective means of mitigation;

- knowledge gained as a result of major accidents either on site or elsewhere;

- lessons learned during the testing of emergency plans.

Emergency planning is an essential element of the COMAH regime. There is some guidance in the consultative document on this but we plan to publish a new guidance document on emergency planning next year which will give much more detail.

Public Access to Information

COMAH contains a new requirement for safety reports to be made available to the public. Certain information can be withheld on the grounds that it is commercially confidential, it is personal, or it might compromise national security or public safety and we are developing guidance on what these exempt categories might mean in practice.

We propose that safety reports should be made available for the public at operator's premises and at the offices of the Environment Agency and SEPA. In addition, we will recommend that operators try and reach agreement with local authorities on putting safety reports in libraries and town halls.

Finally there is the question of how safety reports should be presented where some information is to be withheld. We have left operator's the option of deleting parts of the report, or undertaking a more substantial re-edit of the document.

Environmental Requirements

All of the requirements of the new regime apply to protection of the environment, but COMAH has more explicit requirements for environmental protection than CIMAH Some of these have been mentioned. One of the generic categories of dangerous substances which will attract the regime is substances dangerous substances for the environment. Emergency plans must address remediation and clean up of the environment after a major accident. In addition the content of safety reports regarding the environment has been made clearer.

DETR have commissioned important research in these areas and an update on the outcomes of this work will be given at the symposium.

Land Use Planning

The Seveso II Directive contains, for the first time a requirement for member states to have land use planning or other policies which take account of the location of new sites, and modifications to, and developments around, existing sites. The UK has had a land use planning system in place for some time but this will need to be amended to apply such controls to all sites subject to COMAH, and to ensure that environmental protection is considered in the siting of new establishments. The land use planning aspects of the Directive will be implemented through changes to planning law.

ENFORCEMENT OF THE REGIME

HSE are the sole competent authority under Seveso I/CIMAH. This will change. Under COMAH HSE and the Environment Agency and Scottish Environment Protection Agency will be the competent authority. HSE and the agencies are developing detailed working arrangements for this at the moment. Central features of the arrangements are:

- submission of documents such as safety reports to a single office;

- HSE and the agencies working together as a team on the assessment of safety reports providing the operator with a single set of conclusions;

- HSE and the agencies sharing inspection programmes to identify the need for joint or co-ordinated inspections.

CONCLUSIONS

COMAH is a development of the existing regime but it introduces significant changes. This paper identifies some of the key ones. HSE, DETR, and the Scottish Office have published a consultative document which presents proposals for Regulations and interpretative guidance. The consultation period ends on 4 September. After that, the views of consultees will be analysed, changes to the Regulations will be made and a new draft submitted to Ministers and the Health and Safety Commission. In order to comply with the Directive the new Regulations have to be in force by 3 February 1999.

THE REGULATORS APPROACH TO ASSESSING COMAH SAFETY REPORTS

T. J. Britton
Health and Safety Executive, Chemical and Hazards Installations Division, University Road, Bootle, Merseyside, L20 3RA

> A recent European Directive requires member states to introduce legislation on the control of major accident hazards involving dangerous substances. Regulations will be introduced in February 1999 and central to these is a requirement for establishments above the threshold quantities of these dangerous substances to produce a written safety report. This paper outlines what a safety report is and when it is required. It explains the principles behind the regulator's assessment process and how they were drawn up. The paper also describes how the regulator, consisting of the HSE and environment agencies and dealing with health, safety and environmental major accident hazards, will assess these reports and outlines the procedures and criteria that will be used.
> Key words: Safety Reports, Assessment, Major Hazards

INTRODUCTION

On 9 December 1996, the Council of the European Union published European Directive 96/82/EC on the control of major accident hazards involving dangerous substances. This Directive is commonly referred to as the Seveso II Directive. UK proposals to implement this Directive were made in a Consultative Document published in early May 1998. The proposals are for a single set of regulations, to be called the Control of Major Accident Hazard Regulations (COMAH), to be made under the Health and Safety at Work Act 1974 and the European Communities Act 1972. COMAH will replace the existing Control of Industrial Major Accident Hazards Regulations 1984 (CIMAH).

COMAH will apply to establishments which have, or foreseeably have, threshold quantities of dangerous substances, including those which might be generated in the course of an accident due to loss of control of a process and without any differentiation between storage and processing. The regulations will be goal setting and will therefore place a duty on operators of establishments to take all measures necessary to prevent a major accident. There will also be more specific duties relating to notification, emergency planning, information to the public and accident reporting.

Central to COMAH is the requirement for establishments with higher thresholds of dangerous substances specified in the regulations (referred to as 'top-tier' establishments) to send a report to the competent authority. Sites with lower threshold quantities will be required to prepare a major accident prevention policy (MAPP), which sets out their policy for preventing and mitigating major accidents and demonstrating that they have a safety management system in place to achieve this.

COMAH will be enforced by the 'competent authority' (CA) which is the HSE and either the Environment Agency or the Scottish Environment Protection Agency (SEPA) working jointly or separately in the enforcement of the regulations. HSE and the Agencies will work together closely in their enforcement.

WHAT IS A SAFETY REPORT?

The safety report is prepared by the operator of a top tier establishment and is a key element in identifying, preventing, controlling and mitigating major accident hazards.

Basically the report is prepared for the CA and should contain certain _information_ and descriptions about:

- the management system and organisation of the establishment concerning major accident prevention

- the site and its locality and the areas where a major accident may occur

- the main activities and products, processes and operating methods

- the dangerous substances, including maximum quantities present

- the hazard and risk identification, analysis and prevention methods resulting in a description of possible major accident scenarios and the extent and severity of their consequences

- technical measures used for the safety of installations

- measures of protection and intervention to limit the consequences of a major accident, including technical measures taken and emergency response arrangements.

Operators are providing this information for the following purposes.

To _demonstrate_ that

- a MAPP and a safety management system for implementing it have been put into effect

- major accident hazards have been identified and measures taken to prevent and limit their consequences for persons and the environment

- there is adequate safety and reliability in the design, construction, operation and maintenance of any installation re major accident hazards

- on-site emergency plans have been drawn up and information supplied to enable off-site plans to be drawn up and that sufficient information has been supplied to

- enable the CA to make decisions about the siting of new activities or developments around existing sites.

In summary, operators will have to present the information in such a way that they demonstrate that they have taken all measures necessary to prevent a major accident. They are presenting a case that the measures they have in place, linked to their major accident hazard processes, do and will continue to control the risks and that the management arrangements will achieve this. The continued operation of an establishment does not depend on the acceptance of the safety report by the CA . However, the CA does have certain duties placed on it, which will be described later.

WHEN IS A SAFETY REPORT REQUIRED?

A safety report is required for each top tier establishment. This requirement is different from the CIMAH Regulations which required a safety report for each hazardous installation. We estimate that there are about 350 top tier establishments in England, Wales and Scotland. The need for a safety report and its timing will depend on the circumstances, which are outlined below.

Existing Major Hazard Sites (ex CIMAH)

There are transitional arrangements for operators of existing establishments who have submitted a safety report under CIMAH.

Establishments which came within scope of the CIMAH top-tier requirements and had submitted a safety report under these regulations, must submit their first COMAH safety report on the date that their CIMAH safety report update would have been due **OR** by 3 February 2001, if that is earlier. However, operators whose CIMAH update was due between 3 February and 3 August 1999 have until the 3 August 1999 to submit their COMAH report. The important feature here is that, in these circumstances alone, COMAH, as drafted in the Consultation Document (CD) will allow submission of a safety report in parts.

CIMAH linked an individual safety report to each hazardous installation, and therefore operators with a multi-installation establishment have the option of submitting, on the due date, either

- an installation report on the date that a CIMAH report would have been due, so long as the first installation report is submitted with with, what is often referred to as, the 'core' report. This should contain information about the establishment's management arrangements and site and locality descriptions.

- a complete report for the whole establishment.

The CIMAH and COMAH schedules for specifying the thresholds for dangerous substances at establishments are very similar, although there is an increase in the number of generic classification categories, which has lead to a major reduction in the extensive list of specific substances named in CIMAH. These classification categories are linked to the Classification, Packaging and Labelling Directives for both substances and preparations and which have been brought into UK law as the CHIP Regulations. In practice, there will be some existing establishments which will be new to COMAH, which are described below, and a few establishments which will no longer require a safety report. This will mean, for example, that a large number of timber treatment sites handling copper chrome arsenate (CCA), will no longer require a safety report.

Existing sites new to major hazards legislation

The relatively minor changes to the thresholds for dangerous substances in the COMAH schedules compared to CIMAH will mean few establishments requiring a safety report for the first time. Warehouse sites look the most likely additions. However COMAH will apply (whereas CIMAH did not) to:

- chemical hazards on nuclear installations which are licensed under the Nuclear Installations Act 1965

- certain explosives sites subject to The Explosives Act 1875.

Operators of these establishments are required to submit a safety report by 3 February 2002.

New Establishments

The following describes the position concerning the submission of safety reports for brand new establishments as envisaged in the COMAH Consultation Document. Requirements concerning this may change after consultation.

Operators of brand new establishments, on which construction starts after COMAH comes into force, must submit a safety report to the CA within a reasonable period before construction. What a reasonable period is considered to be will depend on negotiations between the operator and the CA and whether in particular there has been early dialogue between the two before submission of the safety report. Without this, the CA is unlikely to require less than 3 months.

Operators will not be able to start the construction stage until they have received the conclusions from the CA. There is then a requirement for a second report to be submitted a reasonable period before the start of operations, which is taken to be the time when hazardous substances are used in the hazardous installation for the first time. Again there is a 'statutory hold point' because the operator must not start operations before the CA has given its conclusions on the safety report. In this case, a reasonable period is thought to be up to 6 months to allow the CA to assess the report, but again this may be reduced by early dialogue between the operator and the CA, in which information is provided..

The information provided in the 2 reports together should provide the complete information and provide the necessary demonstration described earlier. The split of information will vary depending on the specific project, but the pre-construction report should contain as a minimum the following information:

- description of the establishment installations and processes

- reasoning in selecting process options and chosen design concept

- description of the environment in and around the establishment which could be effected by a major accident

- identification of major accident scenarios

- information about the prevention, control and mitigation measures to be provided or the principles to be followed, where details are not finalised

- demonstration of an effective safety management system to ensure the quality of procedures for the control of design, procurement, construction and commissioning.

These reports are only required for new establishments. Operators introducing a new installation on an existing establishment will be required to prepare a single modification report, as discussed later. It is therefore unlikely that 'pre-construction' and 'pre-operation' reports will be required very often.

Review and updating of reports

Safety reports should be reviewed by the operator at least every 5 years, or whenever new facts or new technical knowledge mean that such a review is thought necessary, and the revised report submitted to the CA.

Modifications

Operators must review and revise their safety reports and inform the CA of the details of the revision before any modifications are made to

- the establishment or installation,

- the processes carried on

- or the nature and quantity of dangerous substances,

which could have significant repercussions on the prevention of major accidents. These considerations should be incorporated into the operator's plant/process change procedures. However, it should be remembered that changes to an establishment, under COMAH, include introducing a new installation, which would not normally be thought of as just a 'modification'.

For modification reports, there is no requirement to submit these within a reasonable period, but if the operator is to avoid retrospective action required by the CA, then early dialogue is again recommended. The period of early 'warning' that the CA would like will vary, depending on the type of modification, but for example, the CA is likely to take up to 6 months to examine a safety report for a new hazardous installation at an existing establishment.

WHAT IS ASSESSMENT?

Assessment is the structured process by which the CA examines the adequacy of safety reports. This examination will assess whether the safety report:

- contains sufficient information

- meets the purposes of a safety report which is primarily to provide the demonstrations required.

The CA's conclusions of its examination of the report, given to the operator of the establishment concerned, will be based on these 2 requirements. There is however a further statutory requirement on the CA. In examining the safety report and assessing whether the required demonstrations are made, the CA will carefully consider the measures and in doing so will

- prohibit the operator of this establishment, installation or any part, where a serious deficiency is identified.

Assessment is an 'enforcement' activity, using the term 'enforcement' in its wider sense to include the wide range of influencing techniques used by regulators, ranging from advice, letters, notices, licensing through to prohibition and prosecution. Each is successful in the right context, but each has to be used within a recognised framework. For assessment of COMAH safety reports, principles have been devised which fully take into account the Health and Safety Commission's policy for enforcement, that is that it should be transparent, targeted, proportionate and consistent.

However, assessment is also quite a complex process, which requires good management practice to deliver the enforcement policy. Not surprisingly, we decided that there was no more appropriate management model than that adopted by HSE in its own guidance (1) on effective safety management and therefore the principles of the assessment process should have a clear Policy and Organisation to deliver it, arrangements for Planning and Implementation, supported by suitable Monitoring, Audit and Review arrangements (known within HSE as the POPMAR model). With this framework in place, and learning from other assessment processes particularly safety case assessment in HSE's Offshore Safety Division and licensing arrangements in the Nuclear Safety Division, both of which have many similar aims, we were able to set out the principles and procedures for assessing COMAH safety reports.

HOW HAVE THE ASSESSMENT PROCEDURES BEEN REVISED?

A project was set up in February 1997 called SHARPP (Safety Report Handling Assessment Review Principles and Procedures) to develop the guiding principles and procedures for assessment and to prepare criteria to be used by the CA as performance measures for assessing the information in the safety reports and helping to judge whether the necessary demonstrations have been made. The criteria have been divided into 5 parts:

- safety management systems
- descriptive criteria
- predictive criteria
- technical measures for prevention and mitigation
- emergency arrangements criteria

and were prepared by 5 separate teams. Details about these criteria will be given by colleagues in separate papers.

An important feature of the project is that it has been run by a Project Board which consisted of members from HSE and both environment agencies, who agreed their contributions to the various 'products' required by SHARPP. For example the agencies have been involved in the pilot exercise and the Environment Agency has been involved in the development of the assessment criteria.

As part of the project, procedures and criteria were prepared and incorporated into an 'Assessment Manual' for use in a pilot exercise, involving 4 teams of assessors from HSE and the agencies, assessing COMAH safety reports prepared by 4 volunteer companies. This assessment process was carried out over a period of 3 months and was complemented by consultation on the pilot Assessment Manual with trades associations, trades unions and individual companies, who had shown previous interest.

HOW WILL COMAH SAFETY REPORTS BE ASSESSED?

A new Division in HSE, the Chemical and Hazardous Installations Division (CHID), was set up in April 1996 focussed on the chemical and explosives industries and in particular to enforce the new COMAH Regulations. This Division will have the primary enforcement role in HSE for COMAH in conjunction with the Agencies. The principles for assessment have been developed within CHID.

The principles underpinning the assessment process are listed in Appendix 1. There are 10 Guiding Principles which set the policy and 8 Administrative Principles which set some fundamental principles on how the process is organised, implemented, monitored, audited and reviewed (POPMAR). The following paragraphs summarise the main points.

Policy - Operator retains duty

Although the CA will examine each safety report and come to conclusions about it, as well as identifying any serious deficiencies in the measures it describes, the duty to ensure that establishments, and installations within them, are designed, constructed and managed safely remains with operators.

Policy - All measures necessary

The term to take 'all measures necessary' to prevent or mitigate major accidents is not one that has any legal precedence in safety legislation in the UK. We have interpreted this to mean that hazards should be avoided if possible or reduced at source through the application of inherently safe principles. In this case, inherent safety means inherent safety, health and environmental protection (i.e. inherent SHE), in which, for example, the substances used are intrinsically less harmful or processes are used in which the consequences of loss of containment are reduced. Where risks remain, then the recognised principle of ALARP (as low as reasonably practicable) will be used by the risk assessor for health and safety issues and BATNEEC (best available technology not entailing excessive costs) for environmental matters. We recognise that the application of inherent SHE principles is economically more viable for new installations and these issues should be considered as early as possible during the design of the installation or any modification. Consideration of inherent SHE will be a particular feature of assessment for pre-construction reports. It is not intended to require major retrospective action on the basis of inherent SHE for currently existing installations.

Policy - Enforcement strategy

Assessment of safety reports will be part of an overall enforcement strategy for COMAH top tier establishments. It will not be an isolated or 'one-off' process. Information gained from assessing the safety report will be used to inform a subsequent inspection plan by the CA. Similarly, inspection will help the CA to continue to build its knowledge and experience of an operator and a particular establishment, which will help it to assess each subsequent report.

The assessment of a safety report is looking at a snapshot in time. It will be based on the documentary evidence in the safety report, or referenced by it. There will not be any site visits to check the accuracy of information, other than following up on suspected serious deficiencies, but conclusions will also be based on other intelligence such as from previous inspections, investigations, reference books and other sources of information. Assessors may seek further clarification from the operator on assessment issues, which exceptionally could include site visits.

The safety report will be used as a fundamental source of information for future inspection. After the assessment is completed, the assessors will make recommendations which will be developed into an inspection plan for the site. The inspection plan for each establishment will form part of the CA's wider inspection programme for the next 5 years This period has been chosen because each safety report must be reviewed by the operator every period of 5 years. The contents of each safety report will be subject to verification as part of the inspection plan and the CA's conclusions may be subject to subsequent scrutiny as a result.

Policy - Selection

It is neither possible nor sensible for the CA to examine every part of each safety report in detail. This is particularly the case for large reports dealing with complex or unusual processes. Instead, parts of a safety report will be selected for full examination. Selection will be guided initially by hazard and by previous assessments both at the particular and related establishments and installations. With this knowledge, account can be taken of plant or system vulnerabilities, or weaknesses in the safety management system and the risk of these contributing to major accidents.

Although a safety report may be selectively examined in detail, it will be read thoroughly at least once by the Assessment Manager (see below) and in practice we have found that all the assessors need to read the report in full to assess selected issues.

Policy - Serious Deficiency

A site visit will be paid where a potential serious deficiency is identified in the measures described to prevent a major accident or limit their consequences during the assessment process. Action, jointly with the agencies where appropriate, will not be delayed to complete the assessment process. Assessors will have to obtain first hand evidence to support prohibition action and check the facts with the operator, before a prohibition notice is issued. Agreement will also be obtained from the inspector's line manager, before any notice is served.

Organising

An Assessment Manager (AM) will be appointed for each safety report to be assessed, who will act as the primary point of contact for dealings on the report. The name of the AM will be agreed between HSE and the Agencies. Normally the AM will be the site inspector working in the HSE inspection group dealing with the site.

Assessment will be by a team composed of the necessary competences eg:

> the local inspector - who will manage the assessment process, assess safety management issues and other matters on which he/she has knowledge and bring the conclusions together

> the topic specialist - who will provide specialist input eg on process safety, mechanical, electrical or civil engineering. Teams will be located in 'local' offices and support will be provided by HSE specialists from outside CHID.

> risk assessors - based in the Major Hazards Assessment Unit, who look at the techniques for identifying and analysing the hazards, consequences and risks and be able to confirm that the major accident scenarios have been properly identified

agency representatives - will look at the above issues, but considering the environmental consequences.

An assessment team will be assembled for each safety report and dismantled on its completion. Some of the team will have more than one task.

Planning and Implementation

A diagram illustrating the flow of the assessment procedures is shown at Fig 1. Critical stages to the smooth working of the assessment team are the drawing up of the assessment plan, its implementation and the assessment outcome meeting.

The AM devises an assessment plan for each safety report to include:

- names of the assessment team

- the resources likely to be expected

- aspects of the report likely to be assessed, including the target agenda

- milestones and timing for particular stages of the assessment.

The Target Agenda sets the items in this safety report to be assessed in full and records the reasons for this. The descriptive parts i.e. that adequate information which is provided to meet Schedule 4, Part 2, the major accident scenarios and the MAPP are always selected for full assessment.

The assessment team members will be allocated their tasks and will be asked to conclude whether the operator has met the purposes required of a safety report and made the demonstrations specified in Schedule 4 Part 1, relevant to what they are examining. In examining the report, they may decide that there is insufficient information in the report itself to come to a conclusion. The assessors will obtain further information from the operator, liaising with the AM, until they are satisfied that they have sufficient to come to a conclusion. At this stage it should be become clear whether there are any serious deficiencies in the measures for preventing major accidents and limiting their consequences, including the management arrangements for delivering these measures. As individual members of the team undertake their examination, they will follow up on suspected serious deficiencies, again liaising with the AM.

During assessment, team members will also be accumulating information from the safety report and forming overall views about how the operator manages his establishment, not just as a result of the arrangements and systems described but also from the conditions described, in other words how they have been put into effect. All these matters will be discussed at an assessment outcome meeting. The prime purpose of the assessment outcome meeting is to produce agreed conclusions and initially send them in draft to the operator. The operator will then have an opportunity to discuss these with the assessment team, before conclusions are sent formally.

Where one or more of the demonstrations written in the safety report has not been made, the assessment team will decide the action they will seek from the operator. We envisage that this will require the issue of an Improvement Notice to remedy the problem/s. There may be deficiencies in the measures described, which are neither seriously deficient nor are they such to prevent the operator making the overall demonstrations. In these cases, the team will decide their relative importance and recommend how they should be addressed in the subsequent inspection plan.

Monitoring, Auditing and Review Arrangements

Key stages of the assessment process have been identified and arrangements for monitoring have been included. When finalised, the assessment procedures will be designed to conform to ISO 9000 quality procedures.

PILOT EXERCISE & CONSULTATION

Four operators of current CIMAH top-tier sites agreed to produce COMAH safety reports for their sites to allow assessment by the CA. The reports were produced for the beginning of April 1998 and were examined by separate teams of assessors who communicated their conclusions about the reports to the operators at the beginning of July 1998. A pilot version of the Assessment Manual was prepared for the assessors and copies sent to the operators. At the same time, copies of the pilot manual were sent to members of the Health & Safety Commission's Sub Committee on Major Hazards, selected organisations and other organisations that expressed an interest, during its development, in receiving a copy.

The pilot and consultation exercise has proved invaluable. A number of important points have been learned as well as detailed points concerning the criteria. Further guidance was required by assessors on how to use the criteria. The criteria are provided as a guiding framework within which professional judgements are made: they are not intended to be a tick list representing a pass or a failure for each criterion. They are 'high level' and will eventually be supported by more detailed guidance to assessors on performance standards that will meet these criteria e.g. what might be expected in safety reports dealing with LPG filling installations or ammonium nitrate fertiliser stores.

Demonstration

It was clear from the pilot reports received that further guidance was required on what should be contained in a safety report and how operators should present this information to help them make the required demonstrations under Schedule 4 Part 1. The CA now plan to prepare an on-sale booklet for report writers providing guidance on how a report might b structured and the type of information to be included. This will be guidance and not an approved code of practice to support the regulations.

In relation to the purposes of a safety report required by COMAH 'demonstration' is thought to mean 'show' or 'make the case/argument' rather than at the stronger end of the meaning of demonstration such as 'prove beyond doubt'. The implication is that prima-facie information should be used and professional judgement should be exercised by the assessor, rather than extensive in-depth scrutiny or exhaustive examination.

Proportionality

Concerns were expressed by assessors and consultees alike that the criteria, although thorough, could overwhelm operators and that there needed to be a sense of proportion. COMAH applies to a wide range of establishments differing in size, numbers of employees, complexity, resources and expertise available, technology, culture and environment surrounding the site. They have one thing in common; they all have major accident potential, although even with this there is a wide variety of hazards. Clearly it would be unreasonable for the safety report for a small ammonium nitrate store to look the same as a report for a multi million pound oil refinery but both must make the same demonstrations.

The demonstration should be proportionate. The depth of the demonstration relates to the hazard but more particularly to whether the process is unusual, innovative, complex or whether there are existing standards/guidance. The size of the establishment or the resources of the operator do not determine of the depth of demonstration required, only the amount of information required to describe what is going on.

CONCLUSIONS

The introduction of the new COMAH Regulations in February 1999 has meant that the regulator has completely revised its procedures for assessing safety reports because of new requirements on the operator, duties on the regulator for the first time and because the regulator is a competent authority consisting of 3 bodies i.e. HSE, the Environment Agency and SEPA. As a result, the procedures have been prepared following a recognised management model for ensuring 'quality'. These procedures and the actions taken will meet the enforcement policy followed by HSE and ensure that the regulator is consistent in its approach, proportionate in its action and targets the right parts of the report for detailed consideration, as well as being transparent. The procedures have been tested in a major pilot exercise and key stakeholders have been consulted, resulting in a number of significant changes. The revised procedures will be made available in a manual to anyone requesting it and we plan to make this document available on the HSE Internet web site.

REFERENCES

1. HSE Booklet (HS/G65) Successful health and safety management, HSE Books

Figure 1 Flow Diagram of Safety Report Assessment Procedures

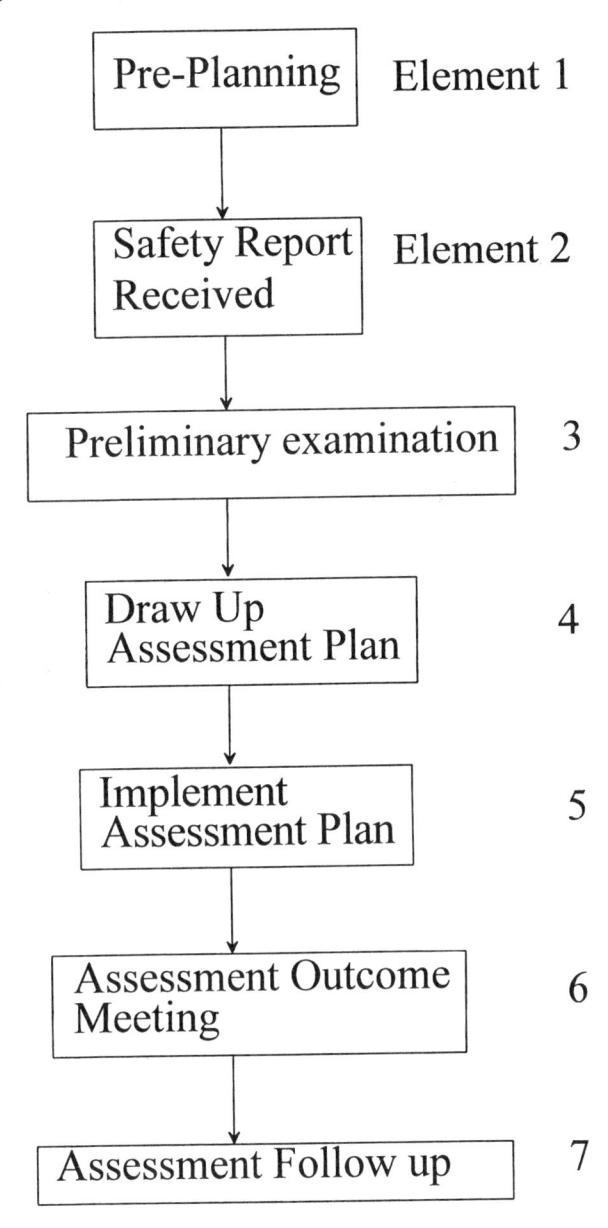

```
Pre-Planning              Element 1
    ↓
Safety Report Received    Element 2
    ↓
Preliminary examination   3
    ↓
Draw Up Assessment Plan   4
    ↓
Implement Assessment Plan 5
    ↓
Assessment Outcome Meeting 6
    ↓
Assessment Follow up      7
```

APPENDIX 1

PRINCIPLES FOR ASSESSMENT

GUIDING PRINCIPLES

GP1 The duty to ensure that necessary measures have been taken to prevent major accidents and to limit their consequences at an establishment remains with the operator and no conclusions reached following assessment of a safety report by the CA diminishes that duty, nor do such conclusions imply that the CA consider the establishment, or parts of the establishment, to be 'safe'.

GP2 Assessment is a structured process by which the Competent Authority (CA) examines the adequacy of safety reports against the purposes set out in Schedule 4 Part 1 of the COMAH Regulations and which contributes to the CA's decision about whether the measures taken by the operator for the prevention and mitigation of major accidents are seriously deficient.

GP3 Hazards should be avoided if possible or reduced at source through the application of inherently safe principles. Where risks remain, then recognised principles such as ALARP (as low as reasonably practicable) for health and safety matters or BATNEEC (best available technology not entailing excessive costs) for environmental matters should be used to determine the extent of the preventive and mitigation measures required.

GP4 The assessment of the safety report is part of an overall enforcement strategy for top tier COMAH sites.

GP5 Conclusions about the safety report will be based on an assessment made against criteria by means of a structured process. The process and criteria will be published.

GP6 The conclusions of an assessment can be based on a full consideration of <u>selected</u> parts of the safety report.

GP7 The CA's assessment conclusions will be based on the evidence in the safety report.

GP8 The contents of the safety report and those conclusions will also be subject to verification and further scrutiny as part of a continuing inspection programme.

GP9 Where a potential serious deficiency is identified, the site before action is taken. Normally the CA's line management action before such action is taken.

GP10 The CA's actions will be confirmed in writing to the operator.

ADMINISTRATIVE PRINCIPLES

AP1 An Assessment Manager will be appointed for each safety report to be assessed and the name made know to the operator to provide the primary point of contact for communications about the report.

AP2 Assessment will be by a team composed of members with the necessary competences.

AP3 The CA's performance standards will be published and met.

AP4 The assessment procedures will be set out in a manual and will follow ISO 9000 quality assurance principles.

AP5 The effort put into assessing a safety report will depend on a number of factors including the hazard, but enforcement will be proportionate to the risk, as identified by the CA.

AP6 Monitoring against published performance standards will be an integral part of the assessment process.

AP7 Procedures will include an in-built mechanism for review.

AP8 The assessment process will be auditable.

THE ASSESSMENT OF TECHNICAL ASPECTS OF COMAH SAFETY REPORTS

R F Evans
Health & Safety Executive, Chemical & Hazardous Installations Division, Bootle

> Under European Union Council Directive 96/82/EC the UK is obliged to introduce new legislation to replace the Control of Industrial Major Accident Hazards (CIMAH) Regulations 1984. To implement this directive new Regulations, the Control of Major Accident Hazards (COMAH) Regulations will come into force on 3 February 1999.
>
> In preparation for implementing the new Regulations HSE has drafted a set of criteria to assist its staff in the assessment of Safety Reports. This paper describes those criteria and a pilot exercise whereby the criteria were tested on safety reports voluntarily submitted by industry.
>
> Keywords: Major hazard, Safety Report, Legislation

INTRODUCTION

Under European Union Council Directive 96/82/EC the United Kingdom is obliged to introduce new legislation to replace the Control of Industrial Major Accident Hazards (CIMAH) Regulations 1984. To implement this directive new Regulations, the Control of Major Accident Hazards (COMAH) Regulations will come into force on 3 February 1999. Among the changes the new Regulations will impose new duties on the Competent Authority[1]. These include the requirements to:

- within a reasonable period of receipt of a safety report, communicate the conclusions of its examination of the report to the operator, if necessary after requesting further information; or

- prohibit the bringing into use, or continued use, of the establishment concerned where the measures taken by the operator for the prevention and mitigation of major accidents are seriously deficient.

It was recognised, therefore, that there was a need to be prepared, before the Regulations came into force, to assess safety reports in a reasonable time and identify any possible serious deficiencies.

PREPARING FOR THE NEW REGULATIONS

---
[1] The Competent Authority comprises the HSE and the Environmental Agency or the Scottish Environmental Protection Agency as appropriate.

HSE in conjunction with the environmental agencies has drawn up a manual called the *Safety Report Handling, Assessment and Review Principles & Processes* manual, known as SHARPP for short. This has undergone widespread consultation and been subject to a pilot exercise to ensure its fitness for purpose.

The manual contains the following parts:

Guiding Principles

These are described in another paper but include such things as the purpose of assessment, the system aims and objectives etc.

Procedures

These are essentially the administrative procedures for dealing with safety reports.

Assessment Criteria

These are the criteria against which safety reports will be assessed and cover the following topics:

- Information on the Safety Management System
- Description of the Establishment
- Predicting the Consequences of Major Accidents
- Technical Risk Reduction Measures

TECHNICAL ASSESSMENT CRITERIA

The Purpose of the Criteria

The technical assessment criteria are summarized in the appendix to this paper. Their purpose is to provide a framework for the technical assessment of safety reports whilst not unduly limiting the discretion of the assessor thus achieving a balance between the conflicting requirements of consistency of assessment between safety reports and the need for flexibility. They are not intended to be regarded as detailed technical guidance although work is in hand to produce such guidance by drawing on existing codes and standards and generating new guidance as appropriate.

The assessment criteria will be made publicly available in the interests of openness and it is expected they will assist authors of safety reports by providing an insight as to how the competent authorities will assess their work. It should however be noted that they are primarily intended to assist assessors and are not intended as a guide to the writing of safety reports.

Drafting the Criteria

The criteria were drafted by a working group. To ensure that the final document reflected the views not only of the Competent Authority but also of those who would have to draft safety reports the working group was composed of members of the Environmental Agency, the HSE, a consultancy and a major petrochemical company. The petrochemical company representative worked on secondment to HSE as a member of the working group but, at the same time, was involved in drafting, on behalf of his parent company, a safety report for the pilot exercise. He was therefore able to give a valuable insight into problems our criteria posed for those writing safety reports.

A separate group consisting of members of the Environmental Agency and the HSE was set up to peer review the document and two consultation exercises were carried out: the first internal only, the second internal and with some informal external consultation.

The process of drafting the criteria was essentially an iterative one starting with a brainstorming session to obtain an initial set of criteria that were then peer reviewed. The document was revised in light of the comments from the review team. The process of review and redrafting was then repeated.

Throughout this exercise we were well aware of the need to draw from past experience of other parts of HSE in safety report and safety case assessment. Therefore, although we drew heavily on the *Assessment Principles for Offshore Safety Cases*[1] and *Safety Assessment Principles for Nuclear Plants*[2], at the same time we also tried to learn from past mistakes. For example from our experience with the manual for assessing Offshore Safety Cases we new that it would not be possible, in the time available, to produce detailed technical guidance of a good quality and we concentrated instead on identifying the main assessment criteria with a view to developing the technical guidance later. By doing this we achieved two things: we were able to produce an assessment manual in advance of the Regulations coming into force and we provided a framework for the subsequent technical guidance.

We also drew on *AVRIM 2*[3], a document drawn up for the Dutch Government to assist in the inspection of Major Hazard Installations.

The Structure of The Criteria

There are 2 fundamental technical criteria and these are that:

- the safety report should show a clear link between the measures taken to reduce risks and the major accident hazards described in the report; and

- the safety report should show how the measures taken will reduce the risks from foreseeable failures which could lead to major accidents.

All other criteria are seen as being subordinate to the above and are classified under the following headings:

Design   This includes conceptual design, plant layout, process design and the detailed design of equipment.

Construction   This includes the manufacture, installation, construction of civil structures, testing, initial inspection and commissioning.

Operation   This includes plant start up, shut down, normal operation and emergency shutdown.

Maintenance   This includes preventive maintenance, repair, replacement, periodic examination by a competent person and the assessment of any defects found.

Modification   This includes all alterations (including decommissioning) which could affect the integrity of the installation.

THE PILOT EXERCISE

It was recognised that no matter how much consultation and peer reviewing were carried out the best way to find out how our assessment criteria would work in practice would be to test them in circumstances as near as possible to those in which they would eventually be applied. A pilot exercise was, therefore, carried out whereby four establishments voluntarily submitted safety reports written to meet the requirements of the draft COMAH Regulations. To assist them in this they were provided with early drafts of the assessment criteria. The establishments in question were:

>The Elf terminal at Flotta
>
>The ISC establishment at Hythe
>
>BP Chemicals at Hull
>
>The British Gas Transco Establishment at Cheltenham

The safety reports were assessed by teams whose composition reflected that of the sort of typical team that would be expected to assess statutory safety reports in practice. They used the draft manual for the basis of their assessments and their conclusions were sent to the operators. Although the reports were assessed as though they were real COMAH safety reports, the primary purpose of the exercise was to obtain information about how the assessment criteria and procedures worked in practice. This information was passed on to the team that was managing the pilot project and, in conjunction with the findings of a formal external consultation exercise that was carried out at the same time, was used as the basis of further revisions of the manual.

The conclusions of the pilot exercise were that some further guidance had to be given as to how assessors should use the assessment criteria but that the technical assessment criteria were about right. The main criticism of the technical assessment criteria arising out of the pilot study was that they gave the impression that all the criteria had to be met whereas in practice not all criteria would apply to all safety cases and the assessor would have to apply considerable judgement in deciding which criteria applied in each case and how much weight to give to them

The consultation exercise generated few comments on the technical assessment criteria and most of them came from within the HSE rather than from external organizations. Most of the comments related to the need to clarify certain points.

The technical assessment criteria were amended to take account of comments received, where appropriate. Other papers presented at this conference deal with how these exercises led to changes in other parts of the SHARPP manual.

CONCLUSIONS

As a result of the consultation and pilot exercises we have a set of assessment criteria that we believe will enable us assess safety reports with the right balance between consistency and flexibility. Initially assessors will have to draw largely on their own knowledge and experience when assessing safety reports but in the long term we hope to be able to provide more technical guidance. It is expected that this guidance will be made publicly available as well as the assessment criteria and will therefore be of some assistance to industry in drafting safety reports

REFERENCES

1. Health and Safety Executive, Offshore Safety Division, 1997, Assessment Principles for Offshore Safety Cases
2. Health and Safety Executive, Nuclear Safety Division, Safety Assessment Principles for Nuclear Plants.
3. Dutch Ministry of Social Affairs, AVRIM 2.

Appendix

TECHNICAL ASSESSMENT CRITERIA

Criterion 1
The Safety report should show a clear link between the measures taken and the major accident hazards described.

The safety report should include a summary of the findings of the hazard identification process. It should show how the identified hazardous events have been ranked on the basis of their perceived likelihood of occurrence and consequences.

For hazardous events that could lead to a major accident, the safety report should show that risk-reduction measures have been put in place to reduce the risks to as low a level as is reasonably practicable. This may be done using qualitative or quantitative methods as appropriate to the circumstances. The report should justify the method chosen.

Criterion 2
The safety report should demonstrate how the measures taken will prevent foreseeable failures which could lead to major accidents

<u>General</u>   The safety report is required to show that the necessary measures have been taken to prevent major accidents and to limit their consequences for people and the environment. The safety report is also required to show that adequate safety and reliability have been built into the design, construction, operation and maintenance of the installation. These demonstrations are closely linked and a single set of assessment criteria has been developed based on the life cycle of the installation.

Criteria 1 & 2 are seen as fundamental. Criterion 2 has been subdivided into lower level criteria, which have been grouped as follows:

2.1 design, which includes conceptual design, plant layout, process design and detailed design of equipment;

2.2 construction, which includes the manufacture, installation, construction of civil structures, testing, initial inspection and commissioning;

2.3 operation, which includes plant start up, shut down, normal operation and emergency shutdown;

2.4 maintenance, which includes preventive maintenance, repair, replacement, periodic examination by a competent person and the assessment of any defects found; and

2.6 modification, which includes all alterations (including decommissioning) which could affect the integrity of the installation.

Appendix

The lower level assessment criteria are designed to be generally applicable, to the full range of installations within the scope of COMAH. However, they may not all be applicable in all cases. The assessor must decide which are appropriate.

2.1 Design

Criterion 2.1.1
The safety report should show that the establishment and installations have been designed to an appropriate standard

This is the main criterion for assessing whether the safety report shows that adequate provision for safety has been included in the design of an installation and covers such matters as containment, redundancy and diversity, and separation and segregation.

Criterion 2.1.2
The Safety report should show that a hierarchical approach to the selection of measures has been used

The design stage in an installation's life presents the best opportunity to reduce risk. The use of a hierarchical approach to the selection of measures will help to ensure that precedence is given to those measures that avoid major accidents, that is through inherent safety and prevention measures. Prevention cannot be guaranteed in all circumstances and therefore it will be necessary to identify other measures to control and mitigate the consequences of any major accidents to reduce risks to as low a level as is reasonably practicable.

Although the design of a new installation offers the best opportunity to apply these principles they may also be applied to the design of modifications and operators of older establishments should be alert to the possibility of taking advantage of technical advances in their industry to improve safety.

The levels in the hierarchy are as follows.

<u>Inherent Safety</u>   Inherent safety is concerned with the removal or reduction of a hazard at source. Examples of inherently safe techniques include; substitution of a less hazardous process, use of corrosion resistant materials of construction, reduction or elimination of hazardous inventory, design for maximum foreseeable operation conditions, fail-safe design principles, and appropriate plant layout, etc..

<u>Prevention Measures</u>   These are intended to prevent the initiation of a sequence of events that could lead to a major accident. They can be management systems or features of the design of the installation, and can apply during design, construction, operation, maintenance and modification.

Appendix

<u>Control Measures</u>   These are intended to prevent a hazardous event from escalating into a major accident. They include measures directed at preventing or limiting small releases that have the potential to escalate to a major accident.

<u>Mitigation Measures</u>   These are measures that are taken to reduce the consequences of a major accident once it has occurred. Examples of this include safety refuges, bunding systems, fire-fighting facilities, emergency response procedures, traverses or mounds for explosives buildings, etc.

Criterion 2.1.3
Layout of the plant should limit the risk during operations, inspection, testing, maintenance, modification, repair and replacement.

Design of the layout of a plant can make a big contribution to reducing the likelihood and consequences of a major accident. The safety report should show that due attention has been given to ensuring safety in the design of the layout of the installation. In particular, it should show how the layout prevents or reduces the development of major accident scenarios.

Criterion 2.1.4
Utilities that are needed to implement any measure defined in the safety report should have suitable reliability, availability and survivability

Failure of a utility, e.g. water, air, steam, electricity, often results in a process upset, and may have effects across the entire establishment. Failure of an emergency facility, e.g. fire water, has the potential to cause an escalation of a relatively small incident into a major accident. The safety report should justify the steps that have been taken in design, construction, operation and maintenance, to ensure that these utilities and facilities will be available when required.

Criterion 2.1.5
The Safety report should show that appropriate measures have been taken to prevent and effectively contain releases of dangerous substances.

The safety report should identify means by which dangerous substances can be accidentally released from the containment and the measures provided to prevent the occurrence. The safety report should demonstrate the suitability of measures to prevent releases.

Criterion 2.1.6
The safety report should show that all foreseeable direct causes of major accidents have been taken into account in the design of the installation

All foreseeable direct causes of loss of containment accidents should be considered at the design stage. The majority of direct causes fall into one of the following categories. The safety report should show that these have been considered and suitable measures taken:

Appendix

a) Corrosion

b) Erosion

c) External loading

d) Impact

e) Pressure

f) Temperature

g) Vibration

h) Wrong equipment

i) Defective equipment

j) Human error

Criterion 2.1.7

The Safety report should show how structures important to safety have been designed to provide adequate integrity.

The safety report should provide sufficient evidence to show that the design of all structures important to safety has been based on sound engineering principles. This includes process and storage vessels, pipework and other items that form the primary containment boundary. Other key structural items such as support structures, bund walls, civil foundations, control rooms, buildings or barriers designed to withstand the effects of accidental explosions should also be included.

Criterion 2.1.8

The Safety report should how the containment structure has been designed to withstand the loads experienced during normal operation of the plant and all foreseeable operational extremes during its expected life.

This assessment criterion is a follow-up to the more general requirements for adequate structural integrity.

The safety report should provide details of the normal operating conditions of the plant and any foreseen operational extremes. Evidence presented in the safety report should include all of the conditions that the containment must withstand, such as external loads, ambient temperatures and the full range of process variations (e.g. normal operation, start-up and shutdown, turndown, regeneration, process upset, emergencies and uncovenanted explosions).

Criterion 2.1.9

The Safety report should show that materials of construction used in the plant are suitable for the application.

Appendix

The safety report should provide evidence that all materials employed in the manufacture and construction of the plant are suitable. Particular attention should be given to the selection of materials used for the primary containment of hazardous substances.

Evidence should be provided to show that materials have been selected with regard to the nature of the environment in which they are to be used. In particular the evidence presented should consider the substances being handled and process conditions such as temperature, pressure and flow. The evidence presented should pay particular attention to possible sources of corrosion and erosion. Evidence presented should also consider the external environment, such as the effects of sea air in coastal areas.

Criterion 2.1.10
The Safety report should show that adequate safeguards have been provided to protect the plant against excursions beyond design conditions

Typically, a plant will be designed to operate within a given range of process variables, the 'normal operating limits'. These are the operating constraints that apply to normal operating conditions. There will also be 'safe operating limits' that are the rated values upon which safety of the plant is based. An excursion beyond the 'safe operating limit' may result in a significant risk of loss of containment, fire or explosion.

Safe operation depends on the measures to prevent excursions from occurring, for example, safety-related control systems, relief systems, shutdown procedures, emergency vent and disposal systems, etc.. The safety report should contain a description of the philosophy underlying the application of these measures, and should describe the foreseeable events that have been taken into account, drawing links between identified hazards, system integrity and the use of suitable standards or good industry practices. It should show how each measure has been designed and constructed and operated so as to be available whenever the plant is operating.

Criterion 2.1.11
The safety report should describe how safety-related control systems have been designed to ensure safety and reliability.

Any safety-related control system that is required to prevent or limit the consequences of a major accident (whether to people or the environment), should be designed in accordance with an appropriate code or standard. This should include an identification and consideration of all the components and devices in the system that need to function to ensure safety. The evidence presented should show that the complete system from sensor to final element, including any software has been considered.

Criterion 2.1.12
The Safety report should show how systems which require human interaction have been designed to take into account the needs of the user and be reliable.

Appendix

An analysis of accidents indicates that most result from human error. The safety report should show how human factors have been accounted for in the design of equipment and operating, maintenance and modification of systems. This should include consideration of how human errors can be reduced and the role of management systems in reducing human errors and identification of the safety implications of human errors and what back up systems are in place.

Criterion 2.1.13

The Safety report should describe the systems for identifying locations where flammable substances could be present and how the equipment has been designed to take account of the risks.

The safety report should identify the system whereby hazardous (flammable and explosive atmosphere) areas have been identified and classified. This may have been via an area classification study in which areas where a hazard exists owing to the normal, occasional or accidental release of process materials to atmosphere have been designated in accordance with recognised standards.

Sources of ignition for flammable atmospheres may include electrical equipment, naked flame or hot surfaces, static electrical discharge, etc. The safety report should indicate how the likely sources of ignition have been considered in the design (e.g. electrical equipment selection for defined hazardous areas, avoidance of hot surfaces or naked flames or sparks associated with equipment such as through the use of spark arrestors, etc., control of static electrical build-up).

2.2 Construction:

Criterion 2.2.1

The safety report should show that the installations have been constructed to appropriate standards to prevent major accidents and reduce loss of containment.

The safety report should show that construction of plant and associated equipment is managed to ensure that it is built in accordance with the design intent. This criterion should be applied by assessors whenever new plant is constructed during the life cycle of the installation. Modifications are covered by a separate criterion.

Criterion 2.2.2

The safety report should describe how the construction of all plant and systems is assessed, and verified against the appropriate standards to ensure adequate safety.

The safety report should provide evidence that suitable assessment and verification of the construction process have been carried out. The evidence presented should show that the construction process has not compromised the design intent.

Appendix

The evidence presented should identify the key assessment and verification activities and the stages at which they are undertaken. The safety report should also provide an explanation of the methods used and show how they will ensure safety. The acceptance criteria for testing and examination programmes should be identified, where appropriate.

2.3 Operation:

Criterion 2.3.1
The safety report should show that safe operating procedures have been established and are documented for all reasonably foreseeable conditions.

The safety report should show that safe operating procedures have been established and documented for all foreseeable normal (including start-up and shut down) and abnormal operating conditions.

The report should identify how reviews of operating procedures are undertaken and recorded to take account of operational experience or changing conditions in the plant.

2.4 Maintenance:

Criterion 2.4.1
The safety report should show that an appropriate maintenance scheme is established for plant and systems to prevent major accidents or reduce the loss of containment in the event of such accidents.

The safety report should show that maintenance procedures are sufficiently comprehensive to maintain the plant and equipment in a safe state. The safety report should also show that maintenance activities will not compromise the safety of the installation and that maintenance staff will not be exposed to unacceptable risks.

Criterion 2.4.2
The safety report should show that there are appropriate procedures for maintenance that take account of any hazardous conditions within the working environment.

The safety report should identify the procedures that are necessary to take into account the working environment and enable maintenance activities to be carried out safely with respect to maintenance staff and to prevent a major accident.

The safety report should show that safe systems of work have been established so that all activities that could result in dangerous situations are or can be identified.

Criterion 2.4.3
The safety report should show that systems are in place to ensure that safety critical plant and systems are examined at appropriate intervals by a competent person.

Appendix

This criterion concerns those activities that are carried out over and above routine maintenance to verify the continuing integrity of safety critical plant and systems. Examinations by a competent person at appropriate intervals may be necessary because certain specialised skills or equipment is required, or because it is demanded by specific legislation (e.g. Pressure Systems and Transportable Gas Container Regulations). For the purposes of this criterion, examination also includes any necessary testing.

Criterion 2.2.4

The safety report should show that there is a system in place to ensure the continued safety of the installations based on the results of periodic examinations and maintenance.

Evidence should be presented to show that defects detected during maintenance or examination are properly assessed by a competent person to determine their significance and appropriate action taken.

2.5 Modification:

Criterion 2.5.1

The safety report should describe the system in place for ensuring modifications are adequately conceived, designed, installed and tested.

The safety report should show that there is a system in place to deal with all modifications on the establishment. Those modifications to a process and its associated equipment, to structures (including warehouses) or to operations and procedures that could affect the safety of the installation should be subject to a formal modification system. This includes both hardware (e.g. pumps, piping arrangements, structures) and software (e.g. control system software, operating systems). Decommissioning of facilities is also included under this heading.

ASSESSMENT OF COMAH SAFETY REPORTS: EMERGENCY RESPONSE CRITERIA

Mrs K.K.McDonald
Health and Safety Executive (HSE), Chemical and Hazardous Installations Division, University Road, Bootle, Merseyside. L20 3RA.

The Control of Major Accident Hazards Regulations (COMAH) will come into effect on 3 February 1999 and will implement the Seveso II Directive in the UK. The Regulations will place a duty on operators of top tier establishments to produce a written safety report to demonstrate that all necessary measures have been taken to prevent major accidents and to limit their consequences for people and the environment. This paper describes the assessment criteria which the Competent Authority, consisting of HSE and the Environment Agencies, will use to assess the emergency response aspects of COMAH safety reports. The paper also outlines the process by which the criteria have evolved.

Key words: Safety Reports, Assessment, Major Hazards, Emergency Response

INTRODUCTION

The Control of Major Accident Hazards Regulations (COMAH) will be introduced in the UK on 3 February 1999. The Regulations will bring into force the requirements of European Council Directive 96/82/EC (the Seveso II Directive) of 9 December 1996 on the control of major accident hazards involving dangerous substances.

The COMAH Regulations, hereafter referred to as the Regulations, require operators of top tier establishments to produce a written safety report to demonstrate that they have taken all measures necessary for the safe operation of their establishment. The Regulations also place duties on the Competent Authority, consisting of HSE and the Environment Agencies, to communicate the conclusions of its examination of the report to the operator; or prohibit the continued operation or bringing into operation of an establishment, installation or any part where the measures taken by the operator for the prevention and mitigation of major accidents are seriously deficient.

HSE has drafted a set of assessment criteria *(the Safety Report Handling, Assessment and Review Principles and Processes manual (SHARPP))*, in conjunction with the Environment Agencies, to provide guidance for inspectors in the assessment of safety reports. This paper describes those criteria intended to assist inspectors in determining whether the safety report adequately demonstrates that all necessary measures have been taken to mitigate the consequences of major accidents and that adequate arrangements have been made for emergency response. The paper also outlines the main stages in the development of the criteria.

THE EMERGENCY RESPONSE ASSESSMENT CRITERIA

Purpose of the Criteria

The emergency response criteria are intended to provide a framework for inspectors in assessing whether a safety report meets the purposes of the Regulations with respect to measures taken by the operator to limit the consequences of major accidents to people and the environment. The criteria will be made publicly available.

Scope of the Criteria

The purposes of a COMAH safety report, as defined by the Regulations, include a demonstration by the operator that necessary measures have been taken to limit the consequences of major accident hazards to people and the environment. An additional purpose is to demonstrate that on-site emergency plans have been drawn up, and that the necessary information has been supplied to Local Authorities to enable the off-site emergency plan to be drawn up.

The Regulations specify the minimum information to be included in a safety report which may satisfy these purposes:

(a) *description of the equipment installed in the plant to limit the consequences of major accidents;*
(b) *organisation of the alert and intervention;*
(c) *description of the mobilisable resources, internal or external;*
(d) *summary of elements described in sub-paragraphs (a), (b) and (c) necessary for drawing up the on-site emergency plan.*

Installed equipment mitigation measures (element (a)) are covered in the technical aspects criteria which are the subject of a separate HSE paper. The emergency response criteria address elements (b), (c) and (d).

The emergency response criteria provide guidance on assessment of information provided in a safety report to establish whether the defined purposes of a safety report have been met in relation to emergency response.

The criteria are concerned with on-site arrangements to respond to a major accident, interface of these arrangements with the off-site emergency plan, and the resources that can be mobilised by the operator to take mitigatory action to minimise the consequences of a major accident. The criteria cover the information required to be supplied by the operator to enable the Local Authority to draw up the off-site emergency plan. The criteria also address those specific aspects of the Safety Management System which are directly relevant to emergency response.

As mentioned above, these criteria are specific to assessment of information submitted in safety reports. In addition, a working group led by HSE (via Safety Policy Directorate (SPD E)) is currently developing guidance on implementing the emergency planning aspects of the COMAH Regulations. Outputs from the SPD E led working group are outside the scope of this paper, however, it should be noted that development of the safety report emergency response assessment criteria took account of any relevant information from this work, through involvement of a member of the SPD E working group.

Drafting the Criteria

The criteria were drafted over a period of five months by a multi-disciplinary working group of six consisting primarily of HSE inspectors with experience of regulating the major hazards sector. The Environment Agencies were also represented. HSE inspectors on the working group were from:

- Chemical and Hazardous Installations Division (CHID) - project manager for development of the emergency response criteria, technical assessor and operational site inspector;
- Nuclear Safety Division (NSD) (whose regulation of emergency arrangements in the nuclear industry is well developed) - a member of NSD's Emergency Arrangements team;
- Safety Policy Directorate (SPD E) - a member of the HSE SPD E COMAH implementation working group on emergency planning.

Development of the criteria drew heavily on the knowledge and experience of the members of the working group. The criteria evolved via an iterative process. An initial set of criteria was drafted by the project manager on the basis of personal regulatory experience of the nuclear industry and on a review of existing literature on the subject. This preliminary draft of the criteria was considered at the first meeting of the working group and revised in light of comments from the working group. The process of review and redrafting was then repeated. The working group invited comments on its agreed assessment criteria from areas of CHID not represented on the working group, namely those areas with responsibility for risk assessment and for regulating the explosives sector. The working group took account of these comments when forming the final version of the criteria. This was sent out in parallel for internal consultation within HSE, for external consultation with industry and for use in the pilot exercise testing the developing criteria against COMAH-style safety reports submitted by industry.

The Structure of the Criteria

The working group made significant changes to the structure of the criteria in light of comments received from the consultation and pilot exercises; information on these exercises and resulting changes to the criteria are covered at a later stage. This section focuses on the revised criteria.

The criteria are representative of the type of measures which the Competent Authority would expect the operator to take to limit the consequences of major accident hazards to people and the environment.

The layout of the criteria is intended to reflect the layout of the Regulations. There are clearly defined links between the criteria and the Regulations to aid transparency of the assessment process. Each criterion is described, an explanation is given of why the criterion helps assessment against the defined purposes of a safety report, and examples of evidence are given indicating what constitutes sufficient information to satisfy the criterion.

Listed below are the headings under which the criteria are arranged. A brief description is given where an explanation of the heading is considered appropriate. In some cases, references have been made to assessment criteria covered in separate HSE papers. The emergency response assessment criteria are listed in Appendix 1 (sections on explanation and examples of evidence are not included).

<u>Organisation of the alert and intervention</u>   *The safety report should describe the organisation of the alert and intervention in the event of a major accident to provide evidence that the necessary measures have been taken on site.*

Typical points covered under this criterion include: functions of key posts; arrangements for controlling and limiting escalation of accidents on-site; arrangements for alerting individuals on-site, neighbouring establishments (where relevant) and the general public to the hazardous situation; arrangements for alerting and mobilising individuals with defined responsibilities under the emergency response; provision for establishing and maintaining communications during the emergency response; location of access routes and emergency control centres; evacuation arrangements; roll call and search and rescue arrangements; the nature and location of any pollution control devices and materials, and arrangements for subsequent environmental clean up and restoration.

<u>Description of the mobilisable resources</u>   *The safety report should describe the on-site and off-site resources which can be mobilised by the operator to provide evidence that the necessary measures have been taken to limit the consequences of a major accident to people and the environment.*

The description should cover human resources, hardware, for example fire fighting equipment, and ancillary equipment required to enable mitigatory action to be carried out, for example, personal protective equipment, vehicles transporting equipment to the site of the accident. This general criterion has been expanded to give ten more specific sub criteria for various types of resource.

<u>Maintenance, inspection, examination and testing of emergency response equipment (aspect of Safety Management System (SMS))</u>   This criterion addresses maintenance (planned and breakdown), inspection, examination and testing of emergency response

equipment and provisions. It expands on a general SMS assessment criterion covering arrangements for safe operation, including maintenance, of plant, processes and equipment; the emergency response criteria cross reference to the SMS criteria. Details of the SMS assessment criteria are given in a separate HSE paper.

Training in the emergency response (aspect of Safety Management System (SMS))
This criterion covers the training of individuals on-site in emergency response procedures. and expands on a general criterion in the SMS assessment criteria covering provision and maintenance of appropriate levels of management and employee competence.

Testing of emergency plans (aspect of Safety Management System (SMS)) This criterion deals with testing, review and revision of on-site emergency plans in light of lessons learned. Interface with the off-site response is also covered.

Information required for off-site emergency plan As mentioned earlier, one of the purposes of a safety report is to include a demonstration that the necessary information has been supplied to Local Authorities to enable the off-site emergency plan to be drawn up. This criterion and its associated guidance identify the minimum information to be included in a safety report to demonstrate that the operator has provided the Local Authority with sufficient information to draw up the off-site emergency plan. This minimum information includes: details of the site, including its location, nearby roads and site access; site plan showing location of key facilities such as control centres, medical centres, location of main process plant and stores, staffing levels; details of off-site area likely to be affected by a major accident; details of dangerous substances on-site; details of technical advice that the company can provide to assist the off-site response; relevant technical details of resources (equipment or chemicals) normally on site which may be available to assist the off-site response; functions of key posts with duties in the emergency response, their location and how they can be identified; outline of initial actions, and procedures in on-site plans, to be taken by on-site staff once an emergency has been declared.

Elements in safety report necessary to draw up the on-site emergency plan This criterion requires the safety report to contain a summary specifying the basis for drawing up the on-site emergency plan and should cover:

- the equipment installed in the plant to limit the consequences of major accidents;
- the organisation of the alert and intervention;
- the on-site and of-site resources that may be mobilised.

SERIOUS DEFICIENCY

The Competent Authority has a duty to prohibit the continued operation or bringing into operation of an establishment, installation or any part where a serious deficiency has been identified in relation to measures taken by the operator to mitigate major accidents.

The Guiding principles for assessment of safety reports are described in a separate HSE paper. These principles state that where assessment of a safety report identifies a potential serious deficiency, assessors need to obtain first hand evidence by a site visit and by checking the facts with the operator, before taking a prohibition decision.

When considering the evidence for a serious deficiency, assessors are advised to take account of the adequacy of the totality of the emergency response provisions. Mitigatory measures which can be mobilised by the operator will be considered by the Competent Authority in the context of the permanent installed (fixed) mitigatory provisions and the resources that can be mobilised by the off-site emergency services. The following are examples of the type of deficiencies associated with the emergency response arrangements which the Competent Authority would regard as serious deficiencies:

- Essential roles and responsibilities not defined in the on-site emergency plans;
- Insufficient staff or other resources to discharge the key functions identified the on-site emergency plans;
- Lack of training for individuals with key roles under the on-site emergency plans;
- Lack of maintenance, examination, testing and inspection arrangements for the mitigatory provisions;
- Failure to supply the information required to enable the off-site emergency plan to be drawn up;
- Effective means of access for off-site emergency services not available.

The working group developing the emergency response criteria concluded that serious deficiencies associated with mitigatory measures would not necessarily attract an immediate prohibition, but could result in a deferred prohibition to enable the deficiencies to be rectified before the use of the plant was prohibited. This recommendation was based on time at risk arguments, that is, on the small likelihood of mitigatory action being needed on an immediate timescale.

CONSULTATION AND PILOT EXERCISES.

Consultation

There was significant consultation outside government departments or related bodies on the SHARPP manual which will form the basis of the Competent Authority's assessment of COMAH safety reports. The main organisations involved in the consultation represented a spectrum of industry considered to have the primary interest in the assessment process:

- the Chemical Industries Association (CIA) - primarily through the CIA COMAH Sub-group which included representatives from major companies such as Shell, BP, Exxon and ICI;

- the CBI (Confederation of British Industries);
- trade associations which represented smaller companies involved in manufacturing or distribution which have been brought together by the British Chemical Distributors and Traders Association (BCDTA);
- British Gas directly because of the large number of top tier sites they will have.

Environmental interest groups (English Nature; Scottish Natural Heritage; Countryside Council for Wales; Green Alliance; Friends of the Earth (Scotland)) were consulted on the assessment criteria.

Particularly relevant to the emergency arrangements criteria, the Society of Industrial Emergency Services Officers (SIESO) held a workshop for discussion of the SHARPP outputs.

The consultation exercise (covering both internal HSE and external comments) identified a number of points for consideration by the working group developing the emergency response criteria. As for other SHARPP criteria, the presentation of the emergency response criteria was to be modified to a common format. Since comments had been made to the effect that there were too many criteria, the working group had to establish whether the number of criteria could be reduced. To show that all criteria were justified in relation to the requirements of COMAH, explanatory text was to be included on how each criterion was rooted in legal requirements; where this justification could not be made, the working group would need to consider whether the particular criterion merited inclusion. The level of detail given in the criteria was to be re-evaluated. For each criterion, examples were to be given indicating what information was required to satisfy the requirements of the Regulations. Overlaps of the emergency response criteria with other criteria were to be considered with a view to eliminating unnecessary overlap.

There were some points specific to the detail of the emergency response criteria. The working group was required to provide guidance on whether emergency plans could be included as part of a safety report or whether the operator had to re-submit this information. It was necessary to clarify that the Competent Authority was only concerned with assessment of the operator's fire fighting and fire protection provisions, not of resources available from local and other fire brigades; although, it was expected that the operator would take account of such resources. Textual amendments proposed by consultees were to be considered in relation to improving the suitability and clarity of the criteria.

Pilot

In order to assist the Competent Authority in evaluating how the assessment criteria would work in practice, the following four establishments volunteered to produce COMAH-style safety reports for assessment:

- Elf, Flotta;
- International Speciality Chemicals Limited (ISC), Hythe
- BP, Hull
- British Gas Transo, Cheltenham.

Correspondingly, the Competent Authority set up four teams, of similar composition to those proposed for assessment of actual COMAH safety reports, to assess these voluntarily submitted safety reports. The pilot version of the SHARPP manual was used for assessment purposes.

The pilot exercise results were in line with those obtained from the consultation exercise. A significant additional point to be considered by the emergency response criteria working group was provision of additional guidance on what constituted a serious deficiency in relation to measures taken by the operator to mitigate major accidents. This was particularly important since the Competent Authority has a duty under COMAH to prohibit where such measures are seriously deficient.

Changes to criteria following consultation and pilot exercises

The working group evaluated the criteria taking into account the results of the consultation and pilot exercises. The criteria were re-structured from a list of 21 criteria with supporting guidance to a format describing each criterion, reasons for inclusion of the criterion, and examples of evidence expected in safety reports to show how the requirements of the Regulations had been met.

Re-consideration of the criteria against legal requirements resulted in the number of criteria being reduced from 21 criteria to 7 key criteria and 10 sub-criteria; the majority of technical information included in the original 21 criteria but not in the revised list was still considered important by the working group and was generally incorporated into examples of evidence under appropriate revised criteria.

Overlaps of the emergency response criteria with other criteria were considered, in particular with those covering Safety Management Systems: maintenance, inspection, examination, testing of emergency response equipment; training in the emergency response; testing of emergency plans. The working group considered that criteria related to these issues was important in consideration of the emergency response aspects of safety reports and supported their retention in the emergency response criteria. These criteria were cross referenced to the Safety Management Systems criteria.

A large number of the textual changes proposed by consultees were taken on board by the working group. These changes were thought to present the criteria in a more clear, concise and user friendly manner. Amendment of a few criteria was considered appropriate to make it clear that the Competent Authority would not be assessing the fire fighting capability of fire brigades.

The working group re-considered measures which could constitute a serious deficiency in terms of emergency response. The types of measures originally suggested remained unaltered. However, more emphasis was given to the point that due account was to be taken of the adequacy of the totality of the emergency response provisions.

The working group concluded that emergency plans could be included in safety reports but a supplementary route map would be required clearly identifying how the safety report purposes were met. Guidance on this point was not to be included in the criteria but would incorporated in an appropriate section of the SHARPP manual.

CONCLUSIONS

HSE, in conjunction with the Environment Agencies, has drafted criteria which will be used by the Competent Authority in assessing the emergency response aspects of COMAH safety reports. The criteria have been developed by a Competent Authority working group with significant experience of regulating the major hazards industry. In developing these criteria, the Competent Authority has consulted industry, Environmental Groups and specialist organisations, for example, SIESO with respect to the emergency response criteria. A pilot assessment exercise has been conducted by the Competent Authority to test and refine the developing criteria using COMAH-style reports submitted voluntarily by industry. Considerable change has been made to the structure of the criteria following the consultation and pilot feedback, however, the technical content remains largely unaltered.

ACKNOWLEDGEMENTS

Development of the criteria was a team effort involving HSE and EA staff. HSE members involved in developing the criteria were: Mrs J Rutherford (team leader to the point at which the criteria were sent out for use in the consultation and pilot exercise), Mr J Carter, Mrs M Wilson, Mr R Cowley, Mr R Hadden, Mr P Rushton. Mr B McGlashan provided the EA input.

REFERENCES

1. HS(R)21(rev) ISBN O 11 8855794, HSE (1990): A Guide to the Control of Industrial Major Accident Hazards Regulations 1984.
2. ISBN 0 11 8820435, HSE (1992): Safety Assessment Principles for Nuclear Plants.
3. HS/G 25: The Control of Industrial Major Accident Hazards Regulations 1984 (CIMAH): further guidance on emergency plans.
4. ISSN 1018-5593, Joint Research Centre, Institution for Systems Informatics and Safety, Guidance on the Preparation of a Safety Report to Meet the Requirements of Council Directive 96/82/EC (Seveso II).

APPENDIX 1

REVISED EMERGENCY ARRANGEMENTS ASSESSMENT CRITERIA FOLLOWING PILOT EXERCISE AND EXTERNAL CONSULTATION.

Organisation of the alert and intervention

1. The safety report should describe the organisation of the alert and intervention in the event of a major accident to provide evidence that the necessary measures have been taken on-site.

Description of mobilisable resources

2. The safety report should describe the on-site and off-site resources which can be mobilised by the operator to provide evidence that the necessary measures have been taken to limit the consequences of a major accident to people and the environment.

    2.1 The safety report should provide evidence that sufficient personnel can be made available within appropriate timescales to carry out the mitigatory actions required by the on-site emergency plans.

    2.2 The safety report should provide evidence that suitable and sufficient arrangements are in place to ensure that the equipment to be mobilised for mitigating the consequences of reasonably foreseeable major accidents will be fit for purpose when called upon for use.

    2.3 The safety report should provide evidence that suitable and sufficient personal protective equipment will be available in the event of a major accident.

    2.4 The safety report should provide evidence that suitable and sufficient on-site fire fighting and fire protection provisions can be mobilised in the event of a major accident, taking account of resources available from local and other fire brigades.

    2.5 The safety report should provide evidence that suitable and sufficient provisions can be mobilised to minimise the release of, and mitigate the consequences of, airborne toxic and/or flammable substances in the event of a major accident.

    2.6 The safety report should provide evidence that suitable and sufficient resources can be mobilised to minimise the consequences of loss of containment of a hazardous substance(s) to ground or water (including Controlled Waters).

2.7 The safety report should provide evidence that suitable and sufficient provisions for monitoring and/or sampling can be mobilised in the event of a major accident.

2.8 The safety report should provide evidence that suitable and sufficient provisions have been made for the restoration and clean up of the environment following a major accident.

2.9 The safety report should provide evidence that suitable and sufficient provisions have been made to mobilise first aid/medical treatment during the emergency response.

2.10 The safety report should provide evidence that suitable and sufficient provisions have been made to mobilise any ancillary equipment which may be required during the emergency response.

Maintenance etc. of emergency response equipment

3. The safety report should provide evidence that suitable arrangements have been made for the maintenance, inspection, examination and testing of the mobilisable resources and other equipment to be used during the emergency response. (This criterion only applies to mobilisable resources and other equipment for which the operator has responsibility.)

Training in the emergency response

4. The safety report should provide evidence that suitable arrangements have been made in the safety management system for training of individuals on-site in the emergency response.

Testing of emergency plans

5. The safety report should provide evidence that procedures have been made and adopted to test and review emergency plans, and to revise the emergency arrangements in the light of the lessons learned.

Information required for the off-site emergency plan

6. The safety report should supply information to enable the off-site emergency plan to be drawn up.

Elements in the safety report necessary to draw up the on-site emergency plans

7. The safety report should summarise those measures of protection and intervention which have been used as the basis for drawing up the on-site emergency plans.

SAFETY MANAGEMENT SYSTEM ASSESSMENT CRITERIA

Dr David Snowball
Health and Safety Executive (HSE), Chemical and Hazardous Installations Division, Sovereign House, 110 Queen Street, SHEFFIELD S1 2ES

This paper describes the safety management system assessment criteria which will be used to assess the information in safety reports relating to Annex III of European Directive 96/82/EC. An operator's safety management system forms only one part of a wider set of management arrangements, reflecting the management philosophy and safety culture of the organisation. The HSE model is used as a template. The paper describes the model and explains how it underpins the assessment criteria.

Key words: Safety Reports, Major Hazards, MAPP, Safety Management Systems

INTRODUCTION

This paper describes how HSE developed the criteria to be used to assess the Major Accident Prevention Policy (MAPP) and the safety management systems (SMS) in safety reports submitted under the Control of Industrial Major Accident Hazards Regulations (COMAH). The criteria did not evolve from a blank sheet of paper but reflect previous HSE experience both in the major hazards sector and under other legislative regimes.

In COMAH the MAPP is the statement of management intent with respect to major accident prevention. This is equivalent to a major hazard version of the health and safety policy which is a *general* requirement under Section 2(3) of the 1974 Health and Safety at Work Act 1974. The Safety Management System (SMS) is the vehicle for delivery and therefore broadly equivalent to the requirement to make arrangements for securing health and safety under Regulation 4 of the Management of Health and Safety at Work Regulations 1992 (MHSWR). The 2 components are complementary: the MAPP provides the strategic thrust behind the requirement to prevent major accident hazards; the SMS provides the framework to implement it.

HISTORIC REQUIREMENTS

The duty to provide management information to describe or demonstrate the adequacy of controls is not novel. Under the 1984 Control of Industrial Major Accident Hazards Regulations (CIMAH), the 'entry level' requirement in Regulation 4 requires all sites subject to CIMAH, not just the top tier ones, to be able to provide evidence for 'Demonstration of safe operation'. CIMAH also required manufacturers to provide certain information in a safety report relating to the management system for controlling the industrial activity. The matters were set out in Schedule 6 to the Regulations and included:

The **staffing** arrangements for controlling the industrial activity;
The arrangements for the **safe operation** of the activity including design, construction, testing, operation, inspection and maintenance;
The arrangements for **training.**

HSE published guidance on what to include under each of the 3 headings, first in 1985 and then in revised form in 1990 (HSE, 1990). This predated the publication of HSE's own guidance entitled 'Successful Health and Safety Management' in 1991 (updated 1997) and usually known more familiarly by its Series number: *HS(G)65* (HSE, 1997).

HSE GUIDANCE ON SUCCESSFUL HEALTH AND SAFETY MANAGEMENT

HS(G)65 conveys a simple message: organisations need to manage health and safety with the same degree of expertise and to the same standards as other core business activities if they are to control risks effectively and prevent harm to people. The guidance is not mandatory and is conceived as a general model for all sectors of work activity, not just major hazards. It draws on the evidence which HSE has gathered about what 'works' in companies who demonstrate they can achieve high standards of health and safety performance.

In HS(G)65, the key elements of successful health and safety management can be broken down into 5 separate stages (see Fig 1) which can be summarised by the acronym POPMAR:

POLICY: Effective health and safety policies set a clear direction for the organisation to follow and contribute to business performance

ORGANISING: An effective management structure and arrangements are in place for delivering the policy. There are structures and systems to:

    establish and maintain management **control**
    promote **cooperation** between staff to make health and safety a collaborative effort
    ensure **communication** of necessary information throughout the organisation
    secure the **competence** of employees

PLANNING: There is a planned and systematic approach to implementing the health and safety policy through an effective health and safety management system which both controls risks and reacts to changing demands.

MEASURING PERFORMANCE: Performance is measured against agreed standards to reveal when and where improvement is needed in two ways: **active** systems which monitor the achievement of plans and the extent of compliance with standards and **reactive** systems which monitor accidents ill health and incidents

AUDITING AND REVIEWING PERFORMANCE: The organisation learns from all relevant experience and applies the lessons through both auditing and review of performance.

The approach bears obvious comparison with other models for planning and decision making. In particular, it incorporates feedback loops to improve performance and emphasises the importance of quality assurance *within* a management process rather than quality control at the end. Many organisations only react to accidents and ill health ('defects') once they have occurred (the quality control approach). But if the 'output' is risk control, then the process has to be properly assured. This can be achieved by designing and implementing an effective proactive health and safety management system. HSE does not promote the POPMAR model as the only way to manage health and safety. *Organisations can manage safety in whatever ways they choose and the model provides only a guide.*

CIMAH did not deal explicitly with 2 elements in the POPMAR model: policy and auditing. MSHWR aligns more closely with the HS(G)65 headings and it is possible for manufacturers to read across from existing requirements under CIMAH to the elements of sound management and to provide a structure for setting out the management information in safety reports. Examples of some of the main links are shown in Table 1.

Table 1: Links between HS(G)65 and CIMAH

HSG65	CIMAH
Policy	Not specifically treated
Organising Control Competence Cooperation Communication	staffing and reporting arrangements training systems securing contributions from all staff up to date information, communication of operating procedures
Planning and implementing	Hazard analysis and risk assessment safe operating procedures application of human factors control of contractors
Measuring	Arrangements for inspection, test and maintenance Quality assurance Checking working methods Investigating accidents and near misses
Audit	Not specifically mentioned, but requirement to ensure the adequacy of the management structure and to audit design
Review	Correcting deficiencies Keeping senior management informed

EXPERIENCE IN OTHER SECTORS

HSE inspectors assess management arrangements in all sectors of industry. Similar approaches exist elsewhere. In the nuclear industry, licensees have to produce and maintain adequate safety cases. In the railway, gas and offshore sectors, operators have to submit safety cases requiring formal acceptance by HSE. These require demonstrations of suitable management arrangements to control risks. In the relevant HSE guidance for offshore safety (HSE 1992), it states:

'The requirement is...to demonstrate the adequacy of the system, and not to show in detail how compliance with all relevant requirements is to be ensured"

Use of the model by HSE inspectors has revealed what successful organisations can achieve. The 1997 edition of HS(G)65 contains a 3 component model illustrating three targets for management effort in controlling risks: These are shown in Fig 2.

Workplace precautions protect people at the point of risk: They include physical equipment such as guards on machines or relief valves on pressure systems and 'software' such as instructions or systems of work.

Risk Control Systems (RCS) produce the appropriate workplace precautions. Organisations need to have a range of RCS which are appropriate to the hazards arising from their activities and which are sufficient to cover all hazards. This means that each organisation has to build up a profile of the risks to which its employees and others may be exposed. The design, reliability and complexity of each RCS needs to be proportionate to the hazards and risks inherent in the operation. For major hazards, typical RCS include control of contractors, permits to work, maintenance, plant and process change and operating procedures.

Management arrangements are necessary to organise, plan, control and monitor the design and implementation of the RCS.

Together, these 3 components can be assembled into a single 'picture' of a health and safety management system to form a framework for planning and auditing.

LINKING THE MODEL WITH THE REQUIREMENTS UNDER COMAH

When the team looked at the requirements under Annex III of the Directive for the MAPP and the SMS, the primary objective was to avoid reinventing the wheel. This accumulated experience could be used to make explicit links between the new requirements in Annex III and the management model in HSG65 (Table 2).

Table 2: Links between Annex III and HS(G)65

ANNEX III (ref to para in brackets)	HS(G)65 ELEMENT
MAPP (a)	Policy
SMS (b)	POPMAR
Organisation and personnel (c)(i):	control
Roles and responsibilities (c)(i)	control
Training (c)(i)	competence
Involvement of employees (c)(i)	cooperation
Identification and evaluation of major hazards (c)(ii)	planning and implementing
Operational control (c)(iii)	risk control system for safe operation, including control and communication
Management of change (c)(iv)	risk control system
Planning for emergencies (c)(v)	risk control system

Monitoring performance (c)(vi)	active and reactive monitoring
Audit and review (c)(vii)	audit and review

The 'fit' between COMAH and HSG65 is not perfect but it is very close. The requirements for a MAPP and for auditing address the omissions from CIMAH. Communication is not mentioned specifically under 'Organising' but is implicit under the other headings as a means of establishing and maintaining control. There are 3 *sets* of risk control systems specified under COMAH for operational control, management of change and planning for emergencies. These are shown in Fig 3 in a modified 'COMAH version' of the 3 component model:

For each criterion included, an explanation was given about why it needed to be there by referring back to the Directive and explaining why individual criteria added value. Some examples of evidence which would satisfy the requirement were also given. 29 separate criteria were identified including the following example:

CONTROL

Criterion: The safety report should show that the responsibilities of everyone involved in the management of major hazards have been clearly defined.

Reason: Unless responsibilities have been clearly defined the operator will be unable to implement the MAPP. Employees and other people involved need to know who is responsible for each aspect of managing the major accident hazards.

Examples of evidence:

Reference to job descriptions or other documents in which responsibilities for the control of major accident hazards are explicitly allocated to line managers.

Descriptions of the responsibilities allocated to key managers and post holders at all levels depending on the management structure of the organisation.

Appropriate references to the way in which the operator has set out how particular jobs should be done eg by using performance standards.

A full list of the criteria is shown in the Annex.

CHALLENGES

The main challenge faced by the team was to agree that the model was appropriate for COMAH requirements and from there to develop a set of workable and robust criteria. It was important for the users to be able to recognise what is required, what evidence might satisfy the criteria and what constitutes an acceptable standard of performance.

The team interpreted 'demonstration' in the following way. Whereas CIMAH required a safety report to contain descriptions (and manufacturers could usually satisfy this requirement by describing a set of outcomes) COMAH requires an explanation of the *process* for delivering specific outcomes. This is equivalent to the difference between 'proof' (which could be an outcome) and 'evidence leading to proof' (a process description which produces the desired outcome).

ASSESSMENT

The purpose of assessment of the MAPP and the SMS is to answer 5 questions:

1 Does the safety report contain a MAPP?
2 Does the information in the safety report demonstrate that there is a SMS for implementing the MAPP?
3 Does the information provided in the safety report as a whole demonstrate that the MAPP and the rest of the SMS have been out into effect?
4 Does the information demonstrate that all necessary measures have been taken to prevent major accidents and to limit their consequences for people and the environment?
5 Has the assessment revealed any serious deficiencies in the measures taken for the prevention and mitigation of major accidents?

The approach to the MAPP and the SMS will be the same as for all the other elements in the safety report. The safety report describes a series of outcomes which are themselves determined or influenced by the SMS. This includes the technical descriptions and predictive elements. The SMS will be assessed as a whole, not as a series of isolated parts.

'SERIOUSLY DEFICIENT'

The following circumstances might arise:

1 the MAPP is seriously deficient or absent
2 the management arrangements are seriously deficient or absent
3 a single element of the SMS or a RCS is seriously deficient or absent
4 there are a number of elements in the SMS or RCS which, taken in isolation, are not seriously deficient, but when viewed as a whole, render the whole SMS seriously deficient
5 there are a number of elements in the SMS or RCS which, in total, are not seriously deficient, but when evaluated together with technical and/or predictive shortcomings, render the report seriously deficient.

CONCLUSIONS

The safety management system assessment criteria closely parallel the elements of sound health and safety management practice set out in HSE's own guidance. However, operators are free to use other management models to suit their own particular situations.

ACKNOWLEDGEMENTS

The task of writing the criteria was a team effort involving staff from HSE and EA. Much of the burden was borne by Graham King, an inspector in CHID 2A. Other HSE staff who contributed were Louise Brearey, Neil Rothwell, Trevor Britton, Ian Travers, David Porter, Andrew Train and David Derbyshire. Adrian Milner provided the EA input. Special thanks also to Julie Hobson and Julie Pilkington at CHID in Sheffield who coped good-humouredly with the numerous typing amendments.

REFERENCES

HSE (1990) A guide to the Control of Industrial Major Accident Hazards Regulations 1984, HS(R)21 (rev) ISBN 0 11 885579 4

HSE (1997) Successful Health and Safety Management HS(G)65, ISBN 0 7176 1276 7

HSE (1992) A guide to the Offshore Installations (Safety Case) Regulations 1992, L30 ISBN 0 11 882055 9

1. **KEY ELEMENTS OF SUCCESSFUL HEALTH & SAFETY MANAGEMENT**

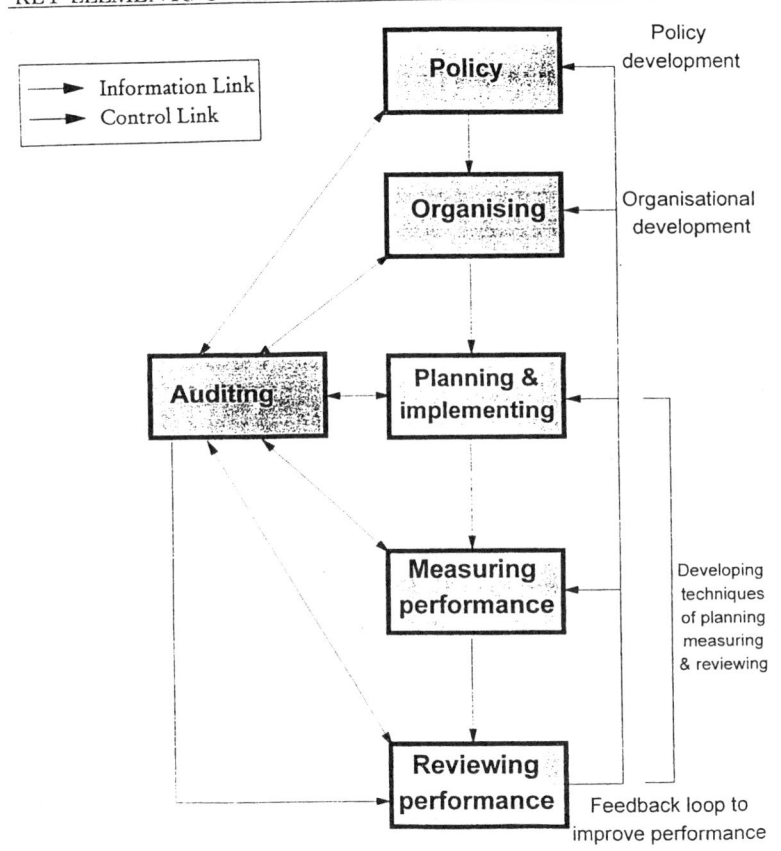

2. HEALTH & SAFETY MANAGEMENT SYSTEM

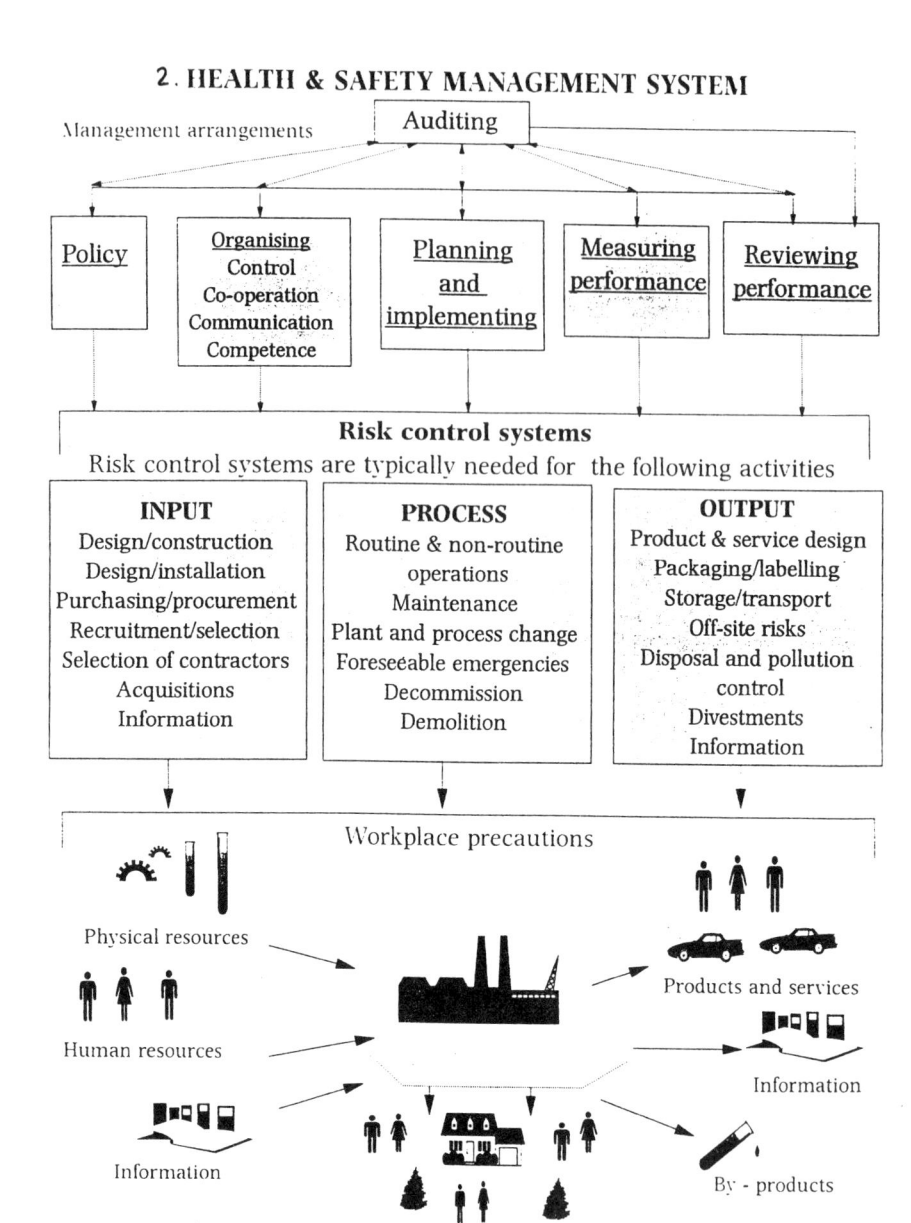

Management arrangements — **Auditing**

- **Policy**
- **Organising**: Control, Co-operation, Communication, Competence
- **Planning and implementing**
- **Measuring performance**
- **Reviewing performance**

Risk control systems
Risk control systems are typically needed for the following activities

INPUT	PROCESS	OUTPUT
Design/construction	Routine & non-routine operations	Product & service design
Design/installation	Maintenance	Packaging/labelling
Purchasing/procurement	Plant and process change	Storage/transport
Recruitment/selection	Foreseeable emergencies	Off-site risks
Selection of contractors	Decommission	Disposal and pollution control
Acquisitions	Demolition	Divestments
Information		Information

Workplace precautions

- Physical resources
- Human resources
- Information

→ Products and services
→ Information
→ By-products

3. MANAGEMENT ARRANGEMENTS

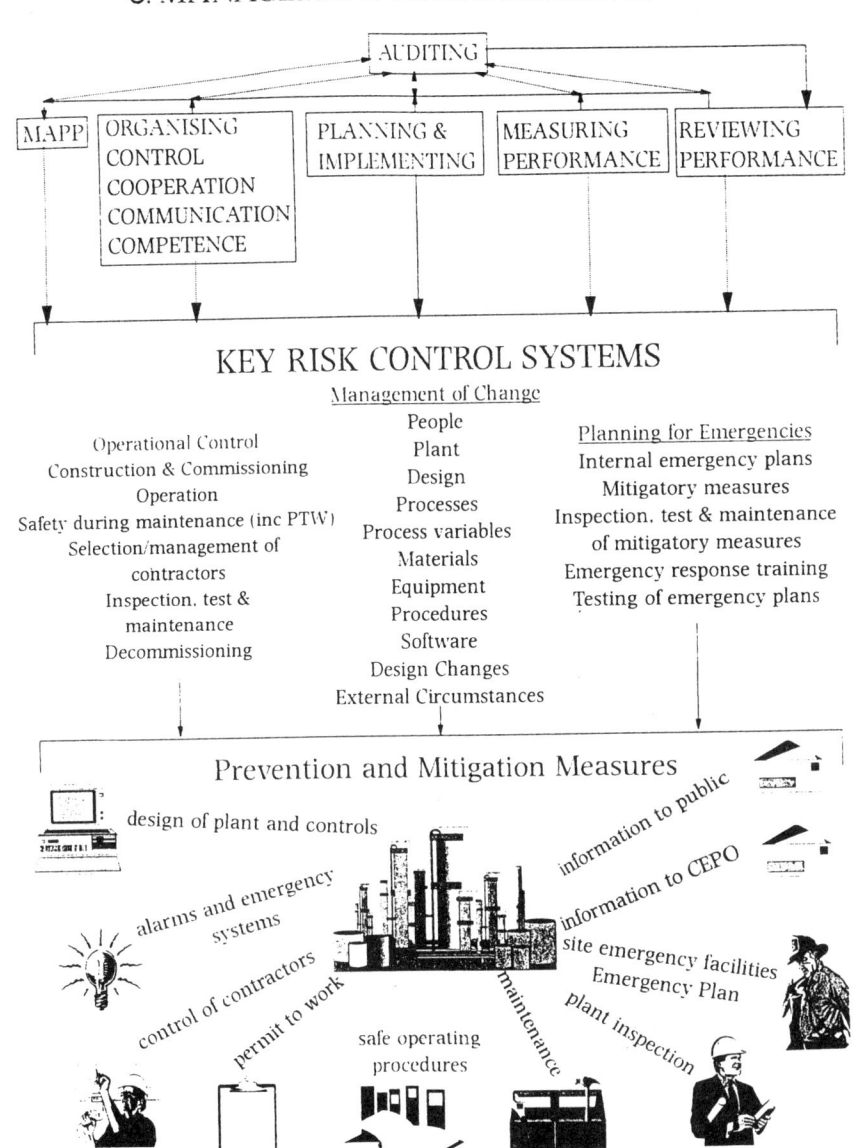

```
                         ┌─ AUDITING ─┐
     ┌──────┬────────────┼────────────┬─────────────┐
     │      │            │            │             │
   ┌────┬─────────┬──────────────┬────────────┬────────────┐
   │MAPP│ORGANISING│ PLANNING &  │ MEASURING  │ REVIEWING  │
   │    │ CONTROL  │IMPLEMENTING │PERFORMANCE │PERFORMANCE │
   │    │COOPERATION│            │            │            │
   │    │COMMUNICATION│          │            │            │
   │    │COMPETENCE│             │            │            │
   └────┴─────────┴──────────────┴────────────┴────────────┘
```

KEY RISK CONTROL SYSTEMS

Management of Change

Operational Control	People	Planning for Emergencies
Construction & Commissioning	Plant	Internal emergency plans
Operation	Design	Mitigatory measures
Safety during maintenance (inc PTW)	Processes	Inspection, test & maintenance
Selection/management of	Process variables	of mitigatory measures
contractors	Materials	Emergency response training
Inspection, test &	Equipment	Testing of emergency plans
maintenance	Procedures	
Decommissioning	Software	
	Design Changes	
	External Circumstances	

Prevention and Mitigation Measures

- design of plant and controls
- alarms and emergency systems
- control of contractors
- permit to work
- safe operating procedures
- maintenance
- information to public
- information to CEPO
- site emergency facilities
- Emergency Plan
- plant inspection

ASSESSMENT CRITERIA FOR THE MAPP/SMS   ANNEX 1

MAPP		The MAPP should:
	1	include a commitment to achieve a high standard of protection for people and the environment
	2	set out the operator's overall aims and principles of action with respect to the control of major accident hazards
	3	include a commitment to achieving high standard of protection for people and the environment and demonstrate that an appropriate management system covering the following elements has been put into effect: the roles and responsibilities of personnel involved in the management of major hazards at all levels in the organisation, including contractors where appropriate, and the provision of training to meet identified training needs the procedures for systematically identifying major hazards arising from normal and abnormal operation and the assessment of their likelihood and severity the procedures and instructions for safe operation, including maintenance of plant, processes, equipment and temporary stoppages the procedures for planning modifications to, or the design of new installations, processes or storage facilities the procedures to identify foreseeable emergencies by systematic analysis and to prepare, test and review emergency plans to respond to such emergencies the procedures for the ongoing assessment of compliance with ht objectives set out in the MAPP and SMS and the mechanisms for investigation and taking corrective action in the event of failing to achieve the stated objectives. The procedures should cover the operator's system for reporting major accidents and near misses, particularly those involving failure of protective measures and their investigation and follow up on the basis of lessons learnt the procedures for periodic systematic assessment of the MAPP and the effectiveness and suitability of the SMS, the documented review of performance of the MAPP and SMS and their updating by senior management
	4	be set at a senior level in the operator's organisation
	5	be in writing
SMS		The safety report should
	6	include sufficient explanation of how the SMS fits into the overall organisational arrangements

62

7	show that all necessary roles in the management of major hazards have been clearly allocated
8	show that the responsibilities of everyone involved in the management of major hazards have been clearly defined
9	show that the operator has allocated sufficient resources to implement the MAPP
10	show that the performance of people having a role to play in the management of major accident hazards is measured and that they are held accountable for their performance
11	show that the operator has in place a system for providing and maintaining appropriate levels of management and employee competence
12	show that the operator has systems for ensuring that employees are actively involved in the management of major accident hazards
13	show that the operator has in place arrangements for co-operating with and securing the cooperation of other organisations
14	show that the operator has arrangements for gathering intelligence needed for the control of major accident hazards from external sources
15	show that the operator has arrangements for communicating information important for the control of major accident hazards within the operator's organisation
16	show that the operator has arrangements for communicating information relevant to the control of major accident hazards to external organisations
17	show that the operator has arrangements for systematically identifying major hazards, assessing the risks from normal and abnormal operations and determining necessary control measures
18	show that the operator has systems for identifying areas for necessary improvement in relation to the control of major accident hazards
19	show that the operator has systems for determining priorities to achieve the objectives of the MAPP and scheduling necessary improvement work in relation to ht control of major accident hazards
20	show that the operator has adopted and implemented procedures and instructions for safe operation, including maintenance, of plant, processes, equipment and temporary stoppages
21	show that the operator has adopted and implemented procedures for planning modifications to, or the design of new installations, processes or storage facilities

	22	show that the operator has arrangements in place to identify foreseeable emergencies by systematic analysis and to prepare, test and review emergency plans
	23	show that the operator has adopted and implemented procedures for the ongoing assessment of compliance with the objectives set by the MAPP and SMS
	24	show that the operator has adopted and implemented a system for reporting major accidents and near misses particularly those involving failure of the protective measures for control of major accident hazards
	25	show that the operator has adopted and implemented mechanisms for investigation and taking corrective action in cases of non-compliance with the objectives set by the MAPP and in relation to major accidents and near misses
	26	show that the operator has adopted and implemented a procedure for systematic independent assessment of the MAPP and the effectiveness and suitability of the SMS
	27	show that the operator has adopted and implemented a review process which uses information from performance measurement and audit
	28	show that results of review are documented
	29	show that the operator has adopted and implemented a system under which the MAPP and SMS is updated by senior management

ASSESSMENT OF THE PREDICTIVE ASPECTS OF COMAH SAFETY REPORTS

Dr Shaun Welsh
Health & Safety Executive (HSE), Chemical and Hazardous Installations Division (CHID), Major Hazards Assessment Unit (MHAU). St Annes House, Stanley Precinct, Bootle, Merseyside L20 3RA

> This paper introduces the 'predictive assessment criteria' and complements other papers which describe the development of the Health and Safety Executive's (HSE) guiding principles procedures and criteria for safety reports submitted under the Control of Major Accident Hazard (COMAH) Regulations 1999. The 'predictive' assessment criteria are needed as part of the demonstration of safety required by Schedule 4 of the draft Consultative Document for the proposed COMAH Regulations (1).

Key Words - COMAH Safety Reports: predictive criteria

INTRODUCTION

The assessment of safety reports submitted under the Control of Industrial Major Accident (CIMAH) Regulations 1984 (2) is not based on any specific evaluation criteria for either the acceptance or rejection of the report. However, in addition to published guidance (3), internal guidelines are used by HSE as the basis for forming judgements about the validity of evidence presented in the report to determine whether or not there are gross omissions or serious deficiencies, when compared against the requirements of CIMAH Schedule 6. In turn, these deficiencies may also be indicative of failures to follow established good practice and safety standards and hence compliance with statutory duties under the Health & Safety at Work etc Act (HASWA) 1974 to control the risks from major chemical hazard installations. Such deficiencies can then be remedied both in terms of requirements for additional information to be submitted for the safety report and if necessary during the later inspections of the installation, for which the safety report is pivotal.

It is important to stress that assessment of the CIMAH safety report is not a measure of the level of compliance with HASWA and relevant statutory provisions. That assessment establishes whether the report contained sufficient information to satisfy the requirements of Schedule 6; a test of the evidence not the activity, A safety report could meet these requirements even though it may reveal areas of weakness in risk control measures. If a report which provided adequate information on Schedule 6 topics, indicated poor compliance on site and is used to determine inspection priorities then it would have fulfilled its purpose. Conversely weaknesses in a report which lead HSE to require extra information from the Manufacturer or even a re - submission of the report does not necessarily imply major defects in on site standards of health and safety. Under CIMAH only when an inspector used Regulation 4 would the Manufacturer be required to satisfy

HSE about the demonstration of safe operation of the activity. Under the COMAH regulations the type and scope of information provided to the Competent Authority (CA namely the Environmental Agency (EA) Scottish Environmental Protection Agency (SEPA) and HSE) is not expected to be significantly different from CIMAH. However, the purpose and use of the information will be significantly different in terms of the operator's 'demonstrations' to the CA.

The essential features of the predictive information required in a safety report are to show that appropriate and systematic analyses of the major accident hazards are carried out and the results presented in the report. These elements comprise:

* Hazard Identification and Analysis: the properties and hazards (fire, explosion and toxicological - including ecotoxicological) of the dangerous substances, nature of process and operating parameters

* Accident Scenario Analyses: the conditions events, both internal and external and mechanisms leading to loss of containment and release of dangerous substances which would have the potential for major impacts on people and the environment - together with broad but justifiable estimates of their likelihood

* Consequence Analyses: the quantification and assessment of likely impacts on people and the environment in terms of their extent and significance

During the assessments of the first submissions of CIMAH safety reports one of the major deficiencies found by HSE in the predictive information, was the failure to consider high consequence low probability accident scenarios. These were often dismissed on the basis of unqualified and unjustified assumptions that such events were considered to be 'non credible' and therefore 'discounted' from the accident scenario and consequence analyses. This was not only a major failure to meet the Schedule 6 requirement for identifying potential sources of major accident hazards, but also led to shortfalls in the information required to satisfy Regulations: 10 (On site Emergency plan) 11 (Off site emergency plan) and 12 (Information to the Public). Other areas included lack of or poor consideration of assessment of the impacts from major accidents on the environment; limited consequence analysis based on assumptions that safety control and intervention measures - such as automatic shut down systems and operator response - would always work effectively and not fail 'on demand'; absence of estimates for the likelihood's of events (even qualitative ranking) and limited evaluation of 'escalation' potential.

In many cases, it was evident through later discussions with HSE, that Companies had carried out a significant amount of work, which in some cases included quantified risk assessments, and had available large volumes of detailed information to cover the predictive aspects, but often missed the opportunity to adequately present this in their safety reports. Fortunately most safety reports, which are now at the stage of their CIMAH '9 Year' review, have addressed these issues to a satisfactory standard. However. it is often the case that these reviews do not take the opportunity to consider the implications of changes to and lessons learnt from operational and accident experience, and technological progress, including the use of up to date mathematical models for consequence analyses.

ASSESSMENT NEEDS UNDER COMAH

Britton (4) describes the essential features of a COMAH safety report, the principles behind the development of the Competent Authority's assessment processes and procedures for dealing with both health safety and environmental major accident hazards. Fundamental to compliance with COMAH will be requirements for the safety report to contain certain information which is presented in such a way that the dutyholder can demonstrate that they have "taken all necessary measures to prevent a major accident", ie presenting a 'case' that the measures they have in place, linked to their major accident hazard processes, do and will continue to control the risks.

It is worth repeating here that the continued operation of an establishment will not depend on the 'acceptance' of the safety report by the Competent Authority. However, the tests for adequacy of the report will need to take due cognisance of the CA's specific duties under COMAH. A key aspect of the Schedule 4 requirements for the minimum information in the safety report concerns the 'hazard and risk identification, analysis and prevention methods resulting in a description of possible major accident scenarios and the extent and severity of their consequences'. This in turn underpins one of 5 elements of the dutyholder's demonstration to the CA, namely that" major accident hazards have been identified and measures taken to prevent and limit their consequences for persons and the environment". The bottom line in safety report assessment by the CA is to ensure that the safety report contains the information required to demonstrate the duties placed on the operator. Other important aspects of the CA's 'assessment' and duties which include communicating the conclusions of the examination of the report (Regulation 17) and prohibiting the operation of the establishment, installation or any part where the measures taken to prevent and mitigate major accidents are seriously deficient (Regulation 18).

PREDICTIVE ASPECTS OF COMAH SAFETY REPORTS

Appendix 1 gives the full text of the revised 'predictive assessment criteria', which first appeared in the COMAH Safety Report Assessment Manual (Pilot Version) issued by HSE on 1 April 1998 (5). The predictive aspects of safety reports covered by this set of criteria form part of the demonstration of safety required by Schedule 4 of the COMAH Regulations, which explains the purpose (Part 1) and contents (Part 2) of the Regulation 7 Safety Report. The duty created by Regulation 4 to take all necessary measures to prevent and mitigate major accidents and the duty to demonstrate it by Regulation 15(1) are based, in part, on information given in the safety report. Operators must provide evidence under Part 1 to '*demonstrate that major accident hazards have been identified and that the necessary measures have been taken to prevent such accidents and to limit their consequences for persons and the environment*'. The specific requirements for the minimum information to be included in the safety report are given in Part 2, paragraph 4 of Schedule 4 which concerns 'Identification and accidental risk analysis and prevention methods':

(a) detailed description of the possible major accident scenarios and their probability or the conditions under which they occur including a summary of the event which may play a role in triggering each of these scenarios, the causes being internal or external to the installation;

(b) assessment of the extent and severity of the consequences of identified major accidents.

The risk analysis and assessment are inextricably linked with all parts of Schedule 4 in terms of the 'demonstrations' (Part 1) and information required from the safety report (Part 2). Because of this pivotal role of risk analysis and assessment there is a fundamental need to ensure that suitably robust assessment criteria are provided for the predictive elements of the safety report. These criteria were developed by external consultants under the close supervision and direction of a HSE Working Group, which included representatives from the Environment Agency, HSE's: - Nuclear Safety Division, Safety Policy Division, Explosives Inspectorate and other CHID Headquarters' Units. The criteria were prepared taking into account the following factors:

- consistency with Guiding Principles for Assessment and Administrative Principles set by the 'SHARPP' Project (Safety Report Handling Assessment Review Principles and Procedures) - see Britton (4)

- need to ensure universal applicability to cover environmental, health and safety risk assessments

The criteria will be applied by the Competent Authority (CA) to assess the fitness-for-purpose of the operator's major accident risk assessment. However, they do not cover other criteria which are being developed separately, such as

1) derogations under Regulation 7 (10) for which the European Commission has published criteria

2) the provision and exchange of information required under Regulation 16 which will be used by the CA to designate groups of establishments where the likelihood or consequences of a major accident may be increased because of the location and proximity of establishments in the group and the dangerous substances present - ie the consideration of the so called 'Domino Effects' referred to in Article 8 of the Seveso II Directive (6).

The criteria should also provide Operators with a clear and definitive steer on the issues which need to be addressed to satisfy their legal obligations. However, whilst these 'high level' criteria include explanatory notes and some examples of the type of information expected to produce a suitable risk analysis and assessment, they are not intended as guidance on the preparation of a safety report. Some Guidance can be found in (7). In due course, HSE will produce and make publicly available (in late 1999), internal guidance - which is outlined in later paragraphs of this paper - for use by MHAU's risk assessors to test the quality and adequacy of the operator's risk assessment. Again this is not intended as guidance on the preparation of safety reports. Similar detailed guidance will be issued by the Environment Agencies in due course.

Structure and Content of the Predictive Criteria

The overall structure of the criteria are shown in Figure 4.1 of Appendix 1 and are based on the 7 main components of the risk assessment process which in outline provide:

1) An understanding of the site operations, the materials involved and the process operating conditions

2) Identification of the hazards to people on-site and off-site and impacts on the natural and wider environment

3) Analyses of the different ways the hazards can be eliminated, reduced in scale, and controlled.

4) For the hazards that remain, predictions about the likelihood of the hazards being realised taking account of the chance of success and failure of possible preventive control and minimisation/ mitigation measures

5) Predictions about the corresponding consequences both when mitigation measures work and fail.

6) Evaluations/ Analyses about the associated risks and the options for reducing them to demonstrate that all measures necessary to make them as low as reasonably practicable (ALARP) have been taken

7) A presentation of the results of the risk assessment to provide the evidence and arguments which demonstrate that all measures necessary have been taken to prevent and mitigate major accidents

For new plant step (3) is particularly important and the hazard analysis of the proposed design should consider the feasibility of:

- eliminating hazards and inherently safer approaches to reducing the scale of the hazards that cannot be eliminated

- reducing the likelihood of realising hazards and

- mitigating the consequences when these measures fail

The same considerations will generally apply to existing plant, but the scope for elimination and reduction in scale of hazards may be less practicable. Work is underway to produce assessment criteria for pre - construction and pre - operation safety reports under Regulations 7(1) to 7(5) based on the predictive criteria in Appendix 1 and other criteria (5).

COMAH does not specifically require the use of quantified risk assessment (QRA). However, the risk assessment whether it is quantitative, semi-quantitative, or qualitative is considered to be the most logical and systematic process for the demonstrations required from the operator. Some of the questions that the risk assessment will need to address (5) for the demonstrations, are whether or not:

- the major accident hazards have been identified?

- the necessary measures have been taken to prevent major accidents?

- the necessary measures have been taken to limit the consequences of major accidents?

The depth of the analysis in the operator's risk assessment needs to be proportionate to the scale and nature of the hazard and the associated risks. To help assessors reach consistent professional judgements on this, the criteria are directly linked to the risk assessment process given above.

The criteria under Test

On 1 April 1998 HSE issued for wider consultation a draft ('Pilot') version of the COMAH Safety Report Assessment Manual (5) to complement the Consultative Draft Regulations and Guidance which were issued in May 1998 (1). Comments on the assessment criteria, inter alia, have been reviewed and modifications made where appropriate. In addition to external consultation a pilot exercise was carried out between April and June. This involved 4 Volunteer CIMAH Manufacturers, who submitted safety reports, for assessment against the newly developed HSE 'COMAH Criteria (5). Further details of this exercise are given in other papers presented at this symposium.

In most cases the responses to the format and content of the predictive criteria were positive and without exception there were no serious concerns about the range and clarity of the criteria. A number of relatively minor amendments have been made to the explanatory text, but the predictive criteria have not been changed. However, as expected, questions were raised about the level of detailed information expected by the Competent Authority. These issues are being addressed as part of the further guidance described in the next Section.

GUIDANCE FOR HSE RISK ASSESSORS

The main objectives for developing internal guidance will be to ensure consistency and proportionality, in the evaluation of information to meet the predictive assessment criteria. This guidance will not be published but it will be made publicly available. It should be stressed that this will not be a guide to the preparation of the risk assessments required in a safety report. However, together with other documents including European Commission's publication (7), and the Agencies' Environmental guidance, it is expected to provide a useful basis for Operators when developing their own internal guidance.

Structure of the Guidance

The original CIMAH guidance on predictive aspects was developed on the basis of HSE's experience in the assessment of the 'first submissions' of safety reports. Specific and important failings were identified which were generally common in the majority of these early reports. This was a valuable means of providing information to assessors, to enable them to make professional judgements about the standard to which a report had or had not satisfied the 'predictive' requirements of Schedule 6, but not as a means for either 'accepting' or 'rejecting' a report.

The main failings found in early CIMAH safety reports included:

(1) Factual Errors with respect to inventories and locations; procedures; process information and descriptions of safety systems

(2) Inadequate identification of or limitation of the sources and sizes of events and the range of initiating mechanisms in all sections of the installation (eg transfer operations) - including off site 'man made' events (eg aircraft impacts) and natural phenomena (eg seismic activity)

(3) Limited or no consequences assessment: large events often dismissed or ignored on the grounds of assumed low frequency - for which suitable estimates were not presented

(4) Consequences of major accidents to the environment

(5) The selection and use of unidentified and sometimes inappropriate mathematical models (source terms, dispersion and vulnerability - for which their limitations, and assumptions were not transparent

(6) Lack of justification and information about the effectiveness, availability and reliability of safety systems including the role of management and procedures - often assumptions that such systems would always be 100% reliable and never fail on demand thus limiting the duration and scale of consequences assessed

(7) Limited consideration of the multiple hazards of dangerous substances including dangerous substances produced during the course of an accident such as combustion products

(8) Limited description and consideration of the links between the consequences of events with specific initiating conditions and events nor consider 'escalation' (on site) and 'domino effects (off site)

(9) Little indication or justification or absence of 'harm criteria' based on suitable vulnerability models for the effects from fire, explosion, toxicological and ecotoxicological hazards

Based on these areas, internal guidance was produced to provide HSE risk assessors with background information about the type of information which should be expected to cover the predictive aspects of the report, namely hazard identification and phenomenology; accident scenario and frequency analyses and consequence analyses - together with information about the uncertainties in the various mathematical modelling parameters (for example see reference 8). In short the guidance was to provide details of the methods and criteria to ensure that inspectors could judge the adequacy and completeness of the information given in the safety report to satisfy the predictive requirements of Schedule 6, taking into account other relevant reference documents and not least, guidelines produced by Industry.

Preparation of COMAH guidance

Risk Assessors based in HSE's Major Hazards Assessment Unit (MHAU) will be required to assess the information given in the safety report about the techniques for identifying and analysing the hazards, consequences and risks and thereby confirm that the major accident scenarios have been properly identified to satisfy legal requirements.

COMAH guidance is being developed to take account of the experience gained in applying current CIMAH guidance and other factors such as:

- the majority of 'top - tier' installations involve relatively 'simple processes' handling flammable gases and liquids (including low pressure liquefiable gases such as LPG, methane and highly and extremely flammable liquids) and chlorine;

- clarity and suitability of information provided for all levels of experience (independent of assessor);

- the relevance of HSE's own risk assessment tools and models for providing land use planning advice to local authorities (8);

- whether or not it would be appropriate for HSE to express its views on the fitness for purpose of the variety of risk assessment methodologies used by industry;

- the availability and suitability of external published guidance on safety reports;

- the need to ensure a careful balance between prescription and professional judgement - and whether or not 'adequacy' statements linked to the criteria would be feasible

- take into account other HSE guidance documents eg assessment principles produced for the nuclear and off shore industries which include specific guidance on for example human factors; consequences models etc

- avoidance of overlap with other COMAH assessment guidance

Four possible structures were considered in terms of the degree to which each could match the factors identified above.

a) Option 1: 'Process Driven'

b) Option 2: 'Criteria Driven'

c) Option 3: Hybrid of Options 1 and 2

d) Option 4: Accident Scenario - Initiating Event Driven

After a detailed review and internal discussion, Option 1 is considered to be the easiest to use because each of the main sections are self contained and apply to a small group of substances with similar properties. The amount of redundant information used when assessing a safety report is minimal, but because each substance group chapter addresses each of the six 'headline' assessment criteria (see figure 4.1 in Appendix 1), a disadvantage will be the large volume of information required in each document.

The guidance will be produced in the form of separate Volumes to cover the most common substance/ process combinations and will include the following :

1. Methane
2. Liquefied Petroleum Gases (Propane and Butane)
3. Flammable Liquids (including Extremely and highly flammable liquids)
4. Chorine
5. Warehouses storing toxic substances
6. Explosives

However, it has been recognised that certain processes and dangerous substances cannot be easily dealt with under Option 1. A good example would be a COMAH petrochemical refinery utilising a Hydrogen Fluoride alkylation Unit and in such cases Guidance based on Option 2 - "Criteria Driven" - will also be prepared.

Conclusions

It is not expected that the information required to satisfy the predictive aspects of COMAH safety reports will be substantially different to that required under CIMAH. Those information needs have been the subject of many published technical papers over the last 12 years (note for example references 9 to 11). However, COMAH will introduce significant changes to the ways Industry and the CA present and use the information required in the safety report. Operators will need to demonstrate that they have 'taken all necessary measures to prevent a major accident', by presenting a case that these measures, linked to the major accident hazard processes do and will continue to control the risks from their activities. A crucial element in this requirement is the need for 'hazard and risk identification, analysis and prevention methods resulting in a description of possible major accident scenarios and the extent and severity of their consequences to persons and the environment'.

This predictive information is clearly an essential feature of the COMAH safety report which must be properly and systematically, addressed by operators, and assessed by the

CA. Detailed and explicit assessment criteria have therefore been developed for use by the CA to ensure that, if met in their entirety, the risk analyses and assessment will be 'fit for purpose'. These 'high level' criteria will require further detailed explanation to ensure that the 'tests for adequacy' applied by the CA can and will be applied consistently and proportionately, to reflect the uncertainties and acceptable variations and uncertainties in risk assessment methodologies. For these reasons HSE will prepare more detailed guidance for its risk assessors, which in due course will be made publicly available and provide further assistance to Industry in the preparation of their COMAH safety reports.

Acknowledgements

The author gratefully acknowledges the assistance from the SHARPP Team who developed the predictive assessment criteria: Dr Peter Kinsman (Consultant) ; Dr Clive Nussey (Consultant); Dr Peter Newman (Environment Agency); Dr Colin Powlesland (Environment Agency); Dr Colin Foan (Environment Agency); Dr Phil Brighton (HSE - Nuclear Safety Division); Peter Sargent (HSE - Safety Policy Division); Dr Roy Merrifield (HSE - Explosives Inspectorate); Trevor Britton (HSE CHID6); Ron Evans (HSE CHID8) and Ian McKay (HSE - Major Hazards Assessment Unit).

References

1. Proposals for Regulations implementing the Directive on the Control of major accident hazards involving dangerous substances: Consultative Document; Health & Safety Commission; The Scottish Office; Department of Environment Transport Regions: HSE Books
2. The Control of Industrial Major Accident Hazard Regulations 1984 (As Amended) (SI 1984/192; HMSO: ISBN 0 11 047902
3. A Guide to the Control Of Industrial Major Accident Hazards Regulations 1984; HS(R)21(Rev) : HMSO: ISBN 0 11 885579 4
4. Britton T J (HSE): 'The Regulators Approach to Assessing COMAH Safety Reports: IChemE - Hazards XIV (November 1998)
5. COMAH Safety Report Assessment Manual (Pilot Version): HSE - CHID Operational Strategy Unit: St Annes House Bootle Merseyside (March 1998)
6. Council Directive 96/82/EC on the control of major accident hazards involving dangerous substances: OJEC No L 10/13 (9/12/96)
7. Guidance on the Preparation of a Safety Report to Meet the Requirements of Council Directive 96/82/EC (Seveso II)
8. Nussey C, Carter D A and Cassidy K: 'The Application of Consequence Models in Risk Assessment: A Regulator's View: International Conference & Workshop on Modelling and Mitigating the Consequences of Accidental Releases of Hazardous Materials - New Orleans 26 to 29 September 1995
9. Cassidy K 'CIMAH Safety Cases - the HSE Approach: Chemical Engineer Loss Prevention Supplement (August 1987)
10. Cassidy K ' CIMAH Safety Cases': Hazard Control and Major Hazards: Loughborough University (18 to 22 January 1988)
11. Cassidy K 'Major Hazards - the Preparation of Safety Reports': Workshop on Major Hazards - IBC Technical Services (8 to 10 March 1988)

Appendix 1

[NB Criteria reference numbers align with the system used in the COMAH Safety Report Assessment Manual (Pilot Version) issued by HSE on 1 April 1998. In this document the Predictive Aspects are given in Part 2 Chapter 4]. Revisions to the text have been made as a result of the comments received during the 'COMAH Pilot' and external consultation which ended on 30 June 1998.

Predictive Aspects

INTRODUCTION

Scope

1   The criteria presented in this chapter of the Safety Report Assessment Manual are applied by staff in the Competent Authority (CA) to assess the fitness-for-purpose of the site operator's major accident risk assessment. The need for the operator to assess the risks stems from the requirement in Schedule 4, Part 1 paragraph 2 of the COMAH Regulations for the safety report to demonstrate that "major accident hazards have been identified and that the necessary measures have been taken to prevent such accidents and to limit their consequences for persons and the environment".

2   A risk assessment (RA) is fundamental to such a demonstration. The need for RA is also recognised in the EU guidance on the preparation of safety reports [1]. The risk assessment may be qualitative, semi- quantitative, quantitative, or a combination of these Operators will need to decide the scope and nature of their RA so that it is fit for-for-purpose in relation to their site specific circumstances and the demonstration required.

RELEVANT SECTIONS OF THE DIRECTIVE

3   The criteria in this chapter assess whether the operator's RA is both suitable and sufficient for the purposes of Schedule 4, Part 1 paragraph 2 (see paragraph 1 above), and Part 2, paragraph 4 'Identification and accidental risk analysis and prevention methods', particularly those under paragraphs 4(a) - accident scenarios, likelihoods etc; and 4(b) - assessment of consequences. It should be noted that Schedule 2 Part 2 defines the <u>minimum</u> information requirement. A suitable and sufficient demonstration of compliance with Part 1 para 2 may require more information and supporting arguments. These supporting arguments should be derived from the results of the risk assessment. Indeed the risk assessment is inextricably linked with all parts of Schedule 4 and the fundamental requirements under the H&S at Work etc Act (1974). The 1974 Act places duties on employers to:

(a)   ensure the health and safety and welfare of their employees; and

(b)   conduct their operations so that persons not in their employment are not exposed to risks to their health and safety.

The Environmental Protection Act (1992) extends these duties on employers to the protection of the environment.

Employers are required to ensure that these duties are met so far as is reasonably practicable. This legal duty is enshrined in the 'as low as is reasonably practicable principle' (ALARP[1]) used in current regulatory practice. RA is the means adopted for demonstrating that risks are ALARP and is a statutory requirement of some enabling regulations, for example the Management of Health and Safety at Work Regulations (1992) address (a) above.

GENERAL GUIDANCE FOR ASSESSMENT OF THE PREDICTIVE ELEMENTS

4   The risk assessment needs to address risks to people both on and off-site and risks to the environment. Regardless of whether the approach to risk assessment (RA) is quantitative, semi-quantitative, or qualitative, a logical and systematic process needs to be adopted. Some of the questions that the RA needs to address are listed in Appendix 1 to Chapter 2 Part 1, 'Guiding principles'. The most relevant questions are those relating to risk analysis ie Q4, 5, 6, and Q11 (domino effects), and the corresponding questions relating to the pre-construction stage and pre-operational stages of the SRs for new establishments.

5   For new establishments the risk assessment needs to include consideration of the elimination of hazards and inherently-safe approaches to reducing the scale of hazards.

6   The risk assessment process for major hazard plant has a number of steps. In outline these are:

   a)   understand the site operations, the materials involved and the process conditions;

   b)   identify the hazards to people on-site and off-site and the environment;

   c)   analyse the different ways the hazards can be eliminated, reduced in scale, realised and controlled;

   d)   for the hazards that remain, predict the likelihood of the hazards being realised taking account of the chance of success and failure of possible preventive measures;

   e)   predict the corresponding consequences both when mitigation measures work and fail;

   f)   analyse the associated risks and the options implicit in (d) and (e) for reducing them.

g) Decide which measures need to be implemented to make the risks to people and the environment as low as reasonably practicable (ALARP[1]);

h) present the results of the risk assessment to provide the evidence and arguments which demonstrate that all measures necessary have been taken to prevent and mitigate major accidents.

7   For new plant step (c) is particularly important. The hazard analysis of the proposed design should pay particular attention to ways of eliminating hazards and inherently safer approaches to reducing the scale of the hazards that cannot be eliminated. Ways for reducing the likelihood of realising hazards and for mitigating the consequences when these measures fail are then analysed. The same applies to existing plant, but the scope for elimination and reduction in scale will be less.

8   The depth of the analysis in the operator's risk assessment should be proportionate to:

a) the scale and nature of the major accident hazards (MAHs) presented by the establishment and the installations and activities on it, and

b) the risks posed to neighbouring populations and the environment ie the assessment has to be site specific.

9   It is recommended that the assessor formulates a view on 'proportionality' at the start of the assessment process. The assessor will need to carefully consider (a) and (b) above when coming to a view. A simple site remote from population and sensitive environments with a single dangerous substance presenting a limited range of hazards may only require a simple qualitative risk assessment to demonstrate that the necessary prevention and mitigation measures are in place. For example a water treatment plant with a total inventory of 30te of chlorine and remote from population and sensitive environments may only need to demonstrate compliance with HSE guidance note HS(G) 28 for the safe handling of chlorine, with supporting statements to demonstrate that the risks to people off-site and the environment are ALARP. (A risk assessment is needed under the Management Regulations and this may form part of the COMAH report). If the qualitative route is adopted for control measures, the operator still has to demonstrate that all MAHs have been identified and that the severity of these has been assessed. In the case of chlorine the guidance published by the CIA and the chlorine producers on emergency planning is helpful here.

10   On the other hand, the same chlorine site in a sensitive location and presenting risks which may be tolerable to people and the environment will require a more detailed analysis to demonstrate that the associated risks are ALARP. Similarly complex site with

---

[1] The ALARP (as low as is reasonably practicable) concept implies that ultimately there is a trade-off between the costs of risk reduction and the benefits obtained. This concept is sometimes referred to as BATNEEC (best available technology not entailing excessive cost) which is often applied in environmental contexts. The political and practical interpretation of 'reasonable' or 'excessive' is the key in the setting of safety standards to be achieved by operators. These and related issues are discussed elsewhere [2].

many processes and several hazardous materials in the vicinity of population and sensitive environments will require a much more detailed assessment and some quantification of the likelihood of hazardous releases and their consequences, and possibly of the associated risks. (NB All sites will require some quantification of the possible consequences to help develop the emergency plan).

11   For explosives facilities and operations which do not meet accepted quantity-safety distances (QDs) the justification that all measures necessary to control the risks will normally require a quantified risk assessment.

12   The fitness-for-purpose of the risk assessment will depend mainly on: the degree to which the expertise of the team conducting it matches the site-specific circumstances; the methods they use; the data and assumptions they adopt; and the time they invest. The safety report should therefore indicate the competence and expertise of the assessment team and describe the process and methods used to conduct the risk analysis and to assess the significance of the risks.

13   In evaluating the results of the operator's risk assessment, assessors will be guided by HSE's and the agencies' approach to risk regulation [3,4,5]. This is based on the concept of risk tolerability which requires duty holders to take measures to reduce the likelihood of hazards and to mitigate their consequences until the associated risks are ALARP. Essential considerations are the scope for hazard elimination and the adoption of inherently safer designs and whether good practice has been, or is to be adopted. Where relevant good practice is not yet established, duty holders will be expected to apply risk-reducing measures. In general, the higher the scale of the hazard and the associated risks the more the balance should tilt in favour of adopting further measures to control risks unless the costs (in money, time and trouble) are clearly disproportionate (excessive in the case of BATNEEC) compared to the benefits gained from the risk reduction. Operators will need to define the basis of their decisions on all measures necessary for controlling major accident hazards (MAHs).

14   In some situations an ethical approach to risk regulation is adopted. This approach may be defined in terms of predetermined levels of safety based on technically achievable standards (eg maximum emission levels (environmental quality standards, EQSs) for particular pollutants), or limits based on historical precedent eg the maximum tolerable level of the risk of fatality from major hazards for a hypothetical member of the public.

15   The TOR framework [3] brings the ethical and cost-benefit approaches together by imposing an absolute maximum level of risk set on the basis of equity. It also applies a lower limit defining broadly acceptable risks below which formal analysis of costs and benefits is not normally required. Residual risks between the two limits need to be made ALARP. Most decisions on whether risks are ALARP are made by exercising professional judgement on whether the risks are reasonable when set subjectively against the cost of further risk reduction. Some companies have adopted this approach and defined their own ALARP bands. In some cases more stringent criteria are set for new plant - typically an order of magnitude lower than the band for existing plant.

16 The concept of tolerability implies that existing control measures should be periodically reviewed to ensure they are properly applied and still appropriate. Whether they are still appropriate will depend on matters such as the availability of new options for reducing or eliminating risks due to technological progress, changes in society's perception of the particular risks, changes in our understanding of the risk analysis, the uncertainty attached to the risk estimates, and new lessons from accidents and incidents etc. Such reviews should figure prominently in safety report updates (see COMAH regulation 8)

17 Some of the risk analyses required to assess the impact on the natural environment and people may already have been documented for other purposes and it may be possible for the operator to re-use some of this information. It is not necessary to repeat the work but the original documentation must be clearly referenced and, normally, copies of the appropriate parts of it attached to the safety report." *(See CD on regulation 7(9).)*

18 The assessment criteria presented below for assessing the quality of the predictive aspects of safety reports are linked directly to the risk assessment (RA) process. The way the criteria are structured and applied is depicted in Fig 4.1. In essence the main steps of risk assessment translate into 6 top level criteria represented by the 6 large boxes - (steps (a) and (b) above are combined into box 1; step (c) may occur in each of boxes 2 to 5 or a combination of these). Criterion 4.1 deals with the operator's approach to RA since this will influence what is done at the various stages of the RA. Each of the six top level criteria (ie 4.1 - 4.6) must be met for the predictive aspects to be acceptable. To help assessors judge consistently whether the criteria are met or not, related lower level criteria are defined. The extent to which each of these applies and is met will determine whether the relevant top level criterion has been met.

19 The tests assessors should apply in making judgements about whether the criteria are met or not are not explicitly defined here. Given the complexity and diversity of the major chemical hazards industry it is not possible to define all the tests assessors will apply in a particular case. Professional judgement is required. These judgements will be strongly influenced by the site-specific circumstances described in the SR. To help assessors reach consistent professional judgements on this, the criteria presented below are linked directly to the risk assessment process. The explanatory text linked to the criteria gives insight into how judgement can be exercised. In addition, to help assessors achieve consistent decisions on the adequacy of the predictive aspects of SRs they will be issued with internal guidance, suitable training will be given and QA checks built into the assessment process.

20 The way assessors will work will depend very much on the nature of the safety report and their own discretion. For example if the site is a warehouse the first test an assessor may apply is "does the risk analysis consider the consequences of a fire in high wind speed conditions?" - as these produce the worst consequences. Alternatively the fire plume may be modelled as a ground level passive release in lower wind speeds. If neither if these apply the assessor should identify a significant omission in the report. The assessor should then give a quick review of the quality of the risk analysis so that any other omissions can be addressed by the operator when the report is revised.

21   In general assessors will familiarise themselves with the safety report to develop the level of understanding of the site and its processes that is sufficient for the assessment of the risk analysis and the conclusions drawn. For simple sites the criteria may be applied sequentially. Most other assessments are likely to require some iterative application of the criteria before the assessor reaches a conclusion. In carrying out this iteration the assessor will be testing whether the assumptions and judgements made at the various risk analysis stages are consistent with one another and accord with the factual information in the SR. The application of the criteria defined here may also expose weaknesses in the quality of the information supplied; such weaknesses, if any, are likely to become apparent under the stages linked to criteria 4.4 (ie event probabilities and sequences) and 4.5 (event consequences). The assessor will also be forming a view on whether the quality of the arguments supporting the company's view that 'all control measures necessary have been taken' are suitable and sufficient. The depth of the RA underpinning the demonstration should therefore be proportionate to the scale and nature of the hazards and the associated risks.

22   If the risk assessment demonstrates that particular dangerous substances present at an establishment are not capable of producing a MAH the operator may apply for a derogation under the EU harmonised criteria developed for this eventuality (regulation 7(10)).

ASSESSMENT CRITERIA

23   Risk assessment is fundamental to the demonstration that all measures necessary have been taken to control risks. Operators therefore needs to present their approach to risk assessment. The approach and the depth of the analysis will be influenced by site specific circumstances.

Criterion 4.1   *The safety report should clearly state the operator's policy on the use of risk assessment to aid decision-making on the measures necessary to prevent major accidents and to mitigate their consequences.*

24   The policy should include a summary of the <u>methods</u> used to analyse risks and the <u>criteria</u> used to judge the significance of the residual risks when control measures have been implemented. The approach to demonstrating that these risks are ALARP is fundamental to the justification that all measures necessary have been taken. This includes the consideration of ways of eliminating hazards, reducing their scale, and other means for reducing the associated risks(ie reducing event likelihoods and mitigating the associated consequences). The approach should embrace current thinking on inherently safer design options, on relevant good practice and on engineering and procedural standards.

25   The basis on which the operator makes decisions on all necessary measures should be clearly stated.

26   The summary should make clear how the operator scopes (ie defines what is and what is not addressed) the risk assessment so that it is both suitable and sufficient. "Suitable" means that it is valid and appropriate for the operators situation and circumstances. "Sufficient" means that the supporting information and arguments are well developed and presented, and do not require further elaboration in order to provide a valid input to the demonstration that all measures necessary have been taken ie the risks to site personnel, people off-site and the environment are, in each case, ALARP. The depth of the analysis also needs to be proportionate to the scale and nature of the hazards, and the associated risks (see General Guidance above). The level of detail will depend on the site specific circumstances eg size and nature of installation and the proximity of population or sensitive environments. For example the off site risks at an LPG facility (provided the vessels are not mounded or fitted with passive fire protective coatings) are usually dominated by the fireball event scenario following whole tank failure; the contribution from the VCE scenario being much less significant. This means that the treatment of the drifting cloud scenario and possible VCE need not be comprehensive. However, the case of explosives facilities and operations, a qualitative approach based on the 'defence-in-depth' principle is appropriate, unless the facility does not comply with QDs - when a quantified risk assessment is needed to demonstrate that all measures necessary to control the risks have been taken.

27   The balance between qualitative, semi-quantitative and quantitative arguments will depend on the nature and complexity of the major accident hazard (MAH) events being analysed in relation to what is at risk. This is considered further under Criterion 4.6 (assessment of the risks).

28   The approach to making the RA a living document should be stated as this supports the periodic review of the safety reports required by the COMAH regulations (Regulation 8).

Criterion 4.1.1   It should be clear that human factors have been taken into account in the risk analysis.

29   Plant personnel are an important part of safety systems. They may also unwittingly contribute to the initiation of a major accident as a result of human error (see Criterion 4.4.4). The role operatives play in controlling hazards and risks therefore need to be identified, and the consequences of failure to carry out such control should be understood so that the various roles can be prioritised. For example an operative may be required to take certain actions following an alarm, the risk analysis will need to make assumptions about the likelihood that the correct action is taken. This task may be critical if a high level of human reliability has to be assumed to make the risks ALARP. If so, automatic control and protection systems may be needed to reduce the reliance on the operator to intervene correctly. The necessary redundancy and diversity should be built into the control systems to achieve the required reliability. This will depend on the scale of the hazards and the associated risk.

30   Operatives need to be well trained, competent and motivated. Equipment and procedures need to be designed to minimise human error (routine unintentional failures,

decision making failures and violation of rules). These and other human factor issues are considered by assessors dealing with criteria in Chapters 2 and 5 of Part 2.

Criterion 4.1.2   *Any criteria for eliminating possible hazardous events from further consideration should be clearly justified.*

31   The justification should be clearly presented and well argued. For example in the case of a plant processing toxic gases, consequence assessment may show that any failure resulting in a release smaller than that equivalent to a 10 mm diameter hole does not produce a hazard to current on - site or off-site populations. This provides a basis for defining major accident hazards. However, operators may need to take account of smaller releases which could trigger other events leading to event escalation. They should also consider any known or foreseeable changes to the sensitivity of the surrounding environment eg to water courses or future dwellings which may be built nearer to the site boundary. Such changes should be also considered whenever the RA is reviewed.

32   In situations where this 'protection' based approach is not sufficiently limiting ie the hazard ranges from very small releases extend into population or sensitive environmental areas, a risk based approach may be needed. This requires the contribution to the residual risk of releases of different sizes to be considered so that a justifiable 'cut-off' can be decided. All contributions to release likelihood need to be taken into account otherwise, the 'cut-off' will be overly optimistic.

33   The criteria should be applied at an early stage to limit the scope of the predictive aspects of the risk assessment. Assessors will assess the validity of the operator's criteria.

Criterion 4.2   *The safety report should demonstrate that the operator has used information and data that are suitable and sufficient for risk analysis.*

34   Schedule 4, Part 2, paragraph 4 of the Directive requires identification of possible major accident scenarios for risk analysis, and the identification and analysis of the adequacy and feasibility of possible prevention and mitigation methods. A suitable and sufficient risk analysis can only be achieved if all relevant information required at Schedule 4, Part 2, paragraph 3 is supplied and the quality of that information is consistent with the needs of risk analysis.

35   A prerequisite is that the safety report has satisfied the criteria developed in Chapter 3 Part 2 which relates to Schedule 4, Part 2, para 1 to 3.

36   However, the information required for risk assessment can be diverse and extensive. For example, weather data is needed to assess the risks of all hazardous materials, but the detail required is process and location specific. Consider lightning: the likelihood of lightning strikes is not a significant issue for LPG facilities but could be the cause of a warehouse fire. On the other hand cold weather is unlikely to pose a threat to a pesticide warehouse, but could cause problems for butane tanks. For many situations involving toxic gas releases an assessment of the consequences in two weather stability/wind speed

combinations may suffice, but for warehouse fires it is the likelihood of high wind speeds and the corresponding consequences that dominates the off-site risk.

37   Similarly, to assess the consequences of hazardous events, a range of harm levels to people and the environment need to be considered, particularly for emergency planning purposes. This requires the use of appropriate harm criteria. Harm criteria for the effects of toxic, thermal, and overpressure effects are generally available but lack accuracy. By comparison, the corresponding criteria for the effects of toxic materials on the environment are relatively scarce and less accurate. Environmental and human impact assessment is therefore an area of considerable uncertainty, and the operator should therefore justify the suitability of the adopted harm criteria. The justification needs to be tested by the assessor. For these types of reasons, assessors considering the predictive aspects will have to be satisfied that the quality of the information supplied and used by the operator is sufficient to support the level of risk assessment required.

Criterion 4.3   *The safety report should identify all potential major accidents and define a representative and sufficient set for the purposes of risk analysis.*

38   Schedule 4, Part 2, paragraph 4 of the Regulations requires identification of all possible major accident scenarios for risk analysis purposes, and the identification and analysis of measures for preventing and mitigating major accidents. To make the risk analysis feasible a representative and sufficient set of major accident scenarios needs to be considered.

Criterion 4.3.1   *The safety report should demonstrate that a systematic process has been used to identify all foreseeable major accidents*

39   The chemical industry is diverse and complex, and presents MAHs ranging from damage to water courses to toxic effects for people downwind of a warehouse fire. A structured approach to hazard identification is therefore required. The process will usually overlap with other stages in the risk analysis.

40   In assessing whether this criterion is met assessors will consider the adequacy of the coverage of different types of MAHs. All MAHs may be broadly classified as loss-of-containment accidents which may be categorised as follows:

   a)   Loss of containment accidents due to vessel or pipe work failures;

   b)   Explosions (batch reactors, tank explosion due to operator error eg wrong contents, BLEVES);

   c)   Condensed Phase Explosions relating to explosives;

   d)   Large fires (Warehouses, pool fires etc);

   e)   Events influenced by emergency action or adverse operating conditions etc (eg allow fire to burn rather than apply water (ie mitigation); dump reactor contents to drain to avoid explosion (ie prevention), abnormal discharge to the environment, etc.;

f) other types of MAH or abnormal discharge.
(Such matters are addressed by criteria in Chapter 6 of Part 2 which deals with emergency response)

41 The coverage of the different types of MAHs needs to be suitable and sufficient for risk assessment purposes. The way the MAHs have been identified should be made transparent. The importance (ie the safety criticality) of each scenario is addressed by subsequent criteria. The potential major accident scenarios need to include the worst case on-site and off-site scenarios both for people and the environment, and be sufficiently comprehensive for assessing the adequacy of methods for preventing major accidents and for limiting their consequences with respect to people and the environment. One way of approaching this would be to:

a) identify the 'worst case events' in relation to people and the environment;

b) assess the consequences. If they are trivial there is no need for further predictions. If they are significant, a range of major accidents needs defining and analysing (see below);

c) the balance between qualitative and quantitative analysis will vary, but in general the level of quantification should be proportionate to the scale and nature of the hazards.

Criterion 4.3.2 **The hazard identification methods used should be appropriate for the scale and nature of the hazards.**

42 The hazard identification methods will vary depending on the type of plant and circumstances. The approach adopted and the expertise of the team involved should be described. This will help the assessor to take a view on the 'completeness' of the list of major accident scenarios.

43 Methods that might be used include:

a) HAZOP (Hazard and Operability Studies)

b) Safety reviews and studies of the causes of past major accidents and incidents

c) Industry standard or bespoke checklists for hazard identification

d) FMEA (Failure Mode and Effect Analysis)

e) Job safety analysis (eg Task Analysis)

f) Human error identification methods.

44   At this stage of the assessment process the focus is on the completeness (but see Criterion 4.1.2) of the event list rather the associated detailed consideration of event initiators and event sequences which is developed under the risk analysis stages linked to Criteria 4.4 (event probabilities and sequences) and 4.5 (consequence analysis). For example whole tank failure into a bund, limited vessel failure, guillotine fracture of a pipe, etc. Foreseeable failure modes leading to each major accident is considered at Criterion 4.4. Scenarios need to cover events when protection and mitigation (actual or proposed for further risk reduction) measures fail to operate. In the case of fires, for example, the events need to take account of any seasonal or operational variations in the range and quantities of stored substances.

Criterion 4.4   *The safety report should contain estimates of the probability (qualitative or quantitative) of each major accident scenario or the conditions under which they occur, including a summary of the initiating events and event sequences (internal or external) which may play a role in triggering each scenario.*

45   Schedule 4, Part 2, paragraph 4(a) requires a detailed description of all possible major accident scenarios and their probability, or the conditions under which they occur, including a summary of the events which may play a role in triggering each of these scenarios, the causes being internal or external to the installation. These are minimal requirements and should not be seen as a choice, though in some straightforward situations one of the alternatives may suffice. In more complex situations a satisfactory demonstration under Schedule 4 may require the consideration of the conditions under which events occur, their likelihood, and how the events interact so the likelihood of certain major accidents can be estimated..

46   The purpose of this criterion is to assess the extent to which the requirement in Schedule 4, Part 2, paragraph 4(a) has been complied with; in particular that the depth of the analysis of the likelihood of realising each major accident scenario under Criterion 4.3 is sufficient relative to the scale and nature of the hazard it presents. The use of operational experience is an important input to the analysis. The operator should bear in mind that the different scenarios may have different levels of significance for employees, people and the environment.

47   An essential feature of the safety report (Schedule 4, part 1, paragraph 2) is the demonstration that the measures necessary for preventing and mitigating major accidents are suitable for their intended purpose and have been applied. (*The off-shore industry refer to these measures as safety critical elements, SCEs. The assessment of the technical suitability of the control measures implemented by the operator and the performance achieved by them is dealt with in Chapter 5 of Part 2, but the quality of the predictive arguments underpinning that justification is considered here*).

Criterion 4.4.1   *The safety report should demonstrate that a systematic process has been used to identify events and event combinations which could cause MAHs to be realised.*

48 All foreseeable causes (initiating events) of the MAH identified under Criterion 4.3 should be considered. Insights gained from the study of previous accidents and incidents can be a useful starting point. The scope of such studies should consider the causes of accidents in other industries which present societal risks. The operator should present evidence to demonstrate that the event sequences triggering the scenarios are correctly identified and clearly justified.

49 Where a sequence or combination of events may lead to a major accident, for example an automatic isolation system fails and the operator fails to respond correctly to an alarm, an assessment should be made of the effects of failure on plant and equipment designed to prevent, detect, or mitigate the hazardous conditions. The purpose of the assessment is to decide whether the event is so hazardous that the reliability of the automatic system is sufficiently high to render the risks ALARP even if the probability of the operative failing to respond is relatively high. Human error should also be addressed as an accident initiating event in addition to intervention activities eg loading wrong reactants into a batch reactor, or wrong operating procedure leading to an abnormal discharge to a water course.

Criterion 4.4.2 *All safety critical events and the associated initiators should be clearly identified*

50 Safety critical events or event sequences are those that dominate the contribution to risk at different distances from the plant. They are relevant to the identification and implementation of suitable control and protection measures for preventing hazardous events or mitigating their consequences.

51 The risk analysis should make clear which events are critical from a safety view point. This requires consideration of the likelihood of the various MAHs and the associated consequences. Operators need to use appropriate methods for assessing the probabilities of each of the listed major accidents.

52 Implementation of control and protection measures should reduce the risk arising from these events. The failure of the control measures to prevent the hazard from being realised or to mitigate the associated consequences then become critical events. The risk analysis should then determine whether the residual risks (determined by the reliability of the control measures etc) are ALARP or whether more needs to be done. This is considered by criteria Criterion 4.6 which consider, among other matters, whether the contribution of each risk reduction measure is then linked to the hazard identification and risk analysis process in a transparent way.

53 If potential control measures are rejected the reasons need to be clearly justified.

Criterion 4.4.3 *Estimates of, or assumptions made about, the reliability of protective systems and the times for operators to respond and isolate loss-of -containment accidents etc need to be realistic and adequately justified.*

54 The quantitative or qualitative arguments presented in the safety report need to be realistic. Significant departure from arguments currently acceptable to risk assessors will need careful presentation and justification, particularly if the scale of the hazards and the associated risks is significant. Well reasoned and plausible arguments backed-up by evidence in the form of credible performance data etc will usually be required.

55 Qualitative arguments will need to be based on currently accepted good standards for engineering and safe systems of work. The assessor will be looking for evidence to support the operator's view on the likely demand on the various control measures and systems and what the consequences might be if these fail.

56 If an operative has to intervene to close an isolation valve manually when automatic isolation fails, the release duration will be determined by the time taken to intervene successfully. In such cases release duration's less than 20 minutes will require realistic justification.

Criterion 4.4.4 *The methods used to generate event sequences and estimates of the probabilities of potential major accidents should be appropriate and have been used correctly*

57 Appropriate methods include the use of relevant operational and historical data, fault tree analysis (FTA) and event tree analysis (ETA), or a combination of these. The methods and assumptions used will therefore need to be described. In particular any failure rate data used for the base events in the FTA will need clear justification in terms of the site-specific circumstances. It will not be sufficient to adopt data from published sources without justification of their suitability, unless it is shown (eg through a sensitivity analysis) that the conclusions of the risk analysis are not sensitive to such data. When the estimates of the likelihoods of the safety critical events are sensitive to the data and assumptions used suitable and sufficient justification is needed.

58 The methods used need to be fit for purpose and used correctly. To enable assessors to judge whether methods have been used correctly, the operator should describe the process and methods (including human error identification and analysis) adopted to generate any probabilities or event sequences, together with assumptions and data sources used. Checks against company benchmarks should be included when appropriate.

59 The sensitivity of the conclusions to the assumptions and other uncertainties may need to be assessed - see also Criterion 4.6. For example, in the case of explosives facilities there is a lack of data on event probabilities leading to considerable uncertainty in the estimation process. Sufficient detail is required to enable an experienced risk assessor assessing the safety report to make a judgement on the quality of this part of the risk analysis.

Criterion 4.4.5 *The safety report should provide adequate justification for event probabilities that are not consistent with historical or relevant generic industry data.*

60 When making judgements about the quality of the estimates of event probabilities HSE will compare the estimates with values commonly used and accepted by experienced risk analysts. In some cases an assessor may perform independent checks to verify that an estimate (qualitative or quantitative) is reasonable.

61 The operator's justification may include quality procedures, plant experience, or other acceptable evidence. The risk assessment assessor will identify the most important parts of the predictive aspects where the justification needs to be further evaluated eg when considering the preventative and mitigation measures in detail later in the assessment process, or for verification during subsequent inspection.

Criterion 4.5 *The safety report should provide details to demonstrate that suitable and sufficient consequence assessment for each major accident scenario has been carried out with respect to people and the environment.*

62 Schedule 4, Part 2, paragraph 4(b) requires an assessment of the extent and severity of the consequences of identified major accidents. (A range of severity's will need to be considered so that corresponding 'hazard zones' defining the extent of affected areas can be mapped out by suitable and sufficient consequence analysis. For people the harms considered should include fatality, serious injury and hospitalisation. A range of potential harms to the environment may also need to be considered.)

63 The purpose of this criterion is to assess the extent to which the requirement in Schedule 4, Part 2, paragraph 4(b) has been complied with; in particular that the severity and extent of each major accident has been properly assessed. The safety report should therefore demonstrate that a systematic process has been adopted for assessing the possible consequences of each major accident hazard.

64 For 'upper - tier' sites, Schedule 4 requires the hazards from all dangerous substances present to be assessed, regardless of quantity. If a substance is present in quantities sufficient to cause a major accident hazard then a detailed consequence assessment is required.

65 The methods used for assessing the consequences of potential major accident impacts on people are now quite mature compared to those for predicting environmental impacts. In applying the methods assumptions need to be made and these should be stated and justified. The criteria below follow the general framework for consequence assessment. Whether these apply, and the extent to which they apply to particular events will depend on the situation. For example, in the case of an LPG facility the risk dominating event will usually (mounded and insulated vessels excepted) be whole tank failure followed by immediate ignition of the BLEVE, resulting in a fireball. If the cloud resulting from the BLEVE event does not ignite immediately it will drift on the wind. Subsequent ignition may result in a flash fire or vapour cloud explosion. If no ignition sources are encountered before the cloud is diluted below the lower limit of flammability, no serious consequences arise. The possible outcomes following an LPG release are usually developed by Event Tree Analysis (ETA). The consequences of each outcome are then assessed using appropriate models. In the case of a loss-of-containment accident resulting in a drifting cloud all the criteria below apply.

66   The worst case scenarios need to be addressed.

67   Operators will need to state which models have been used and justify their suitability. When the scale of the hazards is significant, well validated models should be used throughout the assessment.

Criterion 4.5.1   ***Source term models used should be appropriate and need to have been used correctly for each relevant major accident hazard***

   [Note: Appropriate means 'fit for purpose' - the rationale above defining suitable etc applies. 'Correctly' is described under Criterion 4.4.4 and Criterion 4.5.6.]

68   The source term defines the nature, size, and duration of the release. In the case of releases into the atmosphere, matters such as the influence of obstacles on jets and air entrainment into the release are also addressed. This enables the source term to be defined in terms of the parameters needed by the dispersion model used to predict how the release will disperse. A good introduction to source term models is provided elsewhere [6].

Criterion 4.5.2   ***The material transport models used should be appropriate and need to have been used correctly for each relevant MAH***

69   Releases of hazardous materials can harm people, and pollute the air, water courses, or land. The spatial and temporal variation in contaminant concentration from the release point will depend on the mode of transport. For example there are many competing models capable of predicting the spatial and temporal variation in concentration downwind of a release dispersed in the atmosphere.

70   The choice of model depends on whether a loss-of-containment accident gives rise to:

   a)   a passive (neutrally buoyant) or a heavier-than-air cloud.

   b)   a cloud which contains aerosol which reacts with ambient moisture entrained into the cloud (eg releases of anhydrous ammonia and anhydrous hydrogen fluoride.) It also depends on other factors such as whether the release gives rise to large hazard distances - in which case the validity of the model is an important issue; and

   c)   whether the dispersing clouds will interact with obstacles or terrain features.

71   A range of weather conditions usually need to be considered. For the more significant events it may be necessary to test the sensitivity of the predictions to any assumptions made about the source term.

72   A passive dispersion model may, depending on circumstances, be adequate for a simple plant which releases a heavy gas. This may overestimate the downwind extent of the hazard, but will underestimate the lateral extent. This needs to be borne in mind when justifying the choice of model.

Criterion 4.5.3   *Other consequence assessment models (eg BLEVE, Warehouse fire etc) used should be appropriate and need to have been used correctly for each relevant major accident*

73   The models should be named and described, and their suitability justified.

Criterion 4.5.4   *The harm criteria or vulnerability models used to assess the impact of each MAH on people and the environment should be appropriate and have been used correctly for each relevant major accident.*

74   Sensitivity of the results to the choice of harm criteria or model, or the way it is used may be needed, particularly when the scale and nature of the hazard and risks is significant. It is recognised that harm criteria for the environment are scarce and uncertain. Nevertheless, justification for the approach to environmental impact assessment and data used is needed. An essential requirement is that the operator's controls meet the relevant EQSs.

Criterion 4.5.5   *Assumptions used are justified, realistic, and not unduly optimistic*

75   The sensitivity of the results to assumptions that are pivotal to the analysis should be tested, particularly when the scale and nature of the hazard and risks are significant

Criterion 4.5.6   *Estimates of the severity and extent of each major accident consequence are realistic.*

76   The operator should check that the predictions are realistic by comparison with published assessments and with company benchmarks. If not errors in any of the above steps may have arisen and should be corrected.

77   HSE and the agencies' assessors will exercise judgement in a similar way, but using CA benchmarks and views about the models used by the operator.

Criterion 4.6   *The findings and conclusions from the predictive risk analysis should summarise the relationship between the hazards and risks and demonstrate that the measures adopted to prevent and mitigate major accidents make the risks ALARP.*

78   Schedule 4, Part 2, paragraph 4 of the Regulations requires identification of possible major accident scenarios for risk analysis purposes, and the identification and analysis of prevention methods. Paragraph 4(b) requires an assessment of the severity and extent of each major accident. This needs to be assessed in relation to the functioning and failure of existing control measures and assessment of whether there is a need for further controls to reduce the likelihood of major accidents and the extent and severity of the associated consequences.

79   The purpose of this criterion is to enable a view to be taken on the suitability and sufficiency of the risk assessment for drawing soundly based conclusions. It should be clear that the operator's approach to demonstrating compliance with the 'all necessary measures' requirement, is fit for purpose.

80 The scope and depth of the analysis, and the comprehensiveness of the presentation of the risk assessment therefore will generally be proportionate to the scale and nature of the hazards and the residual risks, and sufficient for demonstrating that all necessary control measures have been taken. There should be clear links between the conclusions and:

   a) the analysis of the risks, including hazardous event likelihoods and the associated consequences; and

   b) the measures (technical or procedural) taken to make the risks ALARP.

81 The ALARP arguments may be qualitative and focus on relevant good practice and sound engineering principles. Several sources of authoritative indications of good practice exist:

   i) Prescriptive legislation

   ii) Regulatory Guidance

   iii) Standards produced by Standards-making organisations

   iv) Guidance agreed by an organisation representing a particular sector of industry

   v) Standard good practice adopted by a particular sector of industry.

82 There is clearly an order of precedence from i) downwards and any conflicts between these sources of good practice should be resolved in favour of the one higher in the list.

83 HSE expects good practices to be followed; but if good practice is used as the sole justification of ALARP, several stringent requirements need to be met. These include:

   i) the practice must be relevant to the operator's situation;

   ii) any adopted standard must be up-to-date and relevant; and

   iii) where a standard allows for more than one option for conformity, the chosen option make the risks ALARP;

84 More complex situations may require the presentation of quantitative arguments coupled with cost benefit analysis in order to provide the justification that all measures necessary have been taken. If quantitative arguments are used the methods, assumptions and the criteria adopted for decision making should be explained. For example in the case of fatality risks to people off-site it is common practice [3] for the maximum tolerable level of individual fatality risk to be set at 10^{-4} per year and for the broadly acceptable level to be set at 10^{-6} per year. For new plant a lower maximum tolerable risk level may be adopted. The use of cost benefit analysis (CBA) enables society's aversion to particular group or societal risks to be considered in a transparent way. Corresponding criteria for judging the significance of environmental impacts have yet to be developed and agreed. Nevertheless operators need to state and justify the benchmark criteria adopted for their environmental impact assessments.

Criterion 4.6.1 *The safety report should demonstrate that a systematic and sufficiently comprehensive approach to the identification of risk reduction measures has taken place.*

85   It is not in the spirit of risk assessment to use it solely to demonstrate that existing controls or the adoption of current good practice make the risks ALARP. Risk assessment is an opportunity to systematically assess the current situation or decide the best option for designing a new facility. It is a chance to take account of technological advance, to seek inherently safer designs, and to take account of improvements in assessment methods and views on good practice etc. Whatever additional measures are identified as being reasonably practicable they should be implemented. The justification for rejecting possible risk reduction measures needs to be well argued and supported with evidence.

Criterion 4.6.2 *The main conclusions on the measures necessary to control risks should adequately take account of the sensitivity of the results of the analysis to the critical assumptions and data uncertainties*

86   The results of any risk assessment will be subject to uncertainty. Uncertainty in qualitative risk assessment arises from the validity of any assumptions made, the 'completeness' of the hazard identification and views on the likelihoods of hazardous events and associated consequences. Uncertainty in quantified risk analysis arises from assumptions, 'completeness', data inaccuracies, and the capability and appropriateness of the models employed. The greater the uncertainty the greater the need for a conservative approach supported by strong qualitative arguments based on sound engineering judgement and relevant good practice. In situations where good practice has yet to be established collateral evidence from analogous situations may be helpful For example if a novel design of storage vessel is adopted, failure modes and likelihoods can be developed by taking account of what is known about these parameters for current designs.

87   The interpretation of 'suitable and sufficient' risk assessment will depend on the complexity of the process, the scale of the hazards, and the degree of associated uncertainty. These factors also influence the balance between qualitative, quantitative and semi-quantitative evidence and arguments.

Criterion 4.6.3 *The conclusions drawn from the risk analysis with respect to emergency planning are soundly based.*

88   The worst case scenarios for people and the environment must be considered. The analysis of these should not be overly optimistic or pessimistic as this could have resource implications for the emergency services. The consequence models and assumptions used therefore need to be appropriate for the scale and nature of the hazards (see also Criterion 4.5). The range of hazardous scenarios considered needs to be representative and suitable for emergency planning purposes. The consequences of catastrophic vessel failure and guillotine fracture of pipework need to be included. The levels of harm considered and the impact criteria/vulnerability models used need to be suitable for predicting the extent of areas where people might be fatally or seriously injured or require hospitalisation. For environmental impact assessment, corresponding levels of harm to the environment should be considered. A considered. For releases resulting in environmental damage a range of representative conditions need to be considered eg to cover the range of flow rates in water courses.

References

1. G A Papadakis and A Amendola (Editors). Guidance on the Preparation of a Safety Report to meet the requirements of Commission Directive 96/82/EEC (Seveso II). Report EUR 17690 EN.
2. HM Treasury 1996. The setting of safety standards
3. HSE, 1992. The tolerability of risks from nuclear power
4. Le Guen, J M, 1997. Incorporating risk assessment and its results in the decision-making process. Proc of the ESREL Conference, Lisbon June 1997
5. Department of Environment, 1991. Interpretation of the major accidents to the environment for the purposes of the CIMAH Regulations. A guidance note by the DoE
6. IChemE, 1995. Source term models. IChemE Monograph

Acronyms

ALARP	as low as is reasonably practicable
BATNEEC	best available technology not entailing excessive cost
BLEVE	boiling liquid expanding vapour explosion
CA	competent authority
COMAH	control of major accident hazards
EA	Environment Agency
ETA	event tree analysis
EQS	environmental quality standard
FMEA	failure mode and effect analysis
FTA	fault tree analysis
HAZOP	hazard and operability study
LPG	liquefied petroleum gas
QD	quantity-safety distance
RA	risk assessment
SR	safety report
SRAM	safety report assessment manual
TOR	tolerability of risk

ICHEME SYMPOSIUM SERIES NO. 144

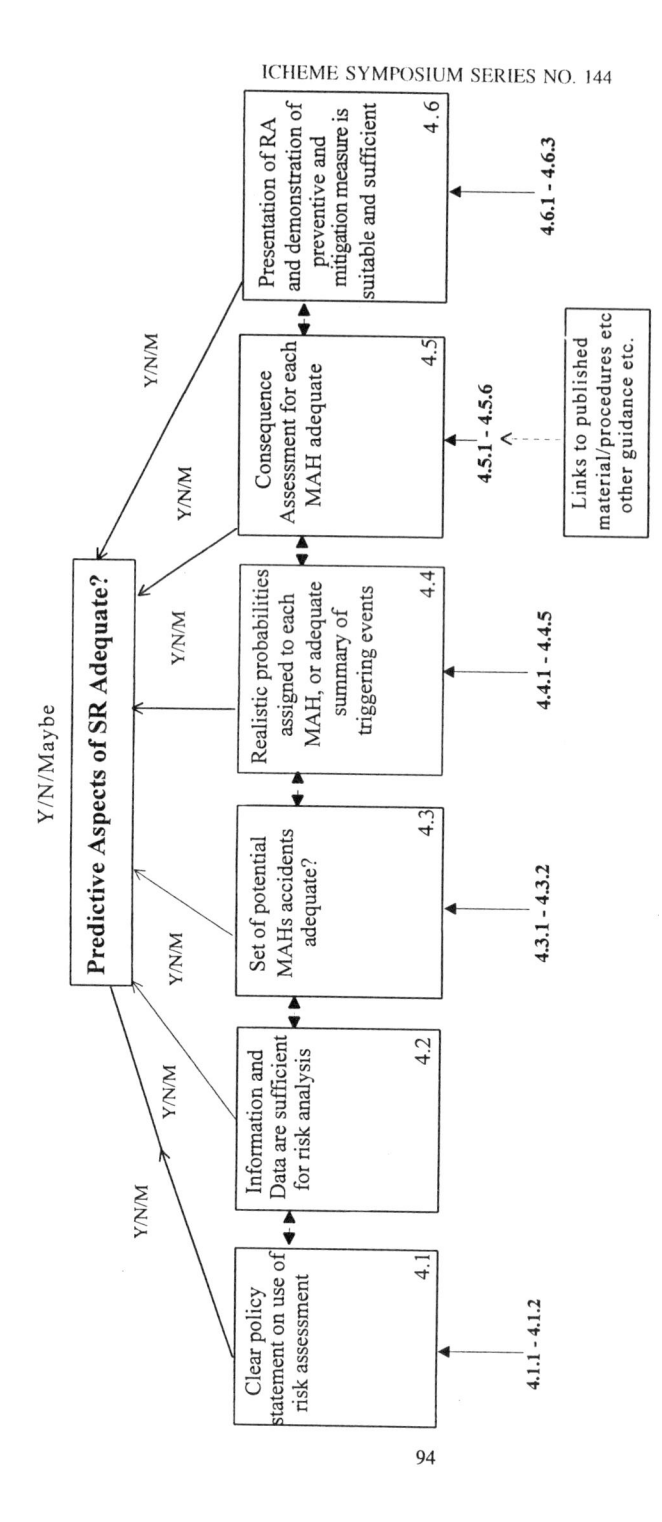

Fig 4.1 Overview of Criteria for Predictive Aspects of Safety Reports.

The extent to which the top level criteria (criteria 4.1 - 4.6) are met is determined by the extent to which the lower level criteria are satisfied.

INDUSTRY EXPERIENCE FROM THE PILOT EXERCISE

Ian Hamilton Senior Technical Safety Adviser, HSE Shared Service, BP Grangemouth

This paper will set out the industry experience in producing a safety report to the new COMAH Regulations. It will describe how the assessment criteria discussed in the previous papers were used to develop the new style reports. It will highlight the any problems that were encountered during the exercise. The paper will also discuss the benefits in participating in the exercise.

Keywords: Safety Report, Criteria, Pilot Exercise, COMAH Regulations

INTRODUCTION

When the CIMAH Regulations were introduced in the 1980s BP decided that a common approach to the submission of Safety Reports would be beneficial from both theirs and the Regulators viewpoint. To meet this goal a network of safety professionals from all BPs UK top tier sites and Corporate functions was established. The main purpose of this group was to provide guidance on the structure and content of safety reports and review the feedback that individual sites received on their submissions.

In July 1997 BP were approached by the Head of the Technical Assessment and Information Technology Strategy Unit, within the Chemical and Hazardous Installations Division of the Health and Safety Executive to see whether they would be willing to second a safety professional into the unit to assist with the development of assessment criteria to be used by assessors of safety reports submitted under the COMAH Regulations. This work was being undertaken by a project team within the Health and Safety Executive. The details of this project entitled SHARPP (Safety Report Handling Assessment and Review Principles and Processes) have been discussed by the previous presenters.

One of the products of the SHARPP Project was to seek volunteers from industry to participate in a pilot exercise to produce a safety report that would meet the requirements of the COMAH Regulations. During the early part of my secondment HSE were in the process of seeking volunteers . As stated previously BP has a number of current top tier CIMAH sites and the opportunity to participate in the pilot exercise was seen as an ideal way to ensure that a consistent approach was adopted in writing safety reports to meet the requirements of the COMAH Regulations. To ensure that BP achieved the maximum possible benefit from the secondment they sought to participate in the pilot study. Originally Grangemouth was put forward as the BP site to participate because of my previous experience there, however the Scottish site for the pilot had already been chosen. I therefore contacted all the other UK sites to see if any of them would be willing to participate. BP Chemicals at Saltend in Hull asked to be put forward as their Core CIMAH and the DF Complex reports were due for resubmission under CIMAH in 1998. In addition to these reports the Hull site have another eight that they currently submit under the CIMAH Regulations.

The DF process involves the liquid phase oxidation of naphtha using air as the oxidising material in four reactors - two on DF2 and two on DF3 - operating at 185°C and 47 barg. The reaction mixture is processed in distillation columns involving azeotropic, pressure and vacuum distillation and liquid/liquid extraction to produce acetic acid, formic acid and propionic acid. A partially oxidised light ends fraction from the main reaction is fed to the acetone recovery plant where pure acetone is extracted and the remainder fed back to the reaction system. Light residues recycle feed is residues from DF2 and DF3. Useful material is removed by flash distillation and returned to the reaction system. Remaining residues are fed to site boiler house as fuel for steam generation.

In December HSE accepted the proposal to use BP Chemicals at Hull as part of the pilot exercise. However, because of the Christmas break work on the report did not start in earnest until early January. This gave us approximately three months to complete the two reports. Once Hull were accepted it was agreed that I could assist with the preparation of the reports and act as a link between BP and HSE.

PREPARATION OF THE REPORT

Each company participating in the pilot exercise was given a set of all the draft criteria which related to the assessment of Safety Reports (i.e. Management, Descriptive, Predictive, Technical and Emergency Response) and the latest revision of the Consultative Document for the Regulations. The task was then to prepare a report which would meet these requirements.

As stated earlier two reports were submitted one plant specific report the other being the core report. The core report gave information concerning the overall aspects of the site, geographical location, details of the safety management system, general management structure, position of installations, etc. The plant specific report gave information about an individual plant which had been identified as having the potential to cause a major accident.

The first step in the process was to establish a contents list for the report. This was done by extracting the headings from each of the criteria and putting them into sections. Each section could be related back to Annex 2 of the Seveso II Directive and Schedule 4 of the draft COMAH Regulations. This ran to four pages. The existing CIMAH reports were then used to provide the detail for each heading and identify where information was lacking or required to be expanded upon. It was then an iterative process to complete the reports using the criteria to decide on the information required.

COMMENTS ON PROCESS

From an early stage in the preparation of the report it became apparent that if Schedule 4 of the Regulations was followed the information would not be presented in a logical manner and would be difficult for the Assessor to gain an understanding where the site was, what hazardous substances were present and how many people could be affected, before details about the management system were presented. We therefore decided not to follow Schedule 4, but to set out the report in a format that could be easily read. In the end the format adopted was similar to that of Schedule of the CIMAH Regulations

Reports submitted under CIMAH were required to, identify the nature and scale of the use of dangerous substances at the establishment, identify the type, likelihood and consequence of major accidents that may occur, give an account of the arrangements for safe operation of the establishment, for control of serious deviations that could lead to a major accident and for emergency procedures at the establishment. The requirements of COMAH take this a step further in that there is now a requirement to demonstrate amongst other things that, the safety management system is adequate, the necessary measures have been taken for major accident prevention, control and mitigation and the plant has been adequately designed, constructed, operated and maintained.

The purpose of the criteria was to give the Assessors a guide as to the information that they should be looking for which demonstrates that all necessary measures have been taken. Each set of Criteria was written in a different style, this lead to difficulty interpreting the level of detail required to demonstrate. The most helpful style was that adopted by the Management Criteria. This listed each criterion, then gave a reason for it and finally gave examples of evidence on how it could be met. This was most helpful as it gave an indication of the level of detail required to ensure that information provided would be sufficient. Some of the other criteria left the writer wondering how much information to include. For example the criteria being used to assess the properties of dangerous substances was essentially asking for the Material Safety Data Sheet (MSDS) in that they required information on boiling point, flash point, vapour pressure, explosive limits, International Union of Pure and Applied Chemistry (IUPAC) nomenclature, Chemical Abstract Number, health hazards, effects on the environment, etc. However to include all the MSDS would have been impractical. They would have created at least a further volume to the report. Thus we chose to extract the information from the MSDS and other sources, when the information was not available from the MSDS, and tabulate it. This proved to be a major exercise in itself as more detail was required than had previously been submitted under CIMAH. The preparation of this table alone, which ran to two sides of A4 paper, took approximately two man weeks.

Another example of this appeared in the Emergency Response criteria. The information being sought after here was in essence the site emergency response plan. However, as with the information on dangerous substances had we included the plan this would have created a further volume to the report.

The style and format of the reports took a considerable amount of time to prepare. We estimated that it took approximately seventy man days to prepare the two volumes for the pilot exercise. This was because it was difficult at times to judge how much of a referenced document or procedure to use in the reports. This was partly due to the fact no definition of the requirements of demonstration had been given with the criteria. In some cases the whole document was included and others it was only referenced by content and title. In some cases it was difficult to establish what the actual measures would be which would prove compliance. This may have resulted from the fact that the criteria which were used during the preparation of the reports were relatively early drafts. They were the ones that had been sent out for consultation both internally within HSE and externally. An example of this arose from the Technical Criteria were there was the perception that the report should demonstrate that a hierarchical approach have been used when the plant was originally designed. Something that is almost impossible to demonstrate on a plant which is 40 years old.

A number of the criteria asked for the same information, although it appeared in a different format. This arose because each set of criteria were written by different teams with the idea that they would be used by different assessors. An example of this was in the Emergency Response criteria where details of the site, which had already been requested in the Descriptive Criteria, were being requested. When writing the reports we had to ensure that we did not provide the same information twice.

One of the main differences between CIMAH and COMAH was the perception that environmental issues would be more prevalent. Although some of the criteria discuss environmental issues there was no guidance on what constituted an environmental accident. We therefore had to rely on existing guidance and hope that this met requirements. This was also the case with the Predictive aspects although a further revision of these criteria was sent out during the exercise.

There was a considerable amount time spent cross checking the Assessment Criteria with the Consultative Document to ensure that the correct interpretation had been made. This was especially evident with the dangerous substances as the categories and definitions had changed from CIMAH.

Initially it was thought that timescale to complete the reports was ample (approximately 3 months). It soon became apparent that it was going to be difficult to complete for the reasons discussed above. However the reports were submitted by the required date although it was recognised that all the cross checking with the Assessment Criteria had not been completed and that there were some areas where additional information would be added.

HEALTH AND SAFETY EXECUTIVE ASSESSMENT OF REPORTS

The Health and Safety Executive forwarded the results of their assessment of the reports to BP Chemicals, Hull at the end of June. The assessment team concluded that there were no serious deficiencies in either report. However, a number deficiencies were highlighted in the written demonstration. Some of these were relatively minor and will be easily rectified in future revisions of the report. The ones that will require more time and effort relate to adequacy of the demonstration and the lack of environmental information. This stems from the fact that guidance on what constitutes demonstration was not contained in the criteria nor was there sufficient guidance on environmental issues.

The most serious concern centred around the Major Accident Prevention Policy (MAPP) that had been put forward. This was a Corporate document which related to prevention of all accidents not just major accidents and is signed by one of the group managing directors and the group chief executive. The Assessment team did not consider this document to meet the requirements of the MAPP in that it did not specifically address major accidents. However when analysed against Schedule 2 of the Regulations, (Principles to be taken into account when preparing Major Accident prevention Policy Document), it met the requirements. Discussions on this are on going at present as the implications of this could affect all BPs Top

Tier sites.

A meeting was held with the assessment manager and some of the team to discuss the points raised in the letter. At this meeting a number of other points, which had not been included in the original letter, were raised. In addition we were presented with a set of environmental criteria which had been drawn up for the Environment Agency by DNV. Prior to this there had been very little information available from them. We are currently awaiting feedback on the outcome of this meeting. Further details will be given when the paper is presented.

CONCLUSIONS

Participating in the pilot exercise has proved a valuable experience for BP in that we have gained a good understanding of the requirements of the new regulations. This will allow us to develop our own guidance for the preparation of safety report thus ensuring that a common approach is taken for all BP establishments.

The lack of environmental criteria meant that when writing the report we were second guessing the requirements of the Agency. Also the lack of a clear definition of demonstration presented problems

The exercise has shown the importance of providing clear guidance to both the Assessor and the report writer thus avoiding the time consuming process that Industry and HSE went through when CIMAH was introduced. It has also shown that whatever guidance is produced it needs to be in the form of a single set which encompasses all aspects of the regulations.

MEETING THE DEMANDS OF THE REGULATOR AND LITIGATOR – AN INTERNATIONAL APPROACH

E Blackmore, MIOSH, RSP.
Eli Lilly and Company Limited, Speke Operations, Fleming Road, Merseyside L24 9LN

> The impact of health and safety regulations on both the employer and the employee has resulted, in general, in improved working conditions and a healthier workforce. However, accidents are still occurring and whilst the regulator, in the form of the under resourced Health and Safety Executive, can effect prosecution, employees are increasingly turning to the litigator to obtain financial compensation for their injuries or ill health.
>
> Keywords: business risk, litigator, public image, regulator, risk perception, standards.

INTRODUCTION

American multinational companies appear to be more vulnerable from the litigator than from the regulator. The reason for this lies in the fact that actions for damages are often sought not just from within America but also from people in countries outside the United States.

Increased environmental and societal awareness has led individuals and organisations to challenge industry to put its house in order in a way which the regulator may never hope to succeed. Recent examples of such challenges are;

- Bhopal
- Brent Spa
- Multinationals operating in Third World countries

Bearing in mind that where America leads today others tend to follow, how can multinational companies meet the demands of both the litigator and the regulator and still hope to remain viable? This paper will discuss some areas of business risk and possible means of minimising exposure to such risks as considered by a UK subsidiary of an American parent company.

INTERNATIONAL BUSINESS RISKS

Although prevention of injury to people and damage to the environment should be of equal concern to any enterprise as making a profit, multinational companies are also subject to other forms of business risk.

Business risk can be defined as any risk that has the potential to affect the earning capacity of a company. Increasing public and social awareness in terms of the environment, public liability and product liability has opened up further avenues of risk - particularly with respect to company image. Increasingly, the source of this risk now includes both the regulator and the litigator.

The requirement to do 'all that is reasonably practicable' to ensure that the public at large are kept free from harm from both the manufacture and use of goods produced by the company has far reaching consequences for multinational companies.

Direct and indirect financial risks arising from accidents and incidents

Up until recently, actions for damages against a company were taken only in the country where the accident occurred however, things are changing. The disaster at the Union Carbide plant at Bhopal in India is regarded as the worst accident to occur in the chemical industry. The size of the disaster varies according to different reports, but it is generally believed that about two thousand people were killed and hundreds of thousands injured. The reason for such a large scale incident was the proximity of the plant to the local community. In addition to the cost in terms of human life, Bhopal also paved the way for the litigator to seek damages in the country of the parent multinational company.

To ensure that their assets are maintained parent companies are increasingly inclined to audit their subsidiaries. For American multinationals this could mean auditing a spread of countries from Europe through Asia to the Pacific rim. Here lies the dilemma as varying standards will be found from country to country. Third world countries struggle to integrate increasing industrialisation from chemical and petrochemical companies into their ancient culture in which age equates to wisdom. Their legislation designed to protect the health and safety of the workers and the environment is in its infancy but industry and technology within those countries continues to expand.

However, having identified a risk exposure in the form of audit findings, a plan of action will be required from that particular subsidiary. Disclosure of information gathered about accidents, incidents and their means of prevention is often hampered by US lawyers concerned about claims for damages. How then can the parent company be sure that appropriate actions will be taken and standards met if information is not shared and, if action is not taken, how serious are the consequences likely to be and how will they impact the business?

Risks arising from the provision of information. Documentation will exist from an audit highlighting the risk and usually categorising it as high, medium or low. In the event of an accident, awareness by the parent company of the existence and magnitude of the risk clearly places the responsibility and any claims for damages with the parent. This is especially true if the risk is in a third world country where the appropriate skills may not be available and the location in terms of the population, less than desirable. In addition, what if the risk is not recognised by the auditor - where then does the responsibility for this omission and consequences of potential claims lie?

The increasing use of the Internet as a means of communication has opened up new areas of potential exposure. Both electronic and paper copies of documents can be cited as 'discoverable' and used in court against the company. So much so that some American owners are going to great lengths to ensure that, for potentially 'sensitive' information, procedures for handling and communicating such information are developed by their legal departments.

This raises another issue. Information about industrial chemical hazards and their means of control are required to be published for acceptance by the Health and Safety Executive (HSE) and the Environment Agency in order to comply with specific legislation (as discussed later in this paper). These 'safety cases', and also details of any enforcement action taken against a company, are placed in the public domain.

Although information which could be considered to be commercially sensitive is not published there could, in the aftermath of an accident or incident, be sufficient information available for an astute lawyer to make a case for civil action to be laid at the door of the highest payer.

America is notorious for exceptionally large financial settlements and is likely to be the focus for future civil claims from overseas subsidiaries - especially for incidents that are the scale of Bhopal or as politically sensitive as the Pan Am disaster at Lockerbie. The effects on the business may be sufficient to impact the viability of company and all of its subsidiaries.

Risks arising from the loss of public image. Further exposure lies in the easy access to the Internet by organisations such as Greenpeace whose increasing following and newspaper coverage can result in considerable damage to company image. The disposal of the redundant oil platform Brent Spa is recent an example of the power of Greenpeace and the resistance of society to bow to the pressure and actions of multinational companies. Although Greenpeace admitted later that their case had some weaknesses, by then the damage was done.

Multinationals operating in third world countries also attract other forms of criticism from the public. The expansion of industry in these countries has increased the need for manual labour usually from women and children. The apparent 'exploitation' of third world cheap labour and its effect on the health and life span of the workers is being publicised by outside organisations and agencies campaigning for the rights of workers in these third world countries. The results can be seen on the shelves of our supermarkets where products from Traidcraft, an organisation who pay a living wage to third world workers, are sold alongside for example coffee produced by traditional suppliers.

How does this use of third world labour by multinationals sit with the often heard phrase 'People are our most valuable asset' ?

Nearer to home pharmaceutical companies, soft drinks manufacturers and suppliers of every type of food product are particularly at the mercy of the public should their product fail in any way. The brand name associated with a company's product has real value in financial terms as Perrier discovered to its cost. In 1990 Perrier water was found to be contaminated with benzene. This, together with poorly handled publicity, took the company from being the market leader in bottled water to a position whereby Nestle were able to take advantage of the low share price and buy 40% of the company.

Companies can also suffer loss of image and loss of business even when products do not cause physical harm. Nike caused offence to the Arab world by using a logo which resembled the Arabic word for Allah. All of their training shoes bearing this logo had to be recalled.

Risk transfer, by means of insurance and / or the use of contracts will do nothing to save the company image when civil action is taken in the pursuit of damages against the company for injury or death caused during the manufacture or use of one of its products. Unfortunately it is only the catastrophic disasters that get publicity and huge payouts, not the annual death toll of child labour in third world countries.

MINIMISING EXPOSURE TO BUSINESS RISK

One approach that can be used by multinational companies to ensure that their risks are minimised is for them to:-

- look at what the local regulator requires,
- examine what standards exist both from local and international legislation and within industry,
- be aware of what society expects from producers and manufacturers.

Means of minimising risk

The reaction of the regulator to major incidents in the past has been to create more regulations. The Control of Industrial Major Accident Hazards Regulations (CIMAH) was the work of an Advisory Committee on Major Hazards which considered the implications of Flixborough in 1974. This incident resulted in twenty-eight deaths on site and extensive injuries and damage. An equivalent European reaction was the Seveso Directives. In 1976 more than two hundred people and large areas of land were affected by the release of Dioxins at Seveso in Italy. The Seveso directive is implemented in the CIMAH regulations in the UK.

In both Flixborough and Seveso, though interestingly not in Bhopal, the regulator intervened with what many consider to be an over reactive approach largely as a result of public pressure.

Companies are faced with the challenge of meeting the requirements of the regulator and the litigator whilst maintaining a viable business and staying ahead of competitors. American owned companies based in the UK and Europe are required to satisfy their national health, safety and environmental legislation and also to meet corporate standards which are usually based on Occupational Safety and Health Agency (OSHA) Regulations. Examination of the regulations in the country of the affiliate operation can often be seen to overlap many 'in house' standards set by the American parent

OSHA Process Hazard Reviews *v* CIMAH accident scenarios. Companies who have sites which are required to comply with CIMAH Regulations should currently be producing their safety cases. These safety cases detail the risks to the public and to the environment and identify the appropriate control measures to be taken. New legislation in the form of the Control of Major Accident Hazards Regulations (COMAH), which will be in force from February 1999, extends this requirement to those companies whose inventory of chemicals had previously fallen below CIMAH thresholds. The result will be that many more companies will be required to assess the off-site impact of potential accidents involving chemicals listed in the schedule to the Regulations.

Whilst this might seem to increase the burden on American owned subsidiaries who may have minimal resources, it also provides opportunities to apply holistic and cost effective approaches to the assessment and control of these and other business risks. One way forward is through combining these demands by extending OSHA Regulation for process safety management to encompass the requirements of our Management of Health and Safety at Work Regulations and the impending enactment of the Seveso 2 directive i.e. the COMAH regulations.

OSHA requires Process Hazard Analysis (PHA) to be carried out to identify and analyse the significance of hazardous accident scenarios associated with a process or activity. PHA is used to pinpoint weakness in design and operation of facilities that could lead to accidental chemical releases, fires or explosions and the resultant on or off site effects.

It provides a basis for a Process Safety Management (PSM) programme and the techniques used are those that anyone in the chemical industry will be familiar with, e.g. HAZOP, What-If-Analysis etc. The key being correct selection of the technique appropriate to the risk and awareness of the strengths and weaknesses of the that technique.

The objectives in conducting a PHA are to:-

1. Identify those hazards inherent in the process or activity,

2. Identify credible failures, both human and / or equipment that could lead to accident scenarios,

3. Assess the risk of those scenarios in terms of likelihood and consequence,.

4. Mitigate the risk by making changes to the design and / or operation of process conditions,

5. Document the findings and actions.

These objectives reflect the requirements in broad terms of both the Management of Health and Safety at Work Regulations 1992, CIMAH and the proposed COMAH Regulations. The main difference being that, for CIMAH and COMAH, the documentation will be accessible to the public and thereby also to the litigator.

Application of standards. With the UK affiliation to the European Community we are seeing British standards gradually being replaced by European and international standards. These changes are reflected in the 'working' standards such as BS 5304 Code of Practice for Safety of Machinery and its replacement BS EN 292 'Safety of Machinery - Basic Concepts', and in items of machinery and equipment which now carry the CE mark as opposed to the British 'Kite' mark. The quality standard BS 5750 and the environmental standard BS 7750 are mirrored by the international standards BS EN ISO 9000 and BS EN ISO 14001.

All of these standards also have equivalents in America. For example respiratory protection and other personal protective equipment use the National Institute for Occupational Safety and Health (NIOSH) standards. Protection appropriate to the risk of exposure will be used by the American employee but what if any form of protection will be being worn in the Far East by the company employee facing similar risks?

Is it acceptable that 'almost' equivalent but differing standards (or even no standards) are used throughout an organisation? It potentially exposes a multinational company to a litigator arguing that differing 'equivalent' standards in some countries afford a lower degree of protection to the employee and therefore cite negligence or breach of duty of care of those exposed to the risk.

Many companies, because of decreased resources and the need to meet customer demand, are now looking to an integrated approach to quality, health and safety, environmental and even financial management. The new British standard for safety BS 8800 'Guide to Occupational Health and Safety Management Systems' has a similar approach to ISO 14001 thus giving companies no excuse for setting universal standards in these areas.

The advantages of an integrated approach to business risk are;

- the visibility of the major risks from both inside and outside the company
- the use of one management system
- better use of limited manpower and other resources

Risk perception and society. In the western world the public are no longer sitting back and allowing the big names in industry to make a profit without considering the impact on the local populations, the earth's resources and the environment.

In the past any adverse comments went largely unheeded. Now however facts and figures right or wrong, as in the case of the previously mentioned Brent Spa, are being used to support the publics' arguments. Companies are being challenged to demonstrate that their actions will not have harmful long term effects on the environment or on the public. Industry should not underestimate the use of pressure groups as a means of risk control.

Although Environmental and now Health and Safety legislation require risks to be reduced or adequately controlled, the decreasing numbers of regulators means that the first awareness of a problem is the incident itself. At this stage the use of legislation is often too late. Prosecution and fines will do little to mitigate further damage or to reduce the loss of lives. The regulator's focus on plant and product design in particular is very limited and therefore contributes little to the control of business risk

FUTURE CHALLENGES

The Health and Safety at Work etc. Act of 1974 and its relevant statutory provisions promote self regulation and are goal setting in their requirements. Conversely, both OSHA and European legislation lean towards prescriptive requirements. COMAH, although basically goal setting, also tends towards a prescriptive approach. It is certainly much harder for a company to argue the case for compliance when legislation is prescriptive.

With CIMAH (and now COMAH) however, are we seeing a shift towards prescriptive legislation in the UK and is it a move in the right direction? The idea of risk being the basis for action is sensible and logical. It was the intention of the Roben's report, which led to the 1974 Act, that those who create the risks should also control them. This has been reinforced from an unexpected direction. In 1992, the Cadbury report on the financial aspects of Corporate Governance [1] recommended that the Board of a corporate body should formally address issues relating to "investments, capital projects, authority levels, treasury policies and risk management policies".

Although Insurance companies have paid out hundreds of millions of dollars in response to major incidents that is not the end of the story. This cost is borne by industry in the form of increased premiums, and in the added cost of rebuilding plant and replacing equipment, to demonstrate to the public and the regulator that chemical plants are safely designed, operated and managed. It seems that when one chemical plant has a major incident the whole of the chemical industry suffers financially.

As discussed, Globalisation of companies for manufacturing and production introduces varying standards (in design, equipment and construction etc.), in the perception of risk and in cultural behaviour and expectations. Can multinationals keep costs down and their shareholders happy by using cheap labour and at the same time apply universal standards ensuring the safety of their employees, the public and the environment?

Setting standards is an integral part of the risk assessment process i.e. the provision of control measures. The means of control must be appropriate to the risk therefore any standard set internationally should give a universal level of protection whether it be for the design of a building, plant or process or for respiratory protection or interlocks on machinery.

Acceptance of one particular standard whether it is American, European or international may not be the answer. The aim should be for the adoption of the highest standard of controls for the protection of the most vulnerable worker wherever they are located.

1. Cadbury Sir A: *1992 Report of the Committee on the Financial Aspect of Corporate Governance*

SAFETY IMPLICATIONS OF SELF-MANAGED TEAMS
Are self-managed teams compatible with safety?

Ronny Lardner, Chartered Occupational Psychologist
The Keil Centre, 5 South Lauder Road, Edinburgh EH9 2LJ

Abstract
One of the most cost-effective methods of work organisation for manufacturing industry is self-managing teams (SMTs). SMTs have an established track record of improving productivity, job satisfaction and employee involvement. There has been a recent trend for onshore and offshore petrochemicals and process industry operators to implement SMTs. A fundamental aspect of SMTs is devolving day-to-day responsibility and decision-making to employees, and reducing or eliminating the role of the first-line supervisor. However, the suitability of SMTs for safety-critical operations has been questioned, and anecdotal evidence exists of flawed implementation and lapses in safety.
This paper summarises the results of a year-long, three-phase joint industry project, funded by HSE and a major UK petrochemicals operator, which examined the safety implications of self-managed teams. Experience from onshore and offshore industries is summarised, and preconditions for SMTs and best practice in implementation are described. Examples of successful implementation are outlined, with key learning points.

INTRODUCTION

Background to this study

The UK Health and Safety Executive's publication "Successful Health and Safety Management"[1] states that the establishment and maintenance of management control within an organisation is one of the key elements of successful health and safety management. Furthermore, this publication observes that commercially successful companies often excel at health and safety management, as many of the features of successful health and safety management are indistinguishable from other sound management practices advocated by proponents of quality and business excellence.

A recent international review of employee involvement methods which foster organisational success through improved quality, productivity and employee attitudes concluded that self-directed or self-managed work teams (SMTs) were one of the most effective techniques[2]. A fundamental aspect of SMTs is redesigning work to devolve day-to-day control, responsibility and decision-making to employees, and reducing or eliminating the role of first-line supervisor. The review concluded that "SMTs appear to be the primary employee involvement approach of the 1990s".

The potential benefits of motivational job or work redesign approaches such as SMTs are summarised as higher job performance, motivation and job satisfaction; greater job involvement and lower absenteeism. Potential costs are greater likelihood of error, mental overload and stress and increased training time[3].

involvement and lower absenteeism. Potential costs are greater likelihood of error, mental overload and stress and increased training time[3].

HSE policy encourages employee involvement in improving health and safety[4]. Their position on SMTs is less clear. HSE guidance states "such initiatives can have positive benefits if group performance criteria covers health and safety. However, the health and safety implications need to be carefully considered, with specific steps being taken to deal with them." Furthermore, HSE-sponsored research has identified that poor planning or implementation of major organisational change can have adverse implications for health and safety[5].

Paradox

There appears to be a paradox. SMTs are a proven method of improving organisational performance. At first glance, the SMT literature is largely silent on the topic of safety[6]. Much of the recent SMT literature has focused on the effect of SMTs on productivity and job satisfaction. However increased errors and stress have been mentioned as possible consequences of implementation.

Many onshore process industries have implemented SMTs. Known examples exist in the petrochemicals and pharmaceutical sectors. Some problems with safety implications have been encountered during implementation of SMTs in these sectors. For example, a recent paper highlighted the problems encountered when SMTs were implemented by consultants with little health and safety knowledge[7], and the present author has personal knowledge of problems being experienced matching existing staff to higher job demands, increased reports of stress-related illness and difficulties with shift handover communication.

Scope of study

This study, joint-funded by HSE and a major UK petrochemicals operator, first examined the scientific literature on self-managed teams and their relationship to safety. The study also conducted three in-depth case studies in UK onshore and offshore process industries which have implemented SMTs, to establish a) reasons for their introduction, b) benefits gained, c) safety implications, and d) lessons learned.

DEFINITION OF TERMS

Self-management

One of the strategic choices open to organisations seeking improved organisational performance through greater commitment and involvement from their employees is job or work design[8,9]. This entails designing or redesigning how work is organised to provide jobs which are broader in scope, involve operative-level employees in managerial tasks such as planning and problem-solving, and where duties are flexibly defined. In short, the emphasis moves from people being told what to do, to self-managing what they do and how they do it, within carefully-defined objectives and boundaries.

In practice, the degree of self-management can vary from (a) making decisions associated with regulating immediate production or work processes, through (b) also determining the order of production to (c) in addition governing how collective decisions are reached[9].

Self-managing teams

A recent trend in work group design has been the widespread application of self-managing teams (SMTs), primarily in manufacturing industries. SMTs typically include all the elements of individual job redesign, coupled with providing the whole team with increased autonomy and responsibility over how they work together to achieve pre-determined outcomes.

Self-managing teams are groups of employees, typically 5 to 15, with the skills and authority to direct and manage themselves. SMTs can vary in the degree of autonomy and scope for self-management afforded to them. A number of different terms have been used to describe SMT variants, including semi-autonomous work groups, autonomous work groups, empowered teams and objective-oriented groups. In this document, the term self-managed team (SMT) will be used throughout, whilst recognising that within this term considerable variety exists.

It is apparent that such teams make decisions which would previously have been made by a supervisor or manager. Such teams often include a former manager or team member who acts as a team co-ordinator or coach. This is a particularly important role during the early stages of SMT implementation, as the team endeavours to come to terms with self-management, and the former manager's tasks and skills are reallocated to competent team members.

Finally, one acknowledged expert[9] has commented that the term "self-managing team" is to some extent a misnomer, as they require active and very careful management, admittedly of a different tenor and quality.

We now turn to the reasons why organisations choose to implement greater self-management.

REASONS FOR INTRODUCTION OF SELF-MANAGEMENT

An understanding of the reasons why organisations choose SMTs can be gained from a US review[10], which identified eleven benefits which may be expected from their introduction. All of these benefits are not necessarily expected by any one organisation, nor may they in fact be realised:

- increased productivity
- improved quality
- more innovation
- faster and better decision-making

- better customer service
- reduced costs
- less managerial bureaucracy
- reduced workforce
- shorter time to market for products and services
- increased employee motivation and commitment
- increased recognition of individual employee's contributions.

Improved health and safety is not explicitly mentioned. Improved mental health of employees is implied by greater job satisfaction, and improved safety may be a consequence of better decision-making and a more committed workforce. Indeed, high levels of stress have been related to work accidents in oil rig workers[11] and low levels of job satisfaction have been related to unsafe driving practices[12].

Another analysis[9] of the reasons underlying the popularity of SMTs points to expectation that one of the first responsibilities often delegated to SMTs is to generate process improvements and improve product or service quality. As decision-making is located near to the source of operational problems and variances, a rapid and effective response to uncertain conditions is possible. It follows that SMTs can be an appropriate choice of work organisation where minimisation of variance under technically complex conditions is important.

None of the reviews cited above explicitly mentions health and safety as a reason why an organisation may choose to implement SMTs. However this does not necessarily mean that SMTs are in any way incompatible with successful management of health and safety. Rather, by promoting more skilled, committed, independent, informed and flexible employees it might reasonably be expected that health and safety would be maintained or enhanced. Alternatively, some industry commentators argue that self-managed team members may take more risks, by taking initiatives without appreciating the full implications of their actions.

HSE'S APPROACH TO MANAGEMENT OF SAFETY

The Robens report established the relevance of a self-regulating or self-managing system to health and safety at work. Two of the guiding principles of this approach were that regulators should set safety goals, rather than determine how those goals should be achieved, and those who create risks are deemed responsible for managing them. These principles informed the regulation of health and safety in onshore organisations from 1974, but were not applied to UK offshore safety until after the publication of Lord Cullen's report into the Piper Alpha disaster in 1992[13]. It seems there is no philosophical inconsistency between this regulatory approach and the principles of self-management at the team level.

There is nothing in HSE's guidance on the management of safety[1] which appears incompatible with the notion of self-managed team working. The main area which requires careful planning is how to allocate responsibility for specific health and safety responsibilities and activities to team members, whilst retaining ultimate management responsibility for policy, supervision, control, audit and review.

SMTs AND SAFETY

We now focus on published literature on the relationship between self-managing teams and safety across a range of industries. It is commonly-accepted wisdom that responsibility for safety should be held by operational staff, rather than a specialist safety function. One reason for this is that many workplace hazards are best uncovered by workers themselves.

Effective management of safety requires active employee involvement, and communication and co-ordination between operational staff and technical specialists across organisational and shift boundaries.

One of the earliest published accounts of SMTs concerned safety[14]. In the early 1950's, UK coal-mining methods were undergoing technological change. Traditional methods involved cohesive teams of multi-skilled, self-managing, interdependent miners working towards common production goals. New mechanised long-wall production technology was introduced, somewhat akin to an underground assembly line. Miners' jobs were redesigned, simplified and de-skilled, thereby reducing variety. Management assumed responsibility for organising production, with a consequent loss of autonomy for miners. A payment system based on common group output was replaced with five different systems.

The result was lowered productivity, reduced co-operation, high absence and increased employee turnover. The changes in work design upset the existing social system, and went against the long-standing tradition of the self-supervising miner who worked within a team responsible for allocation, co-ordination and supervision of their own work.

A modified version of the earlier self-managing work group was reintroduced, and a common production-monitoring and payment system was reinstated. A careful comparison revealed that the new work design led to improvements in output, turnover, absence, accident rates and a reduction in stress-related illnesses.

Improving safety, job satisfaction and productivity was also the focus of a work design intervention at a US mine in 1973[15]. Self-managing autonomous work groups were introduced on a pilot basis into a traditionally-organised small mine. Amongst the reasons for the experiment was a joint concern by management and unions that improvements in safety could not be achieved without increased involvement and training of supervisors and workers. Following the year-long experiment, and evaluation found the experimental autonomous work groups had fewer safety violations, lower overall incidence of reported accidents and showed positive trends towards reduced costs and increased productivity. Employee attitudes showed positive improvements, with considerable enthusiasm for the autonomous work groups.

Similarly, the introduction of SMTs in an Australian heavy engineering workshop[16] led to improvements in job satisfaction and productivity. SMTs used their regular team meetings to address unsafe working practices, rather than rely on a safety representative. During the study period, the SMTs maintained a steady accident rate, whereas traditionally organised teams' accident rates increased.

SUMMARY

These three thoroughly-researched examples help to understand the relationship between self-managing teams and safety.

First, the UK coal-mining study demonstrates that removing self-management unnecessarily (in this case as a result of changing technology) can have unforeseen and adverse effects on safety, and that its reintroduction can restore the damage done. Second, when improving safety is amongst the goals of implementing self-managing teams, safety can be maintained or measurably improved alongside other important organisational outcomes.

PUBLISHED EXAMPLES OF SELF-MANAGING TEAMS IN THE PETROCHEMICALS INDUSTRY

We now examine the implementation of SMTs in the onshore and offshore petrochemicals industry, which includes exploration through refining to manufacture of petrochemicals products. A total of four published accounts of the implementation of SMTs in the petrochemicals industry were identified. These were: (1) a scientific study which sought to measure the effect of SMTs on various aspects of organisational performance, including health and safety and (2) descriptive reports which do not purport to offer the same rigour as the scientific study.

SCIENTIFIC STUDY

This recent study at the UK site of an American-owned chemical-processing company examined the effects of planned strategic downsizing on the well-being of employees who remained in the organisation after a reduction in headcount[17]. Over a four-year period the total number of employees on-site reduced from 455 to 283, a reduction of 40%. In tandem with the reduction in headcount, more efficient technologies and working practices were introduced. An "empowerment" initiative was introduced, which consisted of an increased emphasis on multiskilling, removal of management layers, restructuring of the organisation to create business and support teams and closer integration of production and engineering functions. Greater attention was paid to the development of individual process operators, coupled with the introduction of an annual appraisal process with goal setting and performance review. Training in technical and non-technical skills (e.g. quality improvement techniques) was stepped up. The study authors reported a marked increase in productivity, a substantial decrease in absenteeism and a decrease in lost-time accidents from seven per year to one. Although work demands placed on employees remaining in the organisation increased, this did not lead to an increase in job-related strain, an indicator of mental health. Job satisfaction increased for process operators, and was explained by the introduction of greater participation and clarity to their role.

The study authors certainly do not advocate reducing headcount as a human resource strategy. Rather, they assert that "paying attention to the design of work and the wider context can enhance an organisation's ability to achieve downsizing without incurring severe, negative long-term consequences for employees" (p.299). Furthermore, in this case downsizing was achieved at the same time as improvements were made to productivity and safety performance.

DESCRIPTIVE REPORTS

Three reports of SMT implementation in petrochemicals were identified, which are summarised below.

Table 1

Company / Industry	Improvements reported	Effect on safety
ICI Australia's Botany chemicals plant[18]	• improved "ambience" and "climate" at work • employees working smarter, more flexibly and more co-operatively • Absenteeism dropped by 80%, with resultant cost savings on overtime payments.	No safety performance data reported
Shell Canada chemical plant[19]	• a high level of competency amongst shift team members • efficient plant operations • obvious benefits of multiskilling • widespread participation and learning • efficient problem-solving • excellent industrial relations.	Health and safety outcomes were not explicitly described
Alcoa of Australia's Wagerup refinery[9]	• low levels of blue-collar turnover • low levels of industrial disputes • reduced labour costs and relatively small numbers of managerial, technical and ancillary staff required to run the plant • employees in self-managing teams reported higher levels of satisfaction with their jobs and higher organisational commitment than their counterparts in a sister refinery where jobs were traditionally organised.	No data on safety performance reported.

The scientific study cited was not solely concerned with the introduction of SMTs. SMTs was an important elements in organisational redesign, and was coupled with other changes to organisational structure, reward and training. This strategy proved very successful when productivity, safety, employee mental health and satisfaction criteria were evaluated.

The descriptive reports also paint a positive picture of improvements in productivity and employee satisfaction. Safety performance data is not reported. A more comprehensive assessment of the effects of SMTs on safety was obtained by conducting three in-depth case studies in UK onshore and offshore process industries, which are reported below.

CASE STUDIES

Three UK process industry companies were identified who had implemented self-managed teams, and were willing to describe their experience. Within each company, in-depth interviews were conducted with a senior manager, an operational manager and a self-managed team member to establish a) reasons for their introduction, b) benefits gained, c) safety implications, and d) lessons learned. Relevant documentation was also made available. The three case studies are summarised below:

UK chemical continuous process plant

This company introduced self-managing teams in 1992, to improve productivity and reduce maintenance costs. Each team has ten manufacturing technicians, plus a team leader, who is a working team member and has additional responsibility for emergency response. During implementation maintenance and process staff were combined on shift, and now most team members possess a combination of process and craft skills.

UK offshore oil production platform

This platform did not change its organisational structure to implement self-managing teams. Rather, it radically changed its management style from direction to empowerment. Each team now has a supervisor who assumes a "hands-off" coaching style of management, but is available to help with non-routine problems. An empowerment training package about the attitudinal and behavioural components of effective teamwork was delivered to all platform personnel.

UK chemical batch process plant

To achieve greater measurement and control of quality and performance, this plant redesigned their organisational structure to focus their manufacturing teams on single products. The nature of process operators' jobs were also changed, adding more responsibilities and demanding a higher levels of skill. Self-managing teams now operate supported by a day-based manager. An on-call facility exists, and a permanent on-site incident response team is available.

OUTCOMES OF SELF-MANAGED TEAMWORKING

All three companies reported significant commercial benefits from implementing self-managed teamworking. Some also measured employee morale, motivation and sickness rates and found significant improvements.

Two companies reported that their existing positive safety performance remained unchanged. The third company had detected an improvement in site safety performance, but was unable determine whether this was due to self-managed teamworking or other ongoing initiatives.

By examining output measures of safety (e.g. lost-time accidents) it was not possible to isolate a positive contribution via self-managed teamworking. However, all managers and team members were convinced that self-managed teams were inherently safer, and were able to identify the following mechanisms which they believed led to safer working practices.

Improved production and maintenance operations

- plant uptime significantly improved, resulting in less strain on platform systems due to regular unplanned shutdowns. Smoother operations due to increased uptime allow all staff more time to think ahead, rather than reacting to unplanned events
- completion of safety-critical maintenance on schedule improved from 85% to 100%.

Increased knowledge

- greater knowledge of plant and process - people are able to behave in a safer manner due to better understanding of the plant gained by cross-discipline training.

Changed patterns of communication

- increased involvement and enhanced skill in conducting shift handovers - previously reliant on supervisor
- less scope for communication errors when maintenance work is handed over. Now the people who operate the plant also fix it, significantly reducing the need for cross-disciplinary and inter-departmental communication, a known cause of maintenance-related accidents
- willingness to work on to complete a maintenance job, eliminating the need for handover.

Greater involvement in management tasks, and enhanced management skills

- responsibility for planning of work, and its safety implications
- team members each take greater responsibility for all aspects of their work, including safe working practices
- having and using the discretion to spot and fix problems as they occur.

Greater involvement in safety management and risk assessment

- The openness to involvement fostered by teamworking allowed a switch from a management-driven system of safety auditing to one which involved employees
- greater involvement of team members in risk assessment

- direct team responsibility for housekeeping in a specified area
- individual responsibility for safety tasks
- involvement in monitoring and improvement of safety indicators
- contributing to HAZOPS and design studies
- raising and resolving safety issues at team and safety meetings
- proactive observation and reporting of unsafe acts and conditions.

LEARNING POINTS

Whilst each company believed implementation of self-managed teams had been a positive and worthwhile step, this had not been a straightforward task. With the benefit of hindsight they each offered their key learning points, which include unexpected outcomes which can take the unwary by surprise.

From a senior management view perspective, key learning points were:-

- this type of change only works with top management support, which must be enlisted, maintained and visible to the workforce
- considerable resolve and determination is necessary to see the process through - the workforce will quickly identify if this is not present
- senior managers need to be trained and coached how to maximise the benefits of self-managing teams
- managers and supervisors have to be trained and supported through the changes
- a significant management resource is also required to monitor and coach others during implementation
- do not underestimate the training required for day-based team managers, who have to make a very significant change in their role and management style from that of shift manager
- deselection of existing employees for redundancy required careful and sensitive management. Some first-line managers who were not selected as team leaders found it difficult to revert to being team members, and most left the organisation within two or three years.
- the importance of building-in time for training to manpower planning, and providing sufficient skilled trainers and appropriate training facilities. Manpower plans must also leave sufficient experienced staff on-site to run core operations during training
- the need for a team reward system to recognise team performance
- development of user-friendly team performance indicators and support systems.

Operationally, key learning points were:-

- ensure a suitable organisational structure is in place to support greater self-management
- seek professional advice and guidance
- make a comprehensive, yet flexible plan for implementation
- expect some resistance from people who do not want to change or feel exposed
- make use of benchmarking visits for managers and team members

- prepare the ground by involving all those directly involved several months before implementation
- thoroughly analyse and understand the new role of first-line managers, how this differs from existing first-line managers, and use this information to help them make the transition
- understand the less-visible aspects of the supervisor's former role, e.g. planning, prioritisation and risk assessment, and ensure that these skills are developed in team members prior to implementation
- ensuring sufficient people are available to cover the changing workload, particularly during the early stages of implementation
- provide coaching skills to allow managers to delegate decisions and authority to teams
- carefully specify the function of support staff during transition.

Team members had learned:-

- the need for thorough consultation prior to implementation
- take time and effort required to ensure workforce "buy-in"
- don't rush in, prepare!
- think through implementation thoroughly
- consider designing your own training package to meet local needs, rather than buying in an "off-the-shelf" package
- consider how personalities interact in teams.

DIAGNOSING SUITABILITY FOR SELF-MANAGEMENT, AND BEST PRACTICE IN IMPLEMENTATION

This piece of research also identified key guidance on diagnosing the suitability of self-managing teams for a given task, and best practice in implementation. The final project report containing this information and the full case study text will be published in Autumn 1998, and initial enquiries can be made with the present author at The Keil Centre, 5 South Lauder Road, Edinburgh EH9 2LJ, Tel 0131 667 8059, Fax 0131 667 7946, e-mail keilcentre@compuserve.com

CONCLUSION

This study sought evidence of the effects of self-managed teamworking on health and safety, with particular reference to the petrochemical industries. The scientific studies available from petrochemicals and other industries indicate a positive effect or neutral on health and safety outcomes, dependent on whether improving health and safety was an explicit goal of implementation.

The three UK case studies also identified a neutral or positive trend in health and safety performance, however it was difficult to isolate the contribution of self-managed teamworking from other parallel organisational changes.

Senior and operational managers and team members believed that self-managed teamworking had led to inherently safer working practices, and were able to describe the mechanisms involved.

The study results should prove useful to process industry companies who are considering the introduction of self-managed teamworking, and to those who wish to enhance their existing teams.

References

1. Health and Safety Executive, 1993, Successful Health and Safety Management Suffolk : HSE Books

2. Cotton, J.L, 1996, Employee Involvement, *International Review of Industrial and Organisational Psychology*, 11: 219-241

3. Oldham, G, 1996, Job Design, *International Review of Industrial and Organisational Psychology*, 11: 33-60

4. Health and Safety Executive, 1994, Play Your Part : How offshore workers can help improve health and safety, London : HMSO

5. Health and Safety Executive, 1996, Business re-engineering and health and safety management: literature survey, case studies and best practice model (3 volumes) Suffolk: HSE Books Contract Research Reports 123,124 & 125/1996

6. Parker, S. and Wall, T. (forthcoming) Job and Work Design: Organising Work to Promote Well-being and Effectiveness, Sage

7. Blackmore, E, 1997, Managing health and safety during business process re-engineering, *Proceedings of Hazards X111 - Process Safety - The Future, an IChemE Symposium*, Manchester, UK, April 1997: 183-190

8. Walton, R.E., 1985, From control to Commitment, *Harvard Business Review*, March-April: 98-106

9. Corderey, J.L., 1995, Work design: rhetoric versus reality, *Asia Pacific Journal of Human Resources*, 33(2): 3-19

10. Elmuti, D., 1997, Self-managed work teams approach: creative management tool or fad?, *Management Decision* 35(3): 233-239

11. Rundmo, T., 1995, Perceived risk, safety status and job stress amongst injured and non-injured employess on ofshore pertroleum installations, *Journal of Safety Research*, 26: 87-97

[12] Raggatt, P., 1991, Work stress amongst long-distance coach drivers: a survey and correlational study, Journal of Organisational Behaviour, 12: 565-79

[13] Allison, R., 1995, Nearing the 'sunny uplands' of Robens' self-regulation, Major Hazards Onshore and Offshore II, Institution of Chemical Engineers Symposium, UMIST, Manchester 24-26 October 1995: 1-8.

[14] Trist, E.L. and Bamforth, K.W., 1951, Some social and psychological consequences of the longwall method of coal getting, Human Relations, 4: 33-38

[15] Trist, E.L., Susman, G.I. and Brown, G.R., 1977, An experiment in autonomous working in an American underground coal mine, Human Relations, 30: 201-236

[16] Pearson, C, 1992, Autonomous workgroups: an evaluation at an industrial site, Human Relations, 45(9): 905-34.

[17] Parker, S.K., Chmiel, N, Wall, T.D., 1997, Work characteristics and employee well-being within a context of strategic downsizing, Journal of Occupational Health Psychology, 2(4): 289-303

[18] Anon, 1990, Making people the competitive advantage, Worklife Report Vol 7(5): 10-11

[19] Halpern, N, 1985, Organisation design in Canada: Shell Canada's Sarnia Chemical Plant, Brakel, A. People and Organisations Interacting p.117, John Wiley and Sons Ltd.

ICHEME SYMPOSIUM SERIES NO. 144

THE IMPACT OF THE NEW SEVESO II (COMAH) REGULATIONS ON INDUSTRY

J.C.Ansell, Dr J.R.Mullins and R.Voke
AEA Technology Consulting, Safety Management North, Warrington, WA3 6AT, UK

> The implementation of the new Seveso II Directive or the COMAH Regulations as it will be known in the UK will come into force in early 1999 and will have a major impact on industry. The COMAH (Control Of Major Accident Hazards) Regulations will replace the current CIMAH (Control of Industrial Accident Hazards) Regulations, and place greater emphasis on the demonstration of safe operation, particularly in terms of safety management. This paper will address the issues as to what is required of the operator of an establishment in order to comply with the forthcoming legislation without too much time, money and effort being put into new systems and provision of information beyond that which is required by the competent authority.
>
> Keywords: COMAH, Seveso II, emergency planning, legislation

INTRODUCTION

On the face of it, the new COMAH regulations appear to be just building on the old ones, but there are significant changes that may be feared by some sections of industry, such as public disclosure, the greater requirements on lower tier sites and the testing requirements of emergency plans (see Table 1). However, as a package these regulations must be looked upon as very positive in intention, as it is a mechanism for building on the improvements that CIMAH made in the regulation of on-shore major hazards and is attempting to ensure that these improvements are applied evenly across the industries of Europe. This will lead to an overall improvement in safety to man and the environment, and hopefully a more cost effective approach to safety management systems.

WHO WILL BE SUBJECT TO THE REGULATIONS AND WHEN ?

The regulations will apply to an establishment where the quantity of dangerous substances held exceeds specified inventory thresholds. There is no longer any requirement for the site to be either a defined industrial activity or storage site, indeed this distinction has now been removed. The regulations therefore apply to all activities involving the presence of dangerous substances apart from specified exceptions including military sites, pipelines, temporary storage and waste landfill.

There will still exist named substances which will have a qualifying quantity. However the list will be much reduced as compared to the CIMAH Regulations. This is due to the use of generic categories such as Very Toxic, Toxic, Oxidising, Explosive, Flammable, Highly Flammable, Extremely Flammable, Dangerous for the Environment and Any Classification which does not enter into the above categories but possesses risk phrases such as R14 and R29. As with CIMAH, there are lower and upper tier threshold quantities with different associated duties (these are summarised in Figure 1).

A note of importance is the aggregation rule which may lead to some establishments not subject to CIMAH being required to comply with COMAH. Inventories of individual substances in the same or related generic categories must be divided by their respective threshold quantity and the fractions then added together. If the sum is greater than or equal to one then the regulations apply. An important distinction between COMAH and CIMAH is the fact that under the new regulations, named substances not present in qualifying quantities must also be aggregated under the appropriate categories.

Introduction of the 2% rule means that dangerous substances at an establishment in quantities equal to or less than 2% of the relevant qualifying quantity can be ignored for purposes of calculating the total quantity present if their location within an establishment is such that it cannot act as an initiator elsewhere.

The above assessment will need to be carried out using anticipated inventories of raw materials, by products, intermediates and final products. However, dangerous substances may also be generated during the course of unplanned events. This was the case at Seveso where dioxin was generated when a reaction involving pesticides went out of control. Such establishments where quantities of dangerous substances may be generated during loss of control of a chemical process are also within the scope of the COMAH regulations.

Top-tier establishments which already submit a safety report to the competent authority (CA) under the existing regulations must take action to submit their next updated report under the new COMAH legislation when it would have been due had CIMAH still been in force, or by 3 February 2001 whichever is the earlier. A CIMAH update required between 3 February 1999 and 3 August 1999 can be submitted in the form of a COMAH safety report any time up until 3 August 1999. Other establishments which come under the COMAH regulations have until 3 February 2002 to submit their report.

DUTIES ON OPERATORS

All sites whether lower or top tier have a number of common duties. Notification of activities to the competent authorities and reporting of accidents are similar to the requirements under CIMAH. However, where COMAH differs from CIMAH is the increased emphasis on demonstrating as opposed to describing the adequacy of the physical and organisational safeguards in place to prevent, control and mitigate major accident hazards. This is reflected in the general requirement for a Major Accident Prevention Policy (MAPP).

Additional duties on top tier sites include preparation of safety reports and testing of emergency plans. The safety report is the means of formalising the demonstration that all necessary measures have been taken to prevent, control and mitigate major accident hazards, whilst the testing of emergency arrangements provides additional assurance that effective plans are in place should a major accident occur.

SAFETY MANAGEMENT SYSTEMS AND MAPP
(MAJOR ACCIDENT PREVENTION POLICY)

The increase in emphasis on safety management systems even in lower tier sites (Major Accident Prevention Plan) is indicative of the way things are going in this type of legislation. The fact that Seveso II lays out the requirements of the major accident prevention plan (MAPP) will provide companies that are new to the Seveso legislation with a framework upon which to base their systems or to rearrange existing systems. This will also give a strong lead to companies that are already subject to CIMAH on how they should use their safety management system and what the regulators will be looking for. It is essential that the MAPP specifically addresses major accident hazards and relates to protection of both man and the environment.

The MAPP requires the safety management system to have mechanisms built into it to that will allow its effectiveness to be easily monitored. This means going beyond the reactive measurements of injuries, downtime, incidents etc. that are the staples of many systems at present. It means that positive proactive measurements will have to be used as well e.g. monitoring and review.

This at first seems to be just more demands on safety, health and environment budgets. However, in the major hazard industries it is cost effective to develop proactive safety systems. In large plants like refineries saving one day of downtime will probably pay for most of the improvements to the system. Although these arguments have been put before, COMAH provides a different opportunity because these new measures will be driven by legislative requirements. This will give management the incentive to achieve a better safety profile by investing in new systems that will be good business and provide compliance with legislative requirements (therefore less negative interaction with the regulatory authorities).

The Crux of this new approach is that safety management systems have to be demonstrably effective - measuring injuries only indicates the relative inadequacy of a safety management system.

To further develop the above theme it is possible to use the example of emergency planning. Sometimes emergency planning is looked upon as something that is 'bolted on' the end of a safety management system. It is looked upon as coming into practice only when somebody or something fails, or when a piece of bad luck occurs. Therefore it is almost natural to think of it as a negative aspect and separate it from the rest of the operational management system which is designed to ensure that everything goes as smoothly and positively as possible. The emphasis on safety management and the new requirements in emergency planning in COMAH may encourage managers to see emergency planning as an integrated part of their overall safety management system. For example adequate training is seen by everyone as a key component of a successful safety management system. Reflecting this, there has been a recent trend in the use of competence based assessments as part of operational training (in the UK the most popular are the National Vocational Qualifications - NVQs). This form of training ensures that the operator proves that he/she is adequately trained on a particular aspect of the job by demonstrating his/her competence. It is a 'quality system' based approach to training. However there is little evidence across industry of this approach being used in the training of managers specifically for emergencies or upset conditions. Emergency exercises often only demonstrate whether management representatives can follow a set of given procedures, they do not assess individual manager's competence to make correct decisions in 'upset' conditions in any depth. Yet it is under these extra-ordinary conditions that managers really need to perform well.

An important part of emergency planning is to try to ensure that correct decisions are made as an unplanned event develops, decisions that either prevent an emergency situation developing or, if this is unavoidable, ensure that all necessary actions have taken place in a timely manner so that the emergency plan has the best chance of success. Success in this aspect of emergency planning has direct and large economic benefits. Some forward-thinking companies have developed 'Situational Analysis' type training that assesses their individual managers abilities in mock stressful situations. This is an approach that has been used by airline companies for some time now. For companies that can afford them, the use of simulators can aid this training immensely, however the careful use of table top exercises can be just as effective. If actions can be taken that prevent an upset condition developing into a shutdown let alone a major accident situation then any money spent developing and testing improved safety management systems will be well spent.

COMAH with its requirements for testing of emergency plans will help emergency planning to be treated as part of the 'whole' of safety management, not just something that is an unfortunately needed add-on. It should also encourage a more auditable approach to the assessment of individual managers' ability to deal with formative and full blown emergency situations.

It will be part of the new COMAH Regulations that any establishment qualifying as lower-tier should have to a document in place displaying their MAPP and demonstrating that a safety management system exists. Top-tier establishments will be able to include their MAPP in their safety report which is submitted to the CA, whereas for lower tier establishments it will be more likely that it will exist as a stand alone document. It is necessary that the length and content of the MAPP is proportionate to the scale of hazards at the establishment. It is not necessary to submit the MAPP to the CA, but it must be available for examination by Inspectors, who may use it to structure and plan their inspections.

It may be necessary in some areas of the MAPP to refer to other supporting documentation e.g. plant operating procedures, training records, job descriptions etc. which may be too lengthy or tortuous to include in the MAPP itself. Hence it does not need to be a lengthy document, only setting out what needs to be achieved and an indication of how it is to be done.

The essential contents of a MAPP are briefly described below;

- roles and responsibilities of personnel involved in management of major hazards at all levels within the organisation including selection criteria for competent personnel and training requirement;
- methods in place for the identification of major hazards and assessment of their likelihood and severity;
- methods in place for ensuring safe operation including maintenance;
- methods in place for controlling plant modifications;

- methods in place for identification of foreseeable emergencies including the preparation, testing and review of emergency plans;
- methods for reporting major accidents and near misses, the means of investigation and follow up action;
- methods in place in place for the monitoring, audit and review of the MAPP.

The Health and Safety Executive's (HSE) publication HS(G) 65 'Successful health and safety management' may help in the preparation of a MAPP as it is the foundation for addressing key aspects of effective management of health and safety such as policy, organising, planning, measuring performance, auditing and reviewing performance.

SAFETY REPORT STRUCTURE

The essential requirements of a safety report are the identification, prevention, control and mitigation of major accident hazards. Operators have to demonstrate that they appreciate the nature and scale of potential major accident hazards and that they have taken all reasonable steps to ensure safe operation of their establishment. From initial interpretation of the legislation, this does not appear to be different to the CIMAH regulations. However, it is stated in the new COMAH regulations that all operators (both lower and top tier) shall take all measures necessary to prevent major accidents and limit their consequences to man and the environment. This requires a greater depth of substantiation of the systems in place to prevent or mitigate a major accident hazard and places a greater duty on operators to demonstrate this in their safety reports. The duty is also extended to the environmental aspect whereby standards should be proportionate to hazard and risk and in line with the Health and Safety Commission's (HSC), the Environment Agency's (EA) and Scottish Environmental Protection Agency's (SEPA) policy. It is difficult to guarantee prevention. Therefore the principles of inherent safety should be the first port of call e.g. reduction in inventory, use of similar substances which create less of a hazard etc. Another important aspect of this process will be compliance with appropriate codes of practice, standards, HSE guidance etc. The guiding principle in determining the level of justification required in the report is that it should be proportional to the level of hazard and risk involved.

The emphasis within COMAH on demonstration of the adequacy of the prevention, control and mitigation measures also has implications for the assessment of hazards itself. In particular, COMAH will require a more structured approach to the identification of hazards and the associated initiating events and conditions. Systematic techniques such as Hazard and Operability studies will be suitable for this purpose.

The remaining sections of a safety report should provide a full description of the establishment highlighting all the installations, processes, storage facilities, pipework etc relevant to major hazards. This will include temperature and pressures, material standards, engineering diagrams, site layout diagrams, quantities of dangerous substances etc. The safety report will also need to provide a description of the land use and sensitive environmental features in the vicinity of the site.

EMERGENCY PLANS

As with CIMAH, there is a requirement on operators of top tier sites to prepare on-site emergency plans. Similarly, local authorities are required to prepare off-site emergency plans.

The essential contents of the on-site emergency plan include:

- details of persons authorised to set emergency procedures in motion and the command arrangements;
- means of issuing warnings and required actions of non emergency personnel;
- description of specific arrangements for dealing with identified accidents;
- staff training;
- procedures for setting in motion the off-site emergency plan.

Many of the above elements of the on-site plan are applicable to the off-site plan:

- details of persons authorised to set emergency procedures in motion and the command arrangements;
- arrangements for receiving warnings and mobilising resources;
- arrangements for providing the public with information on the accident and the actions to take.

Emergency plans have traditionally concentrated on the immediate response to the accident. However these plans now need to also address the longer term clean up and restoration of the environment. This aspect is likely to be new to many companies and may require the use of specialised external expertise.

TESTING OF ON-SITE AND OFF-SITE EMERGENCY PLANS

A new feature of COMAH is the requirement to demonstrate that the emergency plans have been reviewed and tested at suitable intervals not less than once every three years. As a full test of emergency plans can prove to be quite costly, an establishment has the option of the exercise taking the form of a live interactive plan or a table top exercise with the support of some live components. In whichever form the testing arrangements take place, it must fully satisfy the objectives as laid out in the plan itself.

The differences between a live and table top exercise are discussed in more detail below;

- a live exercise incorporates a simulation of an accident and the utilisation of all the appropriate resources. The disadvantage with this type of exercise is that it requires careful planning and organisation and can be quite costly e.g. the use of a fire brigade.

- table top exercises involve all the appropriate resource in one place who work through their roles in an emergency situation. The event has to be carefully selected and it is difficult to ascertain in this type of situation whether the event would run smoothly in a ground simulation exercise or an emergency situation.

For multi - installation sites and where shift teams are present, testing emergency plans may seem a daunting task. However it needs to be borne in mind that what is being tested is the plan itself. Therefore emergency plan testing on different installations need only concentrate on any installation specific arrangements that exist. Similarly there is no specific need to test the plan using each shift. Training can ensure that each shift understands the required actions.

For off-site emergency plans, full scale exercise testing need not take place, but include site visits by all off-site agencies for familiarisation purposes, communication exercises to test the communications procedures during an emergency, and table-top exercises as before. Communication exercises enable the necessary personnel to work through their responses in an emergency situation.

PUBLIC DISCLOSURE

An important new feature of the COMAH regulations is the provision for public access to the contents of safety reports. Under CIMAH, operators had the confidence to disclose information on the basis that confidentiality would be protected. However operators are now understandably concerned about the protection of sensitive information. The COMAH regulations do include the option to withhold information on the basis of commercial secrecy national security etc. Such exemptions will need to be agreed with the competent authorities. Means of satisfying the needs of the competent authorities and ensuring protection of sensitive information from public disclosure include producing two safety reports or including sensitive information in separate documents or removable appendices.

LAND USE PLANNING

Land use planning for major hazard installations has been in existence in the UK for many years. Therefore the inclusion of land use planning in the COMAH regulations will not cause any major changes to existing arrangements. There are however certain implications in relation to hazardous substances consent and substances harmful to the environment. In the first case, hazardous substances consent will be extended to include all lower and upper tier COMAH sites. These additional controls in terms of what substances can be held on site and their location could have significant implications for sites where inventories and locations are subject to change such as warehouses and contract bulk storage sites. The way to overcome such problems include anticipation of future storage needs and notification of worst case substances to cover generic groupings. The inclusion of a specific environmental harm category within COMAH also raises for the first time the need to consider environmental hazards in the context of land use planning. The EA and its Scottish counterpart will therefore need to asses the potential for major accidents to the environment and in certain limited circumstances even set a Consultation Distance.

CONCLUSIONS

The COMAH regulations are very positive in intention and compliance will lead to an overall improvement in safety to man and the environment and hopefully a more cost effective approach to safety management systems. COMAH with its increased emphasis on safety management will promote the incorporation of major hazard controls and accident response in to a company's overall safety management system. There is a wide body of evidence to support the concept that the costs associated with prevention, control and mitigation are far less than the potential financial implications of a major accident being realised. In deciding the level of resources to be applied to COMAH compliance, the guiding principle should be that it is proportionate to the scale of hazard and associated risk.

Table 1 - Main differences between CIMAH and COMAH Regulations

	Seveso I	Seveso II
Scope	• Applies to installations • Distinction between process and storage activities • Application determined by a list of named substances • Explosives and nuclear facilities exempt	• Applies to establishments • No distinction between process and storage activities • Application determined primarily by reference to generic categories of substances • Explosives and chemicals at nuclear sites included.
Lower tier duties	• No specific duties beyond demonstration of safe operation and accident reporting • No clear definition of scope	• New duties include to notify competent authority (CA) and prepare a Major Accident Prevention Policy (MAPP) • Scope of application clearly defined by reference to thresholds of dangerous substances
Top tier duties	• Safety report • Emergency plans	• Safety report requires an expanded content • Greater emphasis required on safety management systems • Greater public access to report • Requirement for CAs to communicate conclusions of the reports to operators prior to construction or operation of new sites and within a reasonable period for existing sites • Requirement for an expanded statement of purpose and content of plans and a requirement to test them

Table 1 (continued)

	Seveso I	Seveso II
		• Requirement for plan to include clean up and restoration of the environment
Inspection systems	• Not dealt with	• Member states to have in place adequate inspection systems to ensure operators implement the directive
Powers of competent authorities	• Primarily normal HSWA duties	• Duty on CAs to prohibit activities if the measures taken by the operator for the prevention, control and mitigation of major accidents are seriously deficient
Land Use Planning	• Not dealt with	• Requirement for land use planning policies taking into account major hazards
Domino effects	• Not dealt with	• CA to identify groups of establishments where there could be a knock on effect and these establishments to share information, particularly on emergency planning

THE ACCIDENT DATABASE: CAPTURING CORPORATE MEMORY

M Powell-Price[*], J Bond[+], B Mellin[†]
[*] Institution of Chemical Engineers, Davis Building, 165-189 Railway Terrace, Rugby, Warwickshire, CV21 3HQ.
[+] 25 Canonsfield Road, Welwyn, Hertfordshire, AL6 0PY.
[†] 7 Mayals Road, Blackpill, Swansea, West Glamorgan, SA3 5BT.

Accident databases are means of capturing corporate memory. The design and structure of data for such software tools has been subject to much research and this paper presents the results of such investigations. The Institution of Chemical Engineers are committed to the dissemination of information in process safety, loss prevention and environmental protection, and as a result have produced a database known as The Accident Database which fulfills a number of key objectives. These include: focusing data sets on previously unpublished materials from company data banks; the inclusion of a 'lessons learned' field which reveal the recommendations and lessons the company learned after the accident or incident and the use of a novel keyword mechanism based on the concept of 'case-based reasoning.'

Case-based reasoning uses formalised rules to try to identify and use information on past cases to provide help with new problems. It softens the all-or-nothing approach used in conventional database matches by retrieving cases that have some degree of relevance to the keywords specified by the user, without matching exactly. In order for the system to know how keywords are related, a domain model is required for each database field.

The paper closes by looking at the future developments of The Accident Database and additional features that users requested from the IChemE.

Keywords: Accident database, lessons learned, case-based reasoning, corporate memory, domain models.

INTRODUCTION

On the 14th December 1785 an explosion occurred in the bakery shop of Senior Giacomelli in Turin (1). The explosion was a dust explosion caused by flour released from a silo. The explosion was not too serious but it was investigated by Count Morozzo who wrote a remarkable report. His explanation of the incident was interesting in an historical sense but it is his final words that are as relevant today as they were 200 years ago:

'Ignorance of the fore-mentioned circumstances, and a culpable negligence of those precautions which ought to be taken, have often caused more misfortunes and loss than the most contriving malice; it is therefore of great importance that these facts should be universally known, that public utility may reap them from every possible advantage.'

More recently safety guru Trevor Kletz pointed out in his book (2) 'Lessons from disaster: how organisations have no memory and accidents recur':

'It might seem to an outsider that industrial accidents occur because we do not know how to prevent them. In fact they occur because we do not use the knowledge that is available. Organisations do not learn from the past, or rather individuals learn but they leave the organisation, taking their knowledge with them, and the organisation as a whole forgets.'

WHY SHARE ACCIDENT DATA?

There are many 'soft' reasons for sharing data in the loss prevention field — improvements to company image, PR benefits, improved employee relations etc. can all be cited — but perhaps one of the most powerful reasons comes from real practical experiences. A member of the IChemE who spent many years working as a safety advisor in a major chemical company put it like this:

'I well remember wishing to circulate information within my own company of an incident I investigated. I had not been aware that some stainless steel flexible hoses were silver soldered to a flange, I assumed they were all welded. When I used a stainless steel hose to off-load a road tanker of 10% nitric acid the hose parted from the flange because of the attack on the silver solder. At the time I was told not to write a note as *we do not advertise our mistakes.*'

In Germany in 1976 and the UK in 1978 two accidents occurred which involved:

- the same chemical;
- the same licensed process;
- built by the same construction company;
- the same start-up stage of the process;
- a contractor doing the same job; and
- on the same piece of equipment.

In both cases a solution of a toxic chemical was discharged to the open settling vessel and the contractor was killed.

One wonders how many identical accidents do we have to have before we share lessons learned from accidents?

CORPORATE MEMORY

Way back in the early 1970's the Institution of Chemical Engineers (IChemE) recognised the existence of the corporate forgettory, the 'well that's never happened before' syndrome and began publishing the Loss Prevention Bulletin. This publication had, and still has, three primary aims — to capture accidents and incidents, to report the sequence of events that led to an occurrence, and to offer recommendations to help organisations learn from the mistakes of others. Since 1974 the Loss Prevention Bulletin has published over 1000 comprehensive case studies and, no doubt, saved many lives and prevented much economic loss.

The shocking impact of events such as Feyzin (1966, 15-18 dead), Flixborough (1974, 28 dead), Seveso (1976, massive contamination), Mexico City (1984, over 500 dead), Bhopal (1984 over 2500 dead), Piper Alpha (1988, 167 dead) and many others have stimulated significant improvements in the safety performance of the process and allied industries. This transformation is arguably down to three main changes:

- the arrival of fully-integrated safety and environmental management systems which engender a culture of continual improvement;
- the recognition that if you cannot measure it, you cannot manage it; and
- improved reporting and investigation of accidents and near-misses — the capturing of the corporate memory.

In the UK the Chemical Industries Association publish, under the Responsible Care programme, a series of annual performance indicators (3). The data relating to Loss Time Accidents (LTA's) has shown continual improvement since the 1980s both for own employees (of member companies) and contractors. Whilst one can argue over the validity of LTA's as an indicator of safety the results are nevertheless impressive — see figure 1 and figure 2.

However, there inevitably reaches a point where the accident rate will start to level off and, worse still, perhaps increase. The next paradigm shift is linked to the sharing of accident data within industry sectors. Whilst many companies have huge amounts of information stored away about minor incidents and near-misses there has not been a culture of sharing information within industry sectors.

Surveys and reports such as that carried out in 1994 by Keller and Pineau (4) have shown that potential users of accident databases wanted a number of key features, these included:

- Identification of accident scenarios;
- Identification of design deficiencies;
- Evaluation of emergency response;
- The accident sequence;
- Chemicals involved;
- Human and management aspects;
- Technological aspects;
- Property and plant loss;
- Environmental impact.

There are a number of accident databases (5) in existence or under development including major initiatives such as:

- Major Hazard Incidents Data Services (MHIDAS) — sponsored by the UK Health and Safety Executive, and managed through AEA Technology.
- Analysis, Research and Information on Accidents (ARIA) — a system set up and operated within the French Ministry of Environment.
- Major Accident Reporting System (MARS) — set up as a requirement under the 'Seveso' Directive.
- Failure and Accidents Technical Information System (FACTS) — a commercially available database of incidents which involve hazardous materials which is maintained by TNO.

Most of these are focused on previously published data sets. Thus the IChemE wished to focus its resources in developing a database that captured the corporate memory of organisations and made that memory available to others, not simply recreate what already existed. Therefore the IChemE Accident Database is concentrating on publishing the data that companies have never made available to the wider audience — so called 'in-company' reports.

THE ICHEME ACCIDENT DATABASE

The IChemE have for a number of years been collecting and publishing accident data. This 'public domain' information provided a very valuable platform onto which to build a large quantity of 'in-company' data. Thus two years ago the IChemE carried out a survey of its safety specialists to discover the features that they would wish to see in a CD ROM based electronic data source and to also ask responders whether they would consider donating data to such a resource. The results of this survey gave the IChemE the information it needed to take the next step to develop a database which had the functionality necessary to meet the needs of a safety professional.

More significantly the survey led to the donation of 1000s of records from companies who shared the same vision of information dissemination. All this data had previously been unavailable to the broader safety community and we now had the opportunity and, as a professional body the obligation, to make this information available to all.

Database structuring

As we have seen, surveys have showed what was wanted from an accident database. IChemE's own survey also started to put figures on the quantity of information that was required. As well as numerous small fields such as date of accident, number injured or killed, location etc., the inclusion of two major field were proposed:

- Abstract — the survey stated that this field should be no more than 500 words long but capture the accident sequence, chemicals involved, human, managerial and technological aspects and where known the actual loss to plant and property. This abstract should be appropriately keyworded so that the database user is most likely to find the accident most relevant to his/her enquiry.
- Lessons learned — this field was suggested again to contain no more than 500 words and capture the stated lessons learned from the original report. This latter point is vital, IChemE were not in the business of the reinventing the lessons learned with the 'benefit of hindsight'; thus 'lessons learned' are taken directly from the original report.

One of the greatest challenges on this project was developing a methodology for the keyword structuring. Traditional databases usually rely on a simple alphabetical listing that the user must scroll through and pick prior to commencing the search. This methodology is prone to error particularly when the keyword sets are large. Originally the IChemE database had the following keyword structure:

- chemicals (over 1000);
- equipment (800);
- type (200); and
- cause (500).

However, many of these keywords were duplicated as they represented synonyms. For example, one accident might be indexed using the equipment keyword 'incinerator' and a different accident might use the keyword 'thermal oxidiser'. It is important when indexing a database to use a consistent approach. This required some rationalisation of the keywords. Coincidentally, a project in the department of chemical engineering at Loughborough University was already looking at how to improve retrieval of information from accident databases. The work of rationalising the database keywords was taken up as part of this project with expert input from the IChemE.

The Loughborough University project is initially looking at the use of case-based reasoning to improve accident information retrieval.

Case-based reasoning uses formalised rules to try to identify and use information on past cases to provide help with new problems. It softens the all-or-nothing approach used in conventional database matches by retrieving cases that have some degree of relevance to the keywords specified by the user, without matching exactly. In order for the system to know how keywords are related, a domain model is required for each database field. Part of the domain model for the activity field is shown in figure 3.

The keywords for the IChemE database were rationalised and organised into domain models (with the exception of the chemical keywords). The number of keywords in the keyword hierarchies is now:

- equipment hierarchy (450);
- cause hierarchy (275);
- consequence hierarchy (65); and
- activity hierarchy (40).

The consequence and activity hierarchies have been developed from keywords in the incident type field.

This is where the Accident Database really comes into its own. If, for example, you were worried about storing a particular substance in a particular place you would start with activity and then select storage. From there you can select the chemical concerned, the type of packaging it comes in, the equipment used to move it and so on. Using a Windows-style drag-and-drop facility you can build up mini-databases based on your own keywords and then save them for ease of future reference. As the Database grows, the number of specific topics could be extended.

Definitions

The hierarchies are defined as follows:

- Substance: the substance could be a raw material, chemical, intermediate or final product involved in the accident. The substance may be a specific chemical or a generic material such as 'plastics', based on Sax's Dangerous Properties of Industrial Materials, but not brand names or commonly used names.
- Activity: the activity being carried out when the accident occurred.
- Equipment: the equipment that is involved in the accident.

- Consequence: the result of the accident including; damage to equipment, injury or harm to personnel, damage to environment, financial loss and will include near-misses.
- Cause: the events which resulted in an accident, incident or near-miss. These may be basic causes and/or underlying causes. Experience shows that underlying and particularly management system failures are seldom reported.

Data quality

The IChemE Accident Database in its first release contains approximately 8000 records, 3000 of these are previously unpublished and approximately 2000 have associated lessons learned. Many of the reports on which the entries are based are lengthy reports which contain excellent well researched data. Some of the data is of a lesser standard. However the management group that controls the database developments believe that all the data included in The Accident Database is useful and will help others to learn from the mistakes of the past.

Confidentiality

One of the hurdles that had to be overcome with regards to the 'in-company' data was the issue of confidentiality. There seems little advantage to be gained from publishing companies names and the exact location of each incident. Quite the reverse in fact; it is unlikely that a company would wish to 'wash its dirty linen' in public. To address this issue the IChemE adopted a management system which meant that each incident from in-company sources went through an abstraction process guided by a panel of experts who form a group run by the IChemE, known as the ADMG (Accident Database Management Group). The process is a simple one and involves taking out specific locations and any trade names and avoiding the inclusion of reference to company procedures. In addition individuals are never named. It has also been agreed that the full report will not be made available to a user unless agreed by the donating company.

This compromise does not reduce the value of the data and enables the IChemE to ensure that each entry contains the maximum of quality information such that the user will be able gain information that satisfies the 'wish list' highlighted earlier in this paper.

Medium of release

Initially The Accident Database is being provided in a CD ROM with an annual update. However the IChemE are presently researching into the viability of using the Internet as the host for the data. This has a number of advantages not least the ability for the data to be kept as up to date as possible without the need to wait for the new release.

THE FUTURE

The IChemE are in the process of contacting representatives of all the chemical companies and allied companies in the UK to invite them to donate records to the database. The Chemical Industries Association Responsible Care programme is so convinced that sharing of data in such a way will lead to further improvements in their members safety performance that they have lent their support to this initiative. The initial survey carried out by the IChemE showed that 65% of responders would consider donating reports to the database. This figure would seem to suggest a cultural change has taken place. As well as inviting companies to donate data IChemE are also providing standarised incident reporting training material (6) which will help in the developments of both its own database but also companies' own reporting systems. The Accident Database will grow year on year with a focus on data that has never been published before. By acting as an 'honest broker' in this way IChemE can bring information that has yet to be seen to the broad safety community and help the process industries to continue improving safety performance.

Finally, remember the old adage 'a wise man learns from his own experience, a wiser man learns from the experiences of others' or put simply, sharing data on accidents saves lives.

FURTHER INFORMATION

For further information about The Accident Database contact the Institution of Chemical Engineers:

John Duffy
Head of Marketing (Safety, Health and Environment)
Institution of Chemical Engineers
165-189 Railway Terrace
Rugby CV21 3HQ
UK

Tel: +44 1788 578214
Fax: +44 1788 560833
E-mail: john@icheme.org.uk
URL: http://www.icheme.org

REFERENCES

1. Bond, J., 1996, The hazards of life and all that, Institute of Physics.
2. Kletz, T., 1993, Lessons from disaster: how organisations have no memory and accidents recur, Institution of Chemical Engineers.
3. Chemical Industries Association, 1997, The UK indicators of performance 1990-1996.
4. Keller, A.Z., Pineau, J.P., October 1994, Initial assessments of the strengths and weaknesses of current accident databases, European Safety and Reliability Data Association Conference.
5. Bardsley, A., Cole, J., Lelland, A., Overview report on the use of data on past incidents on process plant: in particular use to aid hazard identification and control, European Process Safety Centre.
6. Incident reporting, investigation and analysis — training package 026, 1996, Institution of Chemical Engineers.

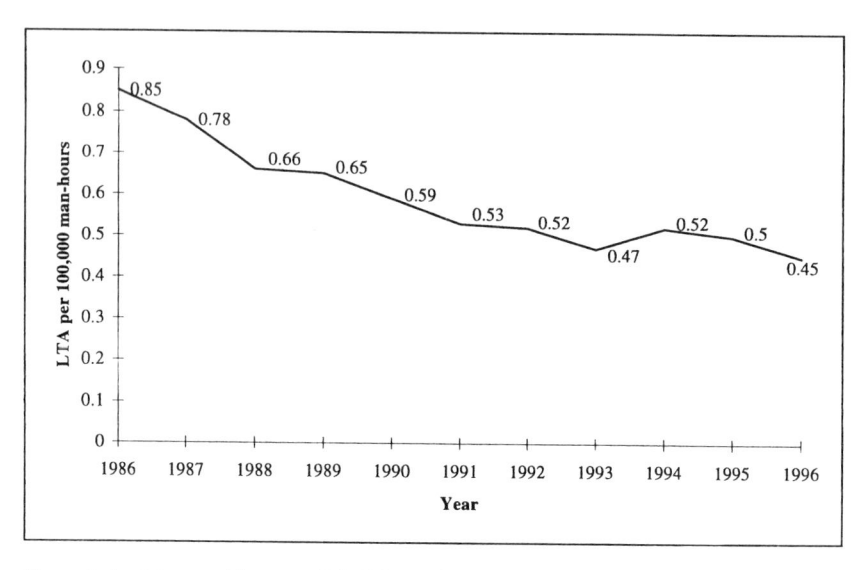

Figure 1: Lost time accidents per 100,000 man-hours — own employees

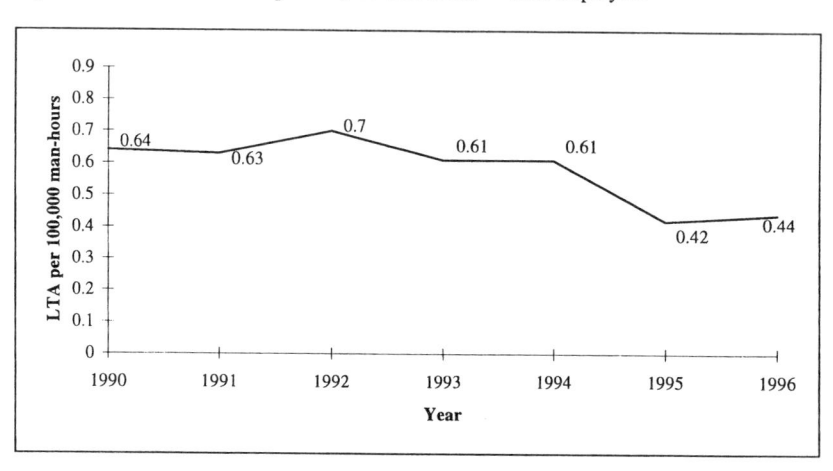

Figure 2: Lost time accidents per 100,000 man-hours — contractors

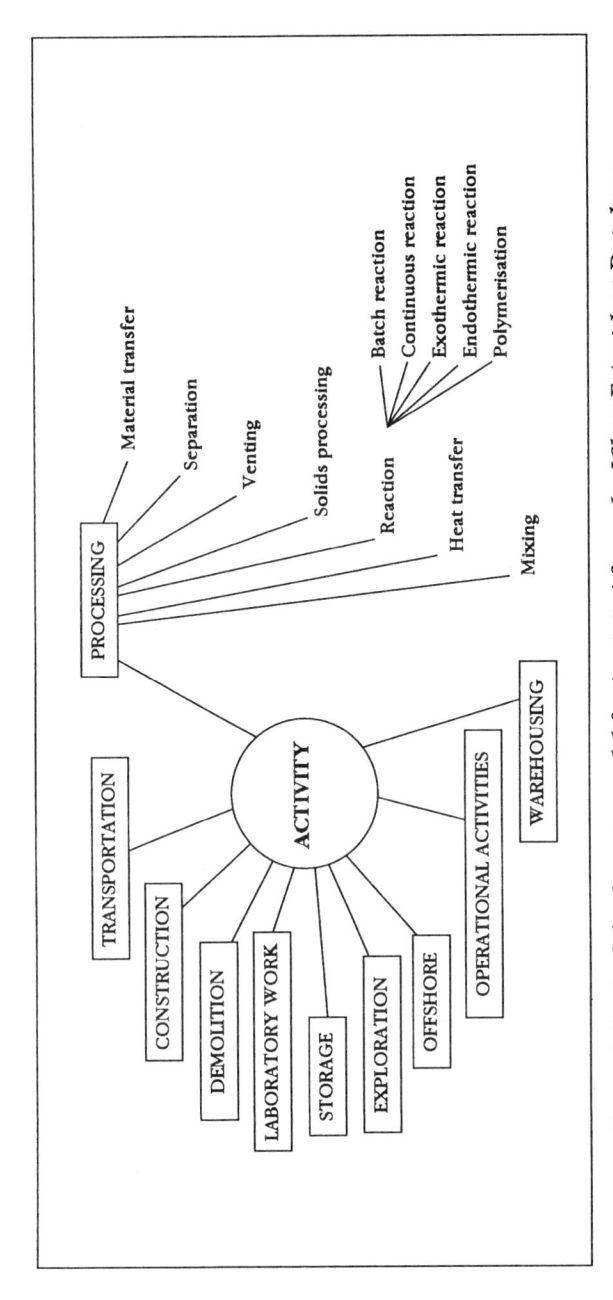

Figure 3: Part of the domain model for 'activity' from the IChemE Accident Database

OPERATIONAL SAFETY REVIEWS

J.K. Broadbent and Dr T. O'Donoghue*
ICI Eutech, Belasis, Billingham
*DuPont Melinex, Dumfries

Operational Safety Reviews, developed during 1997 at Melinex Dumfries, amalgamate features similar to parts of ICI's Process Hazard Review (PHR) technique and the DuPont (ex ICI) Wilton Melinar Plant's Production Task Review. A key element is the use of simple risk ranking to prioritise issues. The actual sessions, conducted using teams with high shop-floor content, encourage an open approach and review what actually happens. The Reviews are suitable for application to safety, occupational health and environmental issues on all kinds of Plant or operation with significant operator interactions. They differ from conventional Hazop and PHR methodology in their ability to focus more effectively on the causes of major and minor injuries and on ergonomic problems to operational staff.

INTRODUCTION

In early 1997, the management team for the Melinex production facilities on the then ICI site at Dumfries were concerned about a levelling off in the trend of continuing improvement in industrial injury performance. Targets being set by the Business and ICI (at that time the owners) were at the same time becoming ever tighter. These concerns were further highlighted in April when a Process Operator sustained serious burns while establishing a Melinex film line. Senior Management within the Business and on Site were aware that some film lines had been built before the creation of the ICI Hazard study process while for other, subsequent lines, only some sections of the equipment had been subjected, at the design and installation phases, to Hazard Studies. On the other hand, it was appreciated on Site that these Hazard Studies (The ICI Studies 1-6, Refs. 1 & 2) with their main focus on the integral safety of the equipment, would probably not in themselves have effectively designed out or prevented this particular injury, or many others where Operator behaviour or action is a major contributory factor.

However, it was, and is, recognised that Hazard Studies have contributed to the inherently safer design of new and modified Film making plant, particularly when carried out by an experienced team at the appropriate stage of the project. Even so, it has been found necessary to develop different guidewords and often additionally use a FMEA (Failure Mode and Effect Analysis) approach to cope with the Film Plant technology which is inherently different from that of the Chemical Process Plants for which the Hazard Study technique was originally developed. One option considered, therefore, was to embark on a programme of comprehensive retrospective Hazard Studies. Reluctance to pursue this course arose from:

- previous experience on another plant of achieving limited value for significant time and effort when Hazard Studies are retrospective, and
- the realisation, as mentioned above, that even if the effort were to be put in, that Hazard Studies would not focus effectively on Operator behaviour, concentrating instead mostly on how the equipment was likely to malfunction.

It was decided that there was a need to carry out some other kind of retrospective study on the older units that would include looking at behavioural aspects. The Plants were already going through a comprehensive series of "Do Not Touch!" studies (Ref.. 3) where Operators under an experienced Study Team Leader reviewed the hazards of entrapment from moving machinery and from film, but it was recognised that the new technique would need to address a wider range of hazards than this.

Eutech were consulted because of their experience with the Process Hazard Review (PHR) technique (Refs. 4 & 5) that had been developed initially by ICI's Teesside Operations to study retrospectively Plants on Teesside that had predated Hazard Studies. It was hoped that PHR could be used flexibly enough to meet the requirement, but on review it was felt that although this technique would contribute significantly to the requirement by considering how the equipment could malfunction, it did not lend itself to effectively reviewing the behavioural risks. Like Hazard Studies, the PHR technique had been developed to identify potential hazards from chemical plant and equipment, reducing the risks of explosions, toxic releases, losses of containment etc., rather than effectively focusing on how individual operational staff could suffer injury from their own behaviour.

In addition the Melinar Plants at Wilton (at the time under ICI ownership, but since transferred to DuPont like their Melinex equivalents at Dumfries) had developed a process that they called Production Task Review (PTR) (Ref.6) that focused entirely on behavioural aspects. PTR involves a Plant team first developing a list of all the tasks carried out by Operators and Tradesmen in the area under study. The Team then assesses the risks (potentially H and E as well as S, though studies to date have concentrated on S) associated with each task and decides whether steps are needed to be taken to reduce them. Activities considered high risk are labelled "Critical Tasks" and are subject to further specific study. This technique was recognised as being capable of providing the basis of the required review of the behavioural aspects that could then complement a PHR style study.

The Operational Safety Review procedure owes debts of gratitude to both PHR and PTR and like a human child that has many of the characteristics of both parents, it also has significant differences from either one parent taken individually.

An important requirement of the Operational Safety Review process is to direct the efforts of the team to the particular topics that require attention. This is achieved by starting from a general overview of the area being studied. Progressively more detail is then introduced but only in conjunction with the use of risk level as a filter. Areas of low or no risk identified in a rapid and subjective filtering process are noted in the records but then no longer addressed. The study team is then able to focus its efforts on detailed consideration of the matters of significance that have not been filtered out. In practice this has been found to be an efficient process, allowing a typical limited area of Plant operations to be reviewed in a single day.

PRELIMINARY STAGES - SCOPING AND INITIAL DATA GATHERING

Required attendees at the Meeting are a First Line Manager with responsibilities for the area together with experienced Operators and perhaps Tradesmen. In some cases a Plant Engineer or Plant Manager may also be able to contribute usefully. The Study Leader should be someone used to chairing Plant level Hazard Study Meetings or similar and have an understanding of the nature of the Plant operations being considered. They need to be able to gain quickly the trust of the operational staff as the Meeting is seeking to consider what actually happens (which may not be quite what is supposed to happen according to the Plant's Operating Instructions).

The Review Meeting starts by agreeing the defined boundaries of the area and activities being considered and will reference drawings, flowsheets or engineering line diagrams that are needed to assist the study. A check is then made of potential or perceived interactions with adjacent areas or activities that could have adverse effects both in the area under consideration and in areas adjacent. An example of this could be that heavy Fork Lift Truck traffic through the area being considered that is not part of the area activities can add to the hazards. Another example is that a spill to drain on one part of a Site could have limited consequences locally, but might have more serious consequences elsewhere on the Site.

The next step is to ensure that Materials Hazard Data is available for all the materials used or encountered in the area and that appropriate COSHH (Control of Substances Harmful to Health) Assessments have been carried out. Typically this would also cover items such as confirming that building and insulation materials in use do not contain asbestos.

Finally, at this stage, the Meeting reviews the Injury Accidents and Dangerous Incidents that have occurred in the area. It is also usually considers engineering failures that have adversely affected process operations even if these have not directly been dangerous. Such failures can often lead to non-standard operation with incumbent increased levels of risks as well as their obvious economic impact.

It will be seen from the above that it is desirable that a certain amount of information gathering occurs prior to the formal meeting. This is usually best done by someone who will be participating in the study. Depending on the nature of the activities, additional work may also be required by Maintenance or Engineering staff.

This stage of the Review often raises Actions to gather further information for subsequent review and it is not uncommon for the Team to consider that the "Actions to prevent recurrence" raised at formal investigations into injury accidents and dangerous occurrences have not been totally effective. The Team may also be prompted to raise Actions suggested in the discussion to consider further or remedy specific situations where unacceptable risk is thought to be present.

Any issues with attendant risk are allocated a "Risk Level" from a simple matrix (see Appendix 1). The "Harm Levels" and "Probabilities" are taken by the Team and benchmarked against the tolerable levels set by the Site or Company. Taking injury as an example, "Severe" would be "potentially disabling or fatal" with the corresponding "Rare" being not foreseeable at a frequency higher than the Company's tolerable frequency for such accidents. "Likely" would be a significant probability (say 0.2) of the accident happening within the Plant's life (say 30 years) and "Unlikely" being between "Rare" and "Likely". The judgements are deliberately kept simply empirical with guidance from the Team Leader to avoid time debating which category applies.

The Risk and Precaution Guidance Table (see Appendix 1) indicates the nature of the Action that will be required to satisfactorily resolve problems at the 3 different Risk Levels.

REVIEW OF EQUIPMENT ISSUES

The second stage of the review is similar to parts of the PHR process with the addition of the simple Risk Rating and considers whether particular problems are possible in the area or activities under consideration. Many of the prompts are derived from PHR prompt sheets. The first 4 topics are compulsory, but the second 4 can be screened out if not relevant. The compulsory topics are (1) Health Issues, (2) Releases to the Environment, (3) Mechanical and other Abnormalities (Weather damage, etc.) and (4) Services Failure. Each topic has a prompt sheet that should be completed even if the response to the "Relevant ?" prompt is "No".

The sheet for Health issues is attached as Appendix (2) to demonstrate the format. Note that a "Risk" rating is applied to each relevant prompt. The hazards or problems are rated by the Team as "High", "Medium", "Low" or "Zero" Risk. Further consideration is given to Medium and High Risk rated issues, but not necessarily to Low Risk issues unless the Team wishes to do so. As in a Hazard Study or PHR, Actions to cure or alleviate the problems are progressed and decided upon outside the meetings and not as part of the Reviews. The Operations Review sessions are charged with addressing current operations and problems but not with identifying the solutions to any issues identified. This clarity of purpose is needed to avoid the Meetings becoming bogged down in detailed debate of optimum solutions.

The optional topics are (5) Web Handling and Cutting (equipment issues relevant to films production processes), (6) Fire and Explosion, (7) Uncontrolled Reaction and (8) Physical Over/Under Pressure. Where relevant these topics are treated in the same way as the compulsory ones.

This part of the study looks at issues arising from the hardware and materials and picks up issues that would probably be found through PHR or retrospective Hazop. By applying the filters of operational experience with the actual equipment and the risk rating, it is possible to focus on the key issues. This has been found in practice to be quite streamlined and certainly an order of magnitude less time consuming than applying the Hazop procedure to the same parts of the Plant. An experienced PHR Team would also be able to skip minor issues and only concentrate on those that are more important, so the time savings against this technique are less clear cut. The authors, though, do believe them to be significant.

The technique is readily adaptable to other technologies than Film Production and Chemical Manufacturing, but if this were to be done, it may well be desirable to draw up further topic specific prompt sheets first. For instance, if bulk transport by road is being considered, a topic sheet with prompts relating to the specific nature of hazards that would be relevant rather than those used in the studies carried out so far. A short brainstorming session between the potential Team Leader and the relevant Operational/Safety management can produce the required prompts.

REVIEW OF BEHAVIOURAL ISSUES

The third stage is the Task Review which focuses on activities performed by the people associated with the operating units. It starts with a prompt sheet that helps decide which kinds of "Activity" need to be studied for a particular segment of operations. For instance, the Team may feel that "Frequent cleaning" and "Routine Control/Electrical on-line maintenance" activities have no specific hazards attached to them and so these activities do not need breaking down into specific Tasks and further study. The sheet prompts at least a brief consideration of activities that could otherwise easily be overlooked such as "Response to process upsets" and Trades activities such as scaffolding and lagging, but this process is rapid and quickly filters out activities that are not hazardous. The Prompt sheet is attached as Appendix 3. Again streamlining of the study arises from using the risk level as a means of focusing attention on the higher risk activities.

Where a type of Activity is deemed to require review, the Team lists all the Tasks associated with it. (A sequence of "Tasks" make up an "Activity") This is where the Team's operational experience is important as it is essential that the description is of what actually happens, not what the Job or Operating Instructions say should happen! It is worth having the formal Operating Instructions available for reference, but the focus must be on what happens in practice, remembering that different shifts may well have developed different methods. When the Tasks have been listed for an Activity, the sheet attached as Appendix 4 is completed by the Team. Team members have in front of them a prompt sheet of potential hazards covering Injury Potential, Spillage Potential and General Hazards. The latter includes reminders that there can be communication failures, Permit to Work misunderstandings, inadequate isolation, etc.. Ergonomic issues that can lead to back strains and other problems are also considered. The risk level for each Task is assessed and any Actions decided upon. As already mentioned for other problems raised, "Medium" and "High" risk Tasks will always be given further consideration, although usually the Action is to see what mitigation measures are possible outside the Meeting. Actions are only raised on Team members so there is no passing of responsibility onto managers or engineers not present. Experience has shown that the Teams become enthusiastic in their desire to reduce the risks and in practice many Actions are also raised for improvements to "Low" risk Tasks.

MANAGEMENT PRIORITISATION OF OUTCOME

Many of the Actions can be pursued to completion without recourse to more senior management approval, but others requiring expenditure or engineering effort will usually need to be fitted into (generally already over-full) Plant programmes. There is thus usually the need for a mechanism to enable review of the Actions raised with more senior Operational management and for priorities to be assessed. This is not part of the formal "Operational Safety Review" procedure as each Plant tends to have its own mechanisms already in place for doing this. All the same, it is essential that this is done or a lot of the effort already put in by the Team will come to nothing.

CONCLUSIONS

At the time of writing (Spring 1998), the technique has been applied to 8 areas on 2 Plants belonging to Melinex Dumfries. The change of ownership (ICI to DuPont) of the Melinex Business announced during the Summer of 1997 and implemented at the end of January 1998 has led to a pause in the implementation programme but it has only been a pause as Reviews are again underway with a year end target date for completion on the largest plant on site. Operations Reviews are also in progress at the Melinex Hopewell site in the USA, following training in the technique of a Hazard Study Leader there.

On the 2 Plants at Dumfries where the technique has been used and at Hopewell the response has been enthusiastic at all levels from the Team Members to the Plant Manager. The participants feel that they are focusing on real issues that affect them and that the reviews provide a realistic route for improving SHE performance. The operating teams are now requesting studies and setting priorities for the order in which reviews should be conducted on the different areas.

ACKNOWLEDGEMENTS:

The Authors are grateful to the Eutech and Melinar Businesses for access to the PHR and PTR techniques during the development of Operational Safety Reviews.

Particular thanks must also be given to the operational staff on the Melinex 5 and 2 Plants at Dumfries who were the guinea pigs during the initial studies while the techniques were still being refined. Their support at this stage was crucial to the successful development of the technique.

REFERENCES:

(1) "Hazop and Hazan" - Book by T.Kletz, 3rd edition published by IChemE in 1992.

(2) "Hazard Studies for Safety, Health and Environmental Protection - Application to Existing Plants and Processes" - Paper by R.D.Turney to the European Conference on Safety and Loss Prevention in the Oil, Gas and Process Industries, London 1991.

(3) "Do Not Touch" - Presentation by D.Shields to Melinex Management Team, Spring 1997.

(4) "PHR - A Programme of Safety Assurancefor Existing Operations" - Paper by R.A.McConnell to AIChemE Symposium, Los Angeles, 1991.

(5) "PHR - Improving Safety, Health and Environmental Protection on Existing Plants" - Paper by R.D.Turney and M.F.Roff to the 8th International Loss Prevention Symposium, 1995.

(6) "Production Task Review Methodology and Training" - Presentation to the ICI European SHE Exchange by S.M.Heppell and R Cheyne, Manchester, Spring 1996.

Appendix 1

Risk Levels

Probability Harm level	Rare (1)	Unlikely (2)	Likely (3)
None (0)	Zero	Zero	Zero
Minor (1)	L (1)	L (2)	L (3)
Moderate (2)	L (2)	M (4)	M (6)
Severe (3)	L (3)	M (6)	H (9)

Risk & precaution guidance table

Risk Precaution	Low (1-3)	Medium (4-6)	High (7-9)
Eliminate or Substitute	✓	✓	✓
Reduce consequences or use PPE	✓	✓	?
Use procedural & training approach	✓	?	X

Appendix 2

Health Issues

Issue	Relevant?	Comment	Risk
Hazardous materials Carcinogens Asbestos Corrosives Asphyxiants Skin sensitisers Respiratory sensitisers			
Manual Handling			
Repetitive activites			
Temperature extremes			
Noise			
Radioactivity			
Infra red and UV radiation			

Appendix 3

Task Review Prompts

Production		Review Required - Y/N
Start-up activities		
Shut down activities		
Routine monitoring	Frequent	
	Infrequent	
Cleaning	Frequent	
	Infrequent	
Process adjustments (incl. Grade changes)		
Process upsets		
Material transfers (incl. Reel changes, scrap disposal, etc)	Arrivals (deliveries)	
	Departures (despatches)	
Maintenance		Mech C/E
Routine	On line	
	Off line	
Breakdown response		
Fault finding		
Schedules work		
Other Trades Work		
Lagging		
Scaffolding		
Civil/Structural		

CARMAN: A SYSTEMATIC APPROACH TO RISK REDUCTION BY IMPROVING COMPLIANCE TO PROCEDURES

David Embrey
Human Reliability Associates, Ltd, 1, School House,
Higher Lane, Dalton, Wigan Lancs WN8 7RP

The first section of the paper discusses the role of procedures in high-risk systems such as chemical processing, aerospace and transportation. The results of a survey, which addressed the factors influencing the use of procedures in the petrochemical industry, are described. This and other evidence shows that there is often a wide disparity between the formal written procedures in an organisation and the ways in which the work is actually carried out. This has major implications for the control of risks and the maintenance of quality. The paper discusses reasons for this problem, and describes a systematic process called CARMAN, (Consensus based Approach to Risk MANagement) to prevent this situation.

Keywords: procedures, training, participation, task analysis, organisational culture

THE ROLE OF PROCEDURES IN ORGANISATIONS

In high risk industries, written procedures are typically subject to considerable scrutiny, since they are intended to represent the way in which the system should be operated, and, at least implicitly, how risks arising from these operations are controlled. For this reason, technical specialists usually write procedures when the system is first set up. If an incident occurs which leads to significant safety, quality or environmental consequences, the organisation will be required to demonstrate that a safe system of operation (as represented in the procedures) existed. Another reason for the existence of written procedures is the need to satisfy the documentation requirements of quality management systems such as ISO 9001. These systems typically require that all working practices that can impact on quality be fully documented in the form of comprehensive written procedures.

It is clear that without procedures and therefore standardised ways of doing things, our complex, interdependent society would rapidly collapse. Heroic interventions have their place in unusual or unique situations which cannot be anticipated and where rapid action is necessary. However, when sitting in an aircraft over Heathrow during the rush hour, most of us would prefer to place our faith in the procedures rather than relying on the creative skills of an individual flight controller who may or may not have developed an improved way to do his job. However in some situations, complying with the 'official' procedure may not be the best or even the safest strategy. A process is needed to ensure that the 'Best Practices' which most effectively control risks are also the preferred practices of the people at the sharp end who actually do the job. In other words, 'Best Practice' and 'preferred practice' need to become identical.

Procedures are the codification of working practices that are seen by an organisation as the best way to get things done. From the perspective of risk analysis, procedures can be seen as the means for achieving the commercial and other objectives of an organisation in the most efficient and cost effective manner, whilst minimising the risks arising from human or hardware failures.

Procedures are not only relevant to safety critical situations. They have a central role to play in quality systems. Without defined working practices it is impossible to achieve adherence to quality standards. Indeed, non-compliance with procedures is probably the most commonly cited causes of product variability in situations where there is significant human involvement in the manufacturing process.

THE PROCEDURES CULTURE SURVEY

As part of our work in a number of industries, we have conducted surveys regarding the attitudes of the workforce to procedures, and the extent to which written procedures are actually used to support operators when they are performing their day to day tasks. In one project, a procedures culture questionnaire was developed and distributed to nearly 400 operators and managers in a particular industry. In this section we summarise the results of that survey.

Extent of Usage of Procedures

The first set of questions related to the extent that procedures were actually used for different categories of task. For tasks perceived to be safety or quality critical, the use of procedures was high (75% and 80% respectively) but by no means universal. For problem diagnosis (regardless of whether a system was safety critical or not) only 30% of the respondents used procedures. In the case of routine tasks, only 10% of the respondents said they used procedures

When a task is described as 'proceduralised' there is an implicit assumption that the procedures will actually be referred to when performing a task. However, the results of the survey indicated that even in tasks where procedures were said to be used, only 58% of the respondents actually had them open in front of them when carrying out the work. These figures imply that the actual average 'on-line' usage of procedures for safety/quality critical, problem diagnosis and routine tasks is quite low, i.e. 43%, 17% and 6% respectively. Later, we shall consider the issue of whether quality or safety will necessarily be compromised if a person does not always use step by step written procedures.

Use of Standardised Working Methods

One of the important functions of procedures is that they can provide the basis for standardised working practices, which should ensure that the objectives of the task are achieved. One of the items in the survey concerned the use of 'Black Books' i.e. personal sets of notes held by individuals as informal job aids. The results indicated a very high usage of Black Books by both operators and managers (56% and 51% respectively). Although there is no reason in principle why such informal job aids should not be compiled by individuals, their existence suggests that there may be considerable individual variation in the way that tasks are actually performed. There are obvious implications for safety or quality critical operations if some of these variations in performance do not achieve the required objectives.

Another dimension assessed by the study was the extent to which procedures should be regarded as being guidelines, or needed to be followed 'to the letter'. Although there was considerable agreement that safety and quality instructions should be followed to the letter (90% and 75% respectively) for most other categories of task about 50% of respondents believed that they were primarily guidelines. This came as a considerable surprise to the management of the companies included in the survey.

Strategies for Improvements

The final part of the survey considered the question of why procedures were not used. Following prior discussions with technicians, seven factors were investigated with regard to their impact on procedure usage. These are set out in Figure 1.

'Procedures are not used because...' (percent agreeing)		
Accuracy	...they are inaccurate	(21)
	...they are out-of-date	(45)
Practicality	...they are unworkable in practice	(40)
	...they make it more difficult to do the work	(42)
	...they are too restrictive	(48)
	...too time consuming	(44)
	...if they were followed 'to the letter' the job couldn't get done in time	(62)
Optimisation	...people usually find a better way of doing the job	(42)
	...they do not describe the best way to carry out the work	(48)
Presentation	...it is difficult to know which is the right procedure	(32)
	...they are too complex and difficult to use	(42)
	...it is difficult to find the information you need within the procedure	(48)
Accessibility	...it is difficult to locate the right procedure	(50)
	...people are not aware that a procedure exists for the job they are doing	(57)
Policy	...people do not understand why they are necessary	(40)
	...no clear policy on when they should be used	(37)
Usage	...experienced people don't need them	(19)
	...people resent being told how to do their job	(34)
	...people prefer to rely on their own skills and experience	(72)
	...people assume they know what is in the procedure	(70)

Figure 1: Reasons for Non-Usage of Procedures

It can be seen from this table that there was a high level of agreement with most of the suggested reasons for lack of usage of procedures. Another part of the survey asked people to indicate the five main reasons that procedures were not used, and the five changes that would be most effective in improving the quality of procedures and their use. The most highly ranked reasons for procedures not being used were as follows:

- If followed to the letter the job wouldn't get done
- People are not aware that a procedure exists
- People prefer to rely on their skills and experience
- People assume they know what is in the procedure

The most highly ranked strategies for improvements were:

- Involving users in the design of procedures
- Writing procedures in plain English
- Updating procedures when plant and working practices change
- Ensuring that procedures always reflect current working practices

There were no significant differences between the reasons for lack of procedure usage, but 'involving users in the design of procedures' was rated significantly higher than any of the other approaches to improvements.

One definition of Best Practice is the performance of a task in a manner which achieves its required objectives whilst minimising any SHE (Safety, Health and Environment), economic and quality risks. As indicated by the survey discussed above, in many cases the formal written procedures do not document Best Practice and also may not correspond to the way in which the work is actually done. The disparity between the formal procedures and Best Practice is due to two main reasons. Procedures are often written by technical specialists or engineers who do not have a high level of hands-on experience with the work environment and may also be unaware of the practical constraints of performing a task in the field. A second reason is that there is rarely a system in place for ensuring that procedures are modified to take into account organisational learning and gradual changes in working practices. These problems gradually erode the credibility of the official procedures, and can give rise to the use of informal undocumented methods that may or may not be effective in controlling risks.

Conclusions from the Survey

The conclusions that emerge from this study are that in the industry surveyed, the majority of tasks were performed without the on-line use of step by step written procedures. There were also variations in the ways in which tasks were performed, between both shifts and individuals, which sometimes differed significantly from the 'official' procedures. People will not follow procedures if they feel they are impractical, and they will not routinely use written procedures if they believe they have sufficient skill and experience to get the job done. However, the existence of 'Black Books' indicates that there is a definite need for some form of on-line support, which is not provided by the existing procedures systems.

Human Reliability Associates have developed a methodology, called the Consensus based Approach to Risk Management (CARMAN), to address some of these problems. The original impetus for CARMAN came from a number of procedures improvement projects where the main focus was on improving the usability of procedures by applying ergonomics design standards to issues such as readability, layout and formatting. However, it was found that even when the usability of procedures was considerably improved, their level of usage was sometimes still low, and procedural violations still occurred. This led to work aimed at understanding the causes of procedural non-compliance, and the development of the CARMAN approach that combined insights from task and risk analysis, group processes, and work on organisational learning. This approach was gradually refined by being applied to a number of petrochemical plants.

In CARMAN, the working practices which are actually used by operators and technicians are examined, using a participative process, which documents the variations that exist, and then attempts to evaluate the practices from the point of view of whether they control the risks. Best Practices are then developed and documented, which take into account the preferences and insights of the people who actually perform the tasks, whilst ensuring that risks are adequately controlled.

DEVELOPING A CONSENSUS BASED PROCEDURES CULTURE

The key to maximising compliance to procedures lies in developing a culture where the preferred practices are identical to the best practices, which are then documented as the 'Reference Procedures'. The Reference Procedures provide the basis for training content and competency assessment criteria, and also guide the development of the most appropriate form of on-line support. In order to achieve these objectives, a consensus process is required which evaluates working practices to achieve agreement about the best methods for performing tasks. It should be emphasised that such a process must not only include shop floor personnel, but also technical specialists who may have insights into why a task should be performed in a particular way.

The process seeks to create a neutral forum for the exchange of information about differing working practices (e.g. between shift teams) and also to allow insights to be gained into the risks associated with different ways of carrying out tasks. Technical specialists contribute to this information exchange process, but do not dominate it. This is because it is essential to ensure that the developers of the revised procedures have a shared sense of ownership. This is a major factor in encouraging compliance, once a consensus has been established amongst the different stakeholders (i.e. shop floor personnel, technical specialists and managers) concerning the working practices that will be adopted. If a working practice has been developed via a consensus process, there are strong group pressures to adhere to the procedures produced by the process. These group pressures partly stem from a desire of a group, e.g. a particular shift that has participated in the consensus process, not be left behind in the implementation of the agreed working practices. Participation in the consensus process involves arguing one's corner and possibly relinquishing the current way of performing a task in favour of a demonstrably superior approach. Given that the consensus process usually involves give and take by the different stakeholders, it is likely that everybody will leave the process with the feeling that at least some of the time their view has prevailed and they have contributed to establishing the best practice. In management theory terms, the objective is to create a 'Win-Win' situation.

Measures to overcome barriers to the acceptance of CARMAN in an organisation can be summarised as follows:

- All participants in the process should feel that they have made a significant contribution to the best practices that have been developed, in order to maximise the sense of ownership
- The process should be driven primarily by the end users of the procedures and job aids, with the assistance of the facilitator
- Sufficient resources must be made available by senior management to ensure that all interested parties have an opportunity to be involved in the development of the best practices via the consensus group
- If incidents still arise, there should be immunity from blame and disciplinary sanctions as long as the agreed best practices can be shown to have been in use at the time. In this case, responsibility is deemed to be shared by all the participants in the process, and there is a continuing role for the consensus group to find out why the practice failed and to develop a revised Best Practice to prevent a recurrence of the problem.

The Role of Task Analysis

One of the important requirements for reaching a consensus is the existence of a common method for describing tasks in a clear, unambiguous manner, such that there is a shared understanding of the alternative ways of performing a task. In order to achieve this, participants in the consensus groups are trained to document their tasks using a form of task analysis called Hierarchical Task Analysis (HTA). HTA was originally developed for use in training operators in the process industries. It has two major advantages: firstly it is easy to learn, and secondly, it allows tasks to be described at varying levels of detail. The level of detail of the description is based on two criteria:

- Can the risks associated with errors be identified at the current level of detail of the description?
- Is the task described in sufficient detail to allow training specifications, job aids and procedures to be developed which will control the risks?

Figure 2 shows first level of an HTA for a booster pump. It can be seen that the task is broken down into a series of subtasks, the text in the bottom left of the box specifying who performs the subtask (OT is the outside technician, and CR the Control Room operator). The preconditions box specifies the starting assumptions of the analysis, and the plan box the conditions governing the execution of the subtasks (e.g. timing, ordering). It should be emphasised that an HTA is not a flowchart, but closer to a functional decomposition into a series of hierarchical goal directed activities. If a line is drawn below a subtask, this means that the task does not need to be decomposed further, since both of the criteria set out above (i.e. risks identified and sufficient detail for training purposes) are satisfied. Figure 3 shows a further breakdown of one of the subtasks. It can be seen that at each level of the decomposition a separate plan is developed. The block diagram form of HTA can be expressed in a more compact textual format, where each level of the analysis is represented by indented text. Our experience has shown that operators have no difficulty in learning HTA given appropriate guidance, and they can use it to represent all types of task. A further advantage of HTA is that it very readily translates into a format suitable for documenting procedures. Breaking complex tasks into a series of subtasks with clearly defined goals facilitates ease of understanding of the overall structure of a task. This is valuable both during training and for on-line use of the procedure.

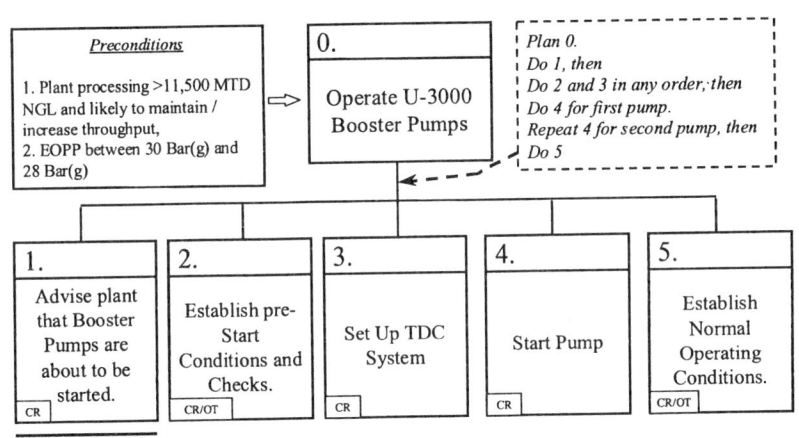

Figure 2: First level of Hierarchical Task Analysis for Booster pump

Integration between Training, Competency Assessment and Procedures

One of the major reasons for lack of compliance with procedures is often that people are simply unaware of the Best Practice for performing a task. This often arises from the absence of a system that provides baseline Best Practices against which to develop training programmes and assess competency. Obviously, unless standardised methods have been agreed with regard to how risks are to be controlled in safety critical tasks, then assessing competency will be extremely difficult. Unfortunately many industries have adopted an approach which essentially relies on providing training in generic skills, with the assumption that task specific skills will be acquired through working with an experienced technician. However, without the existence of a database of Best Practices, there will be no standardisation in the methods transmitted from the trainer to the trainee. The absence of the database also means that competency will probably be assessed against the standards of the trainer, rather than those defined by the Best Practices. In CARMAN, the procedures, training programmes and competency assessments are all based upon the same Best Practices documented as Reference Procedures.

ICHEME SYMPOSIUM SERIES NO. 144

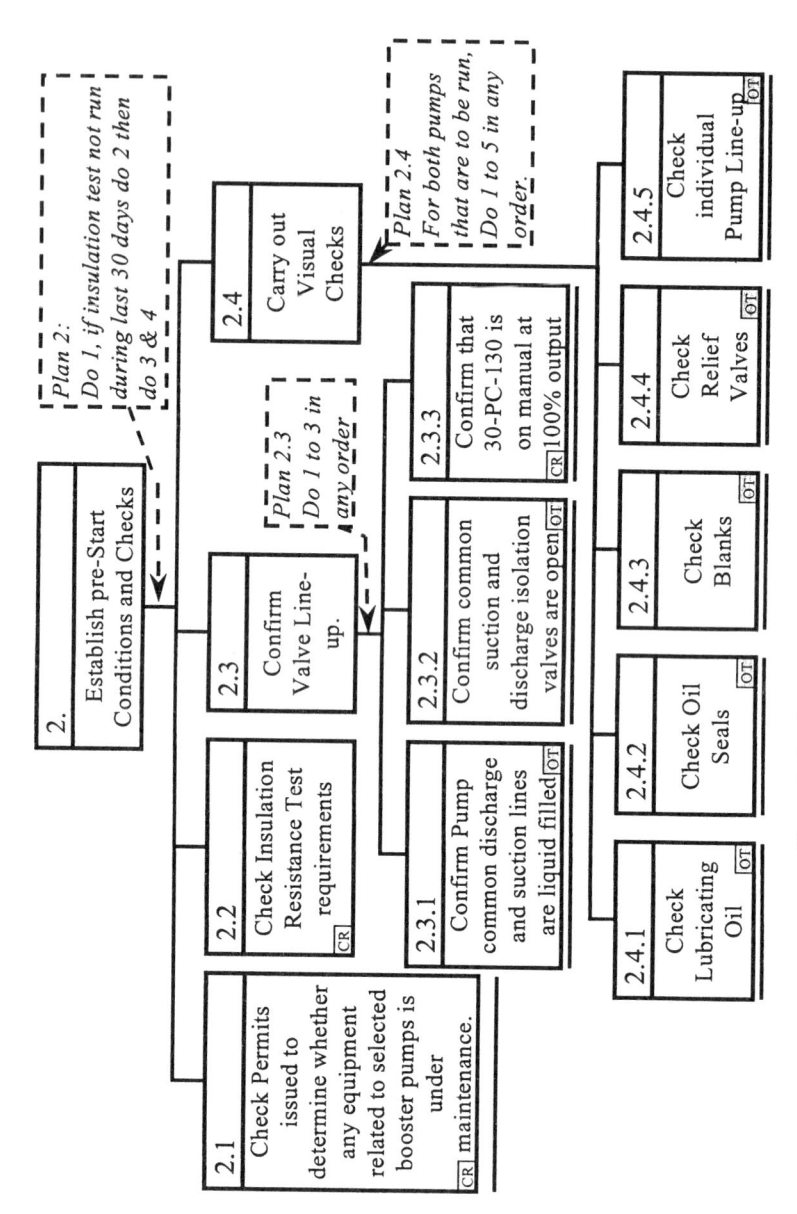

Figure 3: Break down of Subtask 2 in HTA format

Matching the Type of Procedural Support to the Needs of the End User

In many industries, particularly those where formal quality management systems have been applied, it is common to find voluminous manuals containing detailed step by step instructions for performing tasks. However, a close examination of these documents often reveals that they are either in pristine condition, or else are very dusty. This tends to confirm the view that only a small subset of the information in detailed step by step instructions is actually consulted by experienced technicians. The assumption that a large volume of written step by step procedures is the best means for providing the information required to perform a task is based upon a misunderstanding of the role of procedures. They cannot simultaneously function both as formal statements of policy and also as documents to support a person who is performing a task. This is partly because the level of documentation that is needed to support best practice will depend on the level of experience and skill of the worker. However, in almost all cases, we find the documentation of standard practices to be much more detailed than that required for on-the-job support. The Best Practice database generated by CARMAN is essentially for reference purposes, in that it provides the basis for training and competency assessment, and also documents the risks associated with tasks. Only a limited subset of the information in the database needs to be transmitted to the technician in the form of on-line job aids, to supplement the competencies acquired through training. This information may include critical values for parameters such as flow rates or temperatures, sets of rules that will optimise quality or production, or short aide memoirs for handling infrequently occurring but critical disturbances. This information may be in he form of flow charts, checklists or graphics, depending upon how it is best represented for the needs of the end users.

Many tasks will be performed primarily on the basis of skill and experience. However, some form of on-line job aid will often be required, particularly if a task is complex and / or infrequently performed. This situation often arises in organisations where multi-skilling is employed, since this often means that workers only infrequently encounter certain jobs, and hence erosion of skills may arise. The format for such job aids needs to be tailored to technicians' specific needs. Obviously, a trainee will require a more comprehensive set of job aids than an experienced technician. Many of the best job aids are found in technicians' 'Black Books' and it is often a useful exercise to encourage the sharing of this information during the development and documentation of Best Practices. One of the functions of job aids is to provide critical reference information such as dimensions and tolerances in an easily accessible form. Also, written documentation may not be the best way of presenting information, especially for use in unusual situations when a rapid response may be required. Graphical or pictorial representations may sometimes be preferable in this case. The CARMAN process provides decision aids for selecting the appropriate level of support (see Figure 6). Obviously, job aids still need to be subjected to the same levels of change control as other documentation in a quality system.

THE CARMAN PROCESS

CARMAN comprises two stages: the development and documentation of Best Practices, and the production of job aids, competency standards and training programmes based upon the Best Practices.

The overall structure of Stage 1 is shown in Figure 4. Prior to commencing the steps shown in Figure 4, it is first essential to appoint a facilitator, and to provide training in the tools and philosophy of CARMAN. His or her role is to collect information from the various stakeholders, and to assist in the development of consensus regarding Best Practice. It is essential that personnel respect the facilitator, and that he or she has good communication skills. It is also desirable to provide some awareness training for the technicians, and also basic training in task analysis, which is an important technique in the process.

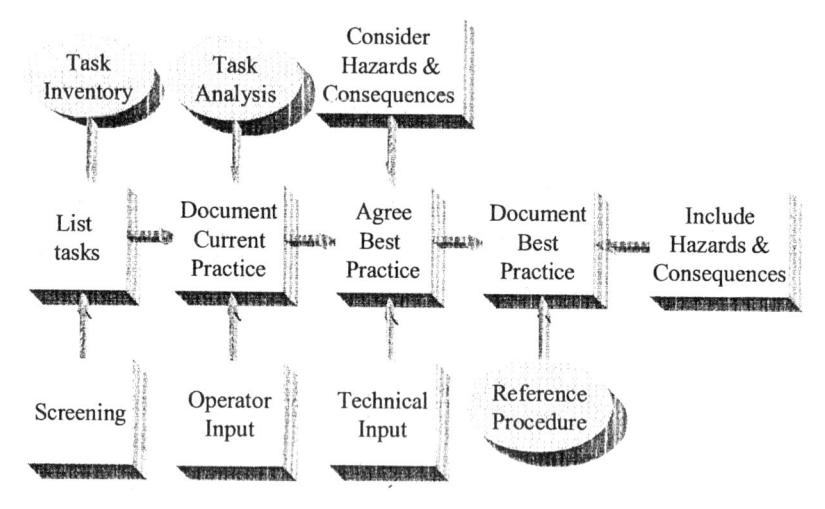

Figure 4: Stage 1 of CARMAN

The first step of Stage 1 is to list the tasks that exist in the system. This list is called a Task Inventory, and is intended to ensure that no important tasks are omitted. Following the development of the Task Inventory, a screening analysis may be conducted to identify all tasks that are considered to be critical. The current practices for the tasks of interest are then documented using HTA. Usually there will be differences between individuals and shift teams regarding how tasks should be performed. These are compiled by the facilitator, and then resolved by convening consensus groups, which examine the similarities and differences between methods. These groups also evaluate the consequences associated with various types of error, and on the basis of these risk assessments and the discussions, consensus is reached on the Best Practice. At this stage, technical specialists are invited to the consensus sessions to comment on the draft Best Practices. Unless the specialists provide specific reasons for modifying the Best Practice, this is then appended to the database in the form of a Reference Procedure together with information concerning the possible hazards and consequences.

In Stage 2 of CARMAN (Figure 5), the Reference Procedures in the Best Practice database are used to develop competency specifications, training programmes and supporting job aids, based upon the level of on-line support required for each task. The primary factors that are considered when determining the level of on-line support are the severity of consequences if the task fails, the frequency with which the task is performed and its complexity. The more severe the consequences, the lower the frequency of task performance, and the greater the complexity, the more elaborate the level of support that is provided.

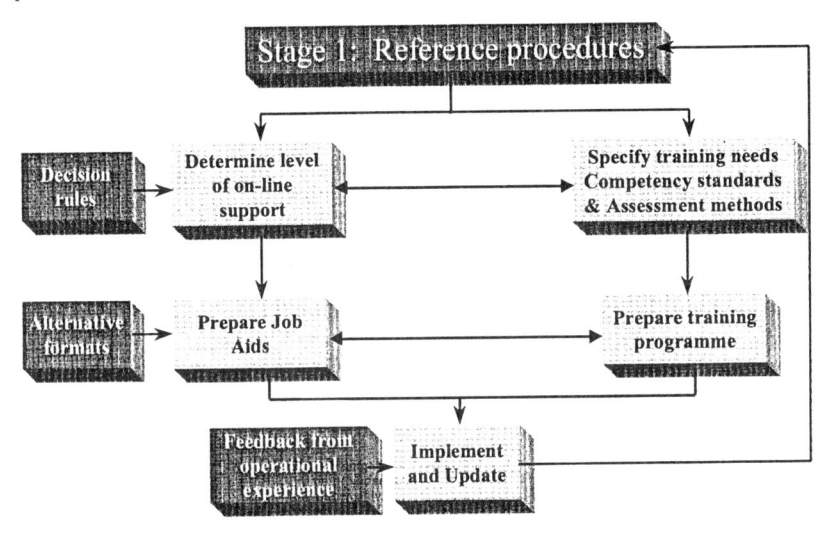

Figure 5: Stage 2 of CARMAN

An example of a decision rule for a particular set of experienced workers is shown in Figure 6. In this figure, it can be seen that the majority of tasks will be performed without written instructions. As the tasks become more critical, complex and infrequent, the level of support increases. However, overall, less than ten percent of the tasks require step by step instructions for on-line support. This is in sharp contrast to many organisations, where nearly all cells in the matrix would specify the use of step by step instructions.

Task Criticality	High			Medium			Low		
Task Familiarity	Freq	Infreq	Rare	Freq	Infreq	Rare	Freq	Infreq	Rare
Task Complexity									
Low	NWI	NWI	JA	NWI	NWI	JA	NWI	NWI	NWI
Medium	NWI	JA	SBS	NWI	NWI	JA	NWI	NWI	NWI
High	JA	JA	SBS	NWI	JA	SBS	NWI	NWI	JA

No Written Instruction required (NWI)
Job Aid required e.g. checklist/memory aid (JA)
Step by Step instruction required (SBS)

Figure 6: Decision Aid for choosing level of Job Aid Support

CASE STUDIES OF THE APPLICATION OF CARMAN

Application to Tasks in a Large Oil Refinery

So far, CARMAN has been most extensively applied in the petrochemical industry. In a project with a major oil refinery, the long-term objective was to apply the process to all operational tasks on the site in order to produce measurable benefits in terms of safety performance. The stages of the project were as follows:

- Pilot project to demonstrate proof of principle and benefits
- Development of task inventory and selection of initial departments within which to implement the process
- Selection of example tasks within departments
- Selection and training of facilitators in tools and techniques
- Developing reference procedures and associated job aids for selected tasks by the facilitators with guidance from the consultants
- Setting up consensus groups
- Initial facilitation of consensus groups by consultants in conjunction with facilitators
- Monitoring of consensus groups to ensure effectiveness of the CARMAN process
- Development of job aids and reference procedures by the consensus groups.
- Repeat of previous steps to develop reference procedures and job aids across the site
- Extension of the process to develop integrated training and procedures system

The pilot project was conducted over a period of six months and included the analysis of a complex process operation and also two non-process but safety critical tasks, i.e. excavations and vehicle entry on site. Both reference procedures and job aids were developed for these areas. The pilot study provided the opportunity to fine-tune the process and to demonstrate its applicability to simple but critical tasks as well as more complex production operations.

Once the pilot study was completed and the results analysed, the next stage involved developing an implementation plan for the site as a whole. The first stage of this plan was to develop an inventory of all the tasks on the site, and select the specific departments and tasks within which the process was to be initially applied. Training programmes were developed and personnel from the selected departments were selected and trained as facilitators. The personal qualities of facilitators are extremely important. They need to be trusted and respected by their peers, and have sufficient social skills to successfully manage the sometimes stormy discussions that can arise during the process of reaching a consensus. If possible, we recommend that facilitators also receive training in how to manage group dynamics. In order to ensure that the facilitators had a comprehensive grasp of the process, they applied it to example tasks from their departments, under the guidance of consultants.

Once the facilitators had demonstrated their competence, by producing acceptable reference procedures for the example tasks, they were involved in setting up consensus groups and facilitating these groups with the guidance of the consultants. Currently the process is now in production mode and is being extended to other units on the site. The next stage will be to extend the project to integrate the site training system into the procedures and job aids. So far the implementation process has taken approximately 18 months. During the second year of the process, the role of the consultants has changed from that of implementation to monitoring, consolidation and the extension of the process to new departments. The measurable benefits to date have included the following:

- Reduced emissions of effluent across the site
- Improved utilisation of steam
- Plant start-up times substantially reduced
- Fewer pump breakdowns arising from incorrect operations

A comprehensive before and after evaluation is currently being implemented, to evaluate key performance indicators in the area of safety and quality. However, it is clear that the project costs have already been recouped in terms of improved efficiency and reduced down time.

Management of plant disturbances

Another significant area of application for CARMAN has been the improved handling of plant disturbances. In a new polymer plant there were concerns regarding incidents which could lead to considerable financial losses as a result of product solidifying in the pipe work following the incorrect handling of disturbances. In the event of a disturbance, in which the heating system usually tripped out, operators were required to diagnose the problem and take appropriate corrective actions to restore the heating system, all within a very restricted time frame. Because of newness of the plant, no effective strategies had been developed for handling the disturbances and for providing appropriate support in the form of training and job aids.

For the application of CARMAN, ten of the most significant disturbances, in terms of potential financial loss, were selected for analysis. The CARMAN process was applied to develop the most effective strategies, by pooling the knowledge that existed both across shifts and between technical specialities. Interestingly, the development of the optimal methods for handling disturbances involved facilitating communication between groups of individuals, e.g. chemical engineers and operators, who would not normally have a forum for discussion. Another result of the analysis was the finding that the process information display system, although very sophisticated in some ways, did not provide the critical information required to handle the disturbances in a way that supported the strategies required for effective response. The CARMAN process can therefore be seen as providing the additional benefit of allowing the process information needs of the operators in critical tasks to be defined in a systematic manner. Applying CARMAN to the polymer plant has led to significant improvements in safety performance as well as reduced losses arising from disturbances.

A similar application of CARMAN to disturbance handling in a refinery has focussed on preventative measures designed to reduce the likelihood of incidents. During a recent insurance evaluation of the plant, the assessors were sufficiently impressed by the results of the process that they retained some of the job aids and other documentation as a model to show other plants what could be achieved. CARMAN is now being applied to develop an improved emergency response system, by identifying the specific needs of the participants in the system by means of task analyses and consensus groups. It is also intended that a new Quality Assurance System being developed at the plant will be based on the principles of CARMAN. The main idea is that the system will provide a participative process so that the practices needed to ensure that quality standards are achieved and maintained are developed by the personnel who have day to day contact with the plant. The intention is that all of the stakeholders in the quality assurance process (managers, plant technicians and technical specialists) are included. In this way, the Quality Assurance system can become a true continuous improvement process rather than a paper driven bureaucracy.

CONCLUSIONS

This paper has described the findings from a study of the use of procedures in the petrochemical industry, and has shown that a major problem exists with regard to the differences that exist between informal working practices and those documented in formal procedures. In order to solve this problem, CARMAN, a consensus-driven approach to identifying risks, and developing Best Practices to control these risks, has been developed. This approach has been successfully applied to several organisations in the petrochemical sector over the last few years. It provides a framework to control safety, quality and other risks by the development of a participative culture, and the integration of training and procedures. It also provides a logical method for selecting the type and degree of on-line support required to complement the skills acquired through training and experience. The process can be seen as a means for achieving many of the objectives of both safety management and quality systems, in addition to providing measurable benefits in terms of improved efficiency.

A METHODOLOGY FOR ASSESSING AND MINIMISING THE RISKS ASSOCIATED WITH FIREWATER RUN-OFF ON OLDER MANUFACTURING PLANTS

Christopher J. Beale, Ciba Speciality Chemicals Plc, Water Treatments Division, PO Box 38, Bradford, West Yorkshire. BD12 0JZ, UK.

> Many manufacturing facilities handle flammable and combustible materials. It is often not practicable to eliminate fire risks on older plants because of their original design features. Fire protection for these plants ultimately depends on some form of active fire protection system, using either dedicated deluge systems or mobile fire appliances. These systems will produce firewater run-off which can cause additional safety, environmental and business impacts away from the incident. These risks must be managed.
>
> Keywords : firewater runoff, fire, risk assessment.

INTRODUCTION

Modern chemical manufacturing plants are often now designed using a structured programme of safety studies. Opportunities for promoting inherent safety are investigated and implemented into the design where practicable. Fire prevention measures can be used extensively to minimise plant fire risks. Firewater containment systems or procedures can be incorporated in the plant design to minimise firewater run-off risks. This allows a balance to be struck between investments in inherent safety, prevention, protection, containment and emergency response facilities.

Contrast this with a typical older plant. It may have been constructed before safety studies were integrated strongly into the design lifecycle or when process safety engineering techniques were less well developed than they are today. Opportunities for fire prevention are likely to be limited and design deficiencies may exist. Typical problems associated with older plants which can cause fire risks to be high include :

- inadequate separation distances between plant and storage areas.

- poor drainage away from high risk equipment.

- limited access for fire fighting in an emergency.

- proximity of plants to populated onsite and offsite populations.

- inadequate firewater supply and firewater run-off containment.

Operating companies are then faced with the dual problem of managing plants with high fire risks and high firewater run-off risks. This paper proposes a methodology for managing firewater run-off risks on older plants supported by practical examples. This allows risks to be identified and minimised using appropriate containment and protection systems.

HAZARDS CAUSED BY FIREWATER RUN-OFF

Although uncontrolled releases of firewater run-off are a well publicised cause of environmental damage to aquatic systems, it is important that all potential hazards are considered. Depending on the local plant design and surrounding area, the following hazards may also exist :

(i) local fire spread to escalate the fire incident to adjacent plant areas. This could occur if the fire extinguishing medium failed to extinguish the fire and acted to increase the pool area of the fire or if the firewater run-off was flammable and re-ignited.

(ii) fire spread to more distant areas. This would be most likely to be caused by re-ignition of flammable firewater run-off but could also occur if the fire extinguishing medium failed to be effective. The fire could start in a plant area where ignition sources were carefully controlled before escalating to an area which would not normally be considered to be a hazardous area. If this occurred, the emergency services would become stretched as they attempted to fight fires in multiple locations simultaneously.

(iii) toxic release in a distant area. The firewater run-off may contain toxic materials from the source of the fire or from other site areas, such as drains, open tanks or Intermediate Bulk Containers (IBCs). The toxic release may occur close to a populated area such as a road, railway, containment lagoon or sewage treatment works, causing risks to offsite populations, office workers or other site workers.

The assessment of firewater run-off risks must consider all of these hazards.

ASSESSMENT OF EXISTING FIRE RISKS

Before firewater run-off risks can be assessed, the fire hazards associated with the plant need to be identified. A Fire Risk Assessment (FRA) needs to be carried out using an appropriate quantitative or qualitative methodology. In practice, it is often most effective to use a semi-quantitative technique, with detailed fire and explosion consequence modelling.

Identification Of Fire Scenarios

Fire scenarios should be clearly identified and are likely to involve the following flammable and explosive events in different areas of the plant.

1. Pool fires at ground floor level, running down a slope or running through a structure from a high floor level.

2. Gas jet fires, liquid jet fires and flash fires.

3. Vapour cloud explosions (VCE) and internal vessel fires and explosions.

4. Proximate fires involving packaging, wood, pallets, vegetation.

5. Local escalations leading to BLEVE (Boiling Liquid Expanding Vapour Explosion) or damage to adjacent combustible materials.

6. Escalations to other plant or storage areas.

7. Fires in warehouses or Intermediate Bulk Container (IBCs) storage areas.

Defining The Current Basis Of Fire Protection

This needs to be investigated and clearly understood and will include :

1. **Inherent safety** features and prevention systems, such as double mechanical seals on pumps, all welded pipes, flange guards and pressure relief systems. These features either ensure that fires cannot start in the first place or reduce the frequency of fires.

2. **Detection systems**, such as operator detection, gas detection systems and fire detection systems (pneumatic, infra-red, heat sensitive cable). On many older plants, fire detection is often dependent on operator vigilance and action. If the plant handles highly toxic materials or is largely unmanned, it may be fitted with a gas or fire detection system.

3. **Active Fire Protection (AFP)**, such as bund foam pouring systems, sprinkler systems, deluge systems and the Fire Brigade response. These systems allow fire fighting water (or foam / water mixtures) to be applied to the fire and / or around critical plant and equipment. Some plants rely on manual intervention by onsite and external fire fighters using fire appliances and portable foam units. Other plants are protected with fixed systems consisting of fixed pipework linking the firewater supply to a matrix of sprinklers, nozzles or foam pourers inside the plant area. Fixed systems can either be manually operated from a safe area away from the fire or automatically operated from fire detection systems and manual call points located around the plant area.

4. **Passive Fire Protection (PFP)** and fire escalation protection, such as firewalls, segregation of high risk areas, PFP for vessels and PFP for structural steelwork. These

systems are designed to protect vulnerable equipment to control the spread and the consequences of the fire. They will not extinguish the fire but will buy time for the Fire Brigade response. There are many different PFP systems. Commonly used technologies include fire insulation boards and lagging, intumescent paint coatings and vermiculite (concrete-like material) coatings. They are all designed to limit the heat transfer from the fire to the vulnerable equipment.

5. **Containment systems**, such as local bunds, drains, slopes and walls. These systems should be designed to direct leaks and fires away from vulnerable areas to a secure area.

6. **Emergency response**, using the onsite services (if they exist) or external Fire Brigade.

Examining The Effectiveness Of The Existing Fire Protection Systems

The effectiveness of the existing systems should be investigated carefully to identify typical problems such as :

1. **Inherent safety**. Corrosion, damaged or missing flange guards, undocumented pressure relief calculations.

2. **Detection systems**. Unreliable gas detection systems. Incorrect detection levels. Systems which have been disconnected. Reductions in manning levels.

3. **AFP**. Areas of the plant not covered by AFP systems because they have been added to the original plant. Damaged or corroded valves, sprinklers, nozzles and foam pourers. Manual deluge valves. Poorly located deluge valves. Poor firewater supply. Degraded foam. Poor system reliability.

4. **PFP**. Damaged PFP. PFP which uses inappropriate materials such as concrete. Unprotected vessels. Extensions to the steelwork which have no PFP.

5. **Containment systems**. Inadequate capacities, blocked or undersized drains, damaged bund walls, areas where large pools may develop due to inadequate drainage. Critical firewater disposal pumps which may fail in a fire situation or may become sludged up. Valves which could isolate or bypass the containment systems.

6. **Emergency response**. Poor access to the plant for emergency services. Slow Fire Brigade response times. Inadequate onsite Fire Brigade manning levels on nights and at weekends.

This stage of the assessment needs to be carried out carefully and tests are often required to verify the system performance. A team based approach is very effective using representatives from the production plant, engineering, process safety and maintenance. An experienced fire fighter should also be present to identify relevant practical fire fighting issues. The assessment should be focused on how each of the identified scenarios would be controlled.

Fire Risk Summary

A frequency and consequence assessment needs to be carried out for each fire scenario. Risks can be summarised using many different techniques. In most cases, a risk matrix is effective for summarising fire risk levels. Frequencies are grouped into a number of defined categories on one axis of the matrix from very low to high. Consequences are grouped into a number of defined categories (covering safety, environmental and asset damage impacts) on the other axis from trivial to catastrophic. Table 1 shows an example of a typical risk matrix.

At this stage, firewater run-off risks are not included in the risk matrix. The risk matrix can be used as a tool to ensure that fire risks are managed to a level which is 'As Low As Reasonably Practicable (ALARP)'. Further information is required before firewater run-off risks can be assessed.

ANALYSIS OF FIREWATER RUN-OFF RISKS

Fire scenarios which pose a high fire risk have now been identified. A quantity of firewater supply will be required for each scenario and this will in turn generate firewater run-off. The potential impacts of the firewater run-off now have to be considered.

Characterisation Of Fire Scenarios

It is very difficult to predict the volume and duration of firewater usage for each of the fire scenarios. In a real fire situation, the Fire Brigade will control the amount of firewater which is used to extinguish the fire and control the incident. An estimate of the firewater application rate and the total volume of applied firewater must, however, be made if firewater run-off risks are to be managed.

In general, four factors tend to dominate firewater usage :

(i) fixed active fire protection systems, such as sprinkler systems, deluge systems and bund foam pouring systems.

(ii) onsite firewater supply capacity through hydrant systems and from fixed storage tanks. This can be used by onsite and external Fire Brigades. The distribution systems will often also feed the fixed active fire protection systems.

(iii) offsite firewater supply capacity such as local lakes, ponds, rivers, canals, water distribution mains. These would probably be used by the external Fire Brigade if the incident was serious.

(iv) rainfall, effluent generation and chemical spills from different plant areas.

Estimates need to be made of the contribution to overall firewater usage for each fire scenario.

Table 1 : Typical Risk Matrix For Assessing Fire Risks.

Frequency Category

	Trivial	Significant	Serious	Major	Catastrophic
High		4a			
Moderate	2a	8			
Medium		11 2b	10 6		
Low		3 1	5 4b	9	
Very Low				7	

Consequence Category

Fire Scenario Summary

- 1   Main transfer line leak.
- 2a   Small leak in day tank area.
- 2b   Large leak over day tank bund.
- 3   Leak in line to adjacent plant.
- 4a   Small pump leak.
- 4b   Day tank drains out through pump.
- 5   Leak from day tank to process.
- 6   Major leak in blending vessel.
- 7   Reactor feed line rupture.
- 8   Reactor vapour space leak.
- 9   Reactor liquid leak.
- 10   Major release through filters.
- 11   Recycle line rupture.

Fixed Active Fire Protection Systems

These systems are normally designed to be compliant with a standard such as NFPA15 (NFPA, 1) or NFPA16A (NFPA, 2). These standards specify minimum required water or foam / water application rates per square meter of protected plant. For example, NFPA15 recommends an application rate of at least 10 LPM / m^2 (litres per minute per square metre) for water spray protection to vessels and NFPA16A recommends an application rate of at least 6.5 LPM / m^2 for foam / water sprinkler protection systems.

The firewater application rate can then be calculated for each fixed system from 1 :

$$LPM = Q \times A \times SF \qquad (1)$$

where LPM is the overall firewater application rate in litres per minute, Q is the NFPA recommended application rate per square metre, A is the surface area of the equipment to be protected in square metres and SF is a Safety Factor. The Safety Factor is required because most systems will be designed to work to NFPA application rates or better when some of the firewater supply system is impaired. The designer of the active fire protection system may be able to advise on actual flow rates, from which the safety factor can be calculated. If this information is not available, a factor in the range 1.1 to 1.3 could be used as a default value.

The LPM could also be obtained by monitoring system water demands when the system is tested or by taking measurements at points on the plant during a test.

Fire Brigade Usage

The Fire Brigade will extract firewater from dedicated firewater supply mains (which may also be supplying fixed active fire protection systems) and from other water sources using mobile pumps. Table 2 shows typical firewater application rates for fires on chemical plants.

It is difficult to predict these usage rates accurately. It may be possible to obtain additional information from experienced fire-fighters in the onsite or external Fire Brigades.

Rainfall, Process Effluent And Chemical Spills

Firewater may be applied at a time when the site effluent systems are fully loaded with process effluent or when storm conditions exist. This will create an additional demand on the site drainage and containment systems. Data can be obtained for stormwater flow rates (eg. a 1 in 25 year 8 hour duration storm event) in mm of rainfall per square metre. This rate has to be adjusted to account for the surface area which feeds into the drains to produce a rate of drainage in LPM.

Process effluent generation rates also need to be assessed in LPM. This should include any unburnt chemicals released during the fire.

Table 2: Typical Firewater Application Rates For Chemical Plants.

Type Of Fire	Typical Usage In LPM From	To	Typical Duration (hours)	Typical Inventories
High Severity	27000	54000	4	Hundreds of tonnes of flammables
Medium Severity	18000	27000	3	Tens of tonnes of flammables
Low Severity	9000	18000	2	Less than 10te of flammables

In most cases, stormwater and process effluent generation rates are much lower than the rates produced by fixed active fire protection systems and the Fire Brigade.

Duration Of Firewater Application

This is again very dependent on the individual fire scenario and the fire fighting strategy adopted by the Fire Brigade.

Typically, firewater may be applied to a large fire for about four hours and to a small fire for about two hours. After this time, cooling or damping down may occur for up to 24 hours at rates of about 4500LPM.

Composition Of Firewater Run-off

In most cases, especially when the fire involves a mixed storage area, it is almost impossible to quantify the likely composition of the firewater run-off. The run-off is likely to consist of a complex mixture of unburnt chemicals, partially combusted chemicals, by-products from the combustion process, fire fighting foam and materials which are stored in the path of the firewater run-off.

If detailed water pollution consequence modelling is being carried out, source terms will need to be derived for each release scenario. This may concentrate on the predicted dominant cause of pollution or contributor to fire and safety risks.

Definition Of Firewater Scenarios

Many of the fire scenarios will produce firewater run-off incidents which have similar safety, environmental and asset damage consequences. These scenarios can be grouped into firewater scenarios. Table 3 illustrates a typical set of firewater scenarios for a plant.

The adequacy of the existing containment systems also needs to be assessed. Calculations need to be performed to understand when each element of the containment system will become overwhelmed (eg. the intermediate bund will fill in 8 minutes, cascade into the main bund and fill this bund in 14 minutes, and then cause an uncontained release. It is also important to calculate the maximum flow rates (LPM) which can be accommodated by relevant plant drains.

Identification Of Firewater Pathways

A pathway is a mechanism by which firewater can be transported from the source of a fire to a vulnerable plant area, populated area or sensitive environment. Pathways need to be identified for each firewater scenario. This can best be achieved by :

1. Identifying the plants, storage areas, offices, offsite populated areas, streams, culverts, rivers and ground which could be susceptible to damage from firewater run-off.

Table 3 : Typical Firewater Scenarios For A Plant.

Scenario	Description	Bund Foam Pouring System	Plant Deluge System	Building Water Curtains	High Level Fixed Monitors	Fire Brigade	Storm Water	Process Effluent	Total Firewater Run-off
FW1	Transfer line leaks	0	0	0	6000	9000	2000	500	17500
FW2	Day tank leaks	1000	0	1100	3000	9000	2000	600	16700
FW3	Low severity fire in process area	0	2900	3300	0	9000	2000	800	18000
FW4	Medium severity fire in process area	0	8800	3300	0	9000	2000	2300	25400

All firewater run-off volumes in LPM (litres per minute)

2. Deciding if each firewater scenario can credibly affect any of these areas.

3. Identifying protection and containment systems which must fail to cause damage to be realised by each scenario. This should include hardware, such as valves, and software, such as emergency procedures.

It is often useful to show each pathway using an event tree format. If a quantitative frequency assessment is being used, branch probabilities can be inserted to calculate the frequency of causing different types of damage. Figure 1 shows how an event tree can be used to define pathways.

Onsite And Offsite Impact Assessment

Each firewater scenario with the potential to cause significant damage should be assessed to determine it's impact on man, the environment and the business. These consequences can either be assessed using a qualitative risk matrix based approach or using quantitative modelling techniques.

If quantitative modelling is used, factors such as composition, pool size, dilution rates and flow rates can be calculated. These parameters can then be fed into the appropriate consequence modelling software for fire damage, groundwater pollution, river pollution or toxic impacts.

Risk Summation

Risks can either be summarised using quantitative techniques or using risk matrices (as shown in Table 1).

RISK ASSESSMENT

Decisions now need to be made about the tolerability of the firewater run-off risks. It is very rare to find that these risks can be eliminated entirely. There is normally some combination of hardware failures, natural events and human error which could lead to a pollution incident, asset damage or impacts to people.

Risk tolerability criteria need to be used to determine whether risks are :

1. So high that they should be considered to be intolerable.

2. Sufficiently low that they can be regarded as broadly acceptable. or

3. Somewhere between these two levels, where risks should be managed to a level which is considered to be ALARP (As Low As Reasonably Practicable).

Most plants fall into this middle ALARP band. It is therefore necessary to examine a series of risk reduction options and assess their cost and the reduction in risk that they produce. An optimal series of risk reduction measures can then be developed to ensure that this ALARP criterion is met.

Figure 1 : Definition Of Firewater Pathways Using An Event Tree Format.

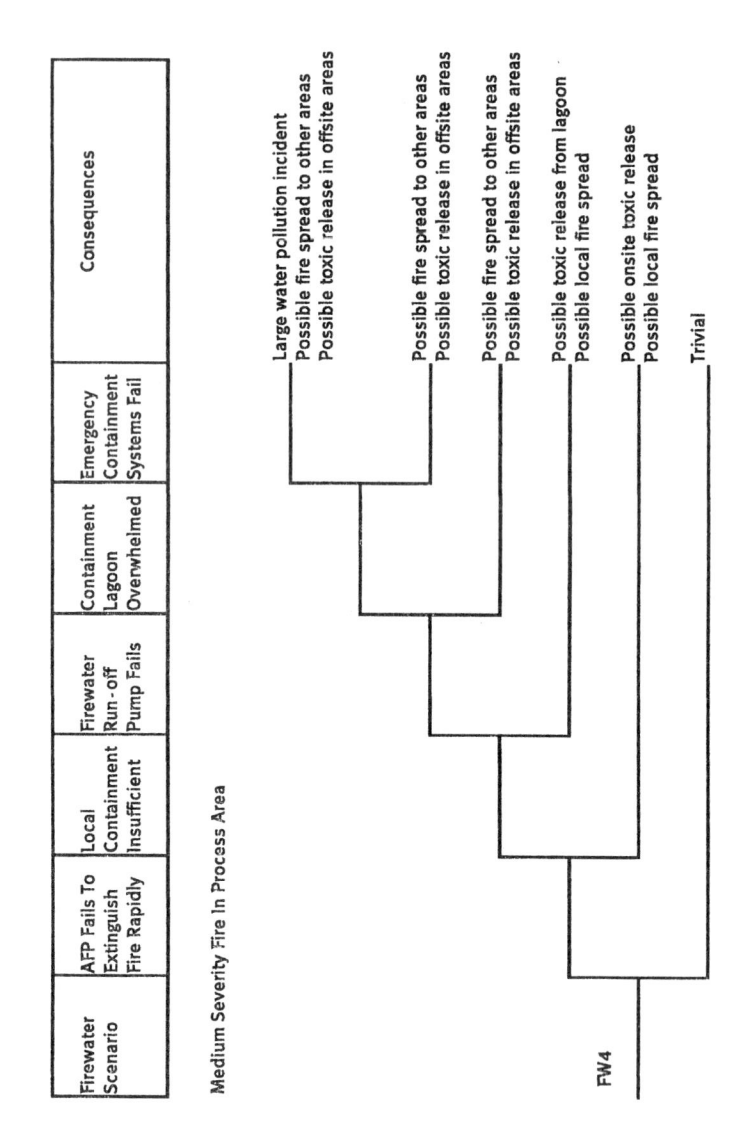

Firewater Scenario	AFP Fails To Extinguish Fire Rapidly	Local Containment Insufficient	Firewater Run-off Pump Fails	Containment Lagoon Overwhelmed	Emergency Containment Systems Fail	Consequences

Medium Severity Fire In Process Area

- Large water pollution incident
- Possible fire spread to other areas
- Possible toxic release in offsite areas

- Possible fire spread to other areas
- Possible toxic release in offsite areas

- Possible fire spread to other areas
- Possible toxic release in offsite areas

- Possible toxic release from lagoon
- Possible local fire spread

- Possible onsite toxic release
- Possible local fire spread

FW4 — Trivial

Risk Reduction

There are a large number of risk reduction measures which could be adopted. These are described in a number of publications including (HSE, 3) and (NRA, 4) and include :

Reduction Of Fire Risks

By eliminating or reducing fire risks, the frequency of producing a firewater run-off incident is automatically reduced. Typical risk reduction may include :

1. Design changes to minimise leak frequencies and measures to control ignition sources in hazardous areas.

2. Passive fire protection to control fire escalation or isolate the plant into smaller fire zones.

3. Fire and gas detection systems to provide early warning that an incident has occurred.

4. Fixed active fire protection systems to allow fires to be extinguished rapidly before they have been able to develop.

5. Trained on-site fire fighting teams equipped with mobile fire fighting equipment.

Improved Containment Systems

These measures are designed to control the flow of firewater run-off and ensure that it is directed to a safe area. Typical measures include :

1. Improved bund designs, increasing bund heights, making greater use of mini-bunds and installing spillways at the top of bund walls to direct bund overtopping spills in a preferred direction.

2. Dedicated firewater run-off drains feeding to a large containment lagoon located in a safe area.

3. Temporary booms and sandbags to direct spills to safe areas or sacrificial areas.

Emergency Response

These measures are aimed at minimising the impact of any releases and include :

1. Training onsite teams to deal with emergencies.

2. Improving links with external bodies who are responsible for fire fighting and pollution control.

CONCLUSIONS

This paper has shown how systematic risk assessment techniques can be used for managing and minimising firewater run-off risks on older manufacturing plants. It can be integrated with quantitative and qualitative techniques and can also be used in conjunction with the risk management systems used by individual operating companies.

REFERENCES

1. NFPA15, 'Standard for water spray fixed systems for fire protection', National Fire Protection Association, 1996.

2. NFPA16A, 'Recommended practice for the installation of closed-head foam-water sprinkler systems', National Fire Protection Association, 1988.

3. HSE Guidance Note EH70, 'The control of fire-water run-off from CIMAH sites to prevent environmental damage', Health and Safety Executive, 1995.

4. NRA, 'Pollution prevention measures for the control of spillages and fire-fighting run-off', National Rivers Authority, 1994.

SUMMARY OF ACRONYMS

AFP	Active Fire Protection
ALARP	As Low As Reasonably Practicable
BLEVE	Boiling Liquid Expanding Vapour Explosion
FRA	Fire Risk Assessment
HSE	Health & Safety Executive
IBC	Intermediate Bulk Container
LPM	Litres Per Minute
NFPA	National Fire Protection Association
NRA	National Rivers Authority
PFP	Passive Fire Protection
VCE	Vapour Cloud Explosion

MORE EFFECTIVE PERMIT-TO-WORK SYSTEMS

R.E.Iliffe, P.W.H. Chung and T.A. Kletz
Department of Chemical Engineering, Loughborough University, Loughborough, Leicestershire, LE11 3TU. United Kingdom. Email: R.E.Iliffe@lboro.ac.uk

> Many incidents in the chemical-industrial workplace are associated with maintenance works, which are typically controlled by permits-to-work (PTWs). Computerised PTWs have advantages of flexibility and informational clarity, and allow closer co-ordination of activities and integration with computer applications. A system has been developed linking computerised PTWs with an incident database: the system examines the nature of the job, equipment and chemicals specified on the PTW and draws users' attention to relevant incident reports without requiring explicit search or further data; unknown or forgotten hazards are thus highlighted when preventative action may still be taken.
> Keywords: Permit-to-work, computer integration, accident database.

INTRODUCTION

According to the Health and Safety Executive[1] (HSE) 30% of the accidents which occur in the chemical industries are maintenance-related. A quick check of the Institute of Chemical Engineers accident database reveals that over 700 accidents of the 5000 listed were maintenance-related. Some of these were due to the way in which the maintenance was carried out but most were due to errors in the way the equipment was prepared for maintenance or handed over. Sometimes the permit-to-work (PTW) system was poor, sometimes it was not followed.

Production of PTWs on a computer would have many advantages which could improve permit systems and make it easier to follow them. There would be less tedious form-filling and greater legibility, the permits could then be inspected by supervisors and managers from any location and analysis of the work carried out would be easier. The computer could refuse to issue more than one permit for the same item of equipment and could remind users of other work in progress nearby. However, much greater advantages would follow by linking the computer to a database of accidents and to other information systems. For example:

- The computer could remind the users of any special hazards associated with the equipment to be maintained and its contents, points highlighted during the hazard and operability study and lessons learned during previous maintenance.
- If a vessel is being prepared for entry, the computer could check that the number of slip-plates (blinds) to be fitted (or pipes disconnected) is the same as the number of connections shown on the drawing.
- Suppose a fitter has to replace a gasket during a night shift. On some plants it is easy; only one sort is used and all he has to do is select the right size. On other plants many types are used. The fitter has to get out a diagram, find the line number and then look-up the details in a bulky

equipment list. It should be possible for him to view the line diagram on a computer screen, select the line and have details of it displayed, including the location of the gaskets and any distinguishing marks such as their colour. The line diagram and equipment list will have been prepared on a computer; all that is needed is a link between the design system and the maintenance system. (Of course, we should, if possible, reduce the number of types of gaskets, nut and bolts required even though we may use more expensive types than strictly necessary on some duties.)

- The computer could look at the nature of the job, the type of equipment and the chemicals involved as entered on the permit and then draw attention to any relevant incidents in the accident database. Note that the user would not need to search the database; the computer would do this for him. The computer would be active, the user passive, while in normal information retrieval the opposite usually obtains. This is both the most original and the most fully developed of the proposals discussed in this paper and is presented in more detail from page 8, onwards.

LIMITATIONS OF PERMITS TO WORK

The HSE defines a Permit to Work as:

> A formal written system used to control certain types of non-routine work, usually maintenance, that are identified as hazardous. The terms 'permit-to-work' or 'permit' refer to the certificate or form that is used as part of the overall system of work. The permit is a written document that authorises certain people to carry out specific work at a specific time and which sets out the hazards associated with the work and the precautions to be taken.[2]

Thus a PTW system incorporates both a written document and a series of rules describing and circumscribing safe methods of working. This said, the specific *purposes* which PTWs seek to achieve are more diverse and more complex. One appreciation of PTWs has it that their purpose is first to ensure that proper consideration has been given to the hazards associated with any given proposed operation, second that they should ensure appropriate precautions have been put in place and third that they should facilitate communication between the various parties involved in the works[3]. An alternative view is to say that PTWs perform at least three notionally distinct functions: first they aid the *identification* of potential hazards together with the concomitant precautions which must be taken; second they aid *co-ordination* of the imposition of precautions, the actual carrying out of the maintenance task and the eventual removal of precautions. Third, they *provide a written record* of what was done, by whom, when and how. This may be of use in the event that something does in fact go wrong, as well as to help monitor the procedures which are in place; it would be a mistake to perceive PTW systems as being set in stone; to obtain the best use of them they should be capable of easy modification to meet changing circumstances and individual user needs.

The HSE study of incidents in the workplace reveals that a large proportion of the maintenance-related incidents involved failures of the PTW system: in some cases, to a greater or a lesser degree, the PTW system was inadequate; in others it was entirely absent. This is a sobering state of affairs: safety in the chemical-industrial-workplace is achieved primarily by placing physical guards between personnel and known hazards; maintenance, by definition, typically involves the removal of these together with the introduction to the work-site of additional hazards, such as flames, which might not already be present. It is, therefore, not excessive to regard maintenance itself as an inherently hazardous activity. Largely as a result of the HSE study considerable further work has been done on PTW systems. One report of work carried out by the HSE itself which was published in 1995 was a survey of PTW systems in medium-sized chemical plants[4]. This identified a number of areas where current PTWs are inadequate: the type and format of PTWs varies widely across the spectrum of plants studied; a majority of plants use at least three different forms to cover a variety of jobs, while a significant minority of plants used as many as ten different forms. At the lower end of the scale many firms copy the permits of other companies without paying sufficient regard their appropriateness to their own needs; at the upper end of the scale too great specificity of permit forms may lead to a confusion of paperwork with a consequent loss of efficiency.

This finding was a symptom of a more general confusion on the part of firms over precisely which jobs should be covered by a PTW and which should not. Further, within many companies there was disagreement between management, foremen and fitters over the general applicability of PTWs and the extent to which these had to be followed in every detail. Fitters in particular were unclear about the extent of their freedom to vary their work in light of developing knowledge of the maintenance situation and the extent to which they were bound to follow the plan.

Administrative difficulties exist too: it is frequently difficult to locate authorised issuers when a permit is actually needed, with the result that permits are commonly issued at a specific time in the morning with actual commencement of the work being left till some time later in the day – or in some cases – until several days later. Similarly, confusion frequently exists over which authorisations, additional to that of the issuer, a permit requires – whether it must be signed-off by a fire-marshal, by management or by some specialist fitter. Of course, when additional authorisation *is* required this further exacerbates the problem of locating those whose signature must be obtained. One example serves to illustrate how serious the consequences of these practices may be:

> In 1989 a take-off branch in a polyethylene plant was dismantled to clear a choke. The 8-inch valve isolating it from the reactor loop was open and hot ethylene under pressure came out and exploded killing 23 people, injuring over 130 and causing extensive damage. Debris was thrown six miles and the subsequent fire caused two liquefied petroleum gas tanks to burst. The valve was opened by compressed air and the two air hoses, one to open the valve and one to close it were connected-up the wrong way around. The two connectors should have been of different sizes or design so this could not occur. In addition they were not both disconnected and a lockout device - a mechanical stop - had been removed. It is also bad practice to carry out work on equipment isolated from hot flammable gas under pressure by a single isolation valve. The take-off branch should have been slip-plated and double block-and-bleed valves should have been provided so the slip-plate could be inserted safely. Another factor in the incident was that the equipment had been prepared for repair and had then had to wait for several days until the maintenance team was ready to work on it. During this time the air lines were reconnected, the lockout removed and the isolation valve opened.[5]

Although proper practice would not have directly addressed the issue of bad design, had the permit only been issued *when it was actually needed* there would have been no opportunity for the proper isolations to be improperly removed.

One variant of the problem of administering PTWs which particularly drew the attention of the HSE involves the problem of ensuring that all those people who need to be informed of maintenance work are, in fact, informed. Although this is of particular significance when maintenance carries-over between shifts – the Cullen Report[6] cites the failure to inform a new shift that safety-critical equipment had been disconnected as one of the proximate causes of the Piper Alpha disaster – this problem is not restricted to shift changes. In complex plants, for example, pipework which is disconnected or otherwise isolated prior to maintenance may pass through a number of areas other than that where the work is actually carried out; it is clearly important that the people in charge of those areas be informed of the isolation; sometimes this is not done.

Quite apart from the specific weaknesses identified by the HSE, current PTWs suffer from three *general* weaknesses. First, they are *uninformative*. All current systems assume that issuers are competent to identify hazards and that they merely need to be prompted to remind them of particular dangers. Unfortunately, this assumption is not always wholly valid. The modern workplace is highly and increasingly complex; while it is reasonable to expect issuers to be aware of the more common hazards, such is this complexity that their failure to guard against *every possible* hazard is not culpable.

A second general weakness is that many permits tend to *lack clarity*: the format of most PTWs is some combination of lists, which the issuer checks as appropriate, and boxes which must be physically filled-in. This latter element is essential for adding specificity to the permit and for ensuring that issuers think when completing permits rather than merely operating on autopilot. A drawback, however, is that what seems clear to issuers frequently unclear to anyone else.

A third and final weakness of current systems is that they tend to be *inflexible*. Permits are – or should be – optimised to the specific needs of a given plant. However, to the extent that a PTW system *is* so optimised it tends to be less than optimal should the plant set-up or production requirements change. Given that firms increasingly see flexibility of production as a route to business success an obvious tension can be seen to exist with good PTW practice.

THE COMPUTER VERSUS THE PAPERCLIP

One pedestrian, though still significant, area where a computerised system scores over older systems lies in simple legibility: experience has shown that the rigors of the workplace impact badly on the legibility of permits; issuers' handwriting is frequently illegible; the permit becomes soiled and crumpled; in extreme cases it may disintegrate entirely with the resultant loss of site-specific data such as atmosphere test readings being lost. A computerised system, conversely, at least starts out legible – though individual copies still become stained. However, being automated, additional copies may be printed-out remotely from any convenient printer just as site-readings may be recorded on hand-held units and then downloaded electronically. This minimises delay, protects data and permits the computer to monitor the data so obtained: this last is significant since a number of incidents have occurred due to workers noting *but failing to act upon* instrument readings indicating the development of dangerous situations.

However, it is in the area of *co-ordination* that a computerised PTW system has the greatest comparative advantage over current methods. For example, in order to achieve greater control over maintenance, the trend has been for maintenance tasks to be broken down into separate stages each of which is subject to a separate permit: typically, in the case of a complex job involving isolation of equipment one permit may be issued to control the application of appropriate isolation and/or the disconnection, a second to control the maintenance task itself and a third to co-ordinate the removal of the isolation and the reconnection of the equipment; traditionally co-ordination between these permits has been by means of a paperclip. However, the system of multiple-paper-permits is not beyond criticism: greater apparent control in such a case is achieved only at an increased cost in bureaucratic complexity; workers may be reluctant to perform 'unnecessary' administration and may not complete the PTWs properly, or may link completed permits improperly; fitters may, as a result of information overload, suffer confusion over what actions the permits actually require them to perform. In such a case redundancy may result in decreased safety.

This problem is compounded when we consider the issue of separate maintenance tasks occurring in close physical proximity. Good practice currently calls for maintenance on neighbouring pieces of equipment to be staggered if this is at all possible: two pieces of neighbouring machinery may require similar but not identical isolations; if one maintenance task is completed before the other the potential for confusion over the removal of isolations is obvious. Alternatively, one job may be fire-sensitive while the other may not: many incidents are on record which occurred due to sparks or flame from one job causing inflammable material from another to ignite.

While computer control of permits cannot eliminate these hazards it does allow much closer control of what goes on: hazards attendant on *each* job can be evaluated *in light of the other*; if problems of common isolation are likely to occur the system can identify potential trouble-spots and require that the isolations, once imposed, are fully labelled and locked to prevent improper removal. Co-ordination of this sort is something that computers are good at dealing with: if the system is properly configured, multiple linked permits may be issued without an unacceptable degradation of safety and also without undue bureaucratic proliferation.

OTHER ISSUES OF INTEGRATION

As was highlighted in the previous section, the imposition of isolations on large and complex plants is frequently difficult and time consuming. Two and three-dimensional plant schematics are in common use as a control tool in process plants. An obvious step would be to integrate a computerised PTW system with such systems: the question of which isolations to impose would be much easier for the permit issuer if such a visual representation were available; further, integration with a plant schematic would allow both the location and the progress of various jobs to be tracked with greater ease: a small refinement might be to have each job flagged and colour-coded to indicate the stage the work has reached; linked jobs might be visibly associated by means of coloured lines.

The visual representation of maintenance tasks would have an additional benefit in achieving greater clarity for the PTW: as has already been noted most PTW systems require a combination of selection-checking and box completion from issuers. While the former permits only a minimum of confusion, the latter is frequently a source of difficulty. A description of a work-site as "all the area to the West of building B" may be clear to the issuer but to no one else. The HSE recommends the imposition of a grid or similar system over a map of the plant area which should be used as a reference; representing maintenance tasks on a plant schematic permits is a useful elaboration of this basic theme.

Computer integration and personnel co-ordination

As has already been noted, PTW systems tend to suffer from a variety of administrative problems with respect to the issuing process itself: in some plants PTWs must be jointly signed-off by both the actual issuer and a safety officer; in other cases they will additionally require the signature of a fire-safety officer. A computerised system would potentially ease if not wholly eliminate confusion by identifying which authorisations are required for any given permit and in what order, basing this assessment on the entries which are made on the permit by the initial issuer.

The problem of physically locating authorised personnel in order to obtain their authorisation would presumably remain even under a computerised system. This might be alleviated to some extent by the integration of the PTW system with a plant intranet: it would clearly be useful to be able to inform personnel of works affecting them electronically and to *view* permits remotely, what is moot is whether the added convenience of being able to *issue* PTWs remotely would be offset by a tendency on the part of issuers to skimp on necessary checks: given that PTWs now *could* be issued from wherever an issuer happened to find himself, might this capability encourage him to be lax in performing necessary checks? Given that a major imperative of any PTW is to encourage issuers to *think* about what they are doing, any development which even *might* encourage one simply to sign-off on permits is to be viewed with extreme caution. Conversely, it is clearly desirable for an issuer to be able to give his authorisation from a work site or safety store having seen that precautions *are* in place and safety equipment *has* been issued. Probably a decision on this point is one which would have to be taken on a case-by-case basis; in any event, integration of a computerised PTW with plant intranets should be accompanied by additional training for both issuers and acceptors to highlight that additional convenience in no way decreases the regulatory force of the permit or obviates the necessity for physical checks by issuers prior to the permit being issued and, by acceptors, prior to work being commenced.

The HSE has identified the problems attendant on work carrying-over between shifts as especially significant. Integration of a computerised PTW system with a plant intranet might reasonably be expected to eliminate or at least reduce the problem which occurs when a whole new set of personnel must be informed of ongoing works. A computerised system should readily be able both to maintain a list of those people who need to be informed of work-in-progress and inform them electronically of what they need to know; further it would be simple matter to have the system log their acknowledgement of receipt of notification.

Clearly the impact of all the possibilities noted so far will be greatest in improving the safety of the system but it should be recognised that the implications go beyond this. For example, it is clearly desirable that decisions relating to maintenance should not conflict unduly with those relating to production; a computerised system controlling both scheduled and unscheduled maintenance will allow management to exercise a greater degree of control over the running of the plant without compromising safety. Improved control over maintenance procedures should result in less 'down-time' for plant which in turn translates into greater production efficiency and, thence, profitability.

Computer integration and personnel evaluation

As has also already been noted one of the basic purposes of a PTW is to provide a permanent record of what has been done, by whom, when and how. This may be of use if an incident actually occurs in order to determine what went wrong but similarly, such records may be used for a variety of other purposes. One of the basic purposes of the Active Database (ADB) under development at Loughborough is to provide the issuers of permits with incident reports relating to their proposed maintenance task even in the absence of a specific request for this information. In order to achieve this the ADB makes its own assessment of potential hazards based upon the information entered on the permit by the issuer. If the user's assessment of risk differs significantly from that of the ADB the system will augment the chances of reports relating to apparently neglected hazards being presented to the user for consideration.

The immediate purpose of this system is, of course, to alert issuers to hazards of which they are unaware or which they have forgotten, but in the longer term the maintenance of *an archive* of permits may be useful as a check on the adequacy of training of issuers: if, for example the issuers of a permit within a plant commonly reveal a blind-spot with respect to a certain class of hazard, this is an indication that the training is inadequate; similarly if a particular issuer commonly neglects a type of hazard, he may need more training, or if he typically fails to note the presence of a variety of hazards this may be an indication of a mismatch of man with job, or indicate an unacceptable slackness on the part of this issuer.

An additional benefit of maintaining a substantial archive of maintenance records is that these may be susceptible to operational analysis which in turn may lead to improvements in the efficiency of the plant: if the records reveal, for example, that a certain type of equipment is off-line for maintenance a disproportionate amount of the time this may be an indication that the equipment in question should be replaced. A further benefit may be to indicate that specific pieces of machinery are breaking down more often than they should: firms could integrate their maintenance logs with a computerised PTW system in order to highlight recurring problems with individual pieces of equipment to issuers at the time a PTW is applied-for.

Further, the maintenance of substantial permanent records is now, in practical terms required by the HSE who include inspection of a selection of PTWs as part of their routine auditing of plant safety procedures. Although it is not yet a legal requirement that firms maintain a permit system, let alone that this be of any specific type, the general requirement that 'reasonable precautions' are taken is such that any firm failing to maintain a PTW system, complete with records, does so at its own risk. In addition to the benefits already noted a computerised system enjoys those generally cited in support of the paperless office: the saving in time, space and expense are substantial.

Computer integration and stock-control

Further efficiencies might be achieved by the integration of PTW and stock-control/warehousing systems: consider, for example, the case of a breakdown requiring the replacement of a component or components. Before this repair can occur someone must first identify what part needs to be replaced, obtain authorisation for its use, locate it and *only then* perform the repair. The situation is complicated still further by the fact that the required part may *not* be held in stock and must therefore be ordered, or, after a part is withdrawn from stock, a replacement may need to be ordered, or at least a record made of the part's use. All of this is possible but may require an inordinate amount of effort on the part of a number of staff; should the breakdown occur at night, when most staff are absent, the repair may need to be postponed till the next day. While there are clearly issues of control over stock involved here there is no *technical* reason why a PTW system should not be integrated with common stock-control and warehousing systems, so that the part required may be identified and logged out without undue delay. In practical terms too, most decisions authorising the use of components can probably be made at the level of the permit issuer.

ACTIVE DATABASE SYSTEM DEVELOPMENT

The focus of this half of the paper shifts from the general applicability of computers to the problems of PTW systems to a specific active database of incident reports drawn from the process industries which is currently under development by the Plant Engineering Group at Loughborough University. A particular feature of note is that where possible generic software already in wide distribution has been utilised: the system generates the queries it makes of the database in standard-format Structured Query Language (SQL) while interconnectivity with any standard database package is achieved by using Microsoft ODBC software. The present version of the incident database has been implemented using Microsoft Access; an earlier version was implemented in PARADOX. The use of popular commercial systems enables easy integration of the ADB by users with varying installed system configuartions.

Database Structure and Report Retrieval

As has been noted, a basic problem of current PTWs is that they tend to be uninformative: if an issuer is actually unfamiliar with a hazard no amount of prompting by a permit will serve to remind him; if he is aware of a hazard but does not think to apply his knowledge to a new situation the system will similarly fail. The fundamental goal of the ADB system is to tell workers about hazards of which they may previously have been unaware, or have forgotten, or have otherwise failed to consider, at a time when they can still do something about it.

What allows the ADB to function is the novel arrangement of data according to number of loosely related classification hierarchies rather than by a more traditional single alphabetical list of keywords. These hierarchies, respectively relating to Equipment, Operation, Chemicals, Cause and Consequence together provide a description of the key elements of an industrial incident and together represent a domain description of the chemical-industrial workplace. The significance of this is that it is less necessary to formulate an exact query in order to retrieve information than is required by conventional databases; data is 'loosely' sorted according to the information that appears in the accident report and thematically related information is then stored 'together' at an appropriate level of the various classification hierarchies. An exact query about a particular case will achieve an exact response; a less precise query, however, will still retrieve *appropriate* information, albeit along with some additional information which may not be *precisely* that which is required. In the context of PTW systems, however, a slightly 'fuzzy' view of the problem is actually required: the variety of *notionally distinct* hazards is actually quite limited; what is a matter of almost limitless variation is the way in which these basic hazards may occur in new situations[7].

Let us consider the organisation of data in the ADB. In figure 1 a section of the Causes hierarchy is displayed with the number of reports primarily indexed according to each node being shown in brackets. In this representation *Electrical equipment cause* is shown as a daughter of *Equipment cause* and a parent of both *Short circuit* and *Lack of earthing*. Reports are indexed according to the specificity of information appearing in the text: if an incident occurred due to equipment failure, but no specifics of what type of equipment, or which sort of failure are

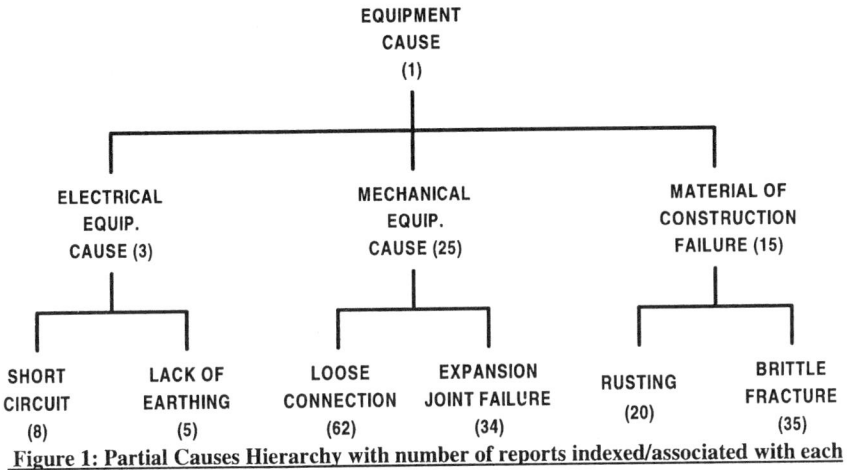

```
                        EQUIPMENT
                          CAUSE
                           (1)
          ┌─────────────────┼─────────────────┐
     ELECTRICAL         MECHANICAL        MATERIAL OF
       EQUIP.             EQUIP.         CONSTRUCTION
      CAUSE (3)         CAUSE (25)        FAILURE (15)
       ┌──┴──┐           ┌──┴──┐            ┌──┴──┐
     SHORT  LACK OF    LOOSE   EXPANSION  RUSTING  BRITTLE
    CIRCUIT EARTHING CONNECTION JOINT FAILURE     FRACTURE
      (8)    (5)      (62)      (34)       (20)    (35)
```

Figure 1: Partial Causes Hierarchy with number of reports indexed/associated with each node.

provided, the report is indexed at a high level of abstraction – further up the hierarchy. If more information is available and it is known, for example, that it was electrical equipment which failed, or that the cause for this failure was a short circuit, the report is indexed at a lower, more specific level. However, since a parent/child link is defined between *Electrical equipment cause* and *Short circuit*, retrieval of related-but-not-identical reports is simplified: if one is interested in cases involving electrical equipment failure then all the children of that node are of potential interest also. To a lesser – but calculable – extent the opposite is also true: interest in failures involving short circuits may make the more general electrical equipment failure of interest also[8].

From this starting point, specific information may be reached by a variety of routes: since every report is indexed under at least four distinct hierarchical heads (primarily these are *Equipment, Operation, Cause and Consequence* with *Chemicals* and *Chemical Properties* possible secondary heads) any report may be accessed by a number of routes. It should be noted, however, that each descriptive acts as a constraint upon the 'hits' achievable by reference to the hierarchy. For example, in Figure 2 each of the various nodes with the exception of *Toxic by skin contact* will result in the retrieval of the report shown; although toxicity by skin contact is a property of Benzene recognised by the system, this characteristic played no part in this incident and hence is excluded from being retrieved. Any combination of the *other* nodes shown in figure 2 would result in this report's retrieval, albeit *some* combinations of nodes might *also* result in the retrieval of a variable number of other more-or-less appropriate reports.

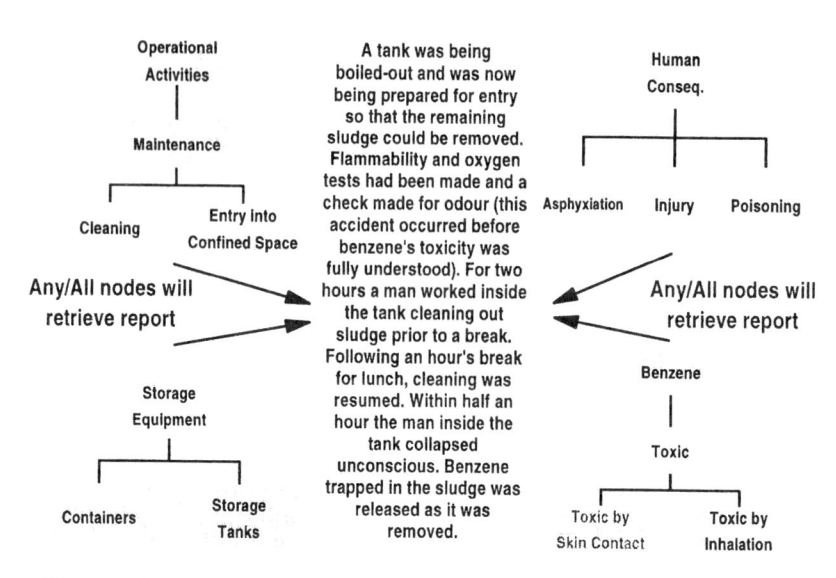

```
Operational                A tank was being              Human
Activities                 boiled-out and was now        Conseq.
    |                      being prepared for entry
    |                      so that the remaining
Maintenance                sludge could be removed.
    |                      Flammability and oxygen
    |                      tests had been made and a
            Entry into     check made for odour (this  Asphyxiation   Injury   Poisoning
Cleaning                   accident occurred before
            Confined Space benzene's toxicity was
Any/All nodes will         fully understood). For two    Any/All nodes will
retrieve report            hours a man worked inside     retrieve report
                           the tank cleaning out
                           sludge prior to a break.
                           Following an hour's break
                           for lunch, cleaning was       Benzene
Storage                    resumed. Within half an          |
Equipment                  hour the man inside the          |
    |                      tank collapsed               Toxic
    |                      unconscious. Benzene
Containers  Storage        trapped in the sludge was
            Tanks          released as it was          Toxic by      Toxic by
                           removed.                    Skin Contact  Inhalation
```

Figure 2: Diagram indicating various routes to retrieve a particular incident report.

Active Database Integration and Permits to Work.

In order for the ADB to function as desired it was necessary to develop a computer front-end in the form of a PTW. In format this is closely based on many existing PTW systems: it was felt that the existing paradigm, despite its various drawbacks, has achieved a fair degree of efficiency overall; in part also it was hoped a familiar format would reassure potential users of the system and minimise the time required for them to familiarise themselves with its use. A specific design goal was that the computerised PTW should be at least as quick and as easy to complete as existing systems. The intent is that the user should complete the permit in a conventional manner and the ADB should then make its determination of the relevance of particular incident reports based on the information so gained without any further action on the part of the user.

Figure 3 shows the first of a number of 'panels' which make up the PTW; the format throughout is the familiar one combining checklists and boxes requiring active completion. In addition to the design principles mentioned above, a 'modular' approach to form design was taken as a result of the HSE finding that firms use different PTW forms for different maintenance tasks. The division of the PTW into a number of panels is intended to improve the logical flow from section to section and facilitate PTW customisation: the choices made by users on the earlier, more general, panels determine which, if any, of the later specific panels are presented for completion. Firms may thus enjoy the advantages of permits tailored to individual and current needs without the drawbacks of bureaucratic proliferation.

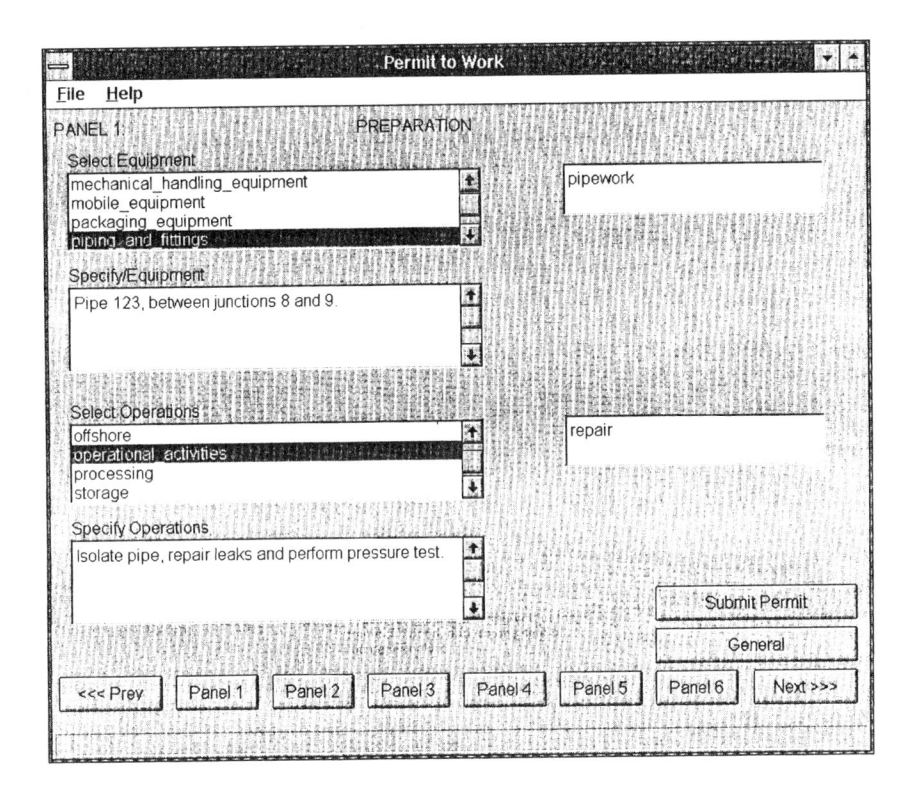

Figure 3: Screen dump of part of Computerised Permit-to-Work: 'Preparation for work: Equipment type and Operations type'.

Appropriateness and 'User Modelling'

One potential problem of the system as it has been presented so far is that it will tend to return too much information rather than too little. The reasons for this are several: as we have seen, the hierarchical organisation of data means that all associated reports will be returned in response to any given query; any which are not desired must explicitly be excluded in one fashion or another. This tendency is exacerbated by a feature of structured taxonomies which has been discovered by experience: the main advantage enjoyed by a hierarchical taxonomy over traditional alphabetical organisations is that proper data placement can quickly be determined from the structure of the taxonomy itself; the drawback is that some data falls outside the pattern prescribed by the taxonomy. Take, for example, an incident involving a solenoid-operated isolation valve which operated too slowly: is this a case involving electrical equipment failure, mechanical equipment failure or safety equipment failure? To what extent can slowness to operate be said to be a 'failure to operate'?

A response to this problem has been to permit incident reports to be indexed by means of multiple descriptives. This makes indexing of data much easier but only does so at the cost of greatly increasing the potential number of 'hits' – with a corresponding degradation of appropriateness of system response. In order to offset this difficulty and re-enhance appropriateness of information retrieval, the system has been developed to accommodate differing needs by different classes of system user, as well as individual needs of individual users. The initial logic guiding this development was the recognition that that plant operators will have different informational needs from those of maintenance personnel and that both will have different needs from those of plant designers. While the former categories of user are likely to require an 'active' database with incident reports being presented without any explicit request being made for them, the latter category probably does not. Further, plant designers are likely to need as detailed as possible an account of an incident to allow them to 'design-out' the possibility of recurrence; conversely, maintenance workers' requirements are accommodated better by a more abbreviated statement of what went wrong together with a statement of what lessons should have been learned. Similarly, plant operators are likely to require a *still more* compressed version of events since their acquaintance with an incident is likely to be made under time pressure as incidents develop.

The necessity for the system to respond differently to different classes of user was further developed by a desire to respond individually to individual users: to prevent information overload, it was decided that a 'cap' should be placed on the number of reports presented to any user at any given time. This cap would vary according to circumstances, and, most notably, in response to what individual users might be expected to know about their own particular situation. If, for example, the job to which the permit pertained was one which is very familiar then less information relating to its hazards need be retrieved; if the job is an unfamiliar one, the need for information is correspondingly greater.

These ends are achieved by maintaining a system-record of the details of past jobs undertaken by each user: if no record exists of a user having undertaken a particular job in the past, then, upon submission of a permit, the system will return the maximum number of appropriate reports, subject only to the arbitrarily imposed 'cap'. If, conversely, the system determines *this* user has performed *that* task several times recently, a much smaller selection of reports – or possibly none at all – will be returned. The intention is that the user not be overloaded with extraneous information and that he should not be bored by seeing again reports that are possibly wearily familiar. Between these extremes, where the system determines *some* degree of familiarity with a job, a less-than-maximum number of reports will be returned. Where possible those reports which are returned will be ones which have not been seen by this user before, or, if this is not possible, are at least those which were reviewed least recently. Figure 4, shows a typical system printout summarising the details entered on a PTW together with report details which were automatically generated.

PERMIT-TO-WORK SECTION A – GENERAL INFORMATION:
A1 - Enter Plant Name: Anytown Chemical Process Plant
A2 - Permit Valid From: 09.00 hrs 11/05/1998   **A3 - Permit Valid Till**: 17.00 hrs, 11/05/1998
A4 - Permit Number: 560   **A5 - Issuer Name**: John Smith   **A6 - Acceptor Name**: Joe Bloggs

SECTION B – PREPARATION
B1 - Equipment Selected: Pipework
B2 - Equipment Specified: Pipe 123, between junctions 8 and 9.
B3 - Operations Selected: Maintenance
B4 - Operations Specified: Isolate pipe, repair leaks and perform pressure test.
B5 - Hazards Identified: Fire and Explosion; Gas or Fumes; Heat; Trapped Pressure.
B6 - Chemicals Present: Liquefied Petroleum Gas (LPG)
B7 - Physical Isolation: Physical Isolation IS appropriate
B8 - Method of Isolation Used: Single/double isolation valve closed off and locked.
B9 - Fire Permit: A fire permit IS necessary.
B10 - Precautions Taken: Pipe 123 isolated by valve; valves locked and marked; protective hoods issued; foreman for area informed of work in progress.
B11 - Factories Act/ Chemical Works Regulations: DO NOT APPLY
B12 - Installed Radioactive Source: No installed radioactive source.
B13 - Electrical Isolation: Electrical Isolation NOT APPROPRIATE
B14 - Master Control Sheet: Master control sheet DOES NOT apply

SECTION C – OPERATIONS
C1 - Type of Job: Job is in NO SMOKING area and involves welding/and/or grinding
C2 - Physical Limits of Fire Permit: Area A sections 3 and 4
C3 - Duration of Fire Permit: 09.00 hrs 11/05/1998 till 17.00 hrs 11/05/1998
C4 - Factories Act/Chemical Works Declaration: Not entered/not appropriate
C5 - Precautions Declaration: All the precautions stated in section B have been put in place and checked by me personally.

SECTION D – SIGNING-OFF
D1 - Status of Permit: Permit current

1 of 3 relevant accident report(s) retrieved as follows: **REPORT NUMBER**: 11
CHEMICAL: LPG   **EQUIPMENT**: PIPELINE   **OPERATION**: HOT WORK
CAUSE: TESTING INADEQUATE   **CONSEQUENCE**: NEAR MISS
DESCRIPTION: Welding was being carried out – during shutdown – on a relief valve tailpipe. It was disconnected at both ends. Four hours later the atmosphere at the end furthest from the relief valve was tested with a combustible gas detector. The head of the detector was pushed as far down the tailpipe as it would go; no gas was detected and a work permit was issued. While the relief valve discharge flange was being ground a flash and a bang occurred at the other end of the tailpipe. Gas in the tailpipe 20m long and containing a number of bends had not been dispersed and had not been detected at the other end of the pipe. **LESSONS**: Before allowing welding or similar operations on a pipeline which has or could have contained flammable gas or liquid (1) sweep out the line with steam or nitrogen from end to end; (2) test at the point at which welding will be carried out. If necessary, a hole may have to be drilled into the pipeline.

Figure 4: Summary printout of Permit with associated incident report

CONCLUSION

To conclude, it is clear that the application of computers to the problems of PTW systems promises a variety of benefits ranging from added legibility to the proactive retrieval of information appropriate to the user's situation but which he has not actually asked for. Collectively, these enhancements promise to make PTWs far more effective; hopefully this should go a long way towards improving plant safety – as well as improving business efficiency – in the chemical-industrial workplace. It is unlikely, however, that any system will be able to render maintenance actually *safe*: the HSE has noted that in many cases workers have failed to do what their permits – correctly – told them they should, either considering the completion of a permit as an end in itself unrelated to actual work practice, or for some other reason. Computerising the process is unlikely to change this singularly human pattern of behaviour.

FURTHER WORK

Further work is also under way to improve the appropriateness of information retrieval from the database by applying case-based reasoning techniques to aid the data recovery across the domain hierarchies as well as up and down. Further work is planned to improve co-ordination between related permits as discussed above as well as in the area of automatically informing workers of the status of current maintenance jobs. The authors would welcome an approach from anyone who might be willing to try-out the system described in an operating plant.

ACKNOWLEDGEMENTS

This work is being carried out with the financial assistance of EPSRC grant (GR/K67502) and with the support of ICI and British Gas. P.W.H. Chung is supported by a British Gas-Royal Academy of Engineering Senior Research Fellowship.

REFERENCES

1. Health and Safety Executive – *Dangerous Maintenance*, p. 2, HMSO, London, (1987).
2. Health and Safety Executive – *Setting up and Running a Permit-to-Work System: How to do it*, p. 1, HMSO, London, (1996).
3. Scott S.J. – *Management of safety - Permit to Work Systems*, Major Hazards Onshore and Offshore, p. 171, IChemE, Rugby, (1992).
4. Riley D.B & Weyman A.K. – *A Survey of Permit-to-Work Systems and Forms in Small to Medium Sized Chemical Plant*, Major Hazards Onshore and Offshore II, p. 369, IChemE, Rugby, (1995).
5. Kletz T.A. – *What Went Wrong: Case Histories of Process Plant Disasters*, p. 2, Gulf Publishing Company, London, (1998), 4th Edition.
6. Cullen W.D. – *The Public Inquiry into the Piper Alpha Disaster*, HMSO, London, (1990).
7. P.W.H. Chung, R.E. Iliffe and M. Jefferson – *Integration of an Incident Database with Computer Tools*, Proceedings of IChemE Research Event 1998, 12148 - 12203.
8. P.W.H. Chung & M. Jefferson – *A Fuzzy Approach to Accessing Incident Databases*. To appear in International Journal of Applied Intelligence.

A QUALITATIVE APPROACH TO CRITICALITY IN THE ALLOCATION OF MAINTENANCE PRIORITIES TO MANUFACTURING PLANT

C.J. Atkinson
Burgoyne Consultants Ltd., Burgoyne House, Chantry Drive, Ilkley,
West Yorkshire. LS29 9HU

> Achieving an effective maintenance regime in a manufacturing plant is a strategic business issue. It is important to focus engineering priorities and effort onto the right areas in order to realise safety in operations and prevent loss. Quantitative techniques for doing this are available but they typically require specialised skills and are time consuming to apply - resources which are frequently unavailable to maintenance engineering departments. A simpler, qualitative approach to assessing criticality, which can be applied both to existing operating units and to new plants at the design stage, is discussed. It provides a simple ranking system and is easily applied by experienced engineering and production staff.
>
> Keywords: Maintenance, criticality, priority, strategic, qualitative approach, simpler.

INTRODUCTION

Why should plant and engineering managers be seriously interested in maintenance as a strategic business issue? One answer to this question is provided by the management consultant, Peter Drucker, who said "The first duty of business is to survive and the guiding principle of business economics is not the maximisation of profit it is the avoidance of loss". Maintenance is about the avoidance of loss in the widest sense of the word.

The purpose of a maintenance system is to improve the reliability and safety performance of a company's capital assets (plant, equipment and structures). The aim is to strengthen the financial performance of the business by maximising the availability of productive capacity and by minimising the risk of unplanned and undesired events. The latter could include injury to people, both employees and the general public; damage to the environment; loss of operating materials and damage to capital assets. Other targeted benefits of an effective maintenance system would be improved service to customers and the avoidance of adverse publicity.

A systematic programme of maintenance can also reduce maintenance costs particularly by avoiding the secondary damage which may result if equipment is allowed to run to failure before anything is done, by avoiding unnecessary maintenance activity and by designing out recurrent problems.

A recent survey by the Institution of Chemical Engineers revealed that two thirds of the companies responding had such a systematic approach. However, it is clear that resources of time and money in implementing such a programme can represent a significant barrier to progress. An objective basis for assessing maintenance frequencies and the risk of "over maintaining" are also common concerns. Of the companies which were operating a system, over half were unhappy with the resources required to sustain it. There are, of course, a variety of approaches to maintenance management and the issues of resources and effectiveness are significant to each of them. An approach which addresses these concerns will be helpful to managers seeking to secure the benefits of a systematic approach to maintenance.

In common with the management of other business activities, part of the function of maintenance management is to provide leadership. By this we mean not only ensuring that the right practices and procedures are followed in carrying out maintenance activities but also that the right maintenance activities are being given priority. The right activities are those which make maximum contribution to the avoidance of loss. Criticality ratings provide an aid to this function of leadership.

It is, of course, quite true that a comprehensive review of the maintenance requirements of a complex plant is a large task; any worthwhile system will require thought and mental application. However, there is more than one approach to analysing the maintenance needs of an operation. The area which this paper will focus upon is the identification, within defined plant boundaries, of critical systems and individual plant items which require priority in any maintenance programme. It develops a methodology for assessing the criticality of those items which is both easy to understand and simple to apply. The criticality ratings derived by the methodology can then be placed in rank order and used for the purpose of determining appropriate maintenance schedules.

The assessment of criticality may lead managers to consider other techniques such as Total Productive Maintenance (TPM) Wilmot (3), Nakajima (4) which focuses on a team approach generated by wider employee involvement and Reliability Centred Maintenance (RCM), Moubray (2) which is aimed at determining the maintenance requirements of physical assets in their present operating context: this is done using an approach similar to failure modes and effects analysis FMEA, Centre of Chemical Process Safety (5), British Standards (6). Managers will also need to consider the range of maintenance strategies available to them from breakdown maintenance to the various planned and preventive maintenance options. However, the primary step in leading the maintenance function to work on the right activities is that of taking an objective view of the relative importance (criticality) of the various elements of a manufacturing plant.

There are two principal methods which can be applied to assessing criticality. One is a rigorous, quantitative approach, typically fault tree analysis (FTA) Lees (1), Barlow et al (7), Shillon & Singh (8), within a Failure Mode Effects and Criticality Analysis (FMECA) (Refs. 1, 5, 6) methodology. This allows other factors such as human error to be taken into account. However, the analysis is both complex and time consuming. It is required for the analysis of major hazard plants but is less appropriate for dealing with the great number of lower risk plants to be found in the process industries. This approach also requires application by staff who are trained in reliability analysis.

A simpler, qualitative approach, still based upon FMECA (Refs. 1, 5, 6), can be applied to these lower risk plants. The methodology is considerably less complex than FTA (Refs. 1, 7 and 8) and, given appropriate training, lends itself to application by local analysis teams who know the plant well. This approach will sit well with a participative, team based company culture and allow the workload to be spread. Qualitative methods are also applicable as an initial screening process for the major hazard plants referred to above. They will allow identification of those elements which require quantitative assessment whilst allowing the simpler qualitative methodology to be applied to the other parts of the operation. Qualitative analysis of criticality can be applied at the process and mechanical design stages of a new project to sharpen the focus on critical systems and plant items at the earliest possible point: key maintenance or design out issues can then be tackled before major costs have been incurred. The principles of assessing criticality at the design stage are the same as those for existing plants and this paper does not treat them separately. This latter qualitative approach for existing plants will now be developed in more detail.

THE QUALITATIVE APPROACH

The qualitative approach to allow criticality to be assessed is outlined in flow chart format in Figure 1. The result is a numerically weighted ranking so that it is possible to prioritise items of equipment in terms of criticality and, subsequently, of maintenance needs. It is also used to define the need for more detailed quantitative assessments.

It allows the benefits of a systematic approach to maintenance to be gained by using a relatively simple methodology. This is easy to apply and can be used by plant staff with no formal training in reliability analysis. The discipline of examining maintenance requirements in the manner described will prompt the investigation of problem areas and stimulate continuous performance improvement within the operation.

The steps in the methodology are set out below:

Define Plant Boundaries for the Study

The plant BOUNDARIES for the study should be defined so that everyone is aware of the extent of the study, which items/processes are to be included and which are not. It is important to ensure that adequate consideration is given to the effect of plant failures on common systems and common system failures on the plant when establishing the boundaries for the study. Similarly, the effect that an incident on the plant could have on other production units will need to be considered.

Define Plant Systems Within the Defined Boundary

The next step is to draw up a block diagram for all the systems of the plant within the defined study Boundaries. At this stage, if the SYSTEM BOUNDARIES are clear, identify any systems that are not critical. These would be systems which, should they fail would not produce a hazard to the public or plant staff or result in a high financial loss. Delete these systems from the study.

Define the Objectives of the Study

The purpose and the objectives of the study should be defined and understood by all the staff involved so that every one is pulling in the right direction. Objectives might take the form of:-

FIG. 1 - FLOW CHART OF THE METHODOLOGY

```
1. Define Plant Boundaries for the Study
            ↓
2. Define the Plant Systems within the Boundaries
            ↓
3. Define the Objectives
            ↓
4. Identify Critical Equipment
            ↓
5. Select First / Next Item of Critical Equipment for Study and Gather Data on the
   design, performance and history of the equipment
            ↓
6. Identify all Potential Causes of Failure (Failures Modes).
   If none, then return to Item 5
            ↓
7. Identify Potential Consequences of the Failure
            ↓
8. Is there redundancy or other systems which will reduce the consequences
            ↓
9. Determine Criticality Rating
            ↓
10. Cross check results versus 'Statutory' & Manufacturers Requirements
    (Then return to item 5 until all equipment has been reviewed)
```

"To identify all critical items of equipment which can have an impact on production, safety and environmental concerns."

"To assess the criticality of such items, evaluate the adequacy of existing maintenance policy and practice and identify the improvements which are required to develop a cost effective maintenance system."

Identify Critical Equipment Within the Defined Systems

This is a first coarse assessment to identify equipment which, if it failed, would cause any of the following :-

- Harm to the public
- Harm to people within the site
- Damage to the environment
- Asset damage
- Lost production

A line diagram would typically be used to go through the process line by line and item by item. This step is not the same as an in depth Hazard and Operability Study (HAZOP), Kletz (9), CIHSC (10). If failure of the item could cause a problem it is simply listed for more thorough consideration in subsequent steps.

The potential for failure should be assessed by a team of people consisting, as a minimum, of the Plant Manager, Plant Engineer and a Safety Engineer. In the real world, the line diagrams may not be up to date and include all modifications. The key selection criterion for team members is that they collectively represent a thorough knowledge of the plant and how it operates in practice.

Select An Item of Equipment for Study

The assessment team should now select an item of equipment which they consider to be of high priority from their initial assessment. They should then gather data on the design, performance and history of the equipment so as to identify the purpose of the equipment, its design conditions and any previous history such as inspection reports, failures and manufacturers' recommendations. This latter activity has two principal benefits:

- it creates an up to date record that can be added to,
- it helps to identify all potential failures/hazards.

It is therefore important to make the best use of all available information. This step also provides valuable background data if excessive maintenance costs and downtime are identified when the criticality rating itself is being calculated.

Identify All Potential Causes of Failure

The assessing team should identify and list all potential failures. If no possible failures are identified the next item can be considered.

Potential failures can be identified by using one or more of the following methods:-
1. The past history of the equipment
2. The experience of people using the same or similar equipment
3. Brainstorming session
4. What if?
5. Hazard and operability study (HAZOP) approach (Refs 9,10).

The method used will obviously depend on the process, the potential for harm and the potential financial loss to the company of a failure. A What if? or brainstorm should be adequate providing that the study team consists of people who have a full understanding of the process, the basis of safety and the engineering aspects. This is a detailed assessment and the Assessment Team must include people who are closely involved with day to day operations; a minimum would be the Plant Manager, an Operating Supervisor, the Plant Engineer and a Maintenance Supervisor. For a more hazardous plant a HAZOP would be more appropriate. Advice should be sought from a Safety Professional on the type of study to be applied. This advice might be obtained internally or externally depending upon the resources available to the business.

The following points should be considered:-
1. Is the vessel or equipment designed to an approved standard or code of practice. If it is, then the chance of failure must be low.
2. Can the equipment be pressured above its design conditions by :-
    a) overfill
    b) runaway reaction
    c) utilities failure
    d) external circumstances such as fire
    e) addition of materials other than those specified
    f) being boxed in
    g) operator error
    h) maintenance error on pressure relief/control systems
3. Are there controls to prevent overpressure, are they adequate and are they maintained?
4. Are all the materials of construction compatible with all the process materials and products?
5. Is it possible to form an explosive mixture in a vessel or a system and if it is are the preventive controls adequate e.g. is there a nitrogen blanket and is the nitrogen supply secure, have all means of ignition been eliminated?
6. For the control of a reaction is the cooling water supply/coolant supply secure?
7. Are the control systems adequate for all potential deviations from design, are they installed and working as designed/intended? Are they understood by maintenance and process? Is there an adequate and meaningful test procedure?
8. Is there a modification procedure and has it been checked that no unauthorised modifications have been carried out?
9. Are the operating instructions and maintenance instructions written and understood?
10. Have staff been given adequate training to be able to identify deviations from the normal and take corrective action?
11. Is the electrical equipment to the required standard and is there a control?
12. Is all maintenance work covered by a permit to work and are all staff and contractors aware of this requirement?

13. Is a hot work permit required for any cutting or welding or the use of sparking equipment in a hazardous area?
14. Is there a procedure to deal quickly with a release of harmful material and have staff been trained in this procedure?

This list is for guidance only and should not be considered comprehensive. It covers some of the potential events that could result in a failure.

Identify Potential Consequences of the Failure

Having identified realistic potential causes for failure of an individual item, the team must now assess the effect of such a failure on the system under examination. These consequences might include fire or explosion, release of harmful materials, material losses, lost production, destruction of assets, creation of a dangerous situation such as a runaway reaction, etc.

The assessment of consequences will be primarily judgemental and hence qualitative. There will be occasions when the team will benefit from commissioning a more detailed quantitative assessment of consequences. Whilst such assessments are outside the scope of this methodology, the variables to be considered are:

1. The inventory that may be released
2. The pressure of the release
3. Whether such a release will be liquid or gas or both
4. The potential effect, toxicity, fire, explosion
5. An estimate of the potential distance of the effect
6. An estimate of the toxic level/discomfort to the public
7. An estimate of potential injury to the public and the numbers that could be affected
8. The potential release to water courses and the environmental consequences of absorption into the ground

Is There Redundancy Or Other Systems Which Will Reduce the Consequences

Protective systems such as alarms, trips, relief valves and non-return valves are installed to prevent hazards from occurring. When there is more than one independent method of performing the protective function, the protective system is said to contain redundancy. An example of such a system would be one where one of two trips with two independent measurement and trip channels will shut down a process safely. Such a system would be more likely to respond to a demand placed upon it than a single channel system. The consequences of the failure of a particular item might therefore be reduced. In considering this reduction of consequences, it is important to consider whether common mode failure such as a failure of instrument air or electrical power could affect the performance of both of the protective systems - are they truly independent?

Determine Criticality Rating

The criticality rating of an item allows it to be placed in rank order when considering the frequency with which it requires to be checked. It forms the basis of the initial test and inspection schedules for that item. This will, of course, be subject to review as the plant history builds up. It is important to recognise that criticality ratings are part of a ranking process. They do not represent absolute frequency values for maintenance schedules but are used to assess the frequency of one item relative to others. This is a qualitative, not an absolute, methodology. Consistent application across similar areas is the key to obtaining meaningful criticality ratings. The work done in identifying the potential consequences of failure will also assist the assessment team in determining the appropriate type of maintenance activity which is required by an item as well as its relative importance.

Criticality rating also serves as a means of prioritising corrective actions identified by the study such as improving means of detection of system failure, minimising the effects on common systems and tackling root causes of excessive downtime.

Criticality must be rated by the assessment team. The ratings are assessed for seven areas which may be affected by the failure of the item. These are:-

Injury to the Public
Damage to the Environment
Injury to people within the factory perimeter
Loss of assets / profit
Effect on common systems
Excessive downtime and maintenance costs
Reliability of detection and control systems

This rating system is targeted at operations where the consequences of failure are potentially significant hazards to people and the environment. It is also important to consider the impact of a failure on the public image of a Company. The perception of the consumer can have a major impact on, say, a food processor. A history of environmental incidents can damage any manufacturer whose clients do not wish to have their image tarnished by association. Perceptions are not always in proportion to reality and should never be underestimated.

The following subsections develop a scoring system based on three rating levels. These are set at 10, 5 and 1. These numbers are intended to serve as a guide. There is no reason in principle why a rating could not be scored as, say 7. However, such nuances are not likely to alter the results obtained in a significant way.

Injury to the Public. The following empirical ratings are probably appropriate :-

Potential effect high and many casualties, people feel sick or suffer serious discomfort	10
Minor irritation to the Public	5
May be aware of a release but suffer no ill effects	1

Damage to the company's public image should also be considered here particularly if this is likely to affect customer attitudes.

Injury to the Environment. Consideration should be given to a failure that could affect the environment such that it would contravene the pollution regulations and cause serious damage to water systems, vegetation, animal life, clean and effluent systems such as storm water drains. Other areas to consider are the potential for flammable liquids and vapours getting into drains and common systems such that they can reach a source of ignition. The potential effects should be calculated or given an empirical rating as follows:-

Major event, clearly perceptible to the public, with for example:

(a) Potential for serious, lasting damage to the environment.
or
b) The possibility that flammable vapours can reach a source of ignition.   10

Minor event, still perceptible to the public, with for example:

Potential for short term damage to the environment but
flammable vapours cannot reach a source of ignition.   5

Minimal effects, containment within the factory and not perceptible to the public.   1

Public perception would relate to either direct observation of the event and/or indirect awareness through reports in the media.

Injury to People within the Factory Perimeter. Consider the potential for injury to people within the factory perimeter fence. This could be fatalities from fire, explosion, asphyxiation or being exposed to toxic materials or various degrees of injury from these effects. Injury because of poor access to the equipment for maintenance or operational reasons should be considered. The potential effects should be given an empirical rating as follows:-

Death or very serious injury/illness   10

Injuries/illness of a less serious nature such that recovery is possible
within a few weeks   5

Minor injuries/illness   1

Loss of Assets/Profit. This will have to be assessed on a plant to plant basis. For the purpose of this exercise the following empirical ratings have been assumed:-

Loss of profits or assets in excess of £100,000   10

Loss of profits or assets in excess of £10,000 but
less than £100,000   5

Loss of profits or assets in excess of £1,000 but
less than £10,000   1

Common Systems. The effects of a failure of systems common to several plants such as services, vacuum, ventilation and liquid and gaseous effluents should be considered. Failure of common systems might include a failure of supply from the common system which results in a hazard on the plant or a failure on the plant which results in damage to the common system and/or a knock on effect to other plants or processes. The potential effects of a cocktail of materials being created in the common system should be considered. If the potential failure could affect the manufacture of other products in the same equipment, this should be considered here. For the purpose of rating the following has been assumed:-

Effects of a service failure could result in a hazardous situation.
An explosive mixture could form in the common system.
A service failure could seriously affect other production units. 10

A service failure would have a minor effect in causing hazards
on the operating units and it is unlikely that an explosive mixture
could form in the common systems.
A service failure would have minor effects on other plants. 5

A service failure would not cause hazard and an explosive mixture
cannot form. There is no effect on other plants. 1

Present Maintenance Costs and Downtime are Excessive. It may be that some equipment is failing on a regular basis because it is not being operated as intended, there is a lack of preventive maintenance, it is not compatible with process materials/conditions or because of its age. Alternatively, the price of a low failure rate may be high maintenance expenditure. In either set of circumstances, whether or not failure could produce a hazard, an investigation should be carried out to determine the cause of the failures/cost of maintenance and a remedy suggested. This may involve replacement especially if the cost of maintenance is greater than the cost of replacement. When rating this element of criticality it is important to take into account not only failure frequencies and the potential consequences of these but also any high cost of preventive measures where failure frequencies are low. In other words, loss of profit arising from high maintenance costs as well as losses from the consequences of high breakdown frequencies should be considered when allocating the ratings defined below:-

Breakdown is occurring more than twice per year and can lead to
a hazardous situation and/or major loss of profit 10

Breakdown is occurring at least once per year and results in
loss of profit but not a hazardous situation 5

Breakdown occurs less than once per year 1

Reliability of Detection/Control Systems. Failures can be prevented if there are means of detection such as regular inspections by qualified staff, control systems to prevent a dangerous failure occurring or operator observation/intervention. The following ratings have been given as follows:-

No means of control or detection 10

Means of control are in place but no schedule to test that it is effective.

Operators not instructed in potential deviations and control 5

Means of control and a meaningful schedule of tests are in place.
Operators are trained to detect and rectify deviations 1

Calculation of Overall Criticality Rating. The overall criticality rating is calculated by multiplying together the ratings scored in each of the seven sections described above. The highest rating is therefore 10^7 and the lowest rating is 1^7. The overall criticality ratings can then be ranked and used for prioritising corrective action or redesign, for scheduling inspections and tests to detect hidden and potential failures, and for scheduling restoration tasks.

Criticality Ranking. A suggested relationship between Criticality Rating and Ranking is set out below.

Criticality Rating		Ranking
Criticality Rating	10^7 to 10^6	Very high
" "	10^6 to 10^5	High
" "	10^5 to 10^4	Medium
" "	10^4 to 10^3	Low
" "	10^3 to 10^2	Very Low
" "	< 100	Very Low

The criticality ratings will always be a clear guide to the priority which one item should receive in comparison to others but engineering management will still have to apply informed judgement according to the situation they are managing when determining the frequency of maintenance activities.

It is important to recognise that modern maintenance systems may call for function checks, visual inspection and incipient fault reporting on a daily or a shift basis. However, like all management systems, maintenance will require formal review activities such as proof testing of safety systems, specialist condition monitoring, statutory inspection, non destructive testing and item replacement routines, it is these formal review activities which are prioritised by criticality ranking.

Items which are identified as having a high criticality rating during a study at the design stage of a new project should be reviewed to see if it is possible to reduce criticality by using an alternative approach.

Cross Check Results Versus 'Statutory' Requirements

Some of the equipment will have had inspection frequencies assigned due to legislative requirements (e.g. Pressure Systems Regulations, IEE Regulations. Other items may be inspected under company procedures such as BS EN ISO 9000 Manufacturers may recommend maintenance schedules for their equipment. It should be checked whether these established activities call for more frequent inspection than required on the basis of the qualitative assessment described above. That dictating the more frequent inspection should be adopted. Statutory requirements will always be the minimum requirement.

The Final Steps

The results should be formally documented in an FMECA record table showing the outcome of the assessment for all items. A specimen table, for some items from a batch organic chemicals plant, is given as Table 1 in Appendix 1.

It is important, as stated above, to choose a maintenance policy appropriate to the plant under consideration. This paper does not set out to provide guidance on the selection of policy. However, in general, it is desirable that maintenance tasks be planned rather than undertaken ad hoc. It is, therefore, necessary to draw up a maintenance schedule table with a system to call items of equipment immediately in advance of their inspection or test and to register that the test has been completed and the results of the test/inspection. Resources must be available to carry out the test/inspection and to repair any defects. Instructions must be written giving the preparation procedure for the test/inspection and the method of carrying out the inspection/test and staff trained in these procedures.

It is important that a competent person monitors the schedule to ensure that it is being operated as intended and to identify any failings so that they can be rectified. As a history is built up it may, on evidence, be prudent to increase or decrease the inspection/testing frequency of a piece of equipment. Change must only take place on the formal authorisation of a competent person designated by the operating company. The maintenance system must be kept under review with a view to continuous improvement in performance.

Having established the criticality ratings of items of plant and equipment, these ratings need to be applied to scheduling within the appropriate maintenance policy. There is no universally applicable maintenance policy and there are considerable differences in the approaches to maintenance which are applied to process plant. This is of course a subject in its own right.

CONCLUSIONS

The Qualitative Assessment of Criticality presented in this training package will provide focus on those items which are critical for Health, Safety, Environmental or commercial reasons and the analysis carried out by the assessment team will assist in deciding which maintenance approach is most appropriate.

REFERENCES

1. Lees F.P. Loss Prevention in the Process Industries 2nd Edition (1996) Butterworth Heinemann.

2. Moubray J., Reliability Centred Maintenance II. Butterworth Heinemann.(1991)

3. Willmot P. Total productive Maintenance - The Western Way.

4. Nakajima S., Introduction to Total Productive Maintenance Productivity Europe Ltd.

5. Centre for Chemical Process Safety, Hazard Evaluation Guidelines (1985/1, 1992/9).

6. B.S. 5760 Reliability of Systems, Equipment and Components, Part 5: 1991 Guide to Failure Modes, Effects and Criticality Analysis (FMEA) and FMECA).

7. Barlow R.E., Fussell, J.B. and Sinpurwalle, N.D. (eds) (1975). Reliability and Fault Tree Analysis (Philadelphia, P.A.: Soc. for Ind. and Appl. Maths).

8. Shillon, B.S. and Singh, C., (1981) Engineering Reliability. New Techniques and Applications (New York: Wiley Interscience).

9. Kletz T.A. (1992) Hazop and Hazan 3rd ed. (Rugby : Institution of Chemical Engineers).

10. The CIA Guide to Hazard and Operability Studies (CISHC, 1977).

TABLE 1

COMPANY:
PLANT:

COMPONENT	PURPOSE	FAILURE MODE	POTENTIAL CAUSE	POTENTIAL CONSEQUENCE	DETECTION/ CONTROLS	CRITICALITY RATING/ RANKING	PRESENT MAINTENANCE SCHEDULE	NEW MAINTENANCE SCHEDULE
Reaction Vessel R1	To react feedstocks 1 and 2 and produce a range of products	Vessel failure. Flange, manhole or agitator gasket	Design error corrosion, erosion, pressure above design. Gaskets, maintenance error.	Large spillage of toxic material. Small leak of toxic material	Inspection for pressure regulations. Flow ratio control/trip. High pressure trip. Low temperature trip.	12,500 Medium	For pressure vessel regulations.	Add an intermediate visual inspection every 6 months.
In line mixer M1	Mixing of reactants in R1 by external circulation.	Seal damage. Bearings fail. Motor fails. Flange fails.	Wrong assembly. Wear and tear. Not suitable for duty.	Small spillage. Loss of production.	None.	250,000 High	None.	Three monthly. Known to fail often. Investigation required
Flow Controllers FIC1, FIC2 and Ratio Controller RCI	Measures flow of feedstocks to reactor & controls the ratio between them.	Fails to safe. One feed valve could fail open but other should close.	Power failure. Element failure Blockage	Fail to danger. In worst case pressure would activate BD/RV and discharge to atmosphere.	Alarm. Valve closes, pump trips on ratio deviation. Valves close and pumps trip on pressure alarm.	12,500 Medium	None	Six monthly.
Non-return valves NRV1 and NRV2 protecting site nitrogen and air systems.	Nitrogen for inert purging. Air for instruments / process.	Fail to danger. Hydro-carbon into site systems.	Failure of NRV flap.	Explosion / Fire.	None.	1,250 Low	None	Annual

EUROPEAN STATE-OF-THE-ART RESEARCH: INTEGRATING TECHNICAL AND MANAGEMENT/ORGANISATIONAL FACTORS IN MAJOR HAZARD RISK ASSESSMENT

Martin Anderson[1], Health and Safety Laboratory[2]
Broad Lane, Sheffield, S3 7HQ

> This paper describes the development and current status of a research project with the objective of providing a model for integrating a broad overview of the organisational and management system of a major hazard site with the initiating events that lead to accidents. Following a desktop trial, this model and the integrated methodology have recently been applied with considerable success at a major hazard installation. This paper outlines the lessons learnt from this exercise and suggests some modifications and considerations prior to a third trial.
>
> **Keywords**: management, organisational, technical, risk assessment, QRA, audit

BACKGROUND AND RATIONALE

The Health and Safety Executive (HSE) provides advice to Local Planning Authorities concerning land-use planning proposals for the siting of new major hazard plant and for the development of housing etc. in the vicinity of existing major hazard installations. The Health and Safety Laboratory (HSL) and HSE have collaborated over the past 10 years in the development of a range of computerised Quantified Risk Assessment (QRA) tools, collectively known as RISKAT[3], to inform such land-use planning advice.

This suite of QRA tools draw heavily on generic failure rate data. In this context, generic data is derived mainly from historical and some theoretical data. These generic failure rates are incorporated into the numerical estimation of risk from a hypothetical release of hazardous materials from a particular installation. Generic failure rates, being derived from historical and theoretical data, include all causes of failure and should thus reflect 'average' conditions or standards.

Even identical plants can, however, be operated, maintained and managed to varying standards. Hurst (2) describes how the differences in safety performance between technically similar installations may be in the region of three orders of magnitude. Some plants thus may warrant a failure rate different from the generic value to reflect site specific conditions. There was, therefore, interest in these generic failure rates and the extent to which they included management, organisational and human factors in addition to engineering and hardware failures.

---
[1] Now at Human Reliability Associates, 1 School House, Higher Lane, Dalton, Wigan, WN8 7RP
[2] An agency of the UK Health and Safety Executive
[3] Risk Assessment Tool, for example, see Hurst et al. (1)

This interest led to research to examine the underlying causes of loss of containment incidents for the main items of equipment on chemical plant (pipework, vessels and hoses and couplings), with the aim of assessing the contribution of human errors, design problems and maintenance errors to generic failure rates (for example, see Hurst et al., 3). Assuming that the quality of safety management will determine standards of plant design, operation and maintenance etc., the inclusion of these factors in QRA will thus more explicitly consider the standards of safety management. In his discussion of what are perceived to be the strengths and weaknesses of QRA, Hurst (2) emphasises the importance of any such assessment being as transparent as possible. Making more explicit the inclusion of organisational and management factors in QRA is a consideration in the proposed 'agenda for risk assessment' in this recent book.

The above research illustrated how previous loss of containment incidents had occurred and identified practical actions to prevent future incidents. Following this work, the HSE developed an audit technique known as STATAS (Structured Audit Technique for the Assessment of Safety Management Systems). This technique has subsequently been included within a set of tools collectively known as 'The FOD[1] Guide to the Inspection of Health and Safety Management' after being modified within the management system framework described in a HSE (4) document 'Successful Health and Safety Management': HS(G)65.

A subsequent research project for the CEC Environment programme 1992-1994 (summarised in Hurst et al., 5) explored a modification of risk methodology whereby an evaluation of the quality of management was used to modify the generic failure rates of QRA. This methodology had been under development since the early 1980s, stimulated by questions from the process industry who wished to have the quality of their safety management accounted for in risk evaluation, and subsequently by the Regulator requiring tools to investigate a site specific Safety Management System (SMS).

This project involved the application of a safety attitude questionnaire and a process SMS audit tool called PRIMA (Process Risk Management Audit) at six major hazard sites in four European countries. The tools provide quantitative measures of safety attitudes and SMS performance respectively. The work compared these quantitative measures with accident performance data for the six sites. The approach allowed the uncertainty in risk estimates due to variation in safety performance at different sites to be explicitly considered within risk-based decision making. The approach adopted in this research was successful and received much support from the Regulator and industry.

In 1995, a new European research project commenced, sponsored by the European Commission, the Dutch Ministry of Social Affairs and Employment and the Health and Safety Executive. This research, 'Development of an Integrated Technical and Management Risk Control Monitoring Methodology for Managing and Quantifying on-Site and Off-Site Risks', has come to be known as 'I-Risk'. This project brings together a multi-disciplinary team of specialists in Europe on two aspects of risk control in major hazards – organisation and management on one hand and technically orientated approaches on the other. The project participants are:

---

[1] Field Operations Division of the HSE

- The Division of Labour Circumstances of SZW (the Netherlands Ministry of Social Affairs and Employment), (project co-ordinator);
- SAVE, Netherlands (consultancy);
- Technische Universiteit, Delft, Netherlands (Delft University of Technology);
- Four Elements Limited (consultancy);
- NCSR Demokritos, Greece (National Centre for Scientific Research);
- RIVM, Netherlands (National Institute for Health and Environment);
- Health and Safety Laboratory (HSL).

This team includes consultants, academics, industry and regulators. This project follows on from the previous EC project, which was co-ordinated by HSL and involved several of the above organisations.

Currently, the relationships between different aspects of the control of risks to people and the environment from major accident hazards are poorly modelled and assessed. Methods of evaluating these risks (e.g. Quantified Risk Assessment), management quality (e.g. audit methods), safety culture (e.g. attitude survey) and organisational structures (by reference to organigrams, responsibilities, manning, emergency plans) are not integrated into a single approach which would enhance an examination and understanding of their inter-relationships.

The requirements for Seveso sites to produce Safety Reports have never clearly indicated how the integration of organisation, management and technical systems in controlling the risks should be considered. This is also true of the proposed COMAH revision of the Seveso Directive. This poses potential problems for assessing the overall sociotechnical system of a major hazard site in its control of risks relating to health, safety and for protection of the environment, and for companies in setting priorities for improvement.

PROJECT OBJECTIVES

Thus, the overall objective of this research is to provide a model for integrating the methods of control of risks. The emphasis is on the chemical and petrochemical industry, focussing on major hazard installations at all stages in the installation life cycle. This will give a basis for controlling the interactions between failures at different levels in the sociotechnical system, as have been repeatedly observed in accidents.

The sub-objectives are:

1. Development of an integrated technical and management risk control and risk monitoring model including both on-site and off-site risks and their variation over time. The model will be developed in the context of considering the requirements of the proposed COMAH Directive.
2. Development of an Integrated Quantitative Risk Assessment (I-QRA) method, based on the integrated model from (1), which takes account of management as well as technical design as an integrated whole and which will produce measures both of the risk level and the rate with which this is expected to change over time. The identification of risk

reduction strategies will then be focussed on the system as a whole, not primarily on hardware, and on a more realistic representation of risk as something which changes over time rather than as a time-based average.
3. Development of management 'corrosion' probes to assist in monitoring the state of the risk management system over time and in setting priorities for improvement.
4. Testing and application of the Integrated Model and I-QRA.

Never before in the chemical/petrochemical industry has the technical risk control model been fully developed in parallel to the management risk control model. Previously, the two have evolved fairly independently, coming together only at certain points. The proposed new model will enable an evaluation of questions about the effects of organisation on risks, which previously may not have been addressed, such as 'what will be the effects of contracting out maintenance compared to keeping it in-house?'

Time varying risk profiles are now viable, giving the opportunity to model how variation in management, plant states etc. over time can affect risk. Questions can be posed such as whether prescriptive versus goal setting regulation have different 'decay rates'. The integrated model would generate key indicators of risk management health equivalent to corrosion monitoring of hardware. It would enable management to target risk control and mitigation measures most effectively to avoid accidents 'waiting to happen' and to respond appropriately in the event of the unforeseen.

THE RISK MODEL

Whereas the previous project linked management weighting factors at a high aggregation level to assessments of loss of containment risk, the current project aims to link management factors to the parameters that govern unavailability of safety related equipment and to the probabilities of human error and hardware failure which appear as initiating events or as failures at lower levels in fault trees. By developing the links between a broad overview of the organisational and management system and the initiating events that lead to accidents, the research will provide a system for directly examining the effect of the management system on risk levels. The I-Risk model consists of two sub-models (technical and management), and an interface between them.

Technical Sub-Model

As the research is concerned with major hazards, within the context of the Seveso Directive, it focuses on loss of containment (of hazardous material) events. These events are represented generically by Master Logic Diagrams, which share the basic features of qualitative fault tree analysis.

The technical model is thus used to identify the plant-specific initiating events that can lead to the release of a hazardous substance; the controls in place to prevent such a release (engineered systems, procedures and actions) and any factors that may mitigate such a release. The Master Logic Diagrams produce a set of hardware failures, hardware and system unavailabilities, initiating events and human errors/recoveries. These features are said to be governed by a set of mathematical parameters which include:

- Frequency of initiating events (both hardware and human errors);
- Time interval between tests;
- The frequency of routine maintenance.

Management Sub-Model

These parameters (or more correctly, the generic data for each of the parameters) are modified by plant-specific management influences which are considered to determine the quality of four main categories of activities:

- Design;
- Construction;
- Operations;
- Maintenance.

These plant-specific management influences are themselves determined by the quality of the systems which specify and deliver the resources and controls to these activities. In the I-Risk model, these are known as the delivery systems. Eight such delivery systems have been developed within this project:

- The availability of personnel with responsibility and authority to carry out the work;
- The competence of these personnel;
- Their commitment and motivation to carry out the work well and safely;
- The resolution of conflicting pressures and demands antagonistic to safety;
- The internal communication and co-ordination of people on the activities;
- The plans, procedures, rules and methods which specify the required level of safety or accepted way to carry out the work;
- The hardware, controls, plant interface etc. on which the activity is carried out;
- The delivery of correct spares for repairs and equipment.

These eight delivery systems are in turn influenced by the system which defines and modifies the safety management system.

The delivery systems have been modelled using a variation of the 'control and feedback learning loop' methodology developed within the previous EC project described above, and taking into account recent research by TU Delft (Hale et al. 6). This control and feedback loop model provides a conceptual basis for characterising a high performance safety management system. In this model, high performance stems from a focus on objectives which are faithfully translated into delivery systems within a closed loop, self-correcting framework. An example of the loop structure for one of the delivery systems is presented in Figure 1. The audit of management quality assesses all of the boxes in this figure and the loops that connect them. The research has identified which aspects of each delivery system are relevant for a particular parameter in the technical model (presented in Bellamy, 7). The delivery systems thus structure the audit and assessment of management influences.

The combination of these influences for each of the aspects to be assessed results in over 8000 potential assessment points for the audit, clearly more than can be addressed in a typical audit. Therefore, the influences are combined with respect to the common modes that exist in the particular company under examination.

The data points are assessed on a scale by the auditor with the help of anchor points at each pole of the scale; these points being textual descriptions of the characteristics of the management control and feedback loops.

The Technical-Management Interface

The interface effectively consists of a table of the parameters, a listing of base events and an audit preparation table. Pre-audit preparation partly consists of determining who (in the company) should be asked what questions and structured documentation enables the recording of information generated in the audit interviews.

Each of the technical parameters has a series of output components, which, when combined, link to the base events in the technical model. For example, the parameter Time to Repair (T_R) comprises of 17 output components grouped under the four headings:

- waiting time prior to repair;
- accessing and replacing time;
- time to do the repair;
- time for return to service.

The management influence is assessed for each of the output components for all of the parameters. The parameter in the technical model is then modified according to this influence. In this way, a plant specific technical model is linked to the plant specific management system by concentrating only on major hazard events and only on those aspects of the technical and management system which are relevant to such events.

MANAGEMENT 'CORROSION'

One of the objectives of the I-Risk project is to consider how the influence of management can lead to a sustained deterioration in safety over time, which if unchecked by regular review and revision processes will eventually lead to an accident. Such accidents generally provoke interventions aimed at restoring an acceptable level of safety, but ideally such intervention would come before, not after, an accident. For this purpose, it is necessary to develop means of identifying when an organisation has become 'an accident waiting to happen', or is heading in that direction.

In the previous project, a measure of the influence of the SMS on risk was achieved by mapping the relevant subsystems onto a conventional control and feedback loop model and developing an audit method for assessing the completeness and performance of the loop by considering each of the elements (comprising both the boxes and the links between them). Depending upon the strengths or the weaknesses of the elements, an overall 'management factor' was determined for the system in question. However, when applied to a plant, the approach does not explicitly distinguish between static and dynamic behaviour, between what the current risk is (as influenced by management) and whether safety is set to improve or deteriorate in the foreseeable future. Providing a methodology for the determination of this future risk is an important aspect of the current research.

The dynamic aspects of the model thus relate to future states, given the current state. These future states are connected to the feedback, monitoring, analysis and revision side of the loop; and the first, second and third order learning loops (see Figure 1). The frequencies of these feedback, monitoring and analysis activities are known as Management Influence Monitors (MIMs) or corrosion monitors. These frequencies, combined with their relevance and quality, are the things which can have an effect on the technical parameters in the future.

PILOT TRIAL OF THE METHODOLOGY

The first test of the methodology was conducted by means of a desktop audit on a major hazard operation in which some of the project team had prior research and production experience. All relevant documentation for this operation was obtained from the company including aspects of the CIMAH Safety Report, procedures, piping and instrumentation diagrams, etc. The technical model adopted in this desktop audit (reported in Papazoglou and Aneziris, 8) expanded upon previous research conducted by the HSE on this operation (Anderson, 9).

Based on the knowledge and experience of the team, a typical SMS was constructed from the site information. In this simulated audit, members of the project team role-played the various site personnel to be interviewed, drawing upon predetermined pen-profiles for these roles.

This pilot study highlighted the strengths and weaknesses of the approach and following this desktop audit several major modifications were made to the methodology, particularly with reference to the complexity of the audit method, the pre-audit preparation phase, support material for the auditors and the nature of the delivery systems.

FIELD TRIAL OF THE METHODOLOGY

Following the pilot study detailed above, preparations were made to audit a section of plant in an oil refinery in the Netherlands. The Regulators were aware that in many respects this site was 'above average' - having a mature and comprehensive safety management system. As such, this site would prove to be an excellent test for the I-Risk methodology.

The technical model was subsequently developed and identified a total of 59 initiating and base events, 10 of which were deemed to relate to human error.

The preparation stage for the audit was comprehensive, requiring three man-days on site, and the assessment of common mode allowed the total number of data points to be assessed to be reduced from over 8000 to less than 500. In some instances, the company was found to have some redundancy in the SMS - where a number of feedback systems mapped onto the same boxes in the model. This preparation also involved drawing up an interview plan and audit data points were assigned to particular individuals to be interviewed.

During the three-day audit, a total of 19 interviews were conducted with a range of personnel in the company hierarchy. These interviews, each involving two auditors, lasted from 15 minutes to 45 minutes depending upon the seniority of the interviewee. Also present in the interviews were three observers recording the process and content of the audit for later analysis.

LESSONS FROM THE FIELD TRIAL

The data from the field trial is still under analysis and it is not yet possible to describe the evaluations of the management influences or estimate their effect on the parameters in the technical model. However, at this stage it is possible to outline several lessons learnt and means of improving the model and the audit methodology:

- There are some aspects that need to be made more explicit, most importantly the emphasis on the fact that the method is specifically in relation to major hazards, within the context of the Seveso Directive. Such an emphasis on major hazards will need to be incorporated in all aspects of the method, including the delivery systems, assessment loops and audit attention points.

- The links (that is, the interface) between the parameters of the technical model and the delivery systems of the management model are considered to require some further consolidation.

- Preparatory work is essential to any successful audit. This preparation involves a visit to the site and discussions with key personnel. It is considered by the audit team that a more structured approach to this stage of the method would be highly advantageous and optimise the limited time available for the audit proper.

- Questions arose during the audit concerning the level of understanding of technical information required by the auditors (for example, relating to specific major hazard scenarios) in order to perform an audit of a management system. This clearly has implications for the level of detail to which the management system is defined and subsequently assessed.

- During the audit it became clear that more common mode existed in the company SMS than originally envisaged, further reducing the amount of audit points to be assessed from 500 to in the region of 120. The method of identifying common mode in the company SMS requires further development, particularly to ensure that it is transparent.

- The audit would have been greatly assisted by an improved matrix of interviewees for each activity within each delivery system, including details of attention points for each interviewee. In addition, these attention points should be prioritised.

- The link between the audit attention points and the interface between the technical and management models needs to be made more explicit.

- Supporting documentation for the auditors should ideally include the facility to record what activity boxes have been assessed and to what extent.

- The importance of time for auditors to consolidate between interviews is stressed, especially if future audits may utilise audit teams working in parallel.

A further field test is planned for later in the year, although it is considered that only fine-tuning will be required at this stage following detailed consideration of the modifications prompted by the above trials. Following the incorporation of lessons learnt in this next field trial, an audit manual will be produced detailing the method and its philosophy.

CONCLUSIONS

The recent field trial of the I-Risk methodology has proved to be an extremely useful exercise. One of the main strengths of the method is that it is based around a collection of concepts (such as the delivery systems, assessment loops and attention points) that are easily grasped by the auditing team and that the links between these components is made explicit. Problems identified with the model and audit in the early desktop exercise have been largely overcome. However, the audit can be greatly improved by attention to the audit strategy and the supporting documentation. This work may include prioritisation of the attention points and their allocation to a specific interviewee in the preparatory phase.

This research project has produced a model and methodology that integrates management and technical factors in QRA in a more highly sensitive manner than has ever been accomplished before. It is intended that the products of I-Risk will enable both industry and Regulators in the participating countries to systematically assess the quality of a safety management system in a site-specific manner and examine its effect on the current and future risk levels, as assessed by the technical modelling. Through explicitly linking the influence of management and organisational factors to the assessment of risk, the methodology will enable improvements to be made to existing systems in a more cost- and safety-effective manner. The audit method has also proved to be a powerful tool in enabling a systematic and critical qualitative examination of an organisation's safety management system.

ACKNOWLEDGEMENTS

This paper reports on the developments and progress of an EC research project involving significant contributions from several organisations. The author acknowledges the work of these participants to the material in this paper. These contributors and their respective organisations are as follows:

- Project Co-ordinator: Joy Oh, Ministry of SZW, Netherlands
- Dr Linda Bellamy, Ingenieurs/adviesbureau SAVE, Netherlands;
- Prof. Andrew Hale, Frank Guldenmund, TU Delft Safety Science Group, Netherlands;
- Helen Shannon, Mark Morris, Helen Walker, Four Elements Limited, London;
- Ben Ale, Jos Post, RIVM, Netherlands;
- Iannis Papazoglou, Olga Aneziris, NCSR Demokritos, Greece;
- Williet Brouwer; Ministry of SZW, Netherlands;
- Andre Muyselaar, Ministry of VROM, Netherlands.

The research work described in this paper is being carried out under the CEC Environment programme 1995-1998 and the author is pleased to acknowledge the support of the following organisations in funding this work:

- European Commission, DGXII;
- Ministry of SZW (Social Affairs and Employment), Netherlands;
- UK Health and Safety Executive.

This paper is dedicated to the late Dr Nick Hurst of the Health and Safety Laboratory who, through his ideas, experience and enthusiasm pushed forward the boundaries of research in this field during a distinguished career in the HSE.

© British Crown Copyright (1998)

REFERENCES

1. Hurst, N. W., Nussey, C. and Pape, R. P. (1989). Development and application of a risk assessment tool (RISKAT) in the Health and Safety Executive. Chem. Eng. Res. Des., 67, 362-372.
2. Hurst, N. W. (1998). Risk assessment: The human dimension. Royal Society of Chemistry, Letchworth, Herts. ISBN 0 85404 554 6.
3. Hurst, N. W., Bellamy, L. J., Geyer, T. A. W. and Astley, J. A. (1991). A classification scheme for pipework failures to include human and sociotechnical errors and their contribution to pipework failure frequencies. J. Haz. Mat. 26: 159-186.
4. HSE (1997). Successful health and safety management. HS(G)65. HSE Books, Sudbury, Suffolk. ISBN 0 7176 0412 8.
5. Hurst, N. W., Young, S., Donald, I., Gibson, H. and Muyselaar, A. (1996). Measures of safety management performance and attitudes to safety at major hazard sites. J. of Loss Prev. Process Ind. 9(2): 161-172.
6. Hale, A. R., Heming, B. H. J., Carthy, J. and Kirwan, B. (1997b). Modelling of safety management. Safety Science. 26: 121-140.
7. Bellamy, L. (1998a). Technical model parameters management systems. Document prepared for CEC I-Risk. SAVE. Apeldoorn, March 1998.
8. Papazoglou, I. A. and Aneziris, O. N. (1997). A technical model for QRA of the chlorine loading to road tankers. Document prepared for CEC I-Risk. NCSR Demokritos, October 1997.
9. Anderson, M. (1997). The development of site-specific failure rates for use in risk assessments of major hazard sites: A case study of road tanker loading operations. MSc Dissertation: University of Sheffield, August 1997.

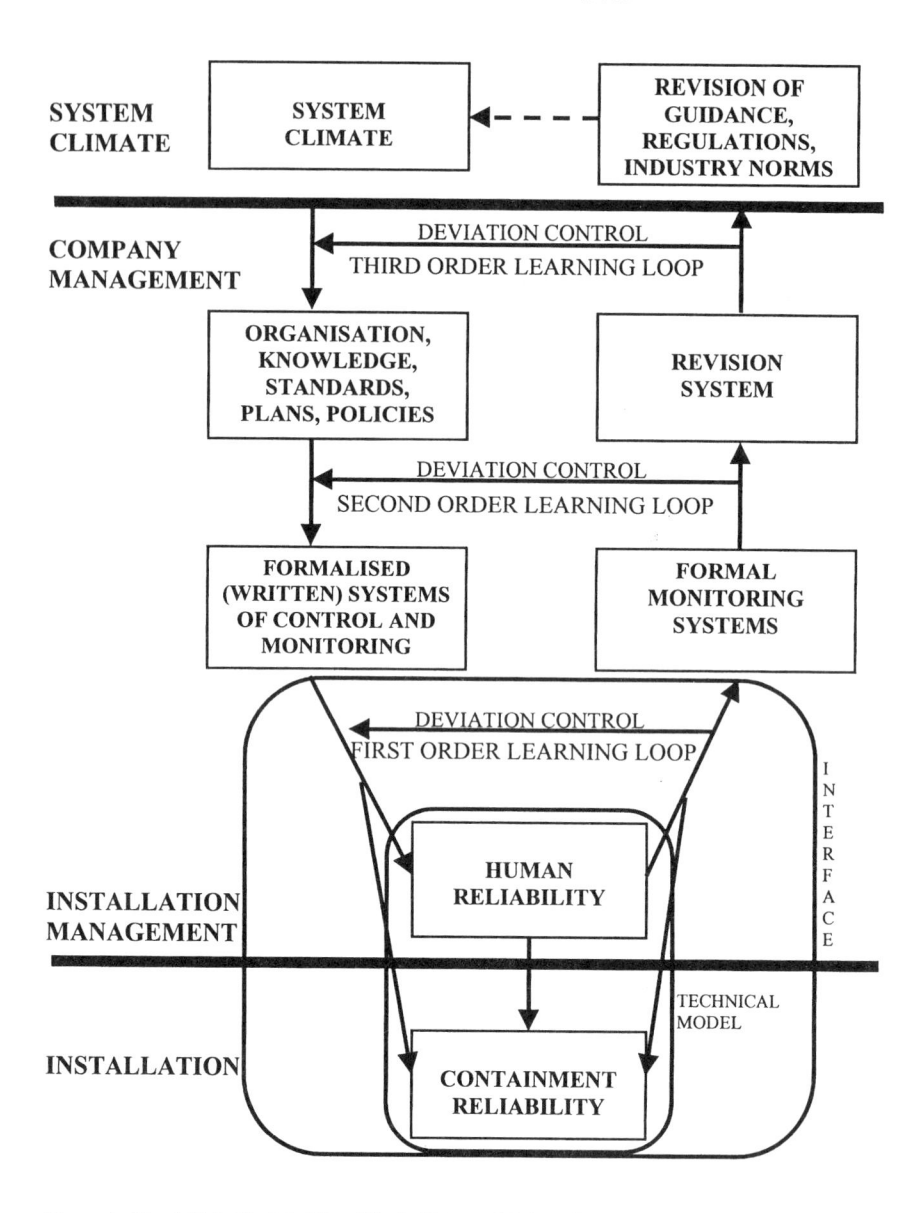

Figure 1: The I-Risk Model. (Simplified) This model describes the development of specific outputs of the safety management system and their modification based on experience with those outputs. The 'loop' runs from the development of a policy, its elaboration into detail, through its implementation and feedback of experiences.

DESIGN FOR SAFETY APPLYING IEC 6-1508
"from the manufacturer's point of view"

S. De Vries, M. Van den Schoor, R. Bours
Fike Europe, Explosion protection systems manufacturer, Herentals, Belgium

> Explosion protection systems are as a well-known technique widely used to prevent process plants from hazardous events and personal injuries. The design of such systems requires knowledge of system design, the user requirements, legislation and hazards. The process of translating between these parameters is often a point of discussion. The IEC 6-1508 document provides a framework for all involved partners in order to communicate and work in a structured and quantifiable way. This paper introduces IEC 6-1508 and handles some of the latest evolutions in explosion protection design techniques and regulations.
>
> Keywords: ATEX 100a, Safety Integrity, Life cycle, E/E/PES, Software design

SAFETY: A QUALITY CHARACTERISTIC

People tend to use the word quality in reference to products, often without even knowing how to define and measure quality. The reason for this is that many different definitions for quality are being used. The definition used for quality and stated in the ISO 8402 dictionary is as follows;

> *"Quality is the totality of characteristics of an entity that bear on its ability to satisfy stated and implied needs".*

Examples of other quality definitions:

> *Quality is the absence of unpleasant surprises*

> *Quality is hard to define, impossible to measure, but easy to recognize*

> *Quality is when the customer comes back and the product doesn't*

Safety can be categorized into quality, as a quality attribute such as cost effective, maintainable, reliable, The safety definition can also be found in ISO 8402.

Safety is a state in which the risk of harm or damage is limited to an acceptable level.

EXPLOSION PROTECTION: ATEX 100a DIRECTIVE

General

On 23/03/1994 Directive 94/9/EC, also known as ATEX 100a, has been approved by the European Parliament. This Directive will become mandatory from 1 July 2003; all previous Directives in contradiction with this Directive will be overruled. ATEX100a is mandatory on all (electrical and non-electrical) equipment for use in explosion hazardous areas. The Directive will define the essential safety aspects necessary for each equipment like for the Machinery-Directive (89/336/EC) and the EMC-Directive (89/336/EC). The individual technical requirements, which are applicable for each device, are to be defined by normative organizations such as CEN and CENELEC. The Directive also describes conformity procedures to be followed by the manufacturer in order to achieve the CE-label. The CE-label is necessary in order to allow for trade of products between the European community states. The CE-label guarantees no quality but states that the product has been made in accordance with all applicable CE-Directives. Electrical equipment for use in an explosive hazardous environment will have to comply with the ATEX 100a Directive as well as with the Machinery-Directive and the EMC-Directive.

Apparatus division

The ATEX 100a-Directive applies to apparatus and systems, which can be used in explosion hazardous environments: mining, gas- and dust-explosion hazardous zones. It applies to all kinds of apparatus and protection systems, electrical and non-electrical. The base for the identification of environments where explosion hazards can be present, is the so-called hazardous zone-classification.

The zone-classification will be mandatory for dust as well as for gas. The Directives are based on a classification of apparatus in groups and categories. The protection level for each apparatus needs to be aligned with the zone-classification of the environment in which the apparatus is being placed and used. For dust explosion hazards there are new obligations for hazardous zone-classification (ref. ATEX 118a, IEC 61241-3 and EN1127-1).

New Zone-classes	Old Zone-classes
gas/dust	gas/dust
zone 0 / 20	zone 0 /Z or 10
zone 1 / 21	zone 1 /Y or 11
zone 2 / 22	zone 2

New apparatus classes
group I: underground
category M1: must remain functional in case of an exceptional fault
category M2: must interrupt energy supply in case of danger
group II: above surface
category 1: must maintain safety level, even in case of exceptional fault (zone 0,20)
category 2: must maintain safety level, even in case of frequent faults (zone 1,21)
category 3: maintain safety level in normal operation (zone 2,22)

Zone 0/20: area in which an explosive atmosphere is continuously present, for a long period of time (more than 1000 hours per year).
Applies to apparatus in group II (above surface) category 1 (very high protection level)
Zone 1/21: area in which an explosive atmosphere can occur occasionally in normal operation (100 to 1000 hours a year).
Applies to apparatus in group II (above surface) category 2 (high protection level)

Zone 2/22: area in which an explosive atmosphere is not likely to occur during normal operation, or for very short periods of time (less than 10 hours per year).
Applies to apparatus in group II (above surface) category 3 (normal protection level)

Examples of zones 20 and 21 are areas where in normal operation an explosive dust/air mixture is continuously present. These areas may include elevators, pneumatic transportation installations, silos, bunkers, mixture systems, sieves, cyclones, dryers, dust-chambers, filters, etc.

PROTECTION AGAINST GAS AND DUST EXPLOSIONS

General

In the design of apparatus and protection systems for use in explosion hazardous environments the explosion protection must be integrated. Therefore measures must be taken in order to:

1. Prevent these apparatus and protections systems from creating an explosive environment.

2. Prevent the ignition of an explosive atmosphere, considering the nature of each potential electrical and non-electrical ignition source.

3. In case an explosion occurs, immediately stop this event and/or reduce the zone, which is affected by the flames and pressure rise due to the explosion, in order to come to a normal safety level.

Corrective protection systems

Explosions can be stopped, and the zone, which is affected by the flames and the pressure, can be reduced, by re-directing the explosion in a safe and known direction, by compartment, extinguishing, pressure relief or a combination of these measures. Therefore different possibilities are available.

Explosion pressure relief
This is a corrective protection method designed to re-direct an explosion into a more safe area. Inside the wall of an apparatus a weak construction is placed deliberately. Explosion vents or explosion rupture disks are specially designed and accepted passive protection systems.

Compartment or isolation
These methods prevent the explosion from propagating to other apparatus, installations or workplaces. Compartment or isolation is obtained by using mechanically or chemical explosion-isolation systems with autonomic functions.

Explosion suppression or extinguishing
By applying these methods the explosion will be extinguished by the quickly injection of a suitable extinguishing agent. The explosion will be detected in the incipient stage and suppressed in order to maintain the explosion pressure under a given acceptable limit.

IEC 6-1508 A SAFETY DESIGN GUIDELINE

General

"Design for safety"; one of the most used slogans in sales meetings concerning protection systems and devices. In the designing of these devices the most difficult questions to answer will be: which level of safety is necessary for implementation, how do we define this safety level in order to have both users and designers understand it's meaning, and how do we validate or measure the safety level in the acceptance of a developed device or system? Different techniques have been used in order to have a clear communication between designer-groups and sales/user-groups. The most common way is to have a project analyst who translates between both groups and who has both technical and commercial expertise and experiences. Without guidelines however the analyst can never succeed in this job. When using tools such as the IEC 6-1508 guideline his goal can be reached more effectively and successfully.
The document IEC 6-1508 defines safety as: *"the freedom from unacceptable risk of harm"*.

Scope

IEC 6-1508:

1. Applies to safety related systems when one or more of such systems incorporate electrical, electronic or programmable electronic systems (E/E/PES).
2. Is mainly concerned with safety to persons.
3. Does not specify those who shall be responsible.
4. Uses Safety Integrity Levels (SIL) for specifying the target level of safety integrity to E/E/PES.

Conformance

Conformance to IEC 6-1508 can be reached as follows:

1. Use as a minimum a quality system, such as ISO 9000 series or similar.
2. If a measure or technique is ranked as Highly Recommended (HR) the reason for not using that measure or technique should be recorded with details

in the safety plan.
3. Compliance shall be assessed by the review of documents required for this guideline and by witnessing tests
4. In general the IEC 6-1508 document provides a framework for a safety life-cycle model, which can be incorporated by its users as an expansion of their quality and safety assessment procedures.

IEC 6-1508 Safety Framework

The framework as shown can be divided into different steps for the design and assessment of a safety device or system. (Fig. 1)

Step 1: Define the system
The first step is to carry out a systematic analysis of the system starting from its functional and qualitative design requirements.
Different techniques can be used depending on the nature of the device or system by using the ISO 9000 quality structure and company internal procedures as a guideline.

Step 2: Analysis of the hazards
Identify the hazards and the events, which could give, rise to them, identifying in particular the way in which an E/E/PES malfunction could contribute to a hazard. There are several formal hazard analysis (HAZAN) techniques, each with particular strengths and weaknesses and fields of application. The most widely used techniques are hazard and operability (HAZOP) studies; failure mode and effect \analysis; event tree analysis; and fault-tree analysis.

Step 3: The risks and risk reduction
After defining any identifiable hazard, a risk analysis must be performed for each individual hazard in order to calculate the risk of hazardous events to occur for the particular device or system in its application. The risk analysis will incorporate frequency of failures from hardware and software devices, frequency at which a person or people are exposed to a risk, the duration of the exposure and the severity of injury after exposure. This analysis will lead to a necessary risk reduction calculation for each type of hazard. The application of the above stated hazard analysis techniques will incorporate risk calculation as well as different standards techniques using checklists and/or mathematical approaches.

Step 4: Safety Functions
In order to have the necessary risk reduction the design has to be equipped with safety functions, each of them preventing hazard events to occur during the operation of the device or system. Each safety function has a measure of safety called the Safety Integrity Level (SIL).

SAFETY INTEGRITY LEVEL (SIL)	Probability of failure on demand <u>or</u> probability of one dangerous failure in one year
4	>= 10E-5 to < 10E-4
3	>= 10E-4 to < 10E-3
2	>= 10E-3 to < 10E-2
1	>= 10E-2 to < 10E-1

Step 5: Safety Allocation

Each safety function requires the implementation of safety function technology.

1. E/E/PE, requiring electrical hardware and/or software techniques.
2. Other technology, such as pneumatically or hydraulically techniques.
3. External safety product implementation, by using standard devices or systems in order to achieve the required level of safety integrity for a specific safety function.

The IEC 6-1508 only takes the E/E/PE techniques into further consideration.

Step 6: E/E/PES Allocation

The considered safety functions will be divided into Safety Related Systems (SRS) with hardware and/or software requirements. These requirements will be formalized in a way to allow designers of hardware and software to clearly understand the specific demands and to measure the required safety integrity level for each safety related specification. The independent test group or person will base the validation of their specific design efforts on the same SRS hardware and software requirements.

Safety Life-cycle

By implementing the Safety Framework into the overall System Life cycle, (as used by ISO 9001) a Safety Life-cycle can be created as the complete Life-cycle model for the design and development of Safety Related Systems. (Fig.2)
This Safety Life cycle can be divided into:

1. Safety Life-cycle for specific E/E/PES allocation (Fig. 3)
2. Safety Life cycle for Software allocation. (Fig . 4)

E/E/PES Safety Life-cycle

General

The safety requirement specification can be divided into: a) The safety functions requirement specification (which identifies and specifies the required safety functions in order to achieve functional safety), the hardware system, operator interfaces and all relevant modes of operation of the equipment under control. b) The safety integrity requirement specification shall contain requirements necessary in the design phase to achieve functional safety of the system, including requirements for the SIL level for each function, techniques for avoiding systematic and random hardware faults during design and development. The next step is the safety validation planning to enable the validation of the safety requirement specifications to take place. To ensure that, in all respects the requirements in the SRS have been included in the design. The design will meet all safety functions and safety integrity as specified. All necessary procedures must be developed to ensure that the functional safety of the safety related E/E/PES is maintained during operation and maintenance. These procedures must allow corrections, enhancements or adaptations to the E/E/PES, ensuring that the required Safety Integrity Level is achieved. Procedures must be developed to test and evaluate the products of a given phase ensuring the correctness and consistency with respect to the product and provided standards.

E/E/PES design considerations

1. At switch-on, or when power is restored following a failure, hardware should ensure to reset the system only from a point that it is safe to do so.
2. Power-supply interruption should not lead to unidentified or unsafe conditions
3. Where all safety related systems are PES based, particular attention is required to minimize the risk of common cause failures (CCF's) such as electrical interference.
4. For systems performing actions on demand, such as emergency shutdown systems or protection systems, some form of on-line proof testing for unrevealed faults may be required.
5. Wherever possible ROM's should be used in preference to RAM's since they are less susceptible to electrical interference.
6. In many industrial applications a specially controlled environment must be provided to promote reliable operation; dust, humidity, temperature and pollution must be taken into account. Corrosive atmospheres for example, often affect printed circuit boards.
7. The level of electrical interference in the environment of the PES must be taken into account, to minimize the effects on the performance of the system, therefore attention should be paid in the design of the equipment and the installation.

Implementation of Safety in E/E/PES

Many different techniques and criteria must be taken into account in order to have safe E/E/PES hardware. Some of the most fundamental criteria and techniques will be given in the following. Depending on the required Safety Integrity Level and the required Functional Safety as specified in the E/E/PES Requirement Specification, one or more of these techniques/ criteria must be applied and will be categorized as being Highly Recommended (HR).

Redundancy	Ventilation-, Heating-System
Systematic Failure testing	Random Failure testing
PE logic configuration	Electrical considerations
Electronic considerations	Processing units
Memories	I/O units and interfaces
Data paths	Power supplies
Watchdog	Clocks
Communication, mass-storage	Sensors
Actuators	Systematic failures
Environmental failures	Operational failures

E/E/PES directives
All hardware related safety products do however have to comply with all applicable mandatory requirements. The most common known Directives in hardware are:

1. Machinery Directive (89/392/EC) which applies to products with at least one element, able to move without applying physical human power.

2. Directive on EMC (89/336/EC)
    EN 50081-2 Generic Emission Standard (1993)
    EN 50082-2 Generic Immunity Standard (1994)
Applies to all electrical products. Specific care must be taken that the Equipment Under Test (EUT) does not generate a level of Electro-Magnetic Energy (EME) which is at higher level than specified in the Immunity Standard, in order not to influence other devices in a way that they will go into a faulty operation mode. Also care must be taken that the EUT will not be affected, and therefore will not perform its normal operation by EME influences generated from external devices. Both EMC-standards provide a safety margin between both immunity and emission levels. Compliance against EMC can be obtained by testing for the following:
    Immunity
        1. conducted via cabling
        2. effects of lightning
        3. electrostatic effects
        4. electromagnetic fields of surrounding devices
    Emission
        1. ground plane
        2. enclosure design

3. The Low Voltage Directive (72/23/EC)
Applies to product safety related to personal for all products with working voltages between 50 V to 1000 V (AC), and 75 V to 1500 V (DC).

4. Explosion protection Directive ATEX 100a (94/9/EC)
Applies to all products for use in explosive atmospheres (ATEX 118a and EN 1127-1). This Directive will be mandatory from July 2003.

Software Safety Life-cycle

Basic Steps
1. Develop the software safety requirement specification in terms of the software safety function requirements and software safety integrity requirements.
2. Develop the software safety function requirement specification for each E/E/PES safety-related system necessary to implement the required safety-functions.
3. Develop the software safety integrity requirement specification for each E/E/PES safety related system necessary to achieve the safety integrity level specified for each safety function allocated to that safety related system.
4. Develop a software safety validation plan to enable the validation of the software

5. Select a software architecture that fulfills the requirements of the software safety requirements, to select a suitable set of tools, including languages and compilers. Design and implement software which achieves the required safety integrity level, and which is analyzable, verifiable and maintainable.
6. Integrate the software onto the target programmable electronics.
7. Provide information and procedures concerning software necessary to ensure that the functional safety of the E/E/PES is maintained during operation and maintenance.
8. Test the integrated system to ensure compliance with the Software Safety requirements specification, at the intended safety integrity level.
Make corrections, enhancements or adaptations to the software, ensuring that the required Safety Integrity Level is achieved.
To the extent required by the safety integrity level, test and evaluate the deliverables of a given phase to ensure correctness and consistency with respect to the deliverables and standards provided as input to that phase.

Implementation of Software Safety

Many different techniques and criteria must be taken into account in order to have Safe Software. Some of the most fundamental criteria and techniques will be given in the following. Depending on the required Safety Integrity Level and the required Functional Safety, as specified in the Software Requirement Specification, one or more of these techniques/ criteria must be applied and will be categorized as being Highly Recommended (HR).

- Structured Methodology for example YOURDON, MASCOT
- Re-try fault recovery mechanisms
- Development tools and programming languages
- Formal Methods
- Modular approach
- Software module testing
- Functional, performance testing
- Data recording and analysis
- Simulation
- Static analysis
- Dynamic analysis
- Checklists

Software standards and guidelines

1. ISO 9000-3
These are a set of guidelines for the application of ISO 9001, to the development, supply, installation and maintenance of (computer) software. We define 'Firmware' as software written for the application into embedded hardware that can not be changed by the user operations, and stored into non-volatile memory. Extensive Firmware is written for protection systems and control applications. The most basic form of Firmware is machine code for micro-controllers and microprocessors, many times written in Assembler-language.

2. ISO/IEC 9126
This provides for the information technology (IT) product evaluation quality characteristics and guidelines for their use. Differentiated into 6 main characteristics and 21 sub-characteristics, each of them requires consideration in the requirement specification phase of the software product. The main software quality characteristics are:

- Functionality
- Usability
- Maintainability
- Reliability
- Efficiency
- Portability

TECHNOLOGY UPGRADE FOR SAFETY PRODUCTS

General

Nowadays technology creates perspectives for miniaturizing and IT-implementation into unlimited ranges of products and devices. Here are some technology upgrades for consideration into safety related systems with protection functionality.

Bus systems

Standard protection systems use 'sensor-PES-actuator' configurations.
In safety applications redundancy of electronics can increase the efficiency and reliability of these systems. But still each type of hardware solution uses single-bit information. By using bus-systems all field devices, which are anyway susceptible to false triggering and data disturbances, can be provided with multiple-bit communication to have: a) complete reliability in their functional safety b) reach a higher level of safety integrity. Due to the bus-protocol (fault and retry-mechanisms), all data will be validated correctly by both communicating devices. Current technology allows implementation for bus-structures even for high-speed protection systems such as active explosion suppression or isolation.
Each device will be addressable having a unique bus-address and can be remotely monitored while it is performing its protection function, without having any delay on its safety function when required to take place. By using special cable types and/or fiber optics the susceptibility for introducing retries into the data transfer will be reduced enormously.

Intelligent active validation

Using software validation techniques on analog signal coming from primary sensor elements prevents false readings. For example, an analog 4-20 mA signal, converted by an 8-bit analog to digital converter, can be sampled and stored multiple times before a validated value is used for further comparison against preset levels. The stored pre-validated values can be software integrated to eliminate spikes and noise of the signal.

For example a sample rate of 200 µSec results in a validation value each 1 mSec after integration of the 5 stored samples.

Measurement is knowledge (monitoring/events)

In most active safety protection systems actuator devices use critical components. All failure modes and frequencies are well known and therefore redundant hardware has been installed for these critical components. However, automatic self-checking to inform the user of any malfunction not always provided, many times sensors are applied which should be inspected by the user at frequent intervals. Redundancy can be lost, or the whole risk and failure analysis can become invalid without the customer knowing. The incorporation of monitoring electronics may solve a lot of these problems because they can immediately affect the working mode of the protection devices and/or process plant. Enhanced techniques shall keep a record of all unsafe conditions and all process characteristics before and after a protection event on demand has occurred. For example an explosion detector will measure and store the pressure after triggering the explosion protection system, providing information on the maximum achieved explosion pressure and response time of the suppression system.

WHY DO WE KEEP MAKING THE SAME MISTAKES?

General

Practical experiences in how to avoid unsafe conditions regarding dust and gas explosions, from an electrical point of view.

Design Considerations

Specifications
In the specification phase the analyst tends to make mistakes by not having all related groups involved in the requirements specifications. (from managers to operators, from designers to testers, from technicians to sales person)

Design Techniques
Many errors are made in the design by implementing unfamiliar techniques. Many designers think they know it all and do not use the skills of specific specialists for certain tasks or problems. Two persons will know more than one and each will have their own area of specialization and limits of knowledge.

Testing
Having a test plan is great, but the implementation of the tests will sometimes be placed in the background due to time limits and project costs. Many project leaders tend to use the testing plan if their still is enough time and money to do the job. A lot of in the field-faults and extra costs are the unfortionally result of such an approach.

Installation Considerations

Documents
Users tend to apply all manuals and prescriptions when things go badly. However by reading these documents in advance many mistakes could be avoided in operating devices for safety applications.

Pre-Commissioning
Cabling, earthing, positioning and handling protection devices require information in advance. Having a knowledgeable person on-site before installing a protection system will avoid maintenance and commissioning problems.

Commissioning
The commissioning part of a protection system is a very critical phase to have a good practical solution for a specific hazardous problem. It is at that very moment that all-normal and abnormal modes of operation of the protection system and process should be tested on-site. The execution of process-shutdown and process-startup will tackle a lot of potential problems (electrical interference). In most cases the normal plant operation will not be the cause of problems.

Maintenance
Frequent maintenance tends to create a lack of specific perception for potential problems. This should be avoided by procedures and job-diversity for the maintenance group or by extended use of self-checking E/E/PES hardware.

RESULTS AND CONCLUSIONS

General

1. By using the IEC 6-1508 Safety Life-cycle in the overall System Life-cycle more specific safety based approach was obtained in our products.
2. We learned not to underestimate the power of metrics and quantitative data.
3. IEC 6-1508 was found to be applicable to large-scale projects in order to absorb the overhead introduced from the initial phases of the project.
4. Applying a Safety Life-cycle model ensures confidence and awareness into Safety Related Systems from the users, the sales group and management.

Conclusion

It is our believe that the use of IEC 6-1508 or a similar approach towards complete 'Design for Safety' will be even more stimulated in the future. As manufacturers we therefore must comply to all related documents, and with all mandatory Directives, in order to take the necessary measures in the design of our future products. It will demand a "Safety Culture" in all phases of the product life by all related person. It is therefore necessary and extremely useful to have different information channels in order to educate all the related persons in the "Safety Culture". The ICHEME symposium helps in the creation of such a "Safety Culture".

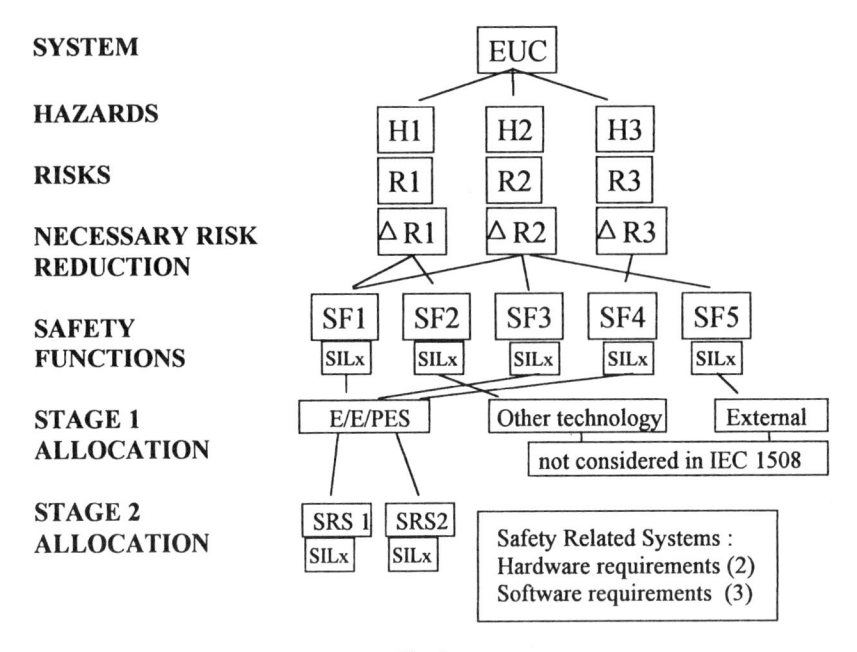

SYSTEM	EUC
HAZARDS	H1 H2 H3
RISKS	R1 R2 R3
NECESSARY RISK REDUCTION	ΔR1 ΔR2 ΔR3
SAFETY FUNCTIONS	SF1 [SILx] SF2 [SILx] SF3 [SILx] SF4 [SILx] SF5 [SILx]
STAGE 1 ALLOCATION	E/E/PES Other technology External not considered in IEC 1508
STAGE 2 ALLOCATION	SRS 1 [SILx] SRS2 [SILx] Safety Related Systems : Hardware requirements (2) Software requirements (3)

Fig. 1

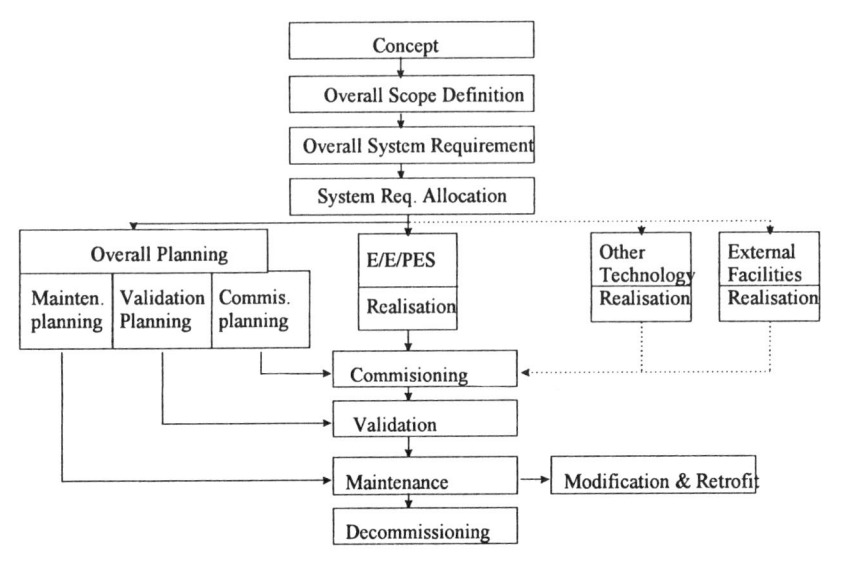

Concept → Overall Scope Definition → Overall System Requirement → System Req. Allocation

Overall Planning: Mainten. planning | Validation Planning | Commis. planning

E/E/PES Realisation

Other Technology Realisation

External Facilities Realisation

→ Commisioning → Validation → Maintenance → Modification & Retrofit
→ Decommissioning

Fig. 2

233

Fig. 3

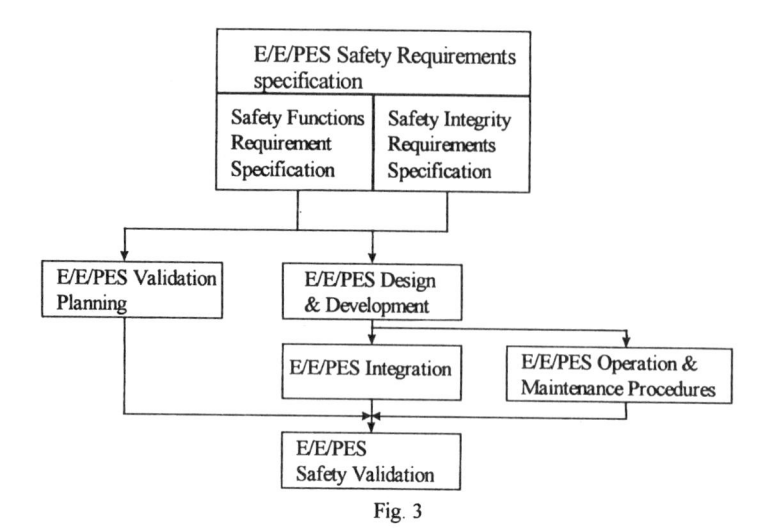

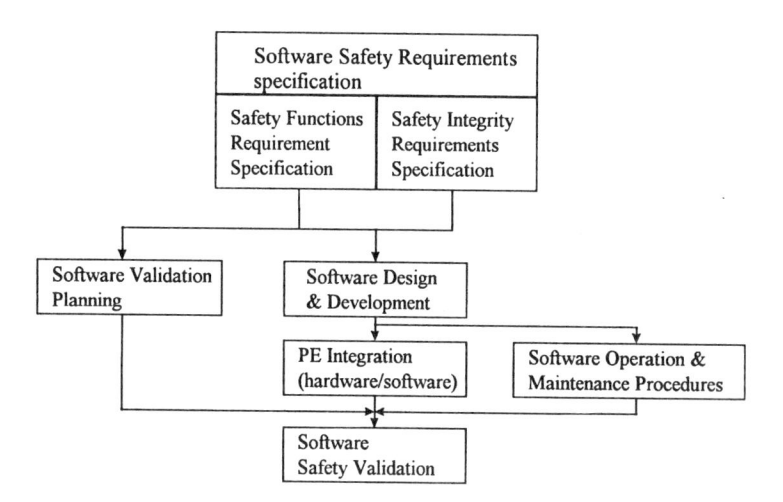

Fig. 4

THESIS: THE HEALTH ENVIRONMENT AND SAFETY INFORMATION SYSTEM - KEEPING THE MANAGEMENT SYSTEM 'LIVE' AND REACHING THE WORKFORCE.

A. Lidstone
EQE International Ltd., 500 Longbarn Boulevard, Birchwood, Warrington, WA2 0XF

> THESIS is a relational database designed to be used by line management and the workforce to collect, analyse and store HSE data for a facility or operation. In addition to compiling information for a HSE Case or HSE Management System Manual, the data held can be output in a wide range of formats to promote dissemination and use amongst the workforce.
>
> The heart of THESIS is the so-called 'Bow Tie' Diagram which provides for a readily understandable visualisation of the links between hazards, the controls preventing their release and the remedial measures available to mitigate the consequences of a release. The use of a relational database permits the linking of these controls to specific tasks, instructions and responsible individuals.
>
> The paper describes the programme key features and some experiences gained during its application world-wide over a number of differing industries and operations.
>
> Keywords: Thesis, HSE Management, Relational Databases, Risk

INTRODUCTION

One of the recommendations arising from Lord Cullen's report on the Piper Alpha disaster [1], was that operators' Safety Management Systems should be based upon quality management principles. In 1991, the Exploration and Production (E&P) sector of the Royal Dutch/Shell Group of companies endorsed and adopted the majority of the Cullen requirements, including the requirement for integrated Health, Safety and Environment Management Systems (HSE-MS) based on quality principles for world-wide application.

A general principle underpinning HSE policy within all Shell E&P companies is that

> '... work should not start before it is confirmed that essential safety systems are in place and that staff are accountable for this requirement. Where safety cannot be ensured, then operations should be suspended...' [2]

A safety, or more generally, HSE Case, in addition to being a regulatory requirement in a number of operating areas, also serves as a formal demonstration of the effectiveness of the facility/operation specific HSE-MS, and that essential controls are in place.

Early experiences in the development of such documents rapidly led to the conclusion that a standardised method was required to ensure the collection, evaluation and presentation of the data necessary to 'make the case'. It was also readily apparent that the preparation of a HSE Case was regarded as a 'paper exercise' with little or no relevance to the management activities conducted on the ground. Part of the problem was that a considerable amount of useful data was relatively inaccessible within the documentation.

To counter these problems, and to increase the perceived relevance of the HSE Case, Shell began the development of an IT tool in 1993 which eventually became THESIS.

BENEFITS OF INFORMATION TECHNOLOGY

THESIS (The Health, Environment and Safety Information System) is a relational database designed for the line management and workforce, to enable the application of the risk control process and the building of management systems. The real benefit of the tool is that it holds relationships between data, for example the hazard, the measures controlling its release and the persons responsible for ensuring each control is effective at all times.

The benefits of IT systems relative to paper-only documents include:

- Quicker activity and hazard analysis
- Provision of a corporate memory to assist the learning ability of the organisation
- Improved accuracy, completeness and consistency
- Visibility of gaps in hazard information
- Automated compilation of areas for improvement
- Easier maintenance and updating of information to keep pace with changing operations
- Cost, time and resource savings
- Presentation of information in multiple formats to suit the end user
- On-line management system information tool
- Implicit relationships between responsibilities and hazards are visible
- Support for multiple users.

THE RISK MANAGEMENT PROCESS

As stated earlier, one of the principle recommendations arising from the Cullen enquiry was that the standards and processes applied in quality management should also apply to the management of safety by operators. Developing on from the Deming Cycle [3] of

PLAN - DO - CHECK - FEEDBACK

Figure 1 shows the key processes in a HSE-MS. At the core of this cycle is what is known within the Shell group of companies as the Hazards and Effects Management Process (HEMP), or more generally known as the risk management process. This is based on the principle that managing risks means that the hazards and effects associated with a particular operation, facility or business process have to be properly managed. HEMP requires the four stages of:

- Identify - Are people, environment, assets or reputation exposed to potential harm?
- Assess - What are the causes and consequences? How likely is the loss of control? What is the risk and is it ALARP?
- Control - Can the cause be eliminated? What controls are needed? How effective are the controls?
- Recover - Can the potential consequences or effects be mitigated? What recovery measures are needed? Are recovery capabilities suitable and sufficient?

i.e. that the hazards and effects should be identified, fully assessed, necessary controls provided and recovery preparedness measures put in place to control any hazard release.

The challenge is then to draw the information recorded during this process into a cohesive structure, rather than a set of items linked in an ad-hoc way (**Figure 2**).

THESIS

THESIS is designed to be used by the personnel of a facility, operation or location to collect, analyse and store HSE data. It provides a structured method to assist, control and check the data collation process. The data held within THESIS may then be presented in a number of custom report formats and may also be used as an on-line information tool, permitting access by multiple users. Later updates to data, to reflect for instance a change in the plant parameters or organisational indicators, is considerably easier given a database approach.

The primary function of THESIS is to identify hazards, the controls preventing their release, the recovery measures required should they be released, and most importantly, to identify and link these controls to the person responsible for their enforcement. THESIS thus guides the user through an Identify - Assess - Control - Recover format, and provides checks to record for example, where data is not available or controls are not defined.

There are many standard techniques available for the identification of hazards, such as HAZOP, HAZID, FMEA, checklists and historical loss analysis. These may then be readily assessed by means of a risk matrix such as that shown in **Figure 3**. This may be used at the outset to qualitatively estimate the people, asset, environment and reputation risks to the business. THESIS contains a comprehensive list of hazards, including the immediately catastrophic (e.g. flammable gas), to long term environmental (e.g. land take, discharge) and chronic health (noise levels >85dBA) effects.

Following on from this comes the important step of identifying the existing controls, assessing their effectiveness and identifying additional controls where current measures are unnecessary or ineffective. At the heart of THESIS is the so called "bow tie" diagram, **Figure 4**, which has been found to be an extremely useful means of presenting data at all levels from the shop floor to the board room.

The terminology which is adopted throughout THESIS may be summarised as:

Hazard - something with the potential to cause harm, including ill health or injury, damage to equipment, product, the environment or reputation, production losses or increased liabilities.

Top Event - the release of the hazard; this is the undesired event arising from the hazard's release e.g. loss of containment, not the ultimate effect e.g. fire or explosion.

Threat - something which may release the hazard to cause an incident e.g. corrosion, operator error etc.

Consequences - the final results of the hazard being released and not being controlled e.g. fire, explosion.

Barriers - the controls that are put in place to prevent a threat from releasing a hazard e.g. isolations, inspections, physical barriers.

Recovery Preparedness Measures - controls that may be applied to limit the chain of consequences arising from the top event e.g. provision of emergency shutdown systems.

Escalation Factors - conditions that could act to reduce the effectiveness of a barrier or recovery preparedness measure e.g. safeguard system not available.

Escalation Factor Controls - the controls that are put in place to prevent the escalation factor from affecting the barrier.

The hazard bow-tie therefore serves to demonstrate for each hazard/top event pairing, the threats that could cause its release, the controls that are in place to prevent the top event occurring and the recovery measures that are, or need to be, identified to mitigate the consequences of a release. Also shown are the escalation factors which may serve to reduce the reliability or effectiveness of any control barrier, and the secondary controls that are in place for their control.

Facilitated workshops, involving the people who are confronted with the risks on a regular basis, supported by specialist technical advisers if appropriate, have been found to be very effective at identifying the components forming the bow-tie. This process also promotes ownership of the final product amongst the workforce, as they have been involved in the identification and development of the information.

However, the identification of hazard controls is insufficient in itself because the analysis is a 'static' record of the risk situation at a particular point in time only. Linkages must be made to those workplace, business or management activities which put in place or maintain the controls shown in the bow-tie diagram, as shown by **Figure 5**.

Such tasks may be deemed "*HSE-critical tasks*" and are a sub-set of all the tasks carried out by employees, management and advisers. The tasks may be design, inspection, operational, financial or administrative tasks, etc. It is only by performing such tasks at the appropriate time that assurance can be gained that hazards are being managed properly.

Furthermore, accountabilities for action must be defined so that risk management becomes a real part of line management activity. Therefore as a minimum it is essential to:

- specify who is responsible for each HSE-critical task;
- the competencies required to ensure the task is carried out properly (in terms of experience, qualifications, training and personal attributes);
- the documentation where the task is defined (e.g. plan, procedure, job description, etc.); and
- how it will be verified that the task has been undertaken properly and at what interval.

It is important to differentiate between *HSE-critical tasks*, which are those which contribute to the control of a hazard, and *hazardous tasks*, which are those which in themselves may present a risk to the person performing the task. Two simple examples illustrate the difference:

- confined space entry is a hazardous activity as it exposes the operator to hazards such as asphyxiation;

- atmosphere checking is a HSE-critical task as it provides one of the barriers necessary to control the hazard of insufficient oxygen atmosphere.

THESIS provides for the identification, assessment and linkage of the HSE-critical tasks necessary to provide the controls to ensure that the major hazards within an operation are controlled. Hazardous activities require to be controlled by the application of local workplace controls such as permit-to-work systems, STOP, etc. The approximate level of distinction is illustrated in **Figure 6**.

The relational database structure of THESIS thus provides for the :

- *identification* of hazards applicable at an operation, facility or area;
- *assessment* of the number and efficacy of the barriers in place to *control* the release of the hazard and also to *recover* from this top event;

- identification, assessment and linkage of the activities necessary to ensure the controls are effective.

This is illustrated in **Figure 7**, which shows some of the information that may be held within the THESIS database for a particular operation or facility and the linkages between them.

Prior to the preparation of the analysis it will become necessary to make a definition of the acceptance criteria to be used. This is generally found to be a combination of:

Quantitative - based upon the position of the hazard within the risk matrix (**Figure 3**), a set number of independent and effective barriers are required between any threat and the top event, the top event and the consequences and also for the control of escalation factors; and

Qualitative - during the preparation and review process of the analysis, experienced judgements are made as to whether the controls identified are suitable and sufficient.

Facilities exist within THESIS for the assessment of barrier effectiveness, for recording and prioritising shortfalls and remedial actions and also for internal error checking e.g. that all barriers are assigned tasks and that all tasks are assigned responsible persons.

BENEFITS OF THE THESIS PROCESS

One of the most significant benefits that has been realised during the development of THESIS based analyses has been increased participation by the workforce in the development, review and implementation of the HSE assessment. The provision of an easily accessible on-line reference tool has helped to remove some of the resentment towards remotely prepared 'paper cases'.

The custom reporter function within THESIS, and especially the graphical bow-tie outputs (**Figures 8** and **9**), mean that specific information can be extracted when required and are easily presented in a readily understandable format.

Some of the uses that THESIS databases have been put to include:

- preparation of HSE/Safety cases, audit plans and emergency planning scenarios;
- assessment of specific high risk operations;
- reviews of operating procedures;
- hazard management awareness training;
- input for contract specifications (following assessment of design concepts); and
- input for workplace hazard management systems.

One notable example of the increased understanding gained from the bow-tie diagram, was an exercise undertaken to take an existing 'paper case' and use the bow-tie diagram to explain to the workforce the levels of safeguards that were in place for their protection and also to show why their contribution to the operation was important.

THE FUTURE

Subsequent to the first version of THESIS being available in January 1996, EQE International Ltd. has partnered with Shell with two principal aims for THESIS:
1. Its continued development as a tool for decreasing the risk faced by organisations; and
2. The development of applications external to the exploration and production oil industry.

The application of THESIS was originally in the oil and gas industry, and HSE Cases have been prepared for operations in a number of countries world-wide, including Europe, South America, Africa and Australasia, some of which have no regulatory requirements for formal risk assessment.

The type of operations that have utilised the method to date include;
- land and marine drilling operations;
- storage and processing facilities;
- logistics activities;
- corporate business risk;
- distribution operations;
- shipping; and
- construction.

The future development of THESIS is continuing with additional features being developed as is the range of industries and applications to which it may be put.

SUMMARY

THESIS leads the user through a structured process to identify and record essential information about each risk management critical task, such as which posts are responsible for executing and supervising these tasks, and how the task is documented and verified. This allows the practical implementation of the risk management process and provides full visibility of risk controls.

THESIS provides for:

- a structured method of activity and task analysis
- a structured method for qualitative analysis of hazards and effects, or for recording results of more formal hazards and effects analysis
- clear references to company standards and procedures
- clear cross referencing between hazards and the activities and tasks providing control
- the ability to set objectives and targets for hazard management
- facilities to check on completeness of data and to highlight conflicts or gaps
- an easily comprehensible graphical presentation of HSE information
- customised hard copy reports.

The relational database permits:

- cost, time and resource savings
- on-line access for multiple users
- easier updates for changing operational circumstances
- automated compilation of areas requiring improvement.

CONCLUSION

'The quality of our individual contributions to the management of safety determines whether the colleagues we work with live or die"

- Brian Appleton, ICI, Technical Advisor to the Piper Alpha Enquiry.

THESIS provides the linkages to develop, demonstrate and enforce these contributions and in doing so provides a useful addition to the process of making HSE management real and live to the workforce.

REFERENCES

1 The Hon. Lord Cullen, 1990, The Public Enquiry into the Piper Alpha Disaster, HMSO, London, Cm 1310.

2 Primrose M.J., Bentley P.D., Van der Graaf G.C., 1996, Thesis - Keeping the Management System "Live" and Reaching the Workforce, SPE 36034.

3 ISO9001

ICHEME SYMPOSIUM SERIES NO. 144

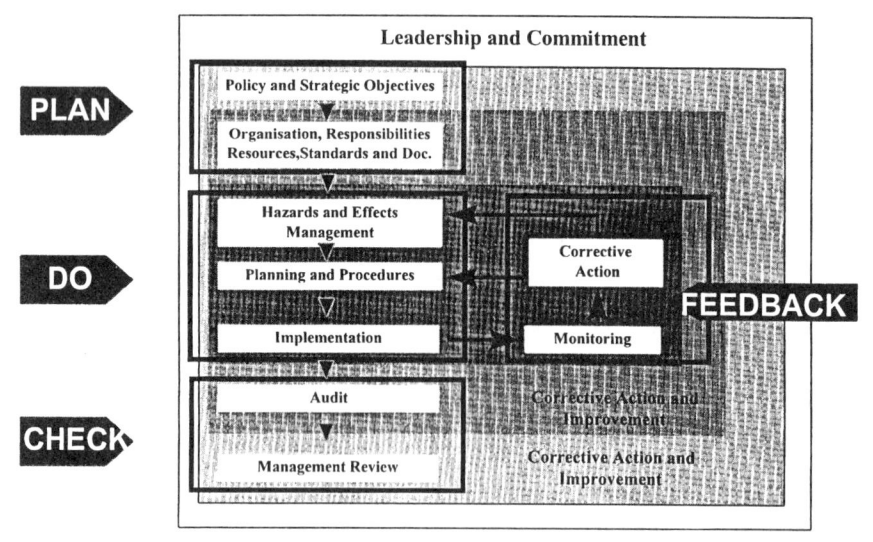

Figure 1 : A structured HSE-Management System, based on the quality management principles of "plan-do-check-feedback".

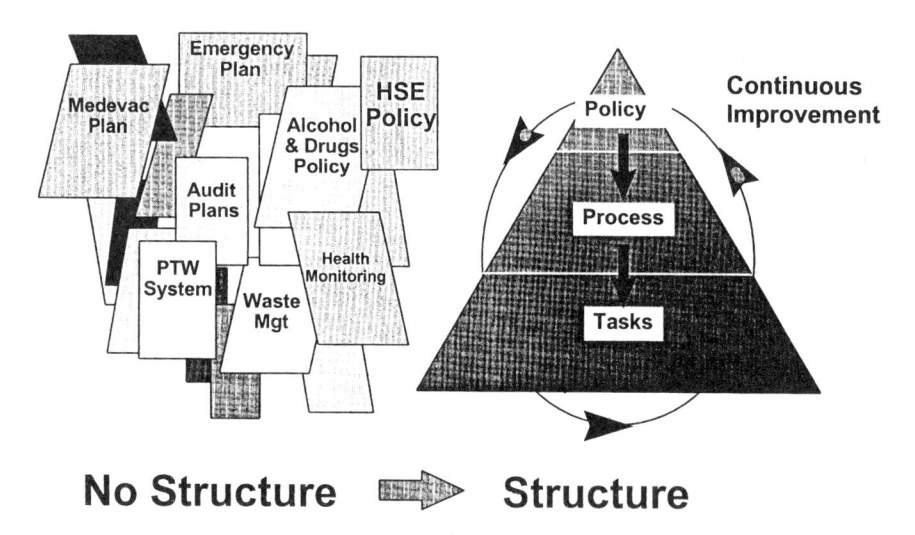

Figure 2 : Developing and documenting the Management System aims to apply a structure to an often ad-hoc approach to risk management.

ICHEME SYMPOSIUM SERIES NO. 144

Rating	IMPACT				INCREASING LIKELIHOOD				
	People	Asset Damage	Environmental Effect	Impact on Reputation	A Never heard of in our industry	B Has occurred in our industry	C Incident has occurred in our company	D Happens several times/yr in our company	E Happens several times/yr at one site
0	No injury	None	None	None					
1	Slight injury	Slight	Slight	Slight	\multicolumn{5}{c}{Manage for continuous improvement}				
2	Minor injury	Minor	Minor	Limited					
3	Major injuries	Localised	Localised	Considerable					
4	Single fatality	Major	Major	National	\multicolumn{5}{c}{Incorporate risk reduction measures}				
5	Multiple fatalities	Extensive	Massive	International					

Figure 3 : A risk assessment matrix may be used to assess the exposure to identified risks and to decide on the most appropriate level of control.

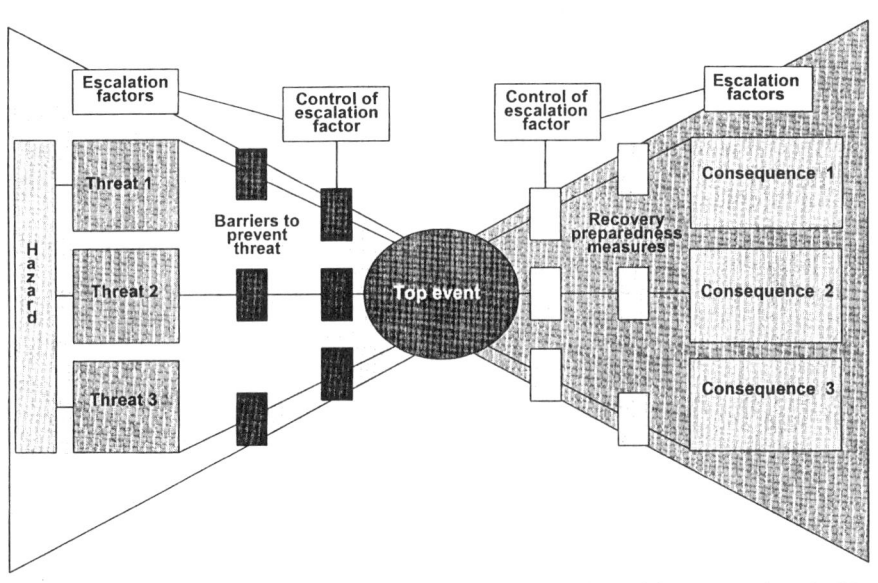

Figure 4 : The "bow-tie" concept has been found to be an extremely useful representation of HSE management and is readily understood at all levels in organisations.

ICHEME SYMPOSIUM SERIES NO. 144

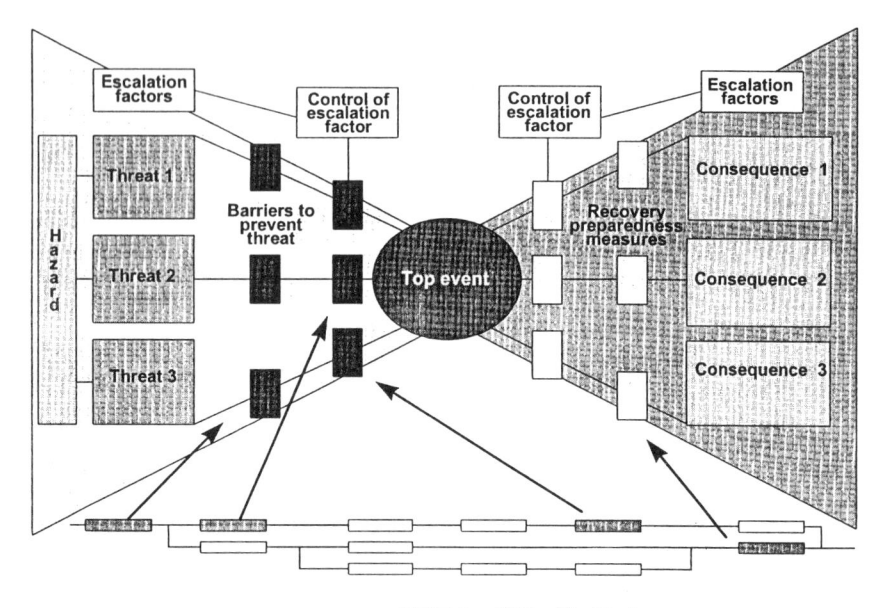

= HSE-critical task

Figure 5 : The identification of the risk controls is insufficient in itself. Linkages need to be made to "HSE-critical tasks" which put into place or administer the controls.

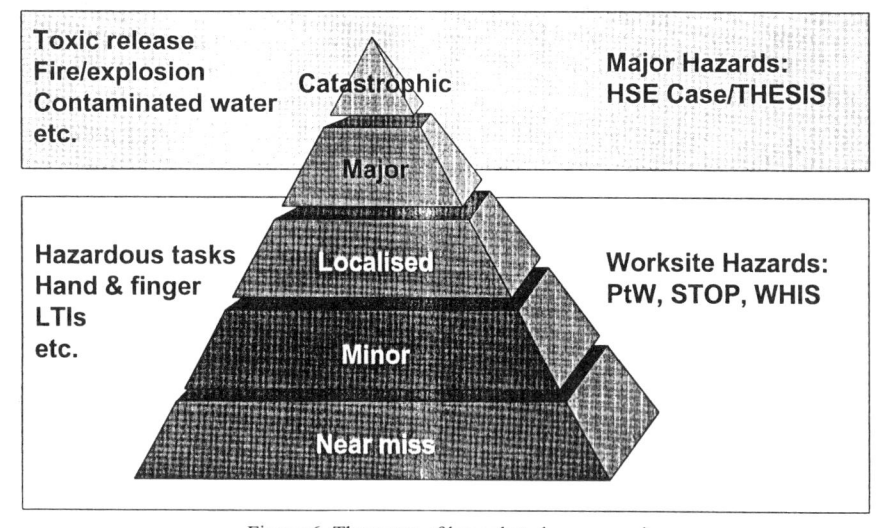

Figure 6 : The range of hazards to be managed

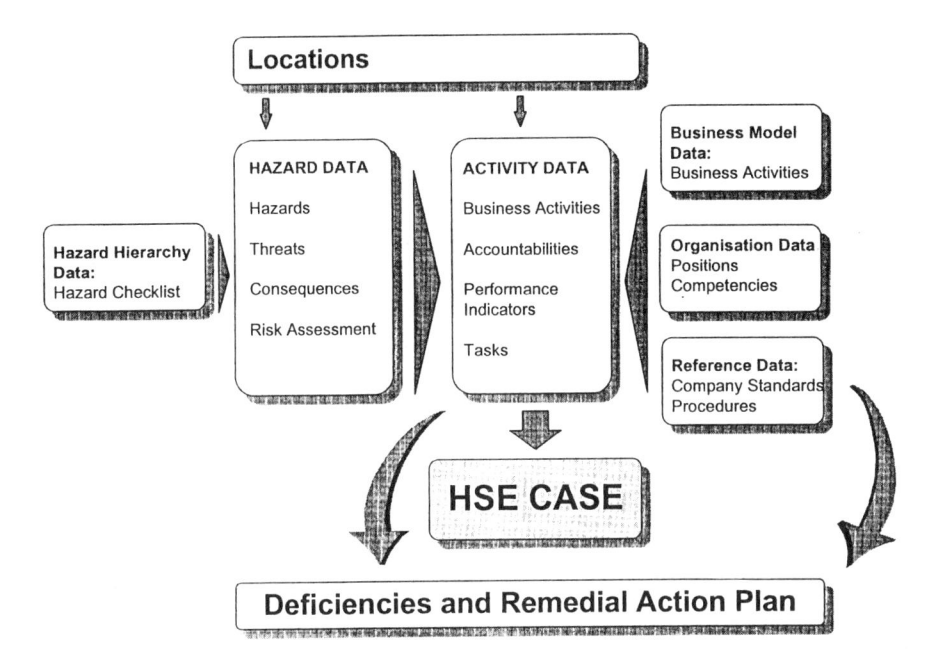

Figure 7 : Illustration of the information that may be held within the THESIS database for a particular operation or facility and the linkages between them.

ICHEME SYMPOSIUM SERIES NO. 144

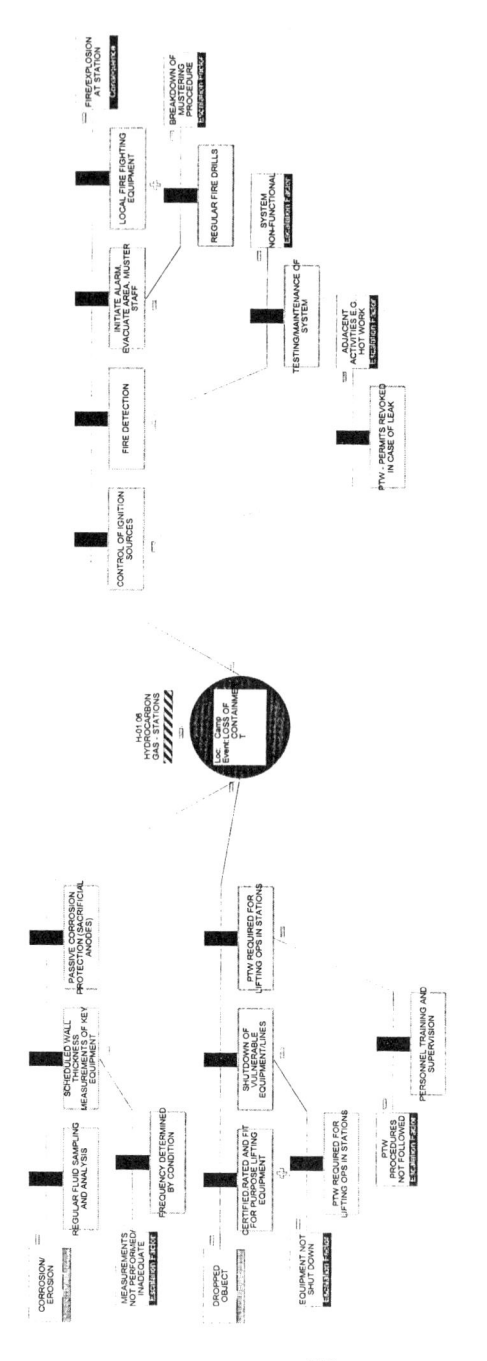

Figure 8 : Example of a Hazard "bow-tie" showing the hazards involved with the loss of containment of hydrocarbon gas.

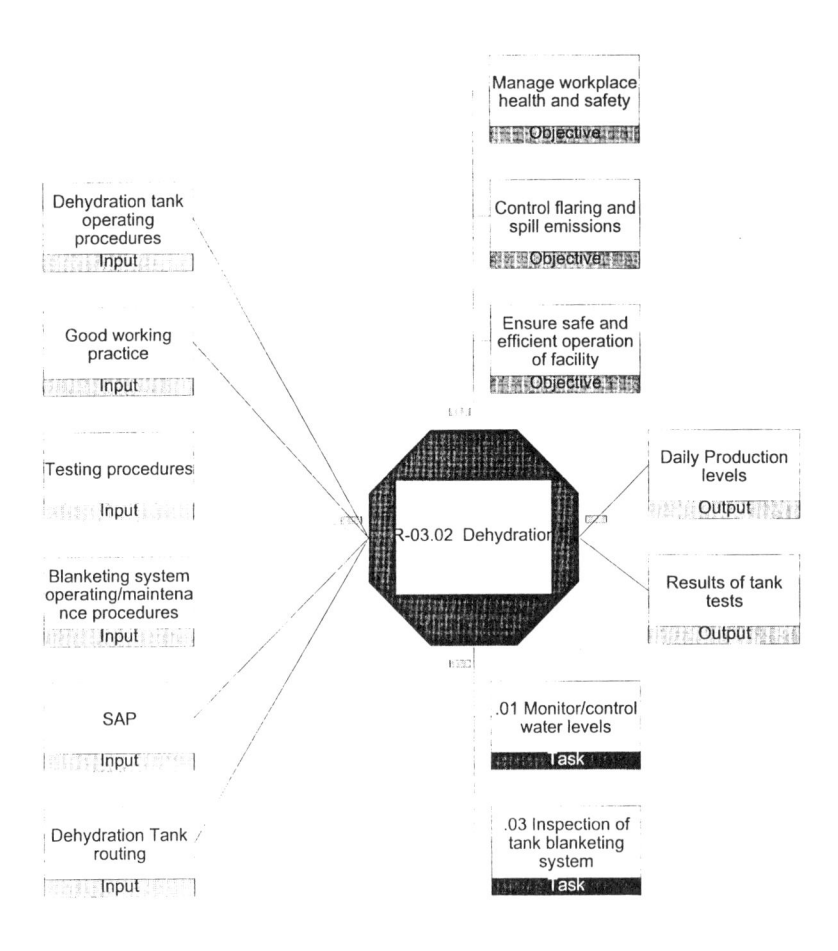

Figure 9 : Example of an Activity "bow-tie" showing the activities involved in the dehydration of hydrocarbons.

SAFETY ISSUES AND THE YEAR 2000

Richard Storey
Allianz Cornhill International, 32 Cornhill, London EC3V 3LJ

Industrial insurers and their clients are increasingly becoming aware of the potential for both property and business interruption losses associated with the millennium bug. This paper is based on the advise which Allianz is currently giving to insured to help minimise exposure to the millennium bug. The paper starts off by describing the nature of the problem along with the types of computer controlled systems which can be affected. Subsequently different loss scenarios are reviewed along with the near disastrous consequences which can follow from a lack of computer based control. The paper aims to provide a framework which allows manufacturers to plan for, and rectify, problems associated with the millennium bug. Ultimately, the advise in the paper aims to reduce the potential for both human and financial loss at manufacturing facilities.

Keywords: millennium bug, Y2K, embedded chips, compliance programme

INTRODUCTION: WHAT IS THE MILLENNIUM BUG

The millennium bug or Y2K problem arises from the potential inability of computer systems, including embedded chips, to recognise correctly the year 2000 as well as other dates both prior and subsequent to the turn of the century. The inability of computer systems to recognise the year 2000 correctly could cause computer systems to malfunction or fail completely resulting in a multitude of serious consequences.

The reason for the possible failure is that until very recently computer programs used computer code which represented year dates as two digits rather than four. For example the year 1963 is encoded as 63 and the year 1901 as 01. The problem occurs when the year 2000 is represented as 00 which may be confused with the year 1900 (or any other century) which is also encoded as 00. Another problematic date is 9th September 1999 which will be encoded as either 99 or 9999 which is often used as a termination sequence within blocks of computer code. Some systems may also not recognise that the year 2000 is a leap year.

Although technically the problem is not difficult to solve, the scale of the problem is enormous due to the reliance of modern society on computer systems. It has been estimated by American computer researchers that the global cost of fixing the problem could be as high as $600 billions US Dollars. Possibly of even more concern is the lack of competent human resources to tackle the problem and the dwindling timeframe available for rectifying any problems.

Allianz Cornhill therefore believes that all companies need to urgently review their year 2000 exposures and take appropriate remedial action to ensure that non-compliant systems are updated. The guidance provided within this paper is aimed at helping companies develop an effective framework for approaching and tackling the year 2000 problem. It is certain that some systems/operations will fail and as such compliance is therefore largely a damage limitation exercise.

What sort of equipment could be potentially affected?

The answer to this is a wide range but in general any equipment containing micro processors will be potentially affected. More specifically a non-exhaustive list would be as follows:-

- Any computer software used within PC and mainframe systems including operating system software
- Security and access control systems
- Alarms including fire alarms
- Process control equipment
- Programmable machinery (including equipment with embedded chip systems)
- Communication systems
- Lifts and other building services equipment
- Bar coding equipment
- Safety critical shutdown systems (major implications for human safety)
- Financial, production scheduling and project planning systems.
- Navigational and satellite positioning systems
- Equipment and machinery testing and monitoring systems

What are the possible consequence of a Year 2000 computer failure?

This is a very difficult question to answer as it will depend on numerous factors including the type of business operations conducted and the degree of reliance of a company on computer systems.

Again a non-exhaustive list of the general consequences is as follows:-

- Computer systems may malfunction or simply shutdown completely
- Production lines may stop or product with an incorrect specification may be produced.
- Alarm systems may fail (both fire and burglary)
- Safety critical systems may fail or malfunction resulting in liability claims
- Financial systems may collapse or malfunction
- Stock control systems may fail
- Building services (heating, cooling, lighting) may malfunction
- Communication systems may fail and associated billing systems crash
- Networked computer systems may not be able to 'talk'
- Supplier and customer chains may breakdown
- Port cargo management systems may fail
- Alignment and navigational systems may cease to operate
- Machinery control may be lost
- Valuable information may be lost

More specifically, with respect to the safety of manufacturing operations such as process plant, possible consequences could be:

Fire

- Failure of computer controlled fire detection systems to recognise fire and raise appropriate alarms. This includes the failure of detection systems to notify remote monitoring stations of alarm conditions

- Failure of computer controlled fire extinguishing systems (such as gaseous or foam systems) to extinguish and control fire

- Fires resulting from process control failures such as overheating of reaction vessels (from loss of coolant), electrical system malfunctions, run-away chemical reaction due to unexpected mixing of reactants, inappropriate handling and processing of flammable liquids

Explosion

- Process control failures resulting in overheating, electrical malfunction, high flammable vapour or dust concentration build-up, failure of monitoring and /or ventilation systems

- Failure of explosion suppression and explosion mitigation systems.

Chemical release

- Release of solid, liquid, or gaseous reactants and/or products into the environment

Nuclear contamination

Release of nuclear material into the environment or reactor meltdown scenario

Equipment start up

- Rotating or other mechanical equipment which starts without warning

- Equipment which is electrically energised without warning

Product spoilage and contamination

- Manufactured with incorrect specification, contaminated or spoiled due to changes in storage conditions.

On a general note, it is clear from the above that there are serious potential consequences to human safety as a result of Year 2000 failures. From a company standpoint there are serious implications with respect to the provision of a safe working environment if a Year 2000 failure occurs. Basically, companies have an obligation to provide a safe working environment under Health and Safety legislation. If a company is shown not to have taken reasonable steps to prevent a Year 2000 injury related incident then it may be culpable under the law.

ADVICE ON TACKLING THE YEAR 2000 PROBLEM

Initial Problem Identification

This is the starting point for ensuring compliance of all computer systems. Companies should consider taking the following action:

- Assess the need to set-up a dedicated year 2000 project team within the company and appoint a competent and proven year 2000 project manager. This individual should hold a senior position within the company to ensure the problem receives exposure at the highest levels of the company so that swift and appropriate action can be taken.

- Generate a detailed inventory of all equipment which is potentially exposed. Where possible try and quantify the effects of the affected systems malfunctioning. Where doubt exists over whether equipment or systems are potentially affected guidance should be sought from equipment manufacturers and suppliers. If in doubt assume the worst and have systems fully evaluated.

- Start a dialogue between yourselves and important customers and suppliers. It is critical to ensure that customers and suppliers are also aware of the problem and taking effective action. Their inability to do so could seriously affect your own business operations. Where a company is either a parent or subsidiary of a large group effective communication lines should be established to ensure that all interested parties are taking action. Particular care is required where computer systems are physically linked within a network as all systems within the network will need to be compliant to ensure that a malfunction or failure does not occur.

- Analyse and allocate sufficient human and financial resources to ensure that the problem can be effectively tackled. Where substantial financial provisions are required these will need to be arranged well in advance.

- Issue clear purchasing instruction to all departments to ensure that all new equipment is year 2000 compliant. Where new equipment and systems are purchased certification should be provided guaranteeing year 2000 compliance and the legal validity of these documents should be assessed.

  Additionally, clear instructions should be issued to all departments who may be responsible for generating new computer code or applications to ensure that these new systems are generated in a year 2000 compliant format.

- The terms and conditions of important contracts including specific wordings should be reviewed to assess whether you are potentially financially or contractually exposed to the year 2000 problem. This should include an assessment of important contracts with suppliers and purchasers.

- You will need to assess the activities of any overseas operations and affiliates with respect to year 2000 compliance

- You will need to review equipment risk assessments conducted under the company Health and Safety policy to ensure that potential year 2000 problems have been considered in relation to possible injury of employees.

Because of the limited timeframe available to achieve compliance, and the potential for many systems to be affected, it will be necessary to prioritise activities. The HSE have developed a scoring system which rates the criticality of systems along with their vulnerability to failure. By using this or similar techniques work can be prioritised in a way which limits exposures to year 2000 failures. It should be remembered that criticality can be defined in a number of ways such as criticality to revenue generation, safety and environmental compliance.

Detailed Implementation and Strategy

Once the scale of the problem has been identified and adequate resources allocated the physical task of making systems compliant can be approached. The key to this vital work is to decide on the approach early on and then stick to it for the duration of the project. In general you should consider the following:-

- The method of achieving compliance needs to be determined. The following overall strategies could be developed:

- Replace existing equipment or programs with new year 2000 compliant systems if this is cheaper than alteration.

- Replace old equipment with new state of the art systems which are year 2000 compliant and highly efficient.

- Modify existing equipment and systems to ensure year 2000 compliance.

- Decide on the implementation strategy. Does the company have sufficient technically qualified in-house resources to tackle the problem? If not, then it will be necessary to hire outside consultants. As the available resources are limited these need to be secured early on to keep costs in check and ensure adequate resources.

- Determine the solution to be applied along with the date format. Basically, there are four different types of solution to the programming problem as follows:-

    - field expansion
    - fixed window
    - sliding window
    - compression.

The most suitable solution should be determined by those involved in the detailed implementation. In general terms a degree of mixing of solutions is possible and often desirable. Suppliers should be closely consulted to assess suitability.

- Where possible automatic conversion methods should be applied. Experience has shown that these are quicker, save on resources and convert in a uniform manner. However, on a cautionary note they will not be suitable for all systems.

- Develop contingencies as part of the conversion process. For business critical systems, contingencies are vital to ensure that operations can continue should problems arise.

CONTINGENCY PLANNING

As previously described, contingency planning forms an important part of the year 2000 compliance process. Contingency plans should be developed to cope with unforseen failures of microprocessor controlled systems or for situations where the failure can be foreseen but the consequences of failure are either unknown or unclear.

Due to the potential for a large number of systems to fail in an unforeseen way (particularly where large numbers of embedded chips are involved), it will not generally be possible to develop contingencies to cover all eventualities. The first task with respect to contingency planning should therefore be to identify and prioritise those systems which are most critical to the continuation of business activity so that contingencies can be developed for these.

Contingency planning in the context of the year 2000 problem will normally involve preparing in advance to either **replace** or **repair** malfunctioning systems or components. This should involve the purchase and stockpiling of substitute components and the development of remedial action plans. Remedial action plans will normally involve assessing the following elements and then documenting the outcome:

- Assessing what probable financial and human resources will be required and ensuring that these will be available. Where outside expertise, usually in the form of IT consultants, may be required, availability should be assessed and contractual arrangements made well in advance. The logistical side of available resources should be carefully analysed to ensure that assistance is available on a twenty four hour continuous basis.

Where in-house human resources are critical it will be necessary to ensure that adequate and appropriate training has been provided. Where possible dry runs of the contingency scenarios should be performed to assess the adequacy of responses.

- An assessment should be made of what tools will be potentially required to alter or repair systems. Where specific tools and systems are required these should be purchased in advance and tested using dummy systems. Tools with respect to the year 2000 problem can have a multitude of definitions including physical tools required to alter machinery and equipment and software tools required to modify computer code.

In addition to an assessment of in-house systems, contingencies may also need to be developed for coping with problems which stem directly from the non-compliance of key customers and suppliers. Generally, these problems can be addressed in advance in a number of ways. Firstly, critical raw materials, components and services can be purchased in advance and stockpiled on site. Secondly, where single sourcing exists an alternative supplier can be sought in advance and preliminary arrangements made to supply materials, components or services in the event that the usual supplier has major problems. Finally, where neither of the above options is seen as being viable, key suppliers and customers should be contacted well in advance and assurances of year 2000 compliance sought.

Obviously, the development of contingency plans will be specific to the operations conducted and the overall company philosophy. A general framework can however be developed and is summarised within the following flowchart:

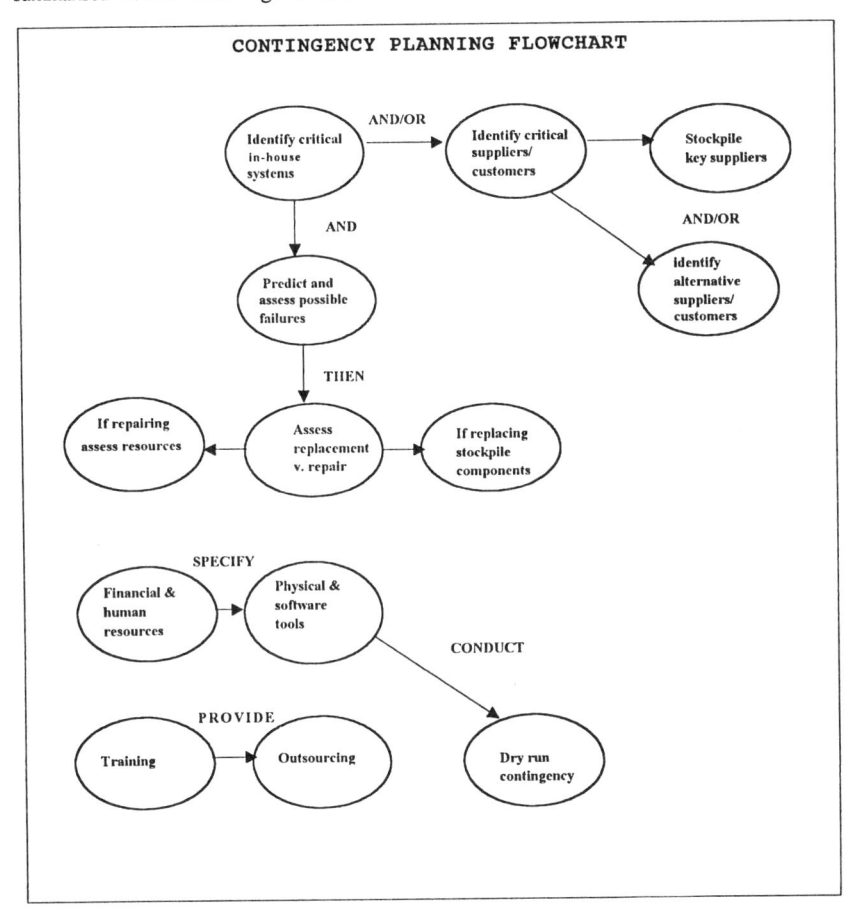

CONTINGENCY PLANNING FLOWCHART

Identify critical in-house systems —AND/OR→ Identify critical suppliers/customers → Stockpile key suppliers

Identify critical in-house systems —AND→ Predict and assess possible failures

Identify critical suppliers/customers —AND/OR→ Identify alternative suppliers/customers

Predict and assess possible failures —THEN→ Assess replacement v. repair

Assess replacement v. repair → If repairing assess resources

Assess replacement v. repair → If replacing stockpile components

SPECIFY: Financial & human resources → Physical & software tools

PROVIDE: Training → Outsourcing

CONDUCT: Dry run contingency

PROCESS VALIDATION

Once equipment has been converted a testing strategy will need to be developed. While it will often not be possible to test all eventualities, the testing regime should be designed to be as rigorous as possible. With respect to testing you should consider the following:-

- Use a separate testing environment from the normal production and/or operational environment. This should ensure that problems can be identified and fixed without putting on-going operations in jeopardy.

- Check that computer licences will not expire and extend or re-new where necessary.

- Monitor all systems and report progress. Problems may not be immediately apparent during testing so that on-going checks will need to be performed.

- Once equipment and/or systems have been successfully monitored certification of equipment can be performed. This would act to benchmark equipment as being compliant and allow progress to be monitored.

CONCLUSION

The year 2000 problem is likely to have major ramifications with potential effects to all aspects of our professional and personal lives. Companies both large and small have an obligation to protect their employees as well as the companies and consumers with whom they interact. Only those companies who can identify the potential for loss and take steps to minimise these potential losses will be able to confidently operate up to and beyond the millennium.

INFORMATION TECHNOLOGY AND TRAINING IN SAFETY

Dinesh Fernando, Head of Product Development - Safety, Health and Environment
Institution of Chemical Engineers, Davis Building, 165-189 Railway Terrace, Rugby, Warwickshire, CV21 3HQ.

During the last decade downsizing in the process industries has led to a number of problems. The removal of experienced employees reduces corporate memory and has implications for safety, health and environmental performance. This reduction in staff numbers leads to increased pressures on trainers to provide materials and courses to train personnel who are constrained by time more than ever before.

The advancement of information technology has now progressed to the point where it is possible to train people remotely or in an open learning situation where personnel can access courses at times convenient to them, fitting around busy work schedules.

This paper describes the benefits and pitfalls of using computer based training (CBT) and illustrates its points by considering a recently launched safety training CD ROM called 'Hazard Spotting'.

Keywords: multimedia, competency, computer based training, distance learning, intranet

BACKGROUND

In today's leaner, flatter organisations, there are less people to do the same job as 10 or 20 years ago. This means that it is increasingly difficult to take people off the job for formal classroom training as there is little cover. While training for performing a particular task can be done on the job, vital issues such as safety, health and environment (SHE) could get sidelined if it means sending people away.

Also, normal staff turnover could mean that even a continuous program of training could omit important personnel until the next tranche of sessions is organised.

There is no getting away from training in SHE: incident prevention and mitigation, control of environmental impact and welfare of personnel all have knock on effects for business performance and training is an integral part of the mix of control measures.

If people cannot be taken off the job for formal training sessions then the next best thing is to make available training that people can access whenever they want and allows a trainer to track the performance of trainees. This can be achieved through training delivered via multimedia CD ROM installed on a computer that appropriate people can have easy access to. The rest of this paper describes some of the main elements of computer based training and, as an illustration, describes 'Hazard Spotting' a new CD ROM providing interactive training in the main hazards from handling chemicals in the process and chemical industries.

COMPUTER BASED TRAINING / MULTIMEDIA

What it is

Multimedia is an imprecise term applied to all kinds of different applications, but a common interpretation is the presentation of information on display screen through the combination of text, images, animation, video, interactive exercises and sound. The information can be presented on demand by the user clicking on-screen buttons or the information can just scroll through like a video.

Common examples of multimedia applications include CD ROM encyclopaedias, web sites, training programs, corporate information CD ROM's, training programs and computer games.

The delivery of these programs can be via CD ROM, computer disk, Internet or company network.

In the training field there are many companies producing material that covers all sorts of different needs such as display screen equipment, manual handling or management techniques. As well as the enforced necessity of companies using computer based training for the reasons outlined previously, there are some benefits to using this medium:

- reduction in travel, time, costs;
- reduction in study time;
- a consistent message is given;
- ensures a longer term retention of information;
- material is self-paced;
- trainees' progress can be recorded and tracked.

This is all based on the premise that the training program itself is effective in conveying the information. Unfortunately there are some programs available that purport to be multimedia training but do not do a good job in driving the messages home.

Too often, the programs are simply electronic books with little or no user interaction other than simply clicking a mouse to get to the next page. Also some multimedia relies heavily on static photographs with a voice-over explaining what is happening.

There is the danger of going too far the other way. Multimedia producers can get carried away with their own cleverness by creating applications that have overly complicated graphics or animations that may look attractive but do not convey information effectively or cannot work on the end user machine.

While elements of the above can be present in successful training programs there is a need for variety in how the information is conveyed. The user also needs to interact with the program for a truly cognitive training experience.

The following elements are common to successful training programs:

- ease of use;
- simple navigation, ie back and forward buttons, exit, bookmark or resume functions;
- meaningful contents;
- quality media - video, animation, graphics and photographs;
- interactive elements where the user can explore concepts or learn by discovery;
- program should provide feedback to trainees and trainer.

To illustrate how multimedia can help with safety training, the rest of the paper considers 'Hazard Spotting' a recent CD ROM training package.

HAZARD SPOTTING

'Hazard spotting' is aimed at teaching personnel the main hazards of handling chemicals in the process and chemical industries:

- fires and explosions;
- chemical reaction hazards;
- occupational hygiene.

The program specifically excludes direct injury incidents such as slips, trips and falls and manual handling.

After logging on to the course new trainees are automatically taken to the Introduction module which describes some common terms for safety and sets the scene with regard to people's responsibility and the scope of the course. There is also a hazard spotting exercise involving a domestic situation which acts as a gentle taster of what is to come.

Once they have completed this section, trainees are presented with the main menu screen where they can choose a topic module (Figure 1).

Each topic module is designed to generate awareness in the subject. The information is conveyed using a mixture of animation, video clips, interactive exercise and information screens. The navigation allows personnel to plough through the course or skip about the main headings to recap on particular points of interest - it is their choice.

Each topic module contains several interactive exercises where familiar principles are brought to life by the power of multimedia. For example the module on chemical reaction hazards explains the principle of why exothermic reactions runaway with the help of an interactive version of the Heat flows versus Temperature graph. Trainees can explore the effects of increasing the reactor set-point temperature by moving a pointer (Figure 2).

As the trainee increases the temperature, the reactor animation gets more vigorous and an explanation is given by the caption. Eventually the reaction runs away and the reactor explodes if the set-point goes beyond the critical temperature.

It is in this type of interactive exercise that the program really comes into it own, helping trainees gain a fundamental understanding of an important principle.

Hazard spotting exercises

Once trainees have been through a topic module they can attempt the multi-choice question set and the appropriate hazard spotting exercise.

There is one hazard spotting exercise for each topic module:

- **fires and explosions** - potential ignition sources;
- **chemical reaction hazards** - hazards that could contribute to an uncontrolled runaway reaction;
- **occupational hygiene** - areas of exposure to toxic materials.

After seeing an animated description of the plant and information about the materials and processes being used, trainees have to click on an area that could be a hazard. There is a specific number of hazards to spot in each exercise and the trainee gets an explanation of each hazard they spot successfully (Figure 3).

The sample plant has a typical array of hazards handling toxic and flammable liquids and powders and the reaction materials can also run away or decompose thermally (Figure 4).

Administration

As well as providing the course material, Hazard Spotting tracks the performance of trainees by writing scores in the hazard spotting exercises and question sets to text files for the course tutor to assess (Figure 5). The program also gives trainees the option of viewing and printing out their scores for each particular session and resuming at the last page they were viewing within the topic modules.

CONCLUSIONS

Computer based / multimedia training will form an important part of future initiatives in SHE training. Although it is hard to beat a good trainer in a classroom training situation computer based training, if done properly, can provide a highly effective alternative.

Effective SHE training is usually an implicit and explicit requirement for complying with laws and regulations. It is also an important element of control measures for eliminating hazards and reducing risk. The advent of effective multimedia training now means that vital training can be provided without disrupting busy work schedules.

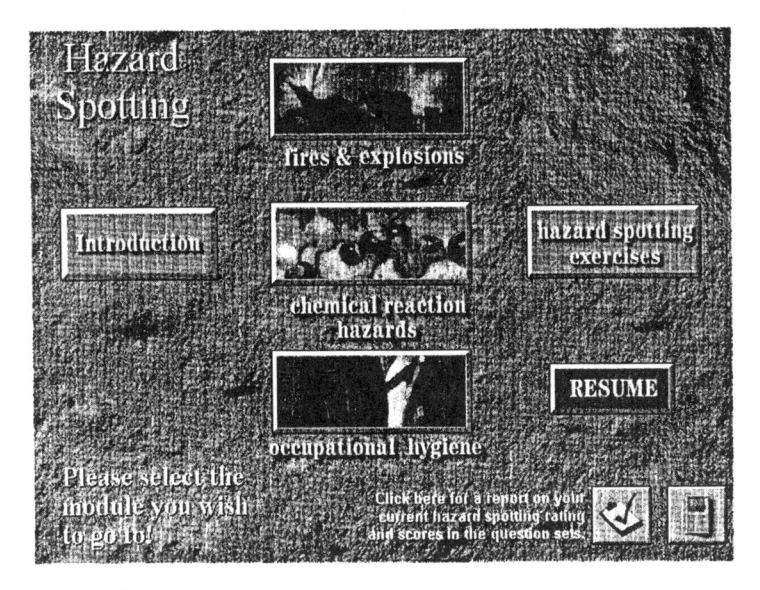

Figure 1: The main menu screen for Hazard Spotting.

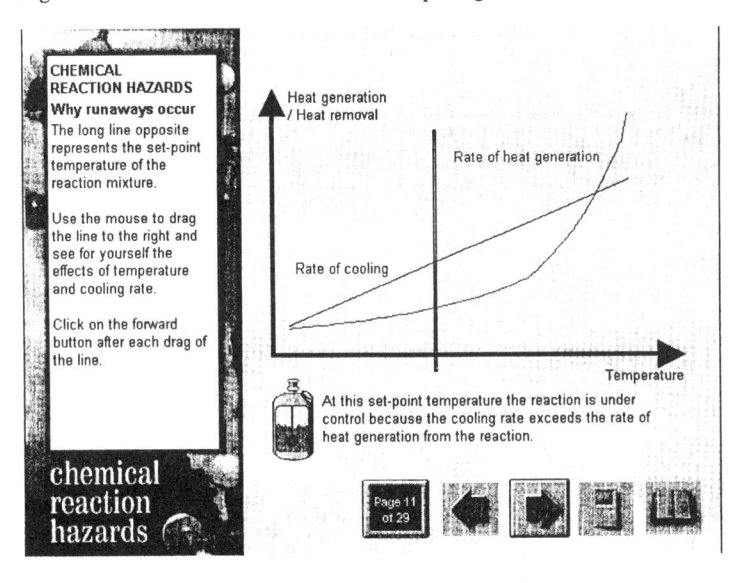

Figure 2: The interactive exercise for why runaway reactions occur.

Occupational hygiene

Exposure to the materials handled in this plant can have acute (short term) and chronic (long term) effects on workers:

POWDERED RAW MATERIAL:
8 hour TWA exposure limit: 15 mg / m³
Suspected asthmagen and probable carcinogen
Oral rat LD50 test: 1800 mg/kg
When heated to decomposition the powder emits toxic fumes
LIQUID RAW MATERIAL:
8 hour TWA exposure: 10 ppm
Vapour is carcinogenic and suspected teratogen
Contact with eyes causes irritation
Contact with skin gives rise to corrosive and defatting action
Acute inhalation of vapour causes dizziness, narcosis and hallucinations
Chronic inhalation could cause loss of ability to concentrate and mild lethargy

Click on forward to go on to a hazard spotting exercise for occupational hygiene. Click on the thumbnail to recap on the plant description.

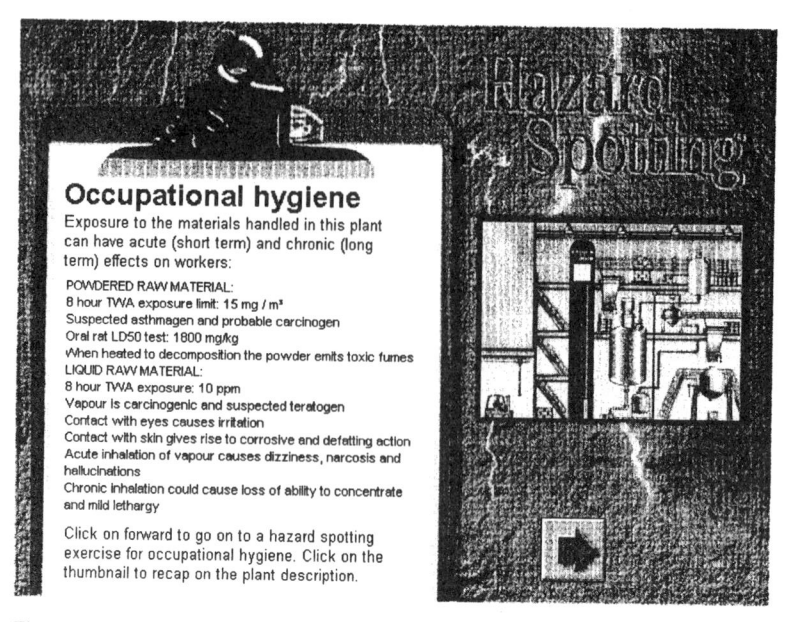

Figure 3: The information screen for one of the Hazard Spotting exercises.

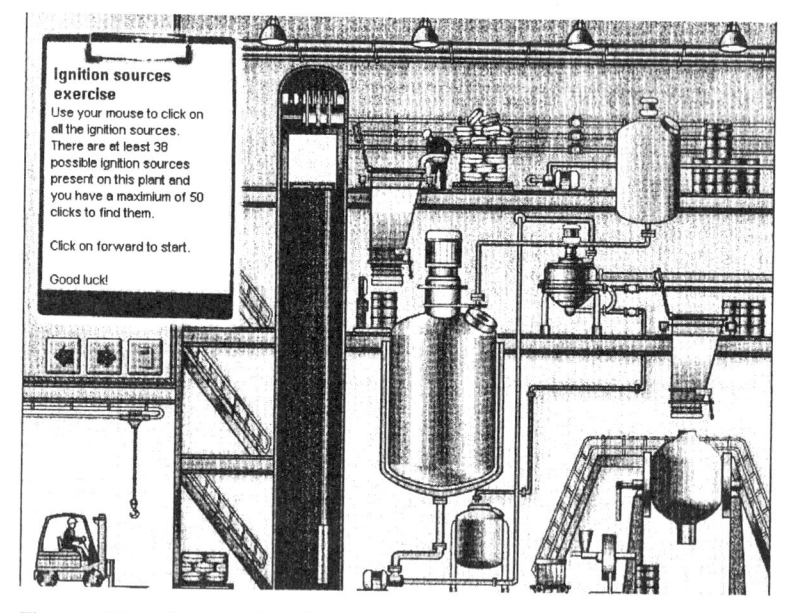

Ignition sources exercise

Use your mouse to click on all the ignition sources. There are at least 38 possible ignition sources present on this plant and you have a maximum of 50 clicks to find them.

Click on forward to start.

Good luck!

Figure 4: The main screen for a Hazard Spotting exercise.

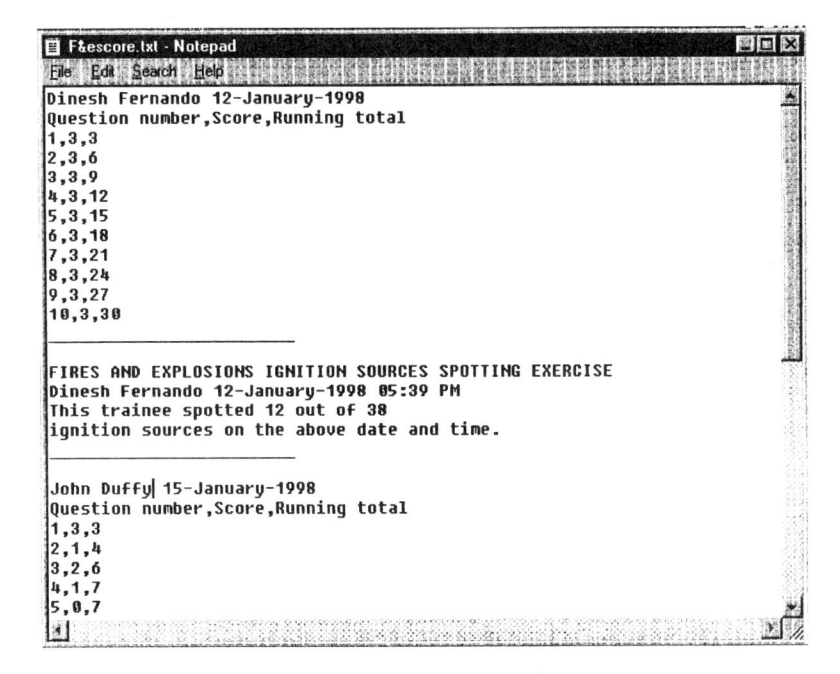

```
Dinesh Fernando 12-January-1998
Question number,Score,Running total
1,3,3
2,3,6
3,3,9
4,3,12
5,3,15
6,3,18
7,3,21
8,3,24
9,3,27
10,3,30
_____

FIRES AND EXPLOSIONS IGNITION SOURCES SPOTTING EXERCISE
Dinesh Fernando 12-January-1998 05:39 PM
This trainee spotted 12 out of 38
ignition sources on the above date and time.
_____

John Duffy 15-January-1998
Question number,Score,Running total
1,3,3
2,1,4
3,2,6
4,1,7
5,0,7
```

Figure 5: Text file containing details of the trainees' performance.

THE SENSITIVITY OF RISK ASSESSMENT OF FLASH FIRE EVENTS TO MODELLING ASSUMPTIONS

P.J. Rew, H. Spencer
WS Atkins Safety & Reliability, UK

T. Maddison
Health & Safety Executive, UK

> Based on reviews of modelling techniques and data from flash fire field trials, models for the prediction of the burn area of a flammable cloud (in terms of fraction of mean LFL of the cloud), flame speed through the cloud and heat transfer from the flash fire have been produced. The validity of the typical assumptions used in the modelling of flash fires has been tested through comparison against these models and a set of revised assumptions is provided relating to the likelihood of fatality of people caught within a flash fire and the mitigating effects of shelter and escape from the cloud. A sensitivity analysis is performed in order to identify which assumptions or aspects of flash fire modelling are the most critical in terms of off-site risk. The analysis suggests that risk assessments are most sensitive to the definition of burn area of the cloud, and therefore also to dispersion modelling, and to the proportion of people within shelter who are fatalities as a result of secondary fires.
>
> Keywords: Flash fires, flammable gas clouds, quantified risk assessment

INTRODUCTION

Flash fire models used for the purpose of risk assessment are usually based on gas dispersion modelling combined with the probability of ignition (e.g. Considine et al (1) and Clay et al (2)), where the boundary of the fire is defined by the downwind and crosswind dimensions of the gas cloud. On behalf of the Health & Safety Executive, a detailed review was conducted (3), which considered incident and experimental data and risk assessment techniques related to flash fire modelling. This review allowed a framework for an improved flash fire model to be produced, which was developed further in a subsequent study (4). Models for the prediction of the burn area of a flammable cloud (in terms of fraction of mean LFL of the cloud), flame speed through the cloud and heat transfer from a flash fire were produced. The study also assessed the validity of typical assumptions used in the modelling of flash fires and a set of revised assumptions was provided relating to the likelihood of fatality of people caught within a flash fire and the effect of shelter and escape from the cloud in reducing fatality levels.

This paper outlines working models for determining the effects of flash fires, which were then used in testing which aspects of flash fire modelling are most critical in calculating off-site risk.

KEY MODELLING ISSUES

When a dispersing cloud of flammable vapour is ignited, its mode of burning will depend on its shape, levels of turbulence present at ignition (or generated during the fire event by the presence of obstacles) and the fuel concentration. Generally, a flash fire is considered to occur when a dispersed gas cloud is ignited within its flammable region, causing a wall of flame to spread throughout the flammable region and back to the release point. In this paper, and as given by CCPS

(5), a flash fire is defined as 'the combustion of a flammable gas or vapour and air mixture in which the flame propagates through that mixture in a manner such that negligible or no damaging overpressure is generated'.

The current approach to flash fire modelling is simplistic in assuming that the burn area covers the dispersion footprint, and that its effects are felt either not at all, or only to a limited extent beyond this area. Current risk assessments for flammable materials which use this simple approach tend to be dominated in the far-field by flash fires and there are a number of possible areas in which the modelling of these events may be over-conservative:

1. *Uncertainties in dispersion analysis of flammable gas clouds.* As suggested by Cracknell & Carsley (6), uncertainties in predicting distance to LFL (or fraction of LFL) may mask other uncertainties in the determination of burn area. Uncertainties may relate to the use of models for complex topographies or the need for simplifying assumptions in defining source properties for events. Even where the scenario fits the type of release for which the dispersion code is designed, uncertainties may still result in large discrepancies between models.
2. *Effect of assumptions regarding burn area.* The fraction of mean LFL taken as the burn area is strongly dependent on the level of mixing (concentration connectivity) within the cloud.
3. *Effects of ignition location and timing on burn area.* Ignition is likely to occur before a cloud disperses to its maximum size. The cloud area at ignition, and thus the risk, is sensitive to ignition modelling. Also, fireball effects may occur if ignition occurs early in the dispersion of the flammable cloud.
4. *Likelihood of fatality for fire engulfment.* At present it is conservatively assumed that 100% of unsheltered people caught within the burn area of a flash fire event are fatalities.
5. *Effect of shelter on likelihood of fatality.* It is generally assumed that buildings provide some form of shelter from a flash fire. The probability of fatality will depend on the integrity of the building within a flash fire, whether gas has entered the building resulting in explosion effects and whether escape is possible from any secondary fires produced.
6. *Possibility of escape from cloud.* Escape from the cloud area may be possible both before and after ignition. Escape will depend on the effectiveness of on- and off-site warning systems and emergency planning and, for large releases, the proximity of shelter.

OUTLINE OF WORKING MODEL

The working model consists of three parts; prediction of burn area (fraction of LFL for which combustion occurs), prediction of flame speed and prediction of fatality due to radiation effects. The development of each of these is described below. Full equations are provided by Rew et al (4).

Definition of burn area

If ignition occurs within the flammable region of a release then it can be argued that the flame will propagate to one of the following three extents;

a. *Localised burning.* Combustion occurs in a small region around the ignition point but fails to spread to the rest of the cloud.
b. *Downwind propagation.* The conditions within the cloud are sufficient to sustain propagation of the flame downwind of the ignition point.
c. *Burn-back.* The conditions within the cloud are sufficient to sustain flame propagation upwind, back to the source of the release, i.e. the mean flame speed is greater than the wind speed.

Burn-back (case c) will result in escalation of the scenario to a pool or jet fire. The extent of both the burn-back and downwind flame propagation (case b) of the flame from the ignition point will determine the 'burn area'. In the Maplin Sands trials (7) and other studies conducted by Shell (8,9), it was found that flames burnt downwind to beyond the mean lower flammable limit (LFL) but not as far as the distance to ½ LFL. However, at present, no clear guidelines exist for defining the fraction of LFL to which flames will propagate. This is partly due to the highly non-homogeneous nature of a dense gas cloud, with the likelihood of flame propagation being dependent on the statistical probability of the flammable pockets within the cloud being connected to each other. Furthermore, the expansion of combustion products as the flame front progresses through the cloud is likely to affect the mixedness of the unburnt cloud ahead of it.

As a first approximation, it is assumed that a flame can travel from one point to another if there is a path between the two points where the fuel concentration is always between the flammability limits, and that the effect of expansion of combustion products is small in the lean region of the cloud. The existence of this path can be modelled using Percolation Theory, which is widely used in many engineering fields and describes how sites or nodes within a random field are connected to each other. Using a 'potential' model of continuum percolation, as described by Sahimi (10), a critical occupied volume (or area) fraction, ϕ_c, can be defined at which the system begins to percolate (i.e. at which a flame path exists). Assuming that the scale of concentration fluctuations of interest within a dense gas cloud are small compared to its width and length, but not its height, then the cloud can be represented as a two-dimensional lattice and ϕ_c can be taken to be equal to 0.44 (Scher & Zallen (11)). Suitable assumptions must then be made regarding the dispersion of the fuel. Using the empirical correlations for concentration fluctuations proposed by Wilson (12), the relationship between ϕ_c and the concentration properties of the cloud can be calculated and the results are illustrated in Figure 1 for $\phi_c = 0.44$. These results can be used to determine whether flame propagation will occur at a certain point within the dense gas cloud and thus to define the fraction of LFL at which combustion can occur.

Using this model, the burn area for a 5 kg/s release of propane from a pool source has been predicted to extend to 0.5 LFL close to the source and to 0.8 LFL on the downwind centreline of the cloud. This latter value compares to that of 0.9 LFL given by Evans & Puttock (8) for sustained burning of a flammable gas cloud at its downwind limit, based on medium-scale trials. However, it should be noted that this method is suitable only for continuous pool releases over flat terrain, as it uses the correlations for concentration fluctuations proposed by Wilson (12), and presently does not include modelling of concentration fluctuations for instantaneous and horizontal jet releases.

Prediction of flame speed

Observations of full-scale flash fire trials suggest that ignition is followed by a transient partially premixed flame passing through the flammable regions of the cloud, followed by a yellow diffusion flame burning through the rich section of the cloud. For the special case of centreline ignition at the downwind edge of the flammable region of the release, a likely flame path is illustrated in Figure 2. Flame speed varies with gas concentration, being at its maximum when the mixture is close to stoichiometry, and is increased by the presence of turbulence. Thus the flame will travel along the path of least resistance (i.e. greatest flame speed) towards the spill point and, initially, when the mean concentration is below stoichiometric, the path of least resistance will be at the maximum concentration, and hence along the plume centreline. The flame will then tend to spread along the stoichiometric contour, around the edge of the rich region of the cloud. This latter stage of the flame propagation matches observations of the Maplin Sands field trials, given by

Jenkins & Martin (7), where the flame front advances along the better-mixed edges of the cloud until a constant velocity, horseshoe shaped, diffusion flame is produced. As suggested by Raj (13), the speed of the diffusion flame front is dependent on the premixed flame speed through the stoichiometric regions of the cloud. The width of the diffusion flame will depend on the rate of air entrainment into this region.

In order to calculate flame speed at a certain position within a cloud, firstly the concentration closest to stoichiometry for which there may be a connected path through that part of the flame is calculated, based on Percolation Theory model discussed above. The laminar flame speed as a function of stoichiometry is approximated by a quadratic fit to experimental data and the turbulent flame speed is defined in terms of the laminar flame speed and the turbulence of the atmosphere using Gulder's (14) correlation. The flame speed with respect to the ground is evaluated in terms of the turbulent flame speed, the wind speed and the effects of gas expansion on a plane flame, noting that the latter may produce a factor of 2 to 3 increase in flame speed.

The model has been shown to produce reasonable predictions of flame speeds observed in the Maplin Sands field trials. Use of the flame speed model for scenarios with turbulence levels typical of industrial plant, or of built-up areas, would suggest that the flame speed is likely to be far greater than those seen at Maplin Sands, where the ground roughness was low (mud flats). Thus for most practical release scenarios it is unlikely that the flame speed will be less than the ambient wind velocity (as was observed for some of the Maplin Sands trials) and burn-back to the source will almost always occur.

Simple modelling of external heat transfer from flash fires

As illustrated in Figure 3, and as originally suggested by Raj & Emmons (15), the external radiation from a flash fire can be calculated by modelling the fire as a wall of flame which moves through the cloud away from the ignition point. The majority of external radiation from the flash fire comes from the rich diffusive burning regions (mean concentration greater than stoichiometric), with a lower intensity of radiation emitted from the 'premixed' burning regions. Due to concentration non-homogeneities in the cloud, the 'premixed' burning region consists of patches of gas at low concentration, burning with a bluish flame and emitting negligible radiation, interspersed with patches of rich gas mixture, burning with a higher intensity yellow flame. It is assumed, for LPG and LNG releases, that the rich region of the cloud emits 30% of its heat of combustion as external radiation ($f_s = 0.3$) and that the lean region emits 10% ($f_s = 0.1$). These values are consistent with surface emissive powers observed in the Maplin Sands trials. The speed at which the wall of flame moves is taken from the model for flame speed within flash fires outlined above.

RISK ASSESSMENT METHOD

Fatality criteria for external radiation and fire engulfment

Two studies (16,17) have recently been completed which review methods for determining the likelihood of fatality for people exposed to thermal radiation from hydrocarbon fires. These studies recommend fatality criteria for an average population, based on a thermal dose, $I^{4/3}t$, where I is the heat flux at the location of the target group and t is the duration of exposure to the flux, as summarised in Table 1. However, these criteria are derived for radiation incident on a target population located outside the boundary of the fire event and require modification if used to predict fatalities due to engulfment within a fire. Based on medical data it was postulated that in order for high levels of fatality to occur, a significant proportion of skin area must receive full thickness burns (approximately 30% body surface area for 50% fatality) and that this severity of burn occurs at 1000 $(kW/m^2)^{4/3}$s. Typically, the unclothed area for an average population is also 30% and, therefore, all of the unclothed area must be exposed to the full thickness burn dose in order to produce 50% fatality. If the population is located outside the boundary of the fire, the thermal radiation can only be incident on half of the exposed skin area at a point in time and so it can be crudely assumed that the cumulative incident dose required to cause 50% fatality is double the dose for full thickness burns, i.e. 2000 $(kW/m^2)^{4/3}$s.

Fatality criteria	Percentage fatality for an average population	Thermal dose $(kW/m^2)^{4/3}$s
Dangerous Dose	1 to 5%	1000
SLOD (Significant Likelihood of Death)	50%	2000

Table 1 Fatality criteria for an average population

If the population is engulfed in the fire event, then all the unclothed skin area will be simultaneously exposed to the heat flux. Thus the cumulative dose for 50% fatality reduces to 1000 $(kW/m^2)^{4/3}$s. The threshold dose for ignition of certain types of clothing may be as low as 1800 $(kW/m^2)^{4/3}$s and fatality is highly likely (50% or higher) if ignition occurs. For engulfment within a fire, this dose will already have caused severe full thickness burns over 30% of the body surface area and this proportion will increase if clothing is ignited. Thus it is assumed that, for fire engulfment, a dose of 1800 $(kW/m^2)^{4/3}$s will result in 100% fatality for an average population.

Probability of fatality for engulfment within flash fires

Flash fire models used for the purpose of risk assessment usually assume that personnel caught within the gas cloud burn area are fatalities and that those outside are not seriously injured. When people are engulfed in a fire, they are subjected to both radiative and convective heat fluxes. For a flash fire event, the radiative flux will come from both adjacent flame areas and also from more distant, but more intense, areas of the fire. The convective flux will be dependent on the temperature and speed of flow of combustion products around the engulfed personnel.

The calculation of heat transfer within the lean regions of a flash fire (where the mean concentration is less than the stoichiometric concentration) requires determination of the temperature of combustion products around personnel caught in the fire, the speed of the combustion products and the duration for which personnel are exposed to the combustion products. These properties vary depending on the position within the flammable gas cloud, and Rew et al (4) use CFD analysis of Maplin Sands Trial 50 to provide such data. All positions within the burn region were found to produce thermal doses far in excess of the dose for 100% fatality, 1800 $(kW/m^2)^{4/3}$s. Close to the stoichiometric contour, the thermal dose is as high as 40,000 $(kW/m^2)^{4/3}$s. However, outside the burn region the thermal dose drops off rapidly. Thus the analysis suggests that the currently used assumption, that all those caught within the burn area of a flammable gas cloud are fatalities, is a reasonable one.

Effect of shelter on probability of fatality

It is generally assumed that buildings provide some form of shelter from a flash fire. The probability of fatality will depend on the integrity of the building within a flash fire, on whether gas has entered the building resulting in explosion effects and on whether escape is possible from any secondary fires produced. Where gas ingress to buildings results in internal flammable concentrations, it would be conservative to assume that the probability of fatality is 100%, due to both burn and explosion effects.

If the flammable concentration within a building does not reach the lower flammable limit, then the probability of fatality of the occupants will depend on whether the building ignites and on the probability of evacuation from the building. Various criteria exist for ignition of wood due to radiated heat, e.g. Lawson & Simms (18). However, there is uncertainty in the application of the above criteria to ignition of buildings caught within a flash fire. This is due to heat transfer occurring by convection as well as by radiation, and also the wide range of materials used in building construction. In order to determine reliably whether ignition of the building occurs, more extensive data is required on ignition of building materials by the thermal dose transferred during a flash fire event. In the sensitivity analysis described in this paper it is assumed that ignition does occur and that a secondary fire is initiated.

Once it has been determined whether a building will ignite during a flash fire event, the likelihood of fatality for building occupants must be estimated. Data taken from the 1993 Home Office Fire Statistics (19) give some indication of fatalities in fires which are similar (in some respects) to a flash fire engulfing a building. For example, malicious fires in dwellings may produce a rapidly growing fire with no warning. However, the percentage of fires where fatalities occur is of the order of 0.5 % for this type of fire. Similarly, for fires caused by bombs, petrol bombs or other incendiary devices, approximately 1% resulted in fatalities. Fires caused by fuel supplies or other fuel-based appliances may also produce fires with little prior warning. Although it is likely that most of these fires originated indoors, the fatality rate is again very low (less than 2%). However, the following should be noted when applying the Home Office data to the determination of the probability of fatality (or percentage being unable to escape) for occupants of a building ignited by a flash fire event:

- people are likely to evacuate a building if they perceive a fire inside it to be serious. However, it is not clear how people would react if the exterior faces of the building were engulfed by fire and, if people attempt to evacuate, whether fatalities may result;
- the initial rate of fire development is likely to be significantly greater for ignition by a flash fire than for the majority of fire events comprising the data given above;
- fire brigade action will have had some impact on the statistics above, but will not be as effective for aiding escape from the large number of dwellings that are likely to be alight after a flash fire;
- flash fire events may produce localised blast effects, and external flames and missiles (glass fragments etc.) may penetrate buildings, increasing fatality levels.

It is evident that there are large uncertainties in estimating the effect of secondary fires on fatality levels. However, as an initial estimate, use of the Home Office Fire Statistics suggests that the proportion of occupants who become fatalities is likely to be small (of the order of 5% say). This assumes that flammable concentrations of gas do not build-up within the building and that external blast effects and flames are unable to penetrate the building boundaries.

Likelihood of escape from flash fires

Escape from the effects of flash fires can be considered to occur either:

a) before the arrival of the flammable cloud, i.e. evacuation as part of an emergency plan;
b) during or after the arrival of the flammable cloud, but before ignition occurs; or
c) after ignition of the flammable cloud, before the arrival of the flame front.

Detailed consideration of evacuation (**a**) has been considered by Purdy & Davies (20), mainly in connection with toxic releases. When a release of gas occurs, the cloud may travel quickly, not allowing sufficient warning for effective evacuation to be implemented. Therefore, as noted by Lees (21), shelter is generally preferred to evacuation, the exception being where escalation of an on-site event threatens loss of containment from a tank or vessel, and where there is sufficient warning of the likely outcome. The likelihood of escape of personnel during or after the arrival of the flammable cloud (**b**) will depend on whether the release is on-site or has travelled off-site. For many on-site cases, it could be expected that personnel will have already vacated the area of the release before ignition occurs, especially where the flammable region is visible (for releases of liquefied gases in humid conditions). For off-site dispersion of flammable gas, the likelihood of escape is lower; the population may be ignorant of the potential hazard or may be trapped within an external area e.g. playground.

Even after ignition occurs (**c**), it could be argued there is some scope for escape of personnel away from the flame front. However, people are unlikely to choose the optimum direction for escape and it could be argued that they would not start to escape until the flame front was within reasonable proximity (of the order of 50-100m say). Even when escape is seen to be necessary there may be a delay before escape commences; a delay of 5 seconds is assumed by Nussey (22) for jet fire events. Therefore, it is assumed that escape from a flash fire once ignition has occurred is unlikely in most situations.

SENSITIVITY ANALYSIS

By using the above assumptions in combination with models for ignition probability, gas dispersion and ingress into buildings and data on population characteristics, the number of fatalities caused by a flash fire event, given that a release of flammable material has occurred, can be calculated. The event tree in Figure 4 illustrates the logic behind such a calculation noting that P_o is the probability that people will not be sheltered (taken to be 0.1 for D weather conditions and 0.01 for F weather conditions (23)). P(A) is the probability that a cloud of area, A, has ignited, noting that it may ignite before it reaches its maximum burn area, and is calculated using the method currently implemented in Flammables RISKAT (2). Ignition is assumed to occur at the edge of the cloud and results from strong, continuous sources, with source density, μ (sources/m^2):

$$P(A) = 1 - e^{-\mu A}$$

Typical values for the conditional probabilities used in the event tree are given in Table 2, which also lists credible variations on these values based on the discussion above. Table 3 gives ignition source data and population characteristics for urban and rural land use areas. The sensitivity of risk is examined for two release scenarios, as defined in Table 4, and for both 2F and 5D conditions. It can be seen from this table that the maximum burn area of the continuous release (for either LFL or ½ LFL) is sensitive to the windspeed used in the gas dispersion modelling. This contrasts with instantaneous releases for which, as shown in Figures 5 and 6, the maximum burn area at mean LFL is not significantly affected by either weather conditions or ground surface roughness (although it should be noted that the effect of initial cloud aspect ratio has not been investigated in this study).

Parameter	Description	Base Case	Variation
A_{max}	Maximum burn area	LFL	½ LFL
P_{fi}	Unsheltered probability fatality	1.0	0.5
P_{fo}	Sheltered probability fatality	0.5	0.05
P_e	Probability escape	0.0	Escape model (4)
P_o	Probability outside	0.01F, 0.1D	0.0
z_o	Ground surface roughness	0.1m	0.2m

Table 2 Variation of parameters considered within sensitivity analysis

Property		Urban	Rural
Population density, ρ_p (people/m^2)		2.5E-3	5.6E-5
Ignition source density, μ (sources/m^2)	Day, 5D	2.5E-5	1.2E-6
	Night, 2F	3.8E-6	1.9E-7

Table 3 Data for Urban and Rural land-use types

Property		Instantaneous	Continuous
Material		\multicolumn{2}{c}{Liquefied Propane}	
Quantity/rate		200 tonne	50 kg/s
Initial concentration		13.4 % (V/V)	100 %
Initial aspect ratio		1.2	-
Pool size		-	2m by 2m
Area at mean LFL	2F, z_o = 0.1m	48 hectares	15 hectares
	5D, z_o = 0.1m	48 hectares	1.2 hectare
	2F, z_o = 0.2m	44 hectares	12 hectares
	5D, z_o = 0.2m	44 hectares	0.9 hectares
Area at ½ mean LFL	2F, z_o = 0.1m	94 hectares	26 hectares
	5D, z_o = 0.1m	79 hectares	2 hectares

Table 4 Release scenario properties

The results of the sensitivity analysis are illustrated in Figures 5, 6 and 7, which show a large variation in fatalities per event between the base case releases. The difference between the instantaneous release over urban land (Figure 5) and the continuous release over urban land (Figure 7) can be attributed to the difference in flammable cloud area. The difference between the instantaneous release over urban land (Figure 5) and rural land (Figure 6) is due to the difference in population densities between these land-use types.

The figures show that, in general, changing assumptions regarding the burn area of the release has a large effect on fatalities per event, noting that burn area is affected by dispersion modelling as well as whether LFL or ½ LFL is used for the flammable region. In general, increasing the burn area increases the number of fatalities, the exception being for an instantaneous release in 5D conditions over urban land, where changing the burn area has no effect on number of fatalities. This results from the modelling of ignition probability and Figure 8, based on analysis conducted by Spencer & Rew (24), illustrates the effect of ignition source density on flash fire risk for an instantaneous release, assuming all other parameters remain constant. It can be seen that the risk peaks at a critical ignition source density, μ_c sources/m^2. This peak occurs where the density is such that there is a high probability that the cloud ignites close to its maximum size. For densities higher than this critical value, the cloud will almost certainly have ignited before it can reach its maximum size. For densities lower than this critical value, the cumulative probability that the cloud will ignite is reduced, even when the cloud reaches its maximum size. Thus, for the 5D instantaneous release over urban land, the source density is greater than the critical value, the cloud ignites before it can reach its maximum burn area, and increasing the maximum burn area to ½ LFL has no effect on fatalities. For all other release conditions studied, the source density is well below the critical value and any increase in the maximum burn area, whether due to changes in the dispersion modelling or use of ½ LFL rather than LFL, will increase the number of fatalities. Thus Figure 8 illustrates the high sensitivity of flash fire risk to ignition probability modelling, and illustrates the uncertainty that can be introduced when two areas of modelling (i.e. of ignition probability with flash fire effects) are combined. It should also be noted that Figure 8 considers risk from flash fire effects only and, due to event escalation and other types of event, the total risk is unlikely to be reduced by increased ignition source density around a site.

For all the releases considered, decreasing the probability of fatality for those indoors, P_{fi}, by a factor of ten, from 0.5 to 0.05, produces a large reduction in the number of fatalities. This is due to the small proportion of people assumed to be outside and not sheltered from the fire. For 2F releases, where only 1% of people are assumed to be unsheltered, the number of fatalities is decreased by a factor of close to 10 from the base case. For the opposite reason, the probability of fatality for those caught outside, P_{fo}, and the probability of escape, P_e, have only a small effect on the number of fatalities per release. Reducing P_o to zero, while keeping P_{fi} equal to 0.05, has a significant additional effect on the number of fatalities (a reduction by a factor of 3) for the 5D cases where P_o was originally 0.1. However, assumptions regarding evacuation or escape have no effect for the 2F cases where only 0.01 of people are assumed to be outside in the first place.

One component of the modelling of flash fires which has not been investigated in the above sensitivity analysis is the effect of thermal radiation external to the burn area. CFD modelling (4) of Maplin Sands Trial 50, confirmed that the probability of fatality dropped rapidly from 100% within the burn region to very low levels just outside the burn region, suggesting risk calculations are insensitive to modelling of external radiation.

SUMMARY AND CONCLUSIONS

The analysis described above suggests that the modelling of flash fire risk is most sensitive to:

- the burn area of the cloud and thus to dispersion modelling assumptions (and, by implication, choice of dispersion model) and the fraction of mean LFL at which combustion is sustained;
- the modelling of ignition probability and, in particular, the assumed ignition source density;
- the likelihood of fatality for those indoors;

It should be noted that the sensitivity studies discussed above consider single flash fire scenarios in isolation and do not consider the importance of flash fire effects in relation to other process fire events, such as fireballs, BLEVEs and VCEs, or in relation to escalation of an incident. Particular risk studies will be sensitive to the relative frequency of these different fire events, as well as the relative frequency of different release modes and sizes. However, the analyses do illustrate the possible large effect on risk calculations of changing modelling assumptions. This has implications not only when comparing risk values, but also when considering which risk reduction measures should be implemented across a site. Based on the work described above, use of the following list of assumptions for flash fire modelling can be suggested:

1. Use of the model for burn area, based on Percolation Theory, suggests that for continuous releases, the mean concentration for burn area is closer to LFL than to ½ LFL, although it should be noted that experimental work and further analysis is currently being conducted to verify this conclusion. Different types of release and variations in topography are also being considered.
2. Those caught within the burn region who are unsheltered will be fatalities. Outside the burn region, the thermal dose drops sharply and it can be assumed for risk assessment purposes that there are negligible fatalities.
3. Approximately 5% of those caught within the burn region who are sheltered will be fatalities. This is based on the analyses presented above and assumes that flammable concentrations of gas do not build up within the shelter and that external blast effects and flames are unable to penetrate the building boundaries. It should be noted that there are significant uncertainties in the estimate of 5% fatalities and that, in light of the sensitivity of risk calculations to this value, further investigation of flash fire effects on buildings is currently being undertaken.
4. People are unlikely to be able to escape from the effects of the flash fire after the cloud has been ignited, i.e. the probability of escape from an ignited cloud is small. For special cases, where the flash fire flame speed is likely to be low, there may be value in considering the effect of escape within a risk assessment. However, as noted above, for most releases into industrial plant or built-up areas the flame speed is likely to be too high for escape to be practical and, in any case, the risk of a flash fire is relatively insensitive to the modelling of escape.
5. Evacuation or escape to shelter of off-site personnel before ignition occurs is considered to be improbable. However, the risk from a flash fire is sensitive to the number of people who are assumed to be indoors and this may warrant further investigation.

ACKNOWLEDGEMENT

The work described in this paper has been undertaken on behalf of the UK Health & Safety Executive. However, the views expressed in this paper are those of the authors and are not, except where the context indicates, necessarily those of the HSE.

REFERENCES

1. Considine, M., Grint, G.C., & Holden, P.L., 1982, `Bulk storage of LPG - factors affecting offsite risk', I Chem E Symposium Series No.71, The Assessment of Major Hazards, April.
2. Clay, G.A., Fitzpatrick, R.D., Hurst, N.W., Carter, D.A., & Crossthwaite, P.J., 1988, `Risk assessment for installations where liquefied petroleum gas (LPG) is stored in bulk vessels above ground', J. Hazardous Materials, 20.
3. Rew P. J., Deaves D. M., Hockey S. M. & Lines I. G., 1995, 'Review of Flash Fire Modelling', HSE Contractor Report WSA/RSU8000/015, HSE Books.
4. Rew P. J., Spencer, H. & Amaratunga, S., 1998, 'Modelling the Effects of Flash Fires', HSE Contractor Report WSA/RSU8000/084.
5. CCPS, 1994, 'Guidelines for evaluating the characteristics of vapour cloud explosions, flash fires and BLEVES', American Institute of Chemical Engineers.
6. Cracknell, R.F. & Carsley, A.J., 1997, 'Cloud Fires - A Methodology for Hazard Consequence Modelling', HAZARDS13, IChemE North Western Branch, Manchester, April.
7. Jenkins, D.R. and Martin, J.A., 1983, 'Refrigerated LPGs Safety Research', Shell.
8. Evans, A. & Puttock, J.S., 1986, 'Experiments on the ignition of dense flammable gas clouds', 5th Int. Symp. on Loss Prevention and Safety Promotion in the Process Industries, Cannes.
9. Hirst, W.J.S., 1986, 'Combustion of large-scale jet-releases of pressurised liquid propane', Heavy Gas and Risk Assessment III, D Reidel Publication Co., Netherlands.

10. Sahimi, M., 1994, 'Applications of Percolation Theory', Taylor & Francis Ltd.
11. Scher, H. & Zallen, R., 1970, 'Critical Density in Percolation Processes', J. Chem. Phys., 53.
12. Wilson, D.J., 1995, 'Concentration Fluctuations and Averaging Time in Vapor Clouds', CCPS, American Institute of Chemical Engineers.
13. Raj, P.P.K., 1977, 'Calculations of thermal radiation hazards from LNG fires - A review of the state of the Art', AGA Transmission Conference, St Louis, Missouri, USA.
14. Gulder, O.L., 1990, 'Turbulent premixed flame propagation models for different combustion regimes', 23rd Intl. Symposium on Combustion, The Combustion Institute.
15. Raj P.P.K. & Emmons, H.W., 1975, 'On the burning of a large flammable vapour cloud', Western and Central States Section Meeting of Combustion Institute, San Antonio.
16. Hockey S. M. & Rew P. J., 1996, 'Review of Human Response to Thermal Radiation', HSE Contractor Report WSA/RSU8000/026, HSE Books.
17. Rew, P.J., 1997, 'LD_{50} Equivalent for the Effect of Thermal Radiation on Humans', HSE Contractor Report, RSU Reference 3520/R72.027, HSE Books.
18. Lawson, D.I. & Simms, D.L., 1952, 'Ignition of wood by radiation', British Journal of Applied Physics, Vol. 3.
19. Home Office, 1995, 'Fire Statistics United Kingdom 1993'.
20. Purdy, G. & Davies, P.C., 1985, 'Toxic gas incidents - some important considerations for emergency planning', Multistream 85, IChemE, 257.
21. Lees, F.P., 1996, 'Loss Prevention in the Process Industries', 2nd Edition, Butterworths.
22. Nussey, C., 1994, 'Research to improve the quality of hazard and risk assessment for major chemical hazards', J. Loss Prevention in the Process Industries, 7, 175.
23. Pape, R.P. & Nussey, C., 1985, 'A basic approach for the analysis of risks from major toxic hazards', The Assessment and Control of Major Hazards, IChemE, 367.
24. Spencer, H. & Rew, P.J., 1996, Ignition Probability of Flammable Gases, HSE Contractor Report WSA/RSU8000/026, HSE Books.

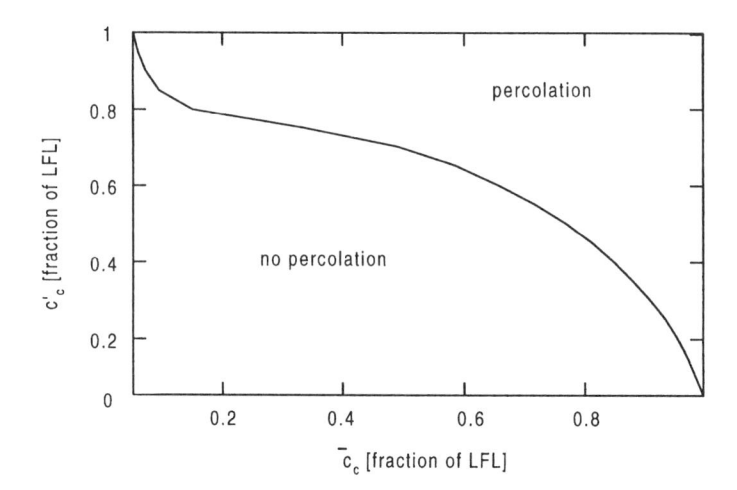

Figure 1 Mean and standard deviation of concentration giving critical occupied area fraction for an LPG release, $\phi_c = 0.44$

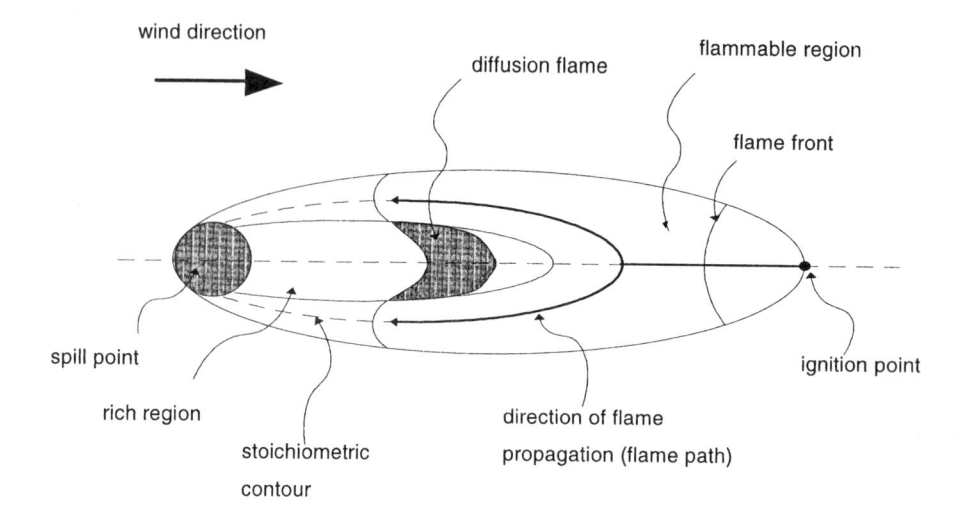

Figure 2 Flame path through a dense flammable gas cloud

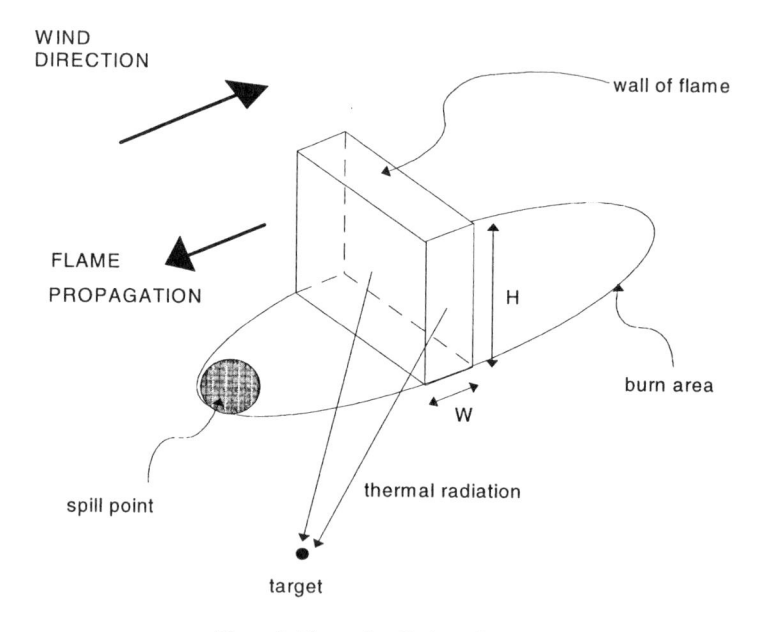

Figure 3 Thermal radiation effects model

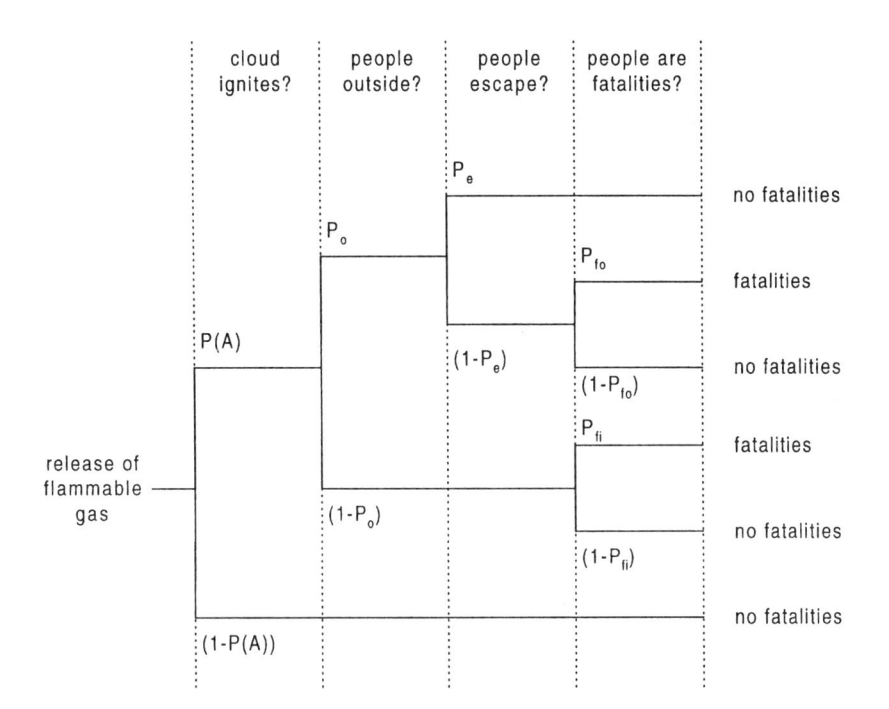

Figure 4 Flash fire event tree

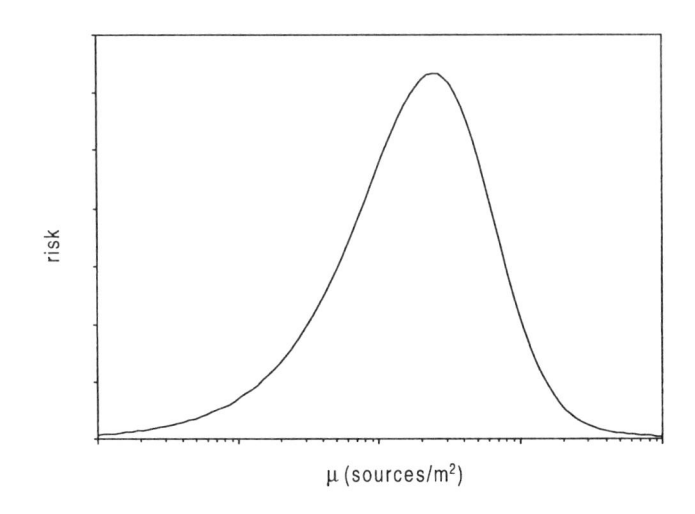

Figure 8 Variation of risk with ignition source density for a flash fire event

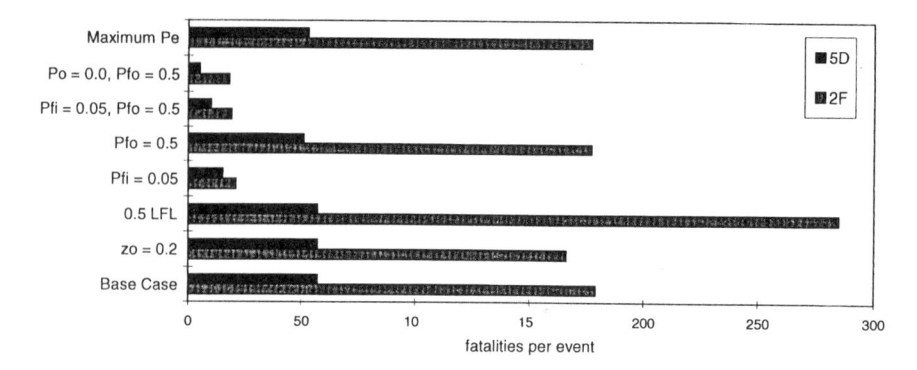

Figure 5 Sensitivity results for instantaneous release over urban land

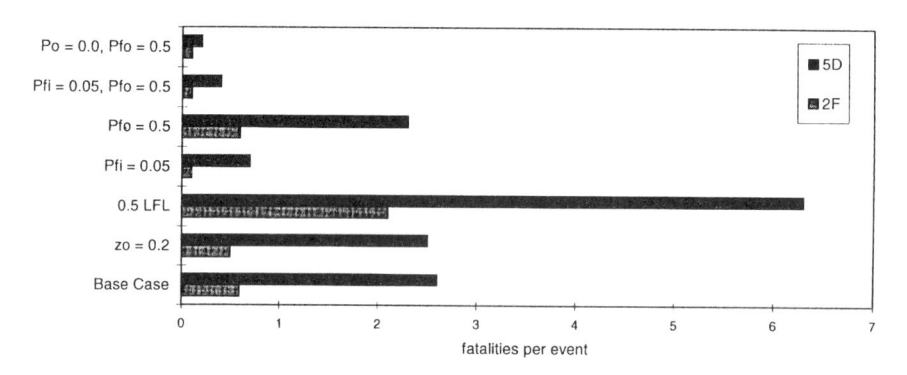

Figure 6 Sensitivity results for instantaneous release over rural land

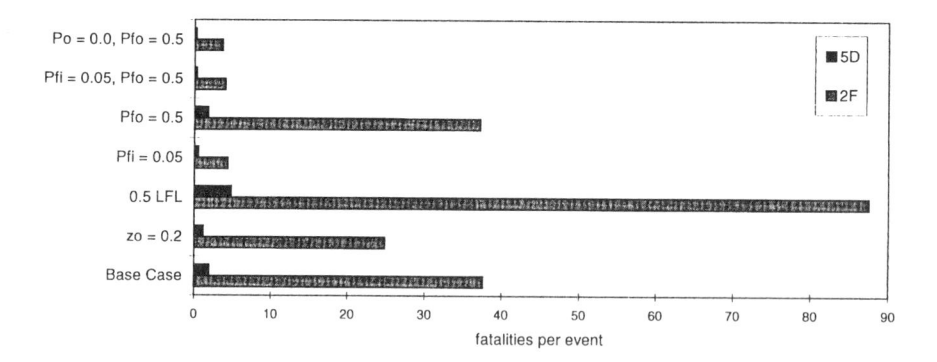

Figure 7 Sensitivity results for continuous release over urban land

TURBULENT-REYNOLDS-NUMBER AND TURBULENT-FLAME-QUENCHING INFLUENCES ON EXPLOSION SEVERITY WITH IMPLICATIONS FOR EXPLOSION SCALING

C.L. Gardner, H. Phylaktou and G.E. Andrews
Department of Fuel and Energy, University of Leeds, Leeds LS2 9JT

Explosion severity is dependent on the turbulent burning rate which is related to the integral-length-scale of turbulence which in turn is a function of the characteristic obstacle scale. Methane/air explosions propagating with an approach flame speed of about 50 m/s, were made to interact with a turbulence-inducing obstacle in the shape of a bar-grid. The scale of these flat-bar grid obstacles was varied by changing the number of bars for fixed blockage ratios. The obstacle scale was taken as the bar width perpendicular to the flow direction of the propagating flame front.

In effect, the turbulent Reynolds number, R_ℓ, was systematically changed from 2,500 to 215,000, by varying both the scale and intensity of turbulence. The maximum overpressure, rate of pressure rise and flame speed increased with an increase of R_ℓ. The Karlovitz flame stretch factor, Ka, was also found to influence the explosion severity. The overpressures and the estimated turbulent burning velocity S_T, were shown to correlate well with R_ℓ, and also shown to separate into two distinct correlations identified by Ka <1 or Ka>1. There was evidence that Ka may be acting as a switch between full-burning and partial/local flame quenching (with associated lower overpressures) at a critical narrow range of Ka around unity. The results show that both the Ka number and the correct dependence on scale are important in providing the fundamental understanding framework for improving explosion scaling approaches used in industry today.

Keywords: Explosion, scaling, turbulent Reynolds number, flame quenching

INTRODUCTION

Current explosion scaling methods [1,2,3] are deduced from considerations of fundamental turbulent combustion models. Such models at present are derived from small scale experiments, at low turbulent Reynolds number (R_ℓ) with little or no variation of length-scale. Catlin & Johnson [2] estimated that in large scale vapour-cloud explosion tests with pipe arrays as obstacles [4,5], turbulent Reynolds numbers of the order of 70000 were induced, while it is estimated that atmospheric explosions can be associated with R_ℓ values in the range of 10^6 to 10^7 [6]. Most experimental flame structure studies and modelling of turbulent combustion have been carried out for regimes with R_ℓ generally below 20000, with the variation of R_ℓ achieved by changing the intensity of turbulence (u') rather than the length-scale (ℓ).

In this investigation, R_ℓ was varied from 4,000 to 215,000 by systematic variation of both ℓ and u'. The influence of these parameters on explosion overpressures, rates of pressure rise and flame speeds, as well as on the derived turbulent burning velocities was investigated.

EXPERIMENTAL

Fifty 10% methane/air explosion tests at 16 different initial conditions were carried out in a 162 mm diameter cylindrical vessel with an L/D (length-to-diameter ratio) of 26. The vessel was mounted horizontally and closed at one end with its open end connected to a large dump vessel (2.5 m diameter) with a volume of 40 m^3, more than 450 times greater than that of the test vessel. This arrangement allowed the simulation of open-to-atmosphere explosions with accurate control of both the test and dump vessel pre-ignition conditions.

A pneumatically actuated gate valve isolated the test vessel prior to mixture preparation. A vacuum pump was used to evacuate the test vessel before a 1 atm, 10% methane/air mixture was formed by partial pressures. The dump vessel was filled with air to a pressure of 1 atm. After mixture circulation, the gate valve was opened and spark ignition was effected at the centre of the test vessel ignition-end flange.

The obstacles were 3 mm thick mild steel single and multi-bar grid plates positioned at 6.2D from the spark. The obstacle characteristics are presented in Table 1. The characteristic obstacle-scale, b is defined as the individual bar width. For any single obstacle configuration each bar had the same width. The obstacles were designed so that the aerodynamic flow areas between the bars were also the same.

An array of thermocouples along the axial centreline of the test vessel was used to record the time of flame arrival. Pressure-time histories were recorded using Keller pressure transducers mounted at the ignition-end flange, halfway along (5D downstream of the obstacle) and at the end (17D from the obstacle) of the test vessel. Two others were located in the dump vessel. Pressure drop and pressure loss measurements were made using differential pressure transducers across the obstacle and this enabled the calculation of the velocity of the explosion-induced gas flow (U) through the obstacle, ahead of the propagating flame. Each test was repeated at least twice.

RESULTS

General effects of obstacles / influence of scale

Figure 1(a) shows the pressure traces recorded at the closed end of the test vessel for tests with a grid-bar obstacle of 30% BR with 1, 3 and 5 bars. The corresponding records of flame position against time are shown in Fig. 1(b). The general phenomena and mechanism associated with explosion development in tubular geometries with and without obstacles have been discussed elsewhere [7].

The turbulence of the fast unburnt gas flow downstream of the obstacle, induced by an initial fast elongated flame propagation, resulted in flame acceleration due to turbulent combustion. This gave rise to a rapid rate of pressure rise (dP/dt). As the flame propagated into the region of turbulence decay the pressure started to fall resulting in the pressure peak P_{max}. No rise in pressure recorded in the dump vessel prior to the flame exit from the test vessel indicated that the the large dump vessel did not influence the explosion development inside the test vessel.

Figure 1(b) indicates maximum flame speeds upstream of the obstacle of the order of 50 m/s for all three obstacles (these flame speeds were similar in all tests). Downstream of the obstacle the average flame speed (approximated by the slope of the dotted fitted lines as 181, 207 and 247 m/s) increased with obstacle scale. This was in accord with the trends in both the rate of pressure rise and the maximum pressure attained.

The thermocouple time-of-flame-arrival data downstream of the obstacle, shown in Fig. 1(b) indicates the flame sometimes arrived at different positions almost simultaneously, while at other times it is shown to arrive at downstream positions before it arrived at neighbouring upstream ones. The high maximum flame speeds in this series of tests (150 to 650 m/s) and the associated high flame acceleration, may have limited the accuracy of the thermocouple technique. However, it will be shown later that most of the present explosions took place in what is described by researchers as the distributed reaction zone or the fragmented-flame regime with possible extinction areas. Therefore, the apparent time-of-arrival anomaly could simply be interpreted as further evidence of fragmented flame zones.

On a practical level this made it difficult to obtain sufficiently resolved and reliable flame speed data from first analysis of the thermocouple records. A simple smoothing procedure was applied to the recorded flame-arrival times so that local minima and maxima points on the derived flame speeds, inconsistent with neighbouring trends, were smoothed out. An example of such a flame speed record, which corresponds to the single-bar test in Fig. 1, is shown in Fig. 2.

In vapour cloud explosions it is usual to assume that the overpressure is proportional to the square of the flame speed [1,8]. A more detailed expression was given by Harrison and Eyre [5], based on the simplified acoustic theory given by Taylor [9], in terms of the flame speed Mach number. If the ambient pressure is atmospheric then the overpressure is given by:

$$P = \frac{2\gamma M^2}{1+M} \quad \text{(baro)} \tag{1}$$

Using a speed of sound of 340 m/s, $\gamma = 1.4$ and applying the <u>averaged</u> flame speeds downstream of the obstacle determined as in Fig. 1(b), i.e. 181, 207 and 247 m/s for the 5, 3 and 1 bar obstacle respectively, Eq. 1 returned overpressure values of 0.52, 0.65 and 0.86 baro. These compare well with the <u>average</u> overpressure after the obstacle shown in Fig. 1(a) for each of the tests.

The implication of this good agreement is that the mechanism of pressure generation in the present tests might be the same as that of vapour-cloud explosions, i.e. the pressure rise is due mainly to the inertia of the gas immediately ahead of the flame, and it is not significantly influenced by the confinement offered by the present geometry.

The pressure signal from the transducer at the test vessel exit showed that in most cases there was no rise in pressure at this position - to correspond to the rise resulting from the fast burning downstream of the obstacle - until some time after the pressure measured by the transducers closer to the obstacle reached a maximum value and began to decay. Since no pressure gradient was measured between the test vessel exit and the dump vessel during this phase, then no gas venting was taking place. While there was no gas movement at the plane of the tube exit, the unburnt gas at a few tube-diameters upstream was being compressed up to pressures of over 3 bara by a flame travelling with speeds of up to 600 m/s. This would indicate a strong blast wave propagation with an associated pressure determined by the inertia of the gas ahead of it that was not influenced by the presence of the vent.

Figure 2 compares the flame speeds predicted by Eq. 1, using the pressure record of the single-bar test (in Fig. 1) with the experimental measurements. For the first 60 ms of the test, during which time the flame was upstream of the obstacle, higher flame speeds than those measured would be needed to predict the overpressure on the basis of the mechanism of Eq. 1. The explosion development in this section of the vessel is effectively a relatively slow explosion, venting unburnt gas through the restriction provided by the obstacle. The mechanism of pressure generation is therefore quite different from that implied by Eq. 1. On the other hand, the good agreement obtained downstream of the obstacle supports the premise that the pressure generation in this region was due to the inertia of the gas ahead the fast accelerating high speed flame. Strehlow et al [10] showed that a constant speed flame and an accelerating flame with the same maximum flame speed would generate equivalent blast waves.

Figure 3 shows a plot of recorded maximum overpressure (for all repeat tests) against the characteristic obstacle-scale. The scale of the obstacle was varied by changing either the blockage ratio (for a fixed number of bars) or the number of bars (for a fixed blockage ratio). The data is grouped for constant blockage ratio and for each set the overpressure increased as the obstacle-scale increased. For the same obstacle-scale the overpressure was generally higher for higher blockage ratios. However, the relative increase in overpressure with increasing blockage ratio decreased as the blockage ratio increased. The overpressures at blockage ratios of 55, 60 and 70 % are shown to effectively be on the same line. At these blockages the overpressure was apparently independent of the blockage ratio and dependent only on scale. Furthermore, the large scale (single-bar, 111 mm) 80% BR tests gave lower overpressures than those given by the lower scale single-bar obstacles at BRs of 70 and 60 and 55 %. As will be quantified below, increasing blockage ratio increases the rms turbulent velocity u' and hence the overpressure increases as the mass burning rate increases. The levelling off of the overpressures (for the same obstacle-scale) for BR>55% and the observed reduction at BR=80% would indicate the onset of a counter-acting mechanism, such as turbulent flame quenching.

Estimation of turbulent combustion parameters

Phylaktou and Andrews [11] presented a method to predict the maximum turbulence levels generated downstream of a grid plate obstacle by an explosion induced flow. For thin sharp obstacles they showed that the turbulence intensity is given by,

$$u'/U = 0.225\sqrt{K} \qquad (2)$$

where K is the pressure loss coefficient and U is the mean velocity of the flow induced ahead of the flame, determined from transient differential pressure measurements across the obstacle [7].

Measurements of the length-scale of turbulence ℓ, immediately on the downstream side of grid plates have been carried out by Baines and Peterson [12] and Checkel [13]. This data was analysed and it was found that at the position of maximum turbulence, ℓ ranged from 30 to 80 % of the characteristic grid-scale, b [14]. Recent experimental measurements of ℓ by Shell Research Ltd. [15] showed that ℓ = b at the plane of maximum turbulence and this was the largest length scale measured in the flow. In view of the uncertainties of the previous measurements it was decided to adopt ℓ = b for the evaluation of the integral length-scale of turbulence in the present work.

Evaluation of u' and ℓ enabled the calculation of the turbulent Reynolds number R_ℓ and the Karlovitz number Ka in the region of maximum flow turbulence downstream of the obstacle just prior to flame arrival, according to Eqs. 3 and 4 [16]

$$R_\ell = u'\ell/\nu \qquad (3)$$

$$Ka = 0.157(u'/S_L)^2 R_\ell^{-0.5} \qquad (4)$$

The Karlovitz stretch factor is a measure of the flame straining. At sufficiently high turbulence levels (and thus high straining) the flame front becomes fragmented and is partially or totally quenched. This number is fundamentally defined as the ratio of chemical to turbulent lifetimes [17]. The turbulent lifetime decreases with increasing turbulence levels and thus Ka increases. In theory, flame quenching occurs when Ka >1, but the actual threshold value may be different, depending on the definition and approximations employed in the quantification of Ka [16,18].

The measured explosion-induced mean gas flow velocities (U) ahead of the flame just prior to its interaction with the obstacle are shown in Table 1, along with the other calculated turbulent combustion parameters for each test condition. Corrections for isentropic compression at the time of flame arrival at the obstacle resulted in minor variation of the mixture kinematic viscosity and laminar burning velocity. The latter was taken to be 0.45 m/s at standard temperature and pressure [19].The turbulent Reynolds number ranged from 4,000 to 215,000 and this covers the range encountered in practical turbulent combustion systems and large scale vapour cloud explosions. In terms of the Borghi flame-structure phase diagram [20], defined by the parameters u'/S_L and ℓ/δ_L, the majority of the present tests lie well within the "thickened-wrinkled flame with possible extinctions" regime or in terms of the terminology of Peters [21], in the "distributed reaction zone".

The influence of Karlovitz stretch factor

As shown in Table 1, for a number of tests the Karlovitz number was greater than 1, which is the theoretical limit above which flame extinction is predicted (see earlier discussion). It should be noted that total flame extinction was not observed in any of the present tests. In all cases the explosion propagated strongly, generating significant overpressures. However, some definite influence of the Karlovitz number was observed and is presented below.

Figure 4 plots the maximum pressure against Ka for a series of tests with approximately constant obstacle scale, as shown. The increase in Ka was achieved by increasing the maximum u' (see Eq. 4) from about 3 to 25 m/s by increasing the blockage ratio and the scale was kept constant by simultaneously increasing the number of bars. Up to approximately Ka = 1 there was an increase in maximum pressure with increasing u' and this may be attributed to the effect of u' rather than to the Karlovitz number directly. However, any further increases in u' which in effect increased Ka to values greater than unity, resulted in slightly reduced, but fairly constant pressures independent of Ka. It is worth noting at this point that in the present system the turbulence was non-uniform, highly anisotropic in the immediate region downstream of the obstacle and decaying further downstream. It would appear that for these high Ka number tests, the flame did not burn in the regions of maximum turbulence until levels of u' had decayed to a lower effective value, ca 13 m/s, (and thus lower local Ka) which would have allowed flame propagation.

For the tests in Fig. 5 both the scale and the turbulence levels were increased simultaneously. The maximum pressure increased strongly to over 2.5 bar, until Ka exceeded unity, at which point a reduction of pressure was observed despite the further increase in both scale and turbulence. In this plot it would appear that the critical Ka was lower than 1, however this might be simply due to the lack of data points for Ka near unity. This plot demonstrates that at larger scales it is possible to induce higher levels of turbulence without entering the flame quenching regime and thus result in significant overpressures which are not possible at smaller scales (cf. Fig. 4), unless the Karlovitz number is maintained below 1, through perhaps an increase in the value of the laminar burning velocity S_L (see Eq. 4). This in fact is the technique used by British Gas [2] and Shell Research [1] in the development of their explosion scaling methodologies, although in the Shell work the increase in reactivity was intended to maintain the compressibility effects at small scale rather than to influence Ka.

Both Figs. 4 and 5 suggest that the Ka number may be acting as a switch between full-burning and partial-burning at a critical value of around unity rather than having a continuous influence over the range of values.

The influence of scale and Reynolds Number

From all the tests in this study, those carried out with the 30% BR and different number of bars (see Fig. 3) effectively isolate the effect of obstacle-scale (or turbulent integral length-scale) on the overpressure. The Karlovitz number was low and approximately constant (0.15 – 0.30) and the rms turbulent velocity was also fairly low and constant (4.4 - 4.7 m/s). The plot of these tests in Fig. 3 suggests a fairly strong dependence of pressure on scale.

In all the tests only u' and ℓ were intentionally changed. The appropriate dimensionless number that might incorporate their combined influence is the turbulent Reynolds number, R_ℓ (Eq. 3). Figure 6 (continuous line curves - LHS axis) is a plot of the measured maximum pressure against R_ℓ. The data separated into two distinct groups identified by their Karlovitz number range. This is effectively a reiteration of the previous observation that at the critical value of Ka = 1 there was a sharp transition to partially quenched combustion with consequent lower overpressures, for the same R_ℓ values.

The equations of the fitted lines for the two combustion regimes are as follows

$$P_{max} = 0.017 R_\ell^{0.43} \quad \text{(baro)} \qquad \text{for Ka} \leq 1.0 \tag{5}$$

$$P_{max} = 0.044 R_\ell^{0.31} \quad \text{(baro)} \qquad \text{for Ka} > 1.0 \tag{6}$$

The single point between the two lines corresponds to a test with Ka=1.00 and it would suggest that it defines the critical transition Ka. Furthermore, the point at the highest R_ℓ corresponds to a Ka=1.05 and this would suggest a very sharp transition boundary between the two regimes. On the other hand, the limited number of data points on this plot means that the above deductions are speculative and it is possible that more points on this plot may reveal a more systematic influence of Ka.

The rate of pressure rise is an important parameter in turbulent explosions as it is related to the mass burning rate, and it is also critical in structural response design. The maximum rate of pressure rise (as indicated in Fig. 1) was measured (by differentiating the smoothed pressure signal) and found to correlate well with maximum pressure according to

$$P_{max} = 0.037 \left(dP/dt\right)_{max}^{0.54} \quad \text{(bar/s)} \tag{7}$$

Graphical correlation of $(dP/dt)_{max}$ against R_ℓ produced the following equations in agreement with derivation of dependence through manipulation of Eqs. 7, 6 and 5.

$$\left(dP/dt\right)_{max} = 0.248 R_\ell^{0.788} \quad \text{(bar/s)} \qquad \text{for Ka} \leq 1.0 \tag{8}$$

$$\left(dP/dt\right)_{max} = 2.45 R_\ell^{0.514} \quad \text{(bar/s)} \qquad \text{for Ka} > 1.0 \tag{9}$$

A TURBULENT BURNING VELOCITY CORRELATION

The applicability of Eq 1 to the present experiments was validated earlier and it was used to obtain the maximum flame speeds corresponding to the measured maximum overpressures. On the assumption that the flame speed is also given by the product of the adiabatic expansion factor and the turbulent burning velocity, $S_f = E\ S_T$, it was possible to obtain S_T (E was assumed constant at 7.5, ignoring any compressibility, flame-thickness and heat loss effects).

Figure 6 (dashed-line curves - RHS axis) shows the derived S_T as a function of R_ℓ, for the two Ka ranges. The correlating equations were

$$S_T = 3.01 R_\ell^{0.27} \quad \text{(m/s)} \qquad \text{for Ka} \leq 1 \tag{10}$$

$$S_T = 5.15 R_\ell^{0.20} \quad \text{(m/s)} \qquad \text{for Ka} > 1 \tag{11}$$

It should be noted that in the present tests the kinematic viscosity of the mixture was not a variable. The implicit dependence on this parameter, indicated in Eqs. 10 and 11 (and in previous equations), was therefore not validated. However, the other two parameters defining R_ℓ (ℓ and u') were the main variables in this study. Additionally their individual influence on S_T was verified by multi-variant regression analysis for the Ka $\leq$ 1 regime (where more data points were available) as

$$S_T = 67.9 u'^{0.25 \pm 0.03} \ell^{0.29 \pm 0.02} \quad (m/s) \quad \text{for } Ka \leq 1 \tag{12}$$

therefore the dependence on R_ℓ indicated in Eq. 10 is an acceptable compromise of the individual dependencies.

By substituting for R_ℓ using Ka (Eq. 4), Eq. 10 can be rewritten as

$$S_T \propto \frac{u'^{1.08}}{Ka^{0.54}} \tag{13}$$

Bradley et al [22] proposed the power law (for Le =1 and Ka = 0.01 to 0.63)

$$S_T \propto \frac{u'}{Ka^{0.3}} \tag{14}$$

In general terms these expressions are similar, however in terms of actual dependencies it can be shown that Eq. 14 gives $S_T \propto u'^{0.55} \ell^{0.15}$ compared to the exponent of 0.27 for both variables in the present correlation.

For combustion in the distributed reaction zone Gulder [23] proposed

$$S_T \propto u'^{0.25} S_L^{0.75} \tag{15}$$

which shows good agreement with Eq. 12 with regard to the dependence on u' but it shows no influence of scale.

Again, for the distributed reaction combustion regime Damkohler [24] proposed

$$S_T \propto u' Da^{0.5} = \frac{u'}{Ka^{0.5}} \tag{16}$$

which is in close agreement with the present correlation - Eq. 13.

Based on fractal theory, Gouldin [25] proposed

$$S_T/S_L \propto R_\ell^{0.26} \tag{17}$$

which is in good agreement with Eq. 10. However, the fractal concept of this model applies to a continuous uniformly-disturbed surface and therefore it could be argued that this concept is not valid in the distributed reaction zone. Nevertheless, Shell Research [1] have employed this model in developing their explosion scaling methodology. The relative success [26] of this approach would indicate that the scale dependence predicted by Eqs 17 and 10 might be correct.

CONCLUDING REMARKS

In this study the turbulent Reynolds number of methane/air explosions was systematically changed by varying both the scale and intensity of turbulence. The maximum overpressure, rate of pressure rise and flame speed increased with an increase of R_ℓ. This trend was also influenced by the Karlovitz number. There was evidence that Ka may be acting as a switch between full-burning and partial-burning at a critical value of around unity rather than having a continuous influence over the range of values.

It was shown that at larger scales it is possible to induce higher levels of turbulence without entering the flame quenching regime. Significant overpressures would result which are not possible at smaller scales, unless the Karlovitz number is maintained below 1, through perhaps an increase in the laminar burning velocity of the mixture.

The overpressures and the estimated burning velocity S_T were shown to correlate well with R_ℓ in the range of 4000 to 215000. The data was shown to separate into two distinct correlations identified by $Ka \leq 1$ and $Ka > 1$. This indicated that at the critical value of $Ka = 1$ there was a sharp transition to partially quenched combustion with consequently lower overpressures, for the same R_ℓ values. There was no evidence of total flame quenching in any of the tests.

This indicates that caution should be used when applying isotropic-turbulence flame-quenching criteria in turbulent explosion modelling. In explosions the turbulence generated by obstructions is highly anisotropic (and transient) and it may be ignited by strong ignition sources (such as jetted flames rather than small sparks). In regions of high turbulence the flame may quench locally, but parts of the flame front could propagate through regions of lower turbulence and therefore the explosion would be sustained and could still result in significant overpressures. Even in the case of complete turbulent flame extinction, the decaying nature of turbulence may allow re-ignition of the mixture by the hot combustion products.

The dominant combustion regime in the present tests was identified as that of a distributed reaction or a fragmented flame front. The S_T correlation (at low Ka) compared well with turbulent combustion models that specifically refer to combustion in the distributed reaction zone.

Figure 6 (and the associated correlations) may provide a basis for predicting maximum explosion overpressures and also for designing scaled explosion experiments for more detailed investigation of the explosion development. They appear to bring together a number of features that have been found to be partially successful in explosion scaling practice to date.

NOMENCLATURE

α	speed of sound
ρ	density of the fluid
γ	specific heat ratio
ν	viscosity of the mixture
(dP/dt)	rate of pressure rise
b	characteristic obstacle-scale / bar width
BR	blockage ratio
D	diameter of tube
Da	Damkohler number (=1/Ka)
E	expansion ratio
K	pressure loss coefficient
Ka	Karlovitz number
ℓ	integral length-scale of turbulence (= b)
L	length
Le	Lewis number
M	Mach number
P	pressure or overpressure
R_ℓ	turbulent Reynolds number
S_F	flame speed
S_L	laminar burning velocity
S_T	turbulent burning velocity
U	mean velocity of the flow
u'	root mean square (rms) turbulent velocity

ACKNOWLEDGEMENTS
We thank the Engineering and Physical Sciences Research Council and the Health and Safety Offshore Division for supporting this work.

REFERENCES
1. Taylor, P.H. and Hirst, W.J.S., *Twenty-Second Symposium (International) on Combustion*, The Combustion Institute, 1989.
2. Catlin, C.A. and Johnson, D.M., *Combust. Flame* 88:15 (1992).
3. Phylaktou, H. and Andrews, G.E., *Trans IChemE*, 73, Part B (1995).
4. Harrison, A.J. Eyre, J.A., *Societe de Chemie Industrielle*, Paris, 1986, 1, p.38.
5. Harrison, A.J. Eyre, J.A., *Combust.Sci.Technol.*, 52:121 (1987).
6. Abdel-Gayed, R.G. and Bradley, D., in *Fuel Air Explosions*. (J. Lee, C.M. Guirao and D.E. Grierson, Eds.), University of Waterloo Press, Montreal, 1982, p.51.
7. Phylaktou, H. and Andrews, G.E., *Combust. and Flame*, 85: 363 (1991).
8. Harris, R.J. & Wickens, M.J., *The Institution of Gas Engineers*, 55th Autumn Meeting, Communication 1408, 1989.
9. Taylor, G. I., *Proc. Roy. Soc. London.* Series A, 186:273-292 (1946)
10. Strehlow, R.A., Luckritz, R.T., Adamczyk, A.A. and Shimpi, S.A., *Combust. And Flame* 35:297-310 (1979)
11. Phylaktou, H. and Andrews, G.E., *25nd Symposium (International) on Combustion*, The Combustion Institute, 1994, p 103
12. Baines, W.D. and Peterson, E.G., *Trans. ASME* 73:167.(1951)
13. Checkel, M.D., 1981, Turbulence Enhanced Combustion of Lean Mixtures, Ph.D Thesis, Cambridge University.
14. Phylaktou, H., 1993, Gas Explosions in Long Closed Vessels with Obstacles, PhD Thesis, University of Leeds
15. Mercx. W.P.M. (project Co-ordinator), Modelling and Experimental Research into Gas Explosions, Overall final report, CEC contract: STEP-CT-0111 (SSMA), 1995.
16. Abdel-Gayed, R.G., Al-Kishali, K.J. & Bradley, D., *Proc. R. Soc. Lond.*, A 391, p.393. (1984)
17. Karlovitz, B., *Selected Combustion Problems* Butterworths, London, 1954, p248.
18. Chung, S.H., Chung, D.H., Fu, C. and Cho, P., *Combust. Flame* 106: 515-520 (1996)
19. Andrews, G.E. and Bradley, D., *Combust. Flame*, 19, p.275 (1972)
20. Borghi, R., *Recent Advances in Aerospace Science* (Bruno, C.& Casci, C. Ed.), Plenum, 1985, p.117.
21. Peters, N., *Twenty-first Symposium (International) on Combustion*, The Combustion Institute, 1988, p 1231
22. Bradley, D., Lau, A.K.C. and Lawes, M., *Phil. Trans. R. Soc Lond.*, A 338:359 (1992)
23. Gulder, O.L., *Twenty third Symposium (International) on Combustion*, The Combustion Institute, 1990, p.743.
24. Damkohler, G., Z., *Elektrochem.* 46: 601(1940) (English translation NACA TM 112, 1947)
25. Gouldin, P. C., *Combust. Flame*, 68: 249-266 (1987)
26. Mercx, W.P.M., Johnson, D.M. and Puttock, J., *Process Safety Progress* 14 No2:120-130 (1995)

Test No	BR %	Number of bars	b=ℓ mm	U m/s	u'/S_L [a]	R_l	Ka	ℓ/δ_L [b]
1	20	1	25.6	23.9	6.0	4787	0.08	793
2	30	1	38.5	23.7	10.3	12299	0.15	1198
3	30	2	20.2	23.1	10.0	6308	0.20	631
4	30	3	13.7	22.5	9.7	4163	0.23	428
6	30	5	8.5	23.0	9.9	2634	0.30	265
5	40	1	51.8	24.1	16.0	25807	0.25	1613
7	40	2	27.0	23.6	15.7	13163	0.34	838
8	55	1	70.0	21.8	25.4	56701	0.42	2235
9	60	1	79.7	20.0	28.5	72766	0.47	2549
11	60	3	28.3	20.7	29.5	26573	0.84	900
10	70	1	94.8	18.7	39.8	125450	0.70	3148
12	70	2	49.3	18.9	40.4	65836	1.00	1631
14	70	3	33.5	18.7	39.9	44261	1.19	1109
15	70	5	20.6	18.3	39.2	26341	1.48	673
13	80	1	111.3	15.9	55.8	214319	1.05	3843
16	80	4	29.6	15.4	54.2	54409	1.98	1003

Table 1. Test conditions, measured and calculated combustion parameters.

[a] The laminar burning velocity was constant (10% CH4/air (v/v) S_L=0.45)

[b] laminar flame thickness, $\delta_L = \nu/S_L$. This gives a thickness that is about 1/30th of the actual experimentally determined value of 1 mm. Nevertheless this approximation is implicit in the evaluation of Ka and a number of turbulent combustion regime diagrams are defined in terms of this approximation.

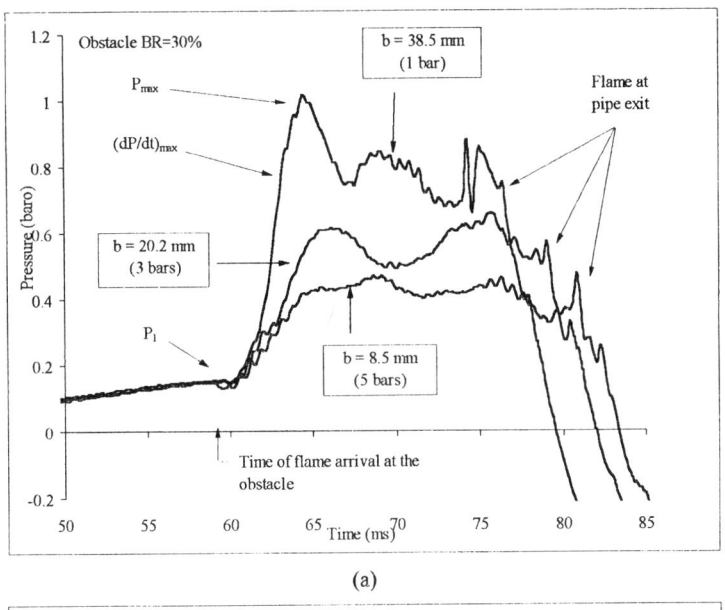

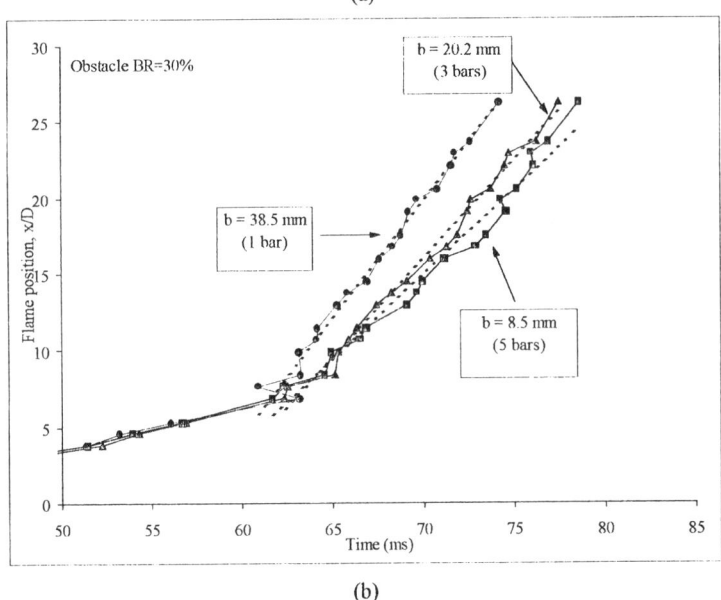

Fig. 1 (a) Pressure-time histories and (b) corresponding flame position against time for tests with obstacles of different scales at constant BR = 30%.

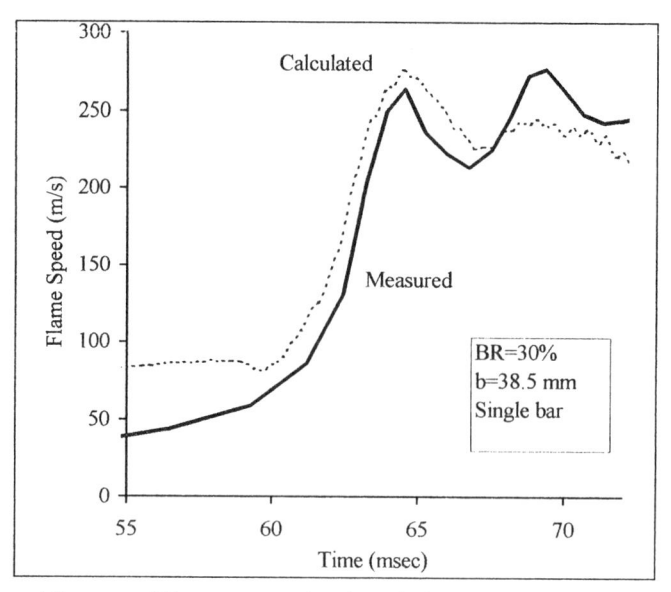

Fig. 2. Measured flame speed history compared to that calculated using Eq. 1 for the single-bar, 30% BR test.

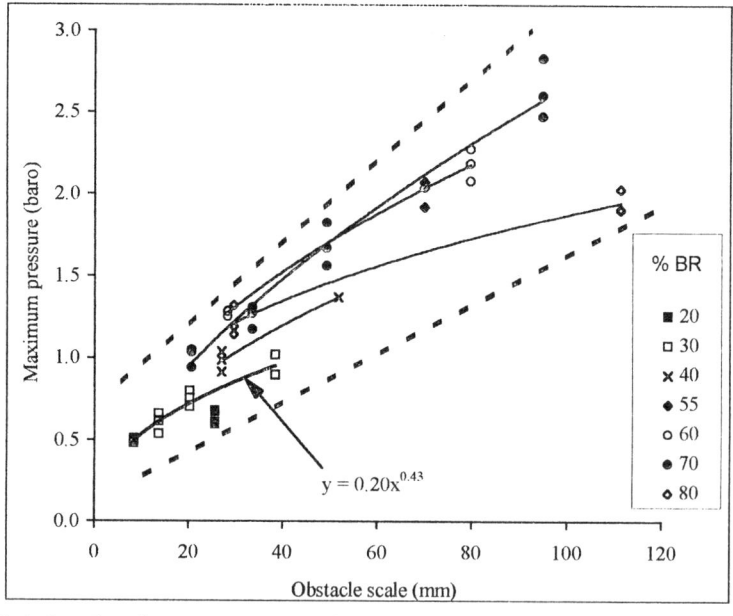

Fig. 3. Variation of maximum overpressure with scale for all blockage ratios tested. (Equation shown is for the line fitted to 30% BR data. See text)

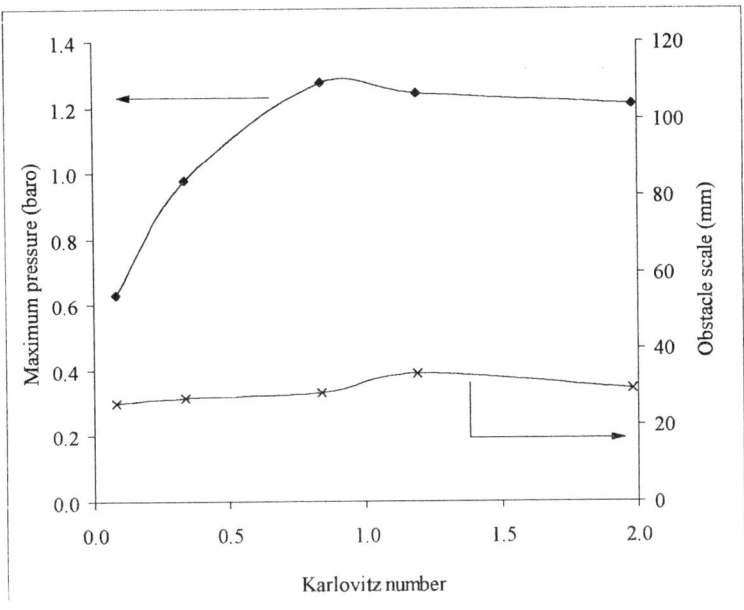

Fig. 4. Variation of maximum overpressure with Ka for approximately constant obstacle-scale (increasing u').

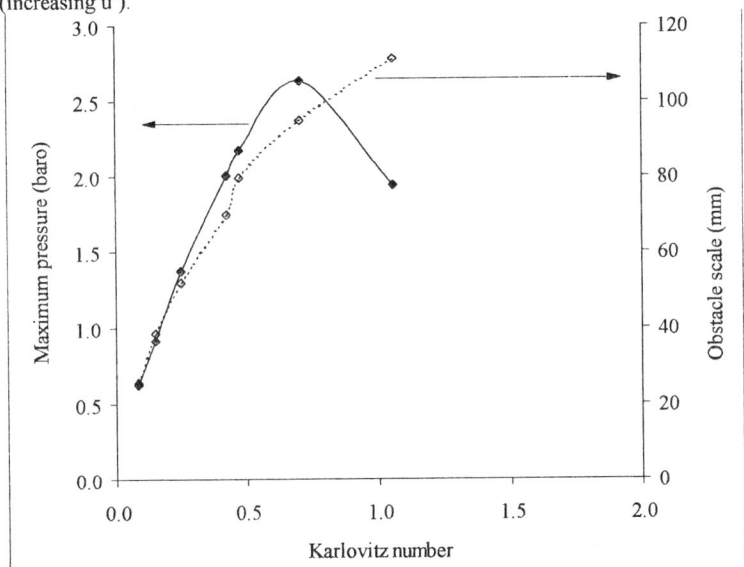

Fig.5. Variation of maximum overpressure with Ka for obstacles of increasing scale. This illustrates the combined effect of u' and scale.

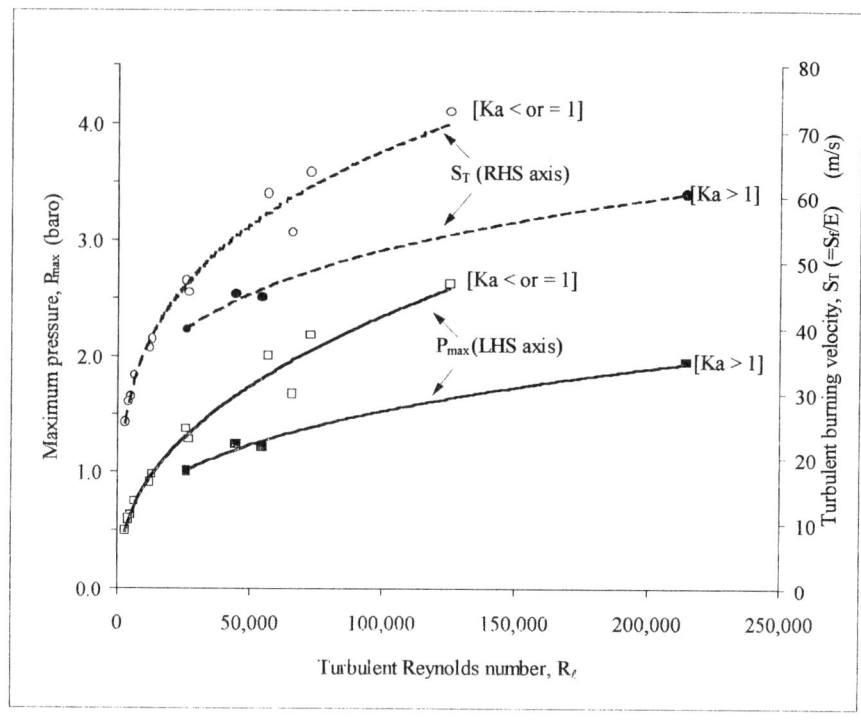

Fig.6. Maximum overpressure (LHS axis) and turbulent burning velocity (RHS axis) plotted against R_ℓ for the data sets, [Ka > 1] and [Ka < or = 1].

EVALUATION OF CFD MODELLING OF GAS DISPERSION NEAR BUILDINGS AND COMPLEX TERRAIN

R. C. Hall
WS Atkins Safety & Reliability, Woodcote Grove, Ashley Road, Epsom, Surrey KT18 5BW

Uncertainties in Computational Fluid Dynamics (CFD) modelling of gas dispersion in the vicinity of buildings and complex terrain have been investigated in two recent studies reported here. The EMU project (Evaluation of Modelling Uncertainty) evaluated the variability of results due to the way in which a CFD code is applied and the accuracy of predictions for realistic scenarios. The second study focused on the applicability of CFD for modelling dense gas dispersion over much larger distances around an industrial site located in complex terrain. These studies have provided useful insights on the modelling strategies which can be used to tackle such applications and the likely accuracy of the predictions.

Keywords: CFD, dispersion, buildings, complex terrain, uncertainty

INTRODUCTION

Currently, a range of 'simple' tools are used to predict the dispersion of gases over flat terrain; however, actual industrial sites are usually much less straightforward and involve complex topography and buildings. CFD techniques offer the capability to simulate realistic industrial problems in greater detail and its use, with the widespread availability of sophisticated commercial software packages and powerful workstations, for safety-related applications is increasing.

EVALUATION OF MODELLING UNCERTAINTY

The validity of CFD predictions is generally uncertain for two important reasons. Firstly, the way in which a CFD code is applied to a specific problem depends on the constraints on staff costs, timescales and computer resources, and can have a critical impact on the final results. Secondly, there is a disparity between the generally simple scope of model validation studies and the complexity of the actual industrial scenarios. To investigate these problems, the European Commission funded the EMU project (Evaluation of Modelling Uncertainty). The specific objectives were:

- to evaluate the spread in results due to the way in which a CFD code is applied;
- to evaluate the accuracy of CFD predictions in large, complex gas dispersion situations.

The EMU project involved a group of four organisations undertaking CFD simulations for a series of realistic near-field dispersion test cases. The organisations were: WS Atkins Consultants Ltd (UK); EnFlo Research Centre, University of Surrey (UK); ARIA

Technologies (France); and NCSR "Demokritos" (Greece). The project was coordinated by WS Atkins. All four partners undertook CFD modelling and EnFlo undertook wind tunnel modelling of certain cases. An overview is presented here of the methodology adopted and the results obtained. Full details are reported by Hall (1).

Methodology

CFD Codes. A key aspect of the project was the use of the same CFD code by all four partners for each particular test case. Actually, two CFD codes were used during the course of the project. Most of the cases were tackled using a commercial CFD code, STAR-CD (Computational Dynamics Ltd (2)). The second code, ADREA-HF (Andronopoulos *et al.* (3)), has been developed specifically for atmospheric dispersion modelling by NCSR "Demokritos" (Greece) and the Joint Research Centre, Ispra (Italy).

Test Cases. The test cases were chosen to provide information on a number of specific factors which have an important influence of the variability and accuracy of CFD results. These factors included: the resources available to the modeller, the problem geometry, gas release conditions (source terms), meteorological conditions, and mesh architecture. The test cases were conceived during the course of the project, rather than being drawn from existing datasets, in order to ensure that each new test case geometry had not been previously encountered by any of the four partners. Furthermore, the test cases were all tackled 'blind', that is without reference to any wind tunnel data.

The test cases are summarised in Table 1. They were specified in three stages of increasing complexity, as noted below.

Table 1 Summary of Test Cases

	Stage A			Stage B							Stage C	
	A1	A2	A3	B1	B2	B3	B4	B5	B6	B7	C1	C2
continuous	✓			✓	✓		✓	✓	✓	✓	✓	
transient		✓	✓		✓		✓					✓
area source	✓											
simple jet		✓				✓		✓	✓	✓		
simple cloud			✓		✓							
'realistic' jet				✓							✓	
'realistic' cloud							✓					✓
buoyant		✓						✓				
neutrally buoyant	✓								✓	✓		
dense			✓	✓	✓	✓	✓				✓	✓
L-shaped building	✓	✓	✓	✓	✓	✓	✓	✓	✓	✓		
additional building						✓	✓		✓	✓		
cliff									✓	✓		
trench										✓		
industrial site											✓	✓
atmospheric conditions	D5	D5	D5	F2	F2	D5	D5	F2	F2	F2	D5	D5
wind direction	0	45	45	45	45	255	45	45	345	345	N	NW
wind tunnel case	✓	✓				✓		✓		✓		
CFD code: STAR-CD	✓	✓	✓	✓	✓	✓	✓	✓	✓			✓
ADREA-HF									✓	✓	✓	✓

- Stage A comprised three cases, A1 to A3, involving a simple building on flat ground. The cases were modelled using STAR-CD. Simplicity was an important consideration because three of the four partners had no prior experience of this code. For this reason, also, only neutral atmosphere and isothermal conditions were considered.
- Stage B incorporated increases in complexity of the geometry (*i.e.* terrain, obstacles and number of buildings), release conditions (*i.e.* two-phase and non-isothermal releases) and meteorology (*i.e.* stability and wind speed). There were a total of 7 test cases. Phase I comprised four cases, B1 to B4, all of which were tackled using STAR-CD. Phase II comprised cases B5 to B7 and involved the ADREA-HF code for the first time.
- Stage C concerned an actual industrial site, featuring numerous buildings and complex local topography. Two scenarios involving chlorine releases were studied.

In addition to the factors described above, it was possible during the course of the project to observe the importance of the user's familiarity with the CFD code. At the start of the project only WS Atkins was familiar with the STAR-CD code and only NCSR "Demokritos" was familiar with the ADREA-HF code. Short training courses were held for the purpose of learning how to run the software packages. Much more time is needed, however, for users to become familiar with such complicated tools. The series of test cases provided the opportunity to assess how much time was really needed for experienced dispersion modellers to switch codes.

Wind Tunnel Modelling. Experiments were performed in EnFlo's large stratified wind tunnel (working section measuring 20m long x 3.5m wide x 1.5m high) at a model scales between 1/133 and 1/250. Measurements were made of mean concentrations using an 18-channel FID gas sampling system. Continuous jet releases of dense, buoyant and neutrally-buoyant gases have been simulated in neutral or stably stratified atmospheres.

Example Results. Figures 1 and 2 show typical meshes used for the simplest and most complex cases respectively. Results were presented in the form of iso concentration surfaces (examples in Figures 3 and 4), cross stream concentration profiles (Figure 5) and vertical concentration profiles (Figure 6). Figures 5 and 6 also show the way in which results were plotted to demonstrate the differences between each partner's results (P1...P4) and the experimental (wind tunnel) data. For transient releases, presentation and comparison of results is less straightforward; Figure 7 shows a typical comparison between cloud shapes predicted by CFD with those predicted by a 'simple' dispersion model, HGSYSTEM (Post (4)).

Analysis of Results. The CFD results were analysed in a number of ways:
- a qualitative assessment of the four partner's results for each release;
- a comparison of overall cloud quantities of practical relevance, such as peak concentration-downwind distance and concentration-width plots;
- a statistical analysis for point-based results, for the cases for which wind tunnel results were available. Two measures were used, namely the geometric mean bias and the geometric variance, as defined by Hanna *et al.* (5);
- a comparison with the results of HGSYSTEM.

Comparison between Different CFD Modellers' Results

CFD requires a very large number of inputs from the user, particularly when designing the mesh, and this inevitably gives scope for variability. During this project, the variability was indeed found to be large. It is difficult to summarise this simply in quantitative terms, but the following gives some indications:

- Considering a concentration level of 1% of the source concentration, for illustrative purposes, the maximum of the hazard ranges predicted by the four partners varied by a factor of 1.5 to 5 greater than the minimum of the predicted hazard ranges. The variation of overall cloud width was of a similar magnitude.

- At higher concentrations, the magnitude of the differences was generally smaller, e.g. for a case involving a continuous dense jet release at a real industrial site, the maximum of the hazard ranges was a factor of 2.4 greater than the minimum at the 10% concentration level and a factor of 3.1 greater at the 1% level.

The particular aspects which contributed most to variability were found to be:

Human factors. Familiarity with the CFD software was an important issue, since the initial learning phase took about 4-6 man-months. User errors were not uncommon, such as the wrong sign for the ground heat flux in a stable atmosphere case (in which case, scatter relative to experiments increased to a factor of 10^4). Interpretation of specifications, conversion of concentrations from mass fractions, used in the CFD models, to other units, and extracting cloud size information caused a surprising number of mistakes.

Mesh design and numerics. This depends on the available computing hardware and project timescales, which force the modeller to consider smaller and coarser resolution meshes than desirable. Each modeller followed their own or published 'design rules' for domain sizes, but there was no consensus between the details of such rules. Crosswind transport was not always adequately accounted for. Preliminary scoping calculations were helpful to determine appropriate domain sizes. Solutions obtained using finer meshes and higher-order differencing schemes showed much more flow structure in the vicinity of buildings, and tended to exhibit reduced scatter, relative to the experimental data, in comparison with the solutions obtained with coarse meshes and upwind differencing. General mesh architecture and the extent of local mesh refinement varied substantially.

Source conditions. These are fundamental inputs to any dispersion model and tuning of other model parameters will not make up for errors here. Variability was greatest for large-scale instantaneous releases for which the initial cloud shape and flow conditions inside the cloud are uncertain for realistic scenarios.

Turbulence model. For STAR-CD, changes to the k-ε model by one partner produced minimal differences in dispersion results relative to the standard model. For ADREA-HF, the differences were more substantial (e.g. hazard range doubling) when the k-ε model was used instead of its default k-l model.

Atmospheric conditions. Inlet velocity profiles were similar, but inlet turbulence conditions varied considerably. Rather unrealistic profiles were used by some modellers; others used empirically-based profiles and checked them for self-consistency. For stably stratified cases, temperature profiles and ground treatment added to the variability.

Representation of geometry. Simplification is often necessary or desirable and depends upon the required and available precision of the geometry data. For the real site, the process involved the building shapes and locations being modified and small details and certain buildings being omitted.

Comparison of CFD and Wind Tunnel Results

The most important point demonstrated by the results was:

- The 'best' CFD solution (that free from any numerical errors) will not necessarily be the same as that from the experiments because of turbulence modelling inadequacies. A coarse mesh model can sometimes give better agreement than a fine mesh model, due to some cancelling out of mesh and turbulence modelling errors. This complicates the definition of accuracy.

Some overall conclusions relating to accuracy for continuous releases are given below:

- For hazard ranges, STAR-CD solutions (using the standard or modified k-ε turbulence model) tended to exceed experimental values by significant amounts, e.g. a factor of 2 to 3 greater at the 1% concentration level. In a few solutions, however, the hazard ranges were under-predicted.

- For STAR-CD, the agreement between predicted and measured concentrations, in neutral conditions, varied between about a factor of 5 and 100, with best performance at high concentrations and worst at low concentrations.

- There was little evidence that solutions based on a large number of cells (typically, 120k to 180k) are much better than those based on more modest numbers (typically, 60k to 80k). With finer meshes and higher-order differencing schemes, the results exhibit less scatter relative to experimental measurements, but the overall bias depends on the capabilities and limitations of the turbulence model.

- For ADREA-HF, there was a slight bias towards over-prediction in one case, but no significant bias in a complex, industrial case. For this case, the agreement between predicted and measured concentrations was within a factor of 10, even though numbers of mesh cells varied from about 35k to 130k.

- The limited evidence available suggests that the k-l turbulence model in ADREA-HF performs better than the k-ε model for atmospheric dispersion applications.

- Agreement for one stable atmosphere case (B5) was generally poor. The results obtained with STAR-CD showed huge spread, biased to under-prediction. One of the ADREA-HF solutions appeared to be much better, although the degree of uncertainty arising from the experiments themselves was significant.

EVALUATION OF A PRACTICAL APPLICATION

This second study considered CFD modelling of continuous releases (44 kg/s) and instantaneous releases (80 tonnes) of chlorine, in neutral (D5) and stable (F2) atmospheric conditions, around an industrial site located in complex terrain. Two wind directions were studied. The problem was much more difficult in modelling terms than those encountered in the EMU project, due to the more complex terrain effects (significant topography) and the greatly increased downwind distances of interest. The EMU-C2 model shown in Figure 2 measured about 1 km long x 660 m wide and the ground height varied by less than ± 25 m relative to the release point. In this study, the main challenge was to balance the need to maintain reasonable near-source detail with a domain measuring some kilometres in length. In view of this, aspects of the CFD modelling strategies which needed close attention included:

- the size of domain needed to establish toxic dose contours around the industrial site;
- the design of the computational mesh and the accuracy with which buildings and structures are represented in a large-scale problem;
- simulation of the wind field over complex terrain, including the specification of realistic atmospheric conditions at the boundaries of the model domains.

Modelling Strategies

The general methodology comprised the following stages:

a) Scoping calculations using HGSYSTEM to predict the dispersion of chlorine releases over unobstructed, flat terrain;

b) CFD simulations of wind flow and chlorine dispersion over unobstructed, flat terrain;

c) CFD simulations of wind flow and chlorine dispersion in a 4 km x 4 km region, using 100 m x 100 m planwise mesh resolution and including topography but excluding buildings and structures (the 'far-field model'). This was intended primarily to provide realistic atmospheric boundary conditions at the boundaries of the more detailed near-field model domain (see below).

d) CFD simulations of wind flow and chlorine dispersion in a smaller region, measuring 2.5 km long x 1.6 km wide, using cells of planwise dimensions up to 32 m x 32 m to resolve the complex terrain, and including the site buildings (the 'near-field model').

To model the continuous two-phase jet release, HGSYSTEM was used to predict the complex initial jet behaviour and these results were then used to define a pure vapour source in the CFD model. This source was defined at the point where the aerosol fraction had dropped to about 1%. Numerical 'sinks' were defined to take account of the entrainment of air by the jet. The presence of tanks and pipework near the gas source were neglected, both in this case and also for the instantaneous release.

Lessons Learnt

For modelling of wind and dispersion behaviour in complex terrain, some key points arising were as follows:

- With a domain extending 2 km downwind of the source, each transient case had to be run for a modelled duration of about 20 minutes (compared to only 5 minutes in the EMU-C2 case). With over 100,000 cells in the near-field model, the computing time required was in excess of 5 days, which was felt to be at the upper limit of acceptability.

- The near-field domains were clearly not large enough to avoid the clouds impacting on the upwind and side boundaries in some cases. Some transient solutions had to be cut short for this reason and, in these cases, it was not possible to determine the relevant dose contours.

- The far-field modelling showed that the predicted wind fields varied rather significantly when 50 m x 50 m cells were used instead of 100 m x 100 m cells. Smaller cells (32 m x 32 m) were used in the near-field model to try to resolve the terrain effects more accurately.
- Using the far-field model results to define the atmospheric conditions at the near-field model boundaries proved to be difficult, mainly because the two meshes were not quite compatible for this operation. The approach eventually adopted was much simpler and only took account of the upwind fetch but not of wind variations across the width of the domain.

Regarding mesh resolution, it was judged that the details of the site buildings and structures would probably not affect the concentration distributions significantly after several hundred metres. Coarse representations were therefore implemented. It was judged that the extra effort involved in defining a more realistic geometry would probably not have yielded a major improvement in accuracy. For example, it was decided not to define storage tanks as being circular, since such details would have been purely cosmetic in view of the coarse mesh resolution. Clearly, it was not possible to model all the important features in accordance with the 'best practice' principles identified in previous studies.

Dispersion Results

For the flat terrain cases, the CFD results were compared with those obtained using HGSYSTEM. The main points arising from this were:

- For continuous releases in neutral (D5) conditions, the peak ground level downwind concentrations predicted by the CFD models generally exceeded the HGSYSTEM values by a factor of 1.5-2. In stable (F2) conditions, the CFD solutions under-predicted concentrations by a factor of 2-3.
- For the instantaneous release, the CFD solutions showed both under- and over-predictions of peak ground level downwind concentrations by a factor of up to 2.
- The results suggest that the performance of the k-ε turbulence model is worse for stable atmospheres than for neutral atmospheres.

For the near-field dispersion cases, the following trends were observed for the effects of the complex site and local terrain, in comparison with the flat terrain cases:

- For the continuous jet release in neutral (D5) conditions, the peak concentrations between 500 and 2000 m downwind of the site were 3-5 times less for the two wind directions considered. In stable (F2) conditions, the peak downwind concentrations were roughly 2-4 times less for both wind directions.
- For the instantaneous release, the peak concentrations at 500 m downwind of the release point were a factor of 1.3-2.4 higher than for flat terrain. Further downwind, however, the peak concentrations were lower than for flat terrain, eg. at 2000 m the peak concentrations were 30-60% less for both wind directions.

CONCLUSIONS

For near-field atmospheric dispersion modelling, the geometric scales of interest range from less than 1m up to 1-2km, forcing the user to compromise between accuracy and computational cost when designing the mesh. It is worth stressing that, for practical 3D simulations of dispersion in the wake of one or more buildings, one cannot hope to attain mesh independent solutions with current work-station based CFD technology. Choice of numerical model and mesh design are thus highly important, and this is a major cause of variation between different modellers' results.

It could be argued that the accuracy required of CFD simulations for near-field dispersion need only be equivalent to the accuracy currently achieved by analytical flat terrain models, (*ie.* to within a factor of 2) or that they should be much lower. However, from the above arguments, a major problem with such an approach is that the potential variability of CFD results between different modellers (and, therefore, organisations) would be great if coarse resolution models were used.

The use of a single CFD model to provide information both for on-site considerations and for evaluating far-field off-site effects is rather over-ambitious for the current generation of 'standard' workstations. However, if targeted more specifically at near source or on-site or far-field effects, then CFD can provide useful information on the combined effects of complex releases, realistic atmospheric conditions and the effects of buildings and complex terrain.

ACKNOWLEDGEMENTS

The EMU project was undertaken jointly with the EnFlo Research Centre (University of Surrey), ARIA Technologies (France) and NCSR "Demokritos" (Greece). The project was a part of the 'Environment' programme of the European Commission (EC), and was also funded by the Health & Safety Executive (HSE) and Electricité de France. Other organisations who contributed to the project included the EC Joint Research Centre at Ispra, through their collaboration with NCSR "Demokritos", Computational Dynamics Ltd (UK), who provided their CFD code STAR-CD to each of the partners, and Associated Octel Ltd (UK), who allowed one of their sites to be used as the basis for two test cases. The site-specific study was undertaken by WS Atkins on behalf of HSE.

REFERENCES

1. R.C. Hall (ed.), Evaluation of Modelling Uncertainty, Final report on contract EV5V-CT94-0531 for European Commission, March 1997.
2. Computational Dynamics Ltd. STAR-CD User's Manual, 1996.
3. S. Andronopoulos, J.G. Bartzis, J. Würtz and D. Asimakopoulos, Modelling the effects of obstacles on the dispersion of denser-than-air gases, J Hazardous Materials, 37 (1994) 327-352.
4. L. Post (ed.), HGSYSTEM 3.0 Technical Reference Manual. Report TNER.94.059, Shell Research Ltd, Thornton Research Centre, Chester, UK, 1994.
5. S.R. Hanna, J.C. Chang and D.G. Strimantis, Hazardous gas model evaluation with field observations. Atmospheric Environment, Vol 27A, No 15 (1993) 2265-2285.

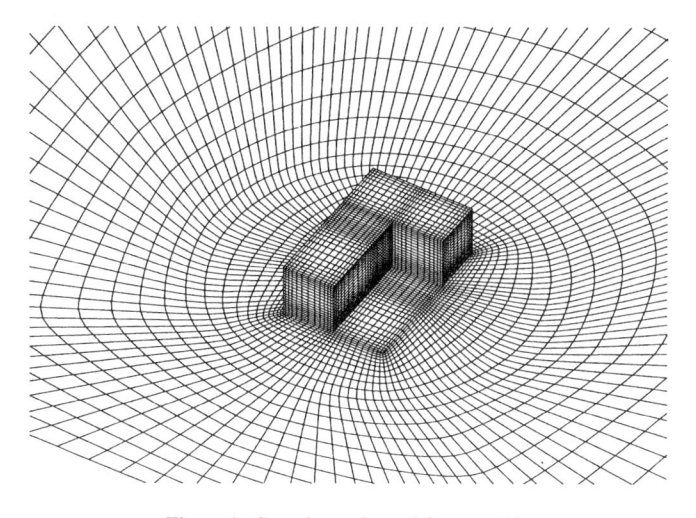

Figure 1 Sample mesh used for case A1

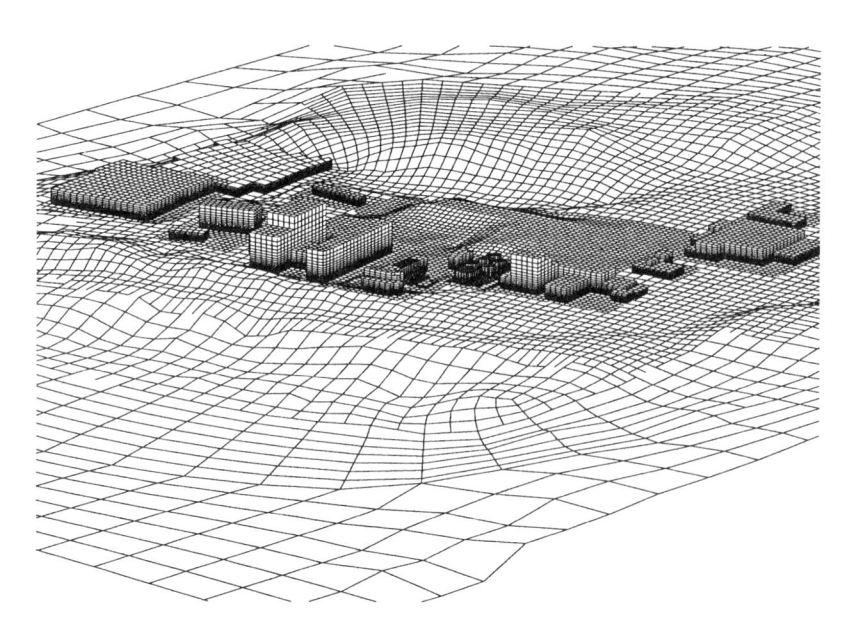

Figure 2 Sample mesh used for case C2

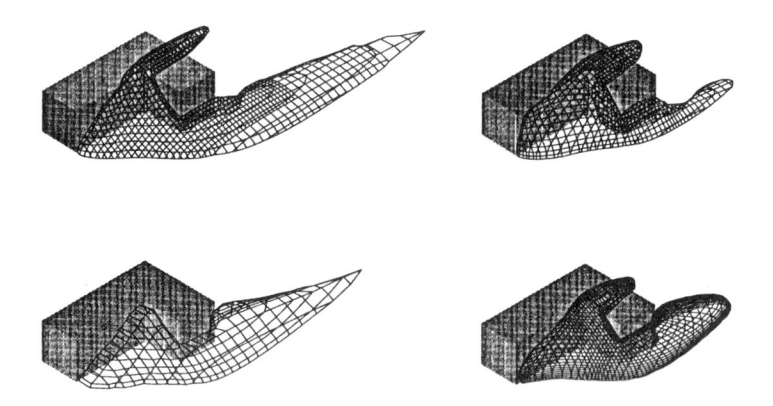

Figure 3   Comparison of partners' 5% iso-concentration surfaces for case A1

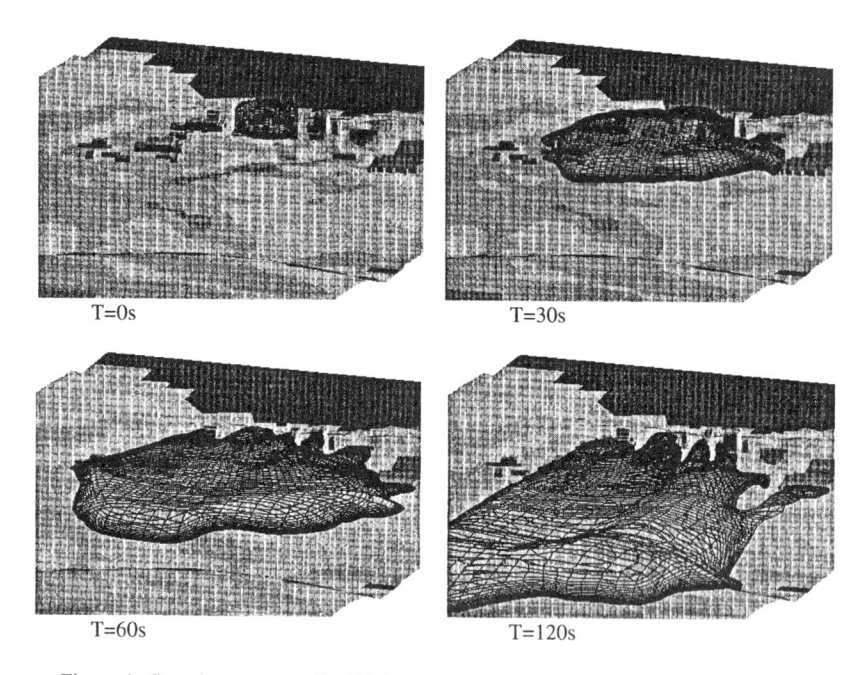

T=0s

T=30s

T=60s

T=120s

Figure 4   Sample sequence of 0.1% iso-concentration surface 'snapshots' for case C2

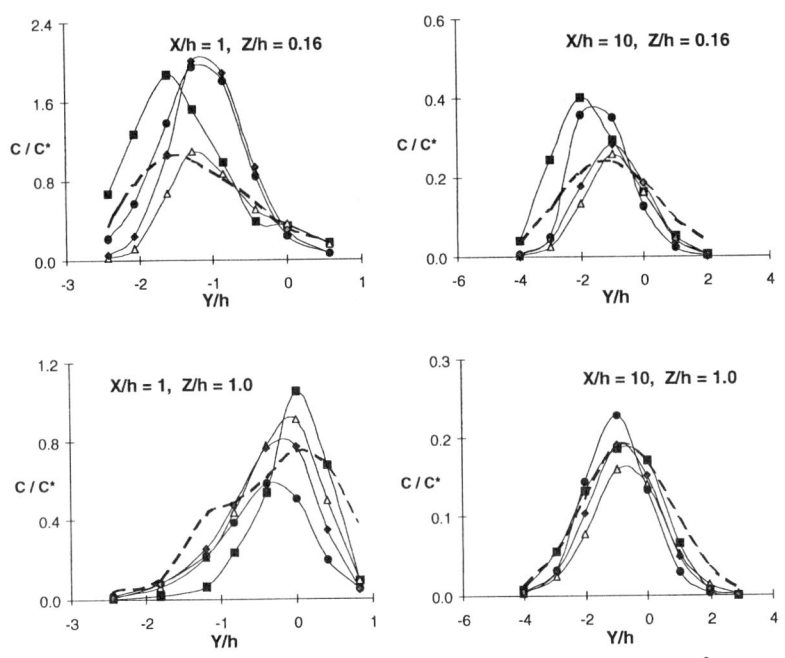

Figure 5 Cross-stream profiles of normalised concentration, $C/C^* = C.U_h.H^2/C_sQ_s$ from CFD (partners P1-P4) and measurements, for case A1

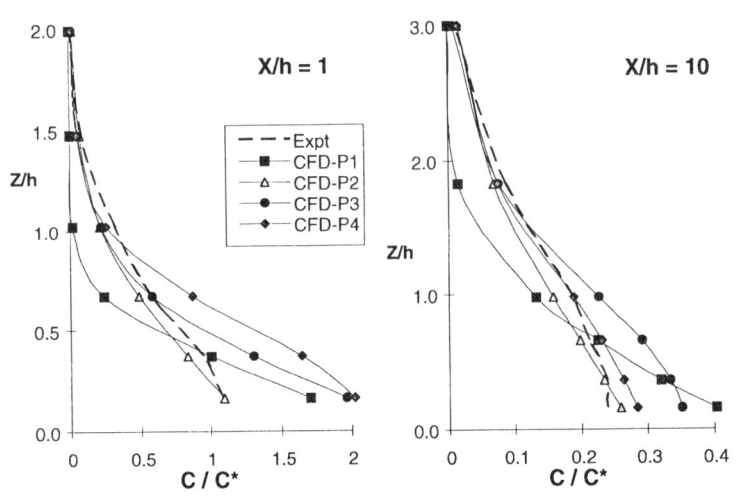

Figure 6 Vertical profiles of normalised concentration, $C/C^* = C.U_h.H^2/C_sQ_s$ from CFD (partners P1-P4) and measurements, for case A1

A	0.1
B	0.01
C	0.001
D	0.0001
E	0.00001

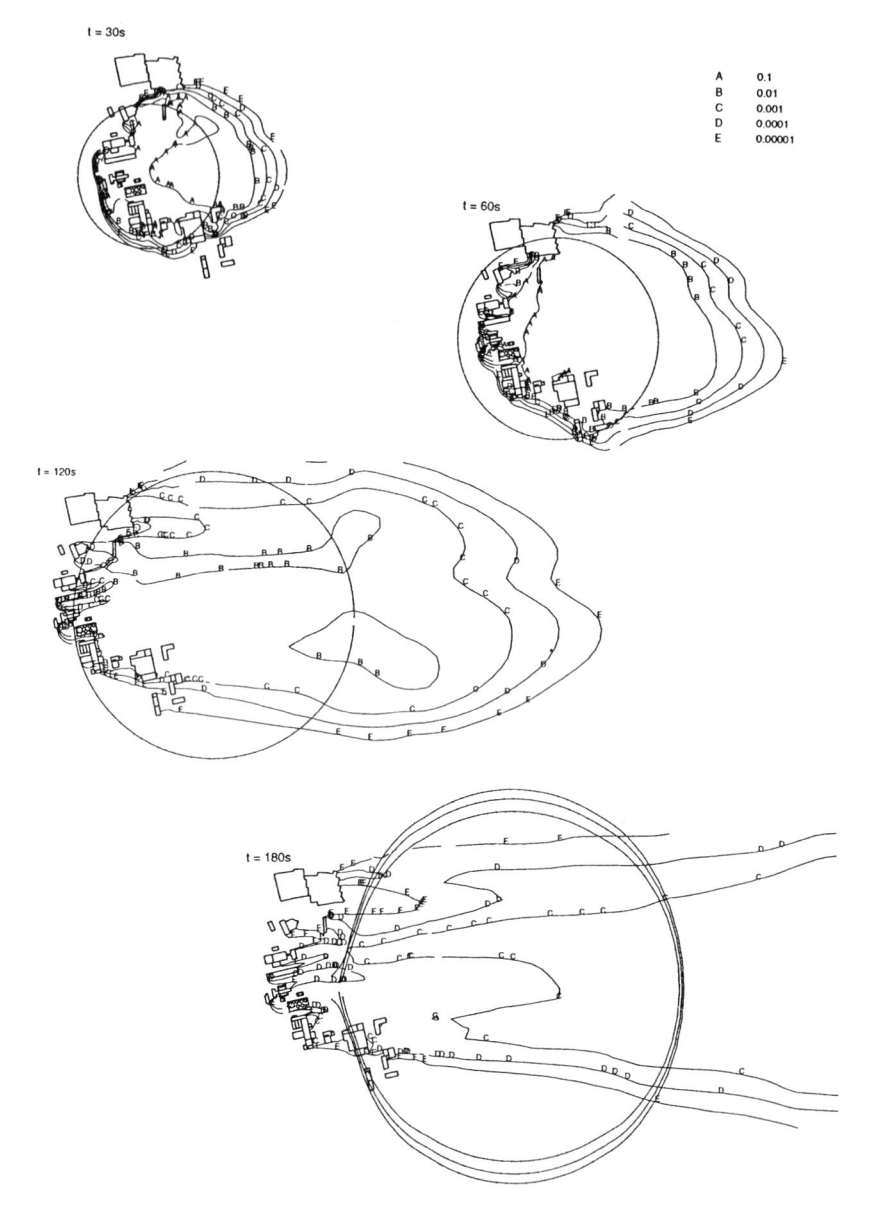

Figure 7 Comparison of ground level concentration contours predicted by a CFD model and HGSYSTEM for case C2
(HGSYSTEM concentrations: 10^{-2} @ t=30, 60, 120s; 10^{-3}, 10^{-4}, 10^{-5} @ t=180s)

EXPLOSION VENTING - THE PREDICTED EFFECTS OF INERTIA

By Steve Cooper - Stuvex Safety Systems Limited

Explosion venting is an established and well used method of primary explosion protection within industry. In recent years there have been various studies into the effect of vent inertia on the venting process. This phenomena may also form a part of the new ATEX directives which may require manufacturers of venting devices to provide vent efficiency data. The paper will present the predicted effects of vent inertia in design as well as the consequence of high inertia vents on weak vessels. There is a need for regular maintenance on all types of vent to ensure that corrosion of hinges and the accumulation of powder deposits does not become a contributory factor to the inertia of the vent once it has been installed.

KEYWORDS: EXPLOSION, VENTING, INERTIA, MASS, DOORS, BURSTING DISCS, GUIDELINES.

INTRODUCTION

Explosion venting is probably the most widely used method of mitigating the effects of an explosion.

The principle can be described as follows. When an explosion is confined it builds up pressure in a short period of time. When the available fuel or oxygen is consumed then the explosion reaches its full potential and realises its maximum explosion pressure (P_{max}). For most types of dusts and gases P_{max}<10 barg. Vessels typically found in process systems, although normally constructed from steel, cannot withstand this level of pressure and would, if sustaining an explosion, burst open, possibly causing local damage and the possibility of a dangerous secondary explosion.

If the vessel is installed with an explosion vent, the explosion flame and pressure can be relieved externally, thus reducing the maximum explosion pressure. This reduced pressure is known as the P_{red}. It is important to recognise that the vent must be directed to a safe location where combustion cannot be supported further, and also away from any personnel, plant or buildings. Explosions should not be vented into the work area unless fitted with special barrier type devices.

DEVICES & DESIGN

The magnitude of the P_{red} is influenced by a number of factors:

- K value
- Vent area
- Vent opening pressure (P_{stat})
- Vessel volume
- Vessel strength

Although this list is not exhaustive, it is possible to design an adequate vent if these characteristics are known and applied using a vent design guide such as is found in the IChemE guides, VDI3673, NFPA68 and Dust Explosion Prevention and Protection Part 1- Venting [4]

The pressure increases rapidly during the early stages of an explosion, therefore it is important that the vent opening is made as large as is practicable and the vent is opened as early as possible.

Another consideration is the type of closure device used to ensure that the vent is sealed during normal process conditions. Very often, bursting discs, rupture panels and bursting membranes are used which can be tested, during manufacture, to ensure accuracy of opening pressure. The devices are lightweight and offer little resistance to the explosion pressure. For vent tests it is normal to use such a device to provide the most effective venting for the vent area available. One disadvantage is that bursting devices such as these can often deteriorate due to fatigue, heat and continuous flexing under pressure pulsing, for instance. In this case the next device that may be considered is the explosion door or pop-out panel (figure 1).

These devices usually comprise a hinged door held down on to a seal with one or more release catches. The catches can be calibrated to release the door at the precise P_{stat} required. Although the release pressure can be controlled, the vent door is usually of a more substantial construction than the bursting membranes, and can tolerate harsh operating conditions. But in constructing a door that will withstand pressure pulsing, erosion etc. is there a penalty to pay when this mass is to be moved out of the way of an oncoming explosion ?

VENT EFFICIENCY

Assuming that the design guidelines mentioned earlier are based on testing with bursting membranes, what would be the effect of substituting the membrane with a hinged, steel door ?

In VDI3673 for example, the vent efficiency is addressed. This compares the door with a bursting membrane in terms of percentage efficiency. If the vent door is assumed to have an efficiency of 80%, compared with a membrane, then the vent area must be increased by 1.25. Also, there are no numerical values presented for the predicted effects of mass and no guidelines on how to test the doors to assess the efficiency.

A vent of greater mass than a bursting membrane will have an inertia effect that must be overcome by the explosion pressure which may (or may not) cause an increase in the P_{red}. Bartknecht [1] suggests a mass of <10kg/m^2, whereas NFPA68 suggests approximately 12kg/m^2. In the UK, values of up to 25kg/m^2 have been acceptable in the past, with some vents being more than 40kg/m^2. So does the absence of inertia effect data mean that there is no problem ? Can the mass values described above, be applied to any vent area, vessel volume or K value ? Do bursting discs and rupture panels operate at 100% efficiency and if not, what effect will the mass of the disc have on the reduced explosion pressure and vent area ?

Apart from mass there are other ways in which the vent inertia may be increased. Hinged doors require attention to ensure that the hinge is free moving and without corrosion. The accumulation of debris, snow, ice and water on top of any vent will increase the inertia effects. The unauthorised modification of vents to prevent nuisance tripping will have an effect and the installation of insulation on low mass membranes or discs will increase the relative mass and thus may influence the venting process. Powder accumulations encrusted on to the vent will not only affect the P_{stat}, but may also increase the mass of the vent.

INERTIA

To try to assess the inertia effects it is first necessary to consider how the pressure is generated within a closed vessel. It is normal to assess the magnitude of a potential explosion by measuring the pressure - time graph. Although the P_{max} is around 10 barg, the rate of pressure rise (dP/dt$_{max}$) is a function of the concentration, explosibility of the material and the vessel volume. The equation for the explosion characteristics is the well known cubic law where the time (in which a device must be effective) can be calculated from:

$$dP/dt_{max} = V^{1/3}/K$$

Figure 2 shows the pressure - time relationship for a K_{st}100 dust in closed vessels of 1, 9, 25 and 60m^3. The larger the volume the greater the time for the vent to become active. Figure 3 however, shows that the time reduces as the K_{st} increases. This pressure-time relationship becomes very important in small volumes.

Firstly, we can assess the effectiveness of low mass bursting membranes by testing and compare these figures with those provided by VDI3673:1992 [3] as prescribed minimum vent areas. Figure 4 shows a summary of tests in the range 1, 9, and 60m^3 and the comparison with VDI3673. The L/D ratio of the vessels was approximately 1.6 which is an important consideration to be addressed later in the paper.

A second series of trials was conducted, this time replacing the bursting disc with a "sandwich" panel of the same size. The panel is constructed from two aluminium sheets with insulating material in between. The density of the material can be altered, thus altering the mass per unit area of the vent. The results for the range of vessels are shown in figure 5. Here we can clearly witness that the P_{red} is increasing as the mass increases although the figures still correspond well with VDI3673.

To investigate further we should attempt to understand why this is happening. If the pressure in the vessel exceeds the P_{stat} of the vent and overcomes the mass m, then the vent will start to move. The acceleration a of the vent is:

$$a = P_{stat}/m$$

where m is the mass per unit area. This movement will cause a gap to appear between the vent and its seat, through which combustion gases may escape. As the gap increases with time, so the venting rate will increase. After a certain time the vent will have moved away from the seat to such a degree, that the vent will no longer present any restriction to the venting process. For a hinged door, the fully developed angle is typically 40°-60°.

From the tests, the time between the beginning of the opening and the attainment of full venting is roughly constant for a certain type of vent.

On the basis of the testing and the relationship with VDI3673 for vents of low mass, we embarked on developing a computer programme to assess the effectiveness of vents with increased mass across the range $10kg/m^2$, $20kg/m^2$ and $40kg/m^2$ and for $K_{st}100$, 170 and 200.

In figure 6 the pressure - time graph for the three K values is shown for a volume of $60m^3$. On to each curve the point is plotted at which full venting is achieved for each of the three vent types. Not surprisingly, as the acceleration properties for each vent are the same there is a straight line proportionality between each pressure time curve. The lines correspond well with P_{red} achieved during testing as shown in figure 6.

The nomogram shown in figure 7 shows the vent area selection from VDI3673 for a $60m^3$ volume and a $P_{red}=0.2$ barg. By transposing the lines from the pressure - time graph we can see a pattern emerging as to the influence of mass on the vent process. For instance, if we were to design a vent to cope with a $K_{st}200$ at 0.2 barg our vent area (A_v) would be approximately $3.2m^2$. By installing a vent with a mass $m=20kg/m^2$ the predicted P_{red} increases from 0.2 barg to >0.4 barg.

The process is repeated for volumes of $25m^3$ and $9m^3$ with almost identical patterns emerging when related to VDI3673. However, when assessing the effects in $1m^3$ a different pattern emerges where the influence of mass is far greater, especially with $K_{st}200$. In figure 8 the nomogram for $1m^3$ shows that for $K_{st}200$, the use of a $12kg/m^2$ vent will increase the P_{red} from 0.2 barg to >0.5 barg.

From these calculations we also find that the influence of mass is reduced when $K_{st}<100$ irrespective of the volume.

Analysis was repeated to ascertain whether there was a benefit in reducing P_{stat} to overcome the penalties of mass. Although marginal benefits were calculated, it did not appear that P_{stat} had very much influence in the range $K_{st}<200$.

LARGE VOLUMES

In an earlier paper by Harmanny [2] it was contended that mass had a diminishing effect as the volume increased. Figure 9 shows the plots for $K_{st}200$ in volumes of 100, 200 and 300m^3. Quite clearly the P_{red} is reducing as the volume increases even though the mass of the vents can vary significantly. From this we can see that mass has only a marginal effect for volumes >60m^3 where the mass is <12kg/m^2. The selection of the vent therefore becomes wider, with lightweight doors becoming acceptable provided the mass is <=12kg/m^2. As the volumes increase beyond 100m^3, doors with a mass of <20kg/m^2 could be employed with little or no penalty on the predicted P_{red}.

L/D RATIO

Although the effects of mass reduce as the volume increases, it is very important to note that the length/diameter ratio begins to influence the vent efficiency. In figure 10 a plot is superimposed to show the effects of an L/D=5 on a volume of 250m^3. The vent area used was for a L/D=2 vessel of 300m^3. It is fair to say that the increase in P_{red} is not significant, but to emphasise the point, the L/D ratio must be taken into consideration at all times when designing venting systems.

CONCLUSION

Very little test data is available on the effect of inertia on vents. From the work undertaken in the preparation of this paper it appears that there can be significant reductions in performance due to the mass (alone) of the vent closure used. The guidelines available do not, currently, provide very much in the way of data to enable accurate assessments of vent performance for any vent with a mass m>1.0kg/m^2.

Efficiency ratings can be subjectively applied unless the closure is purchased from a proprietary manufacture, when such data may be available having been derived from testing. From what we have calculated and seen from testing, the mass may have a considerable impact on vent efficiency and should always be considered as a part of the venting system design.

REFERENCES

1. W.Bartknecht. Explosionen. Springer-Verlag, Berlin, Heidelberg, New York 1978.

2. A. Harmanny. Effect of Inertia on Effectiveness of Explosion Venting. Europex Newsletter April 1993.

3. VDI, 1992. Pressure Release of Dust Explosions - VDI 3673

4. G. Lunn. Dust Explosion Prevention and Protection - Part 1 Venting. Second Edition 1992

POP-OUT EXPLOSION VENT

HINGED EXPLOSION DOOR

Fig 1 Explosion Vents

Fig 2 Pressure/Time (Kst100) 4 Volumes

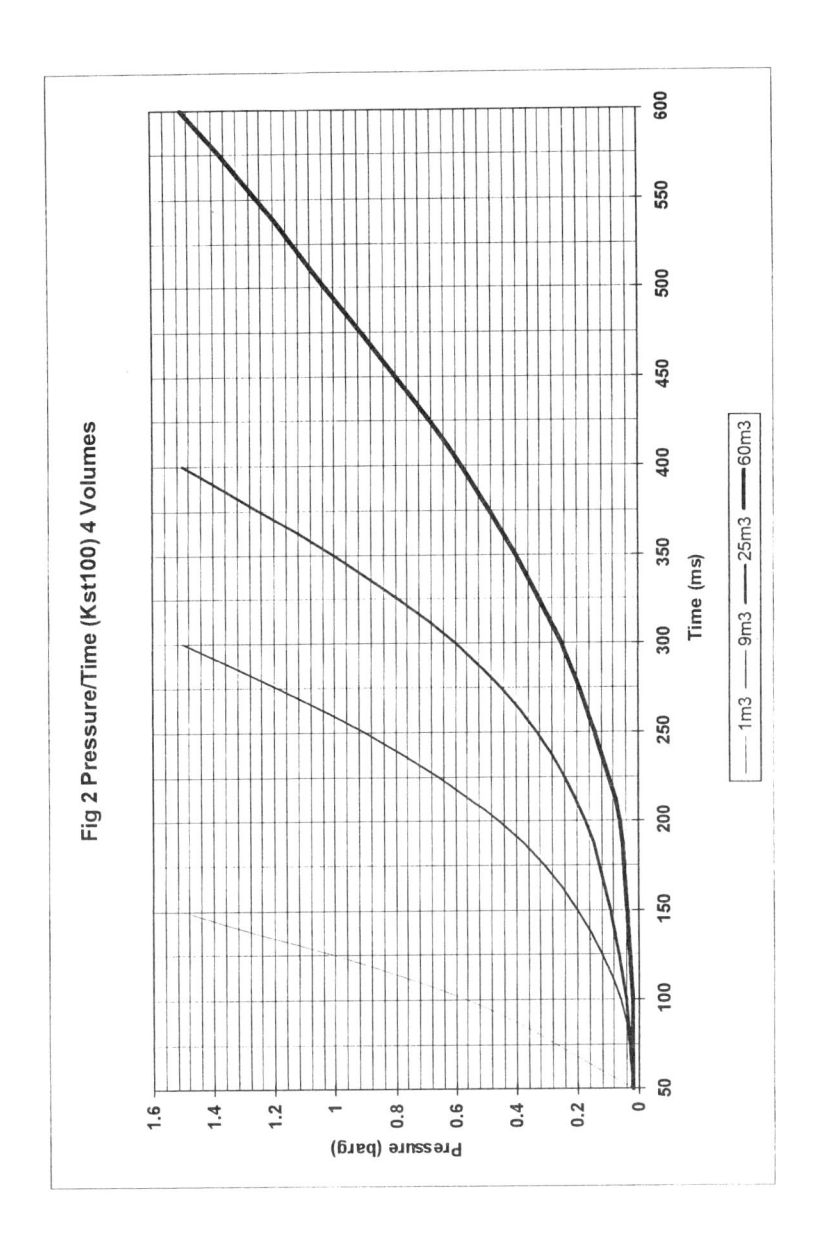

Fig 3 Pressure/Time 9m3 - 4 Kst Values

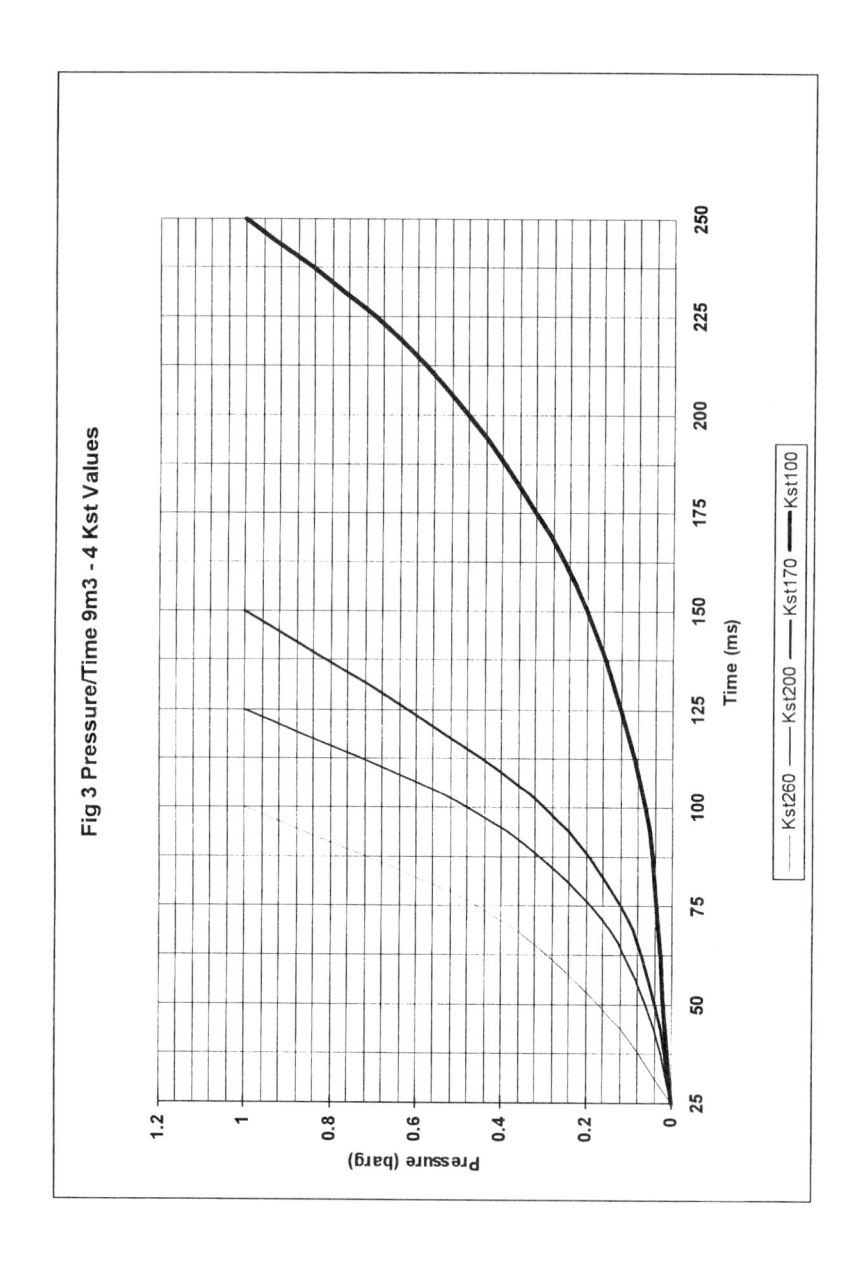

ICHEME SYMPOSIUM SERIES NO. 144

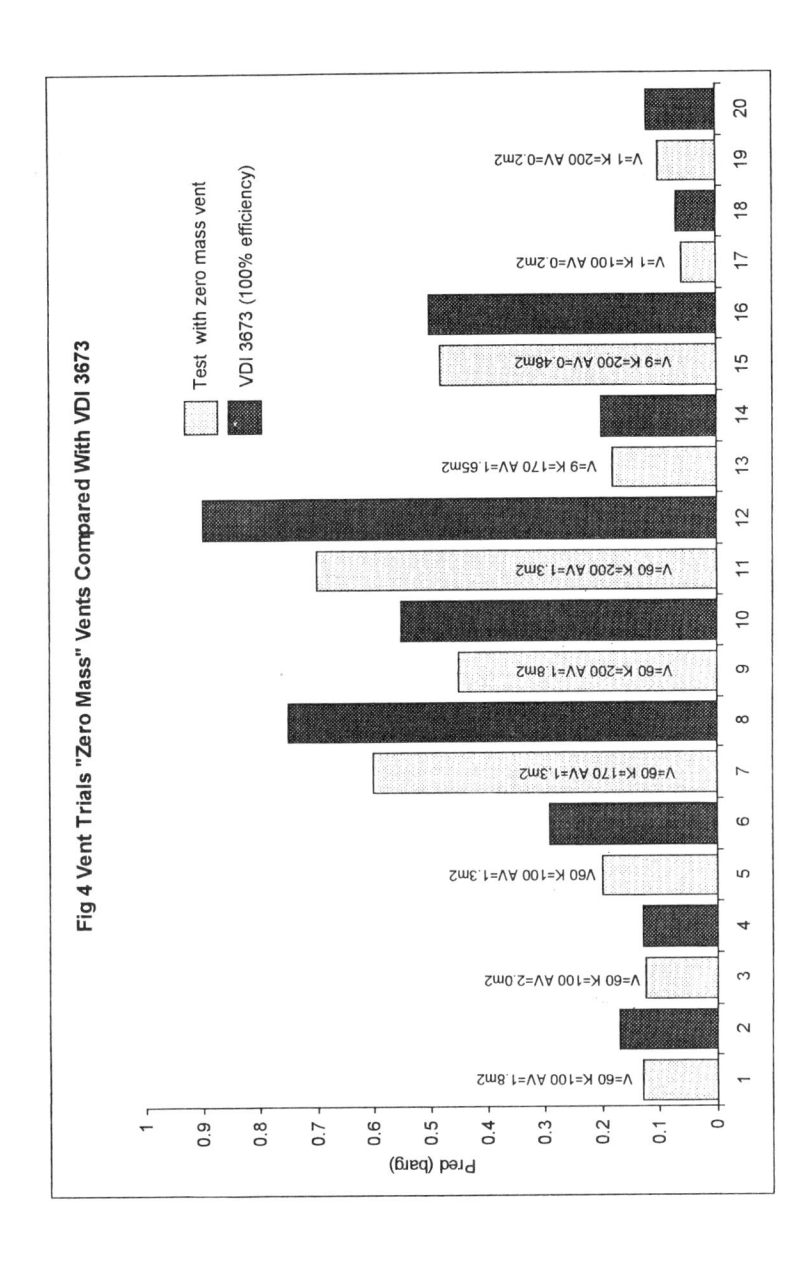

Fig 4 Vent Trials "Zero Mass" Vents Compared With VDI 3673

ICHEME SYMPOSIUM SERIES NO. 144

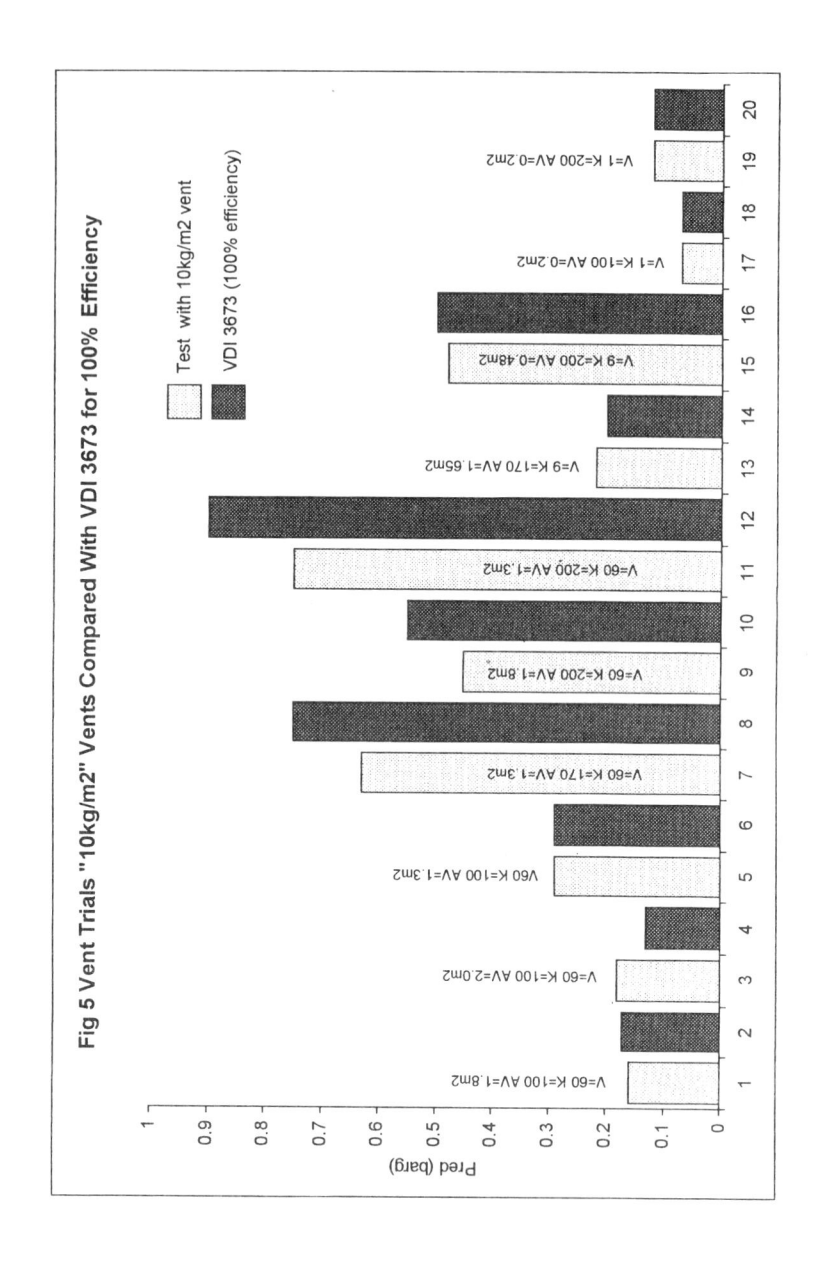

Fig 5 Vent Trials "10kg/m2" Vents Compared With VDI 3673 for 100% Efficiency

ICHEME SYMPOSIUM SERIES NO. 144

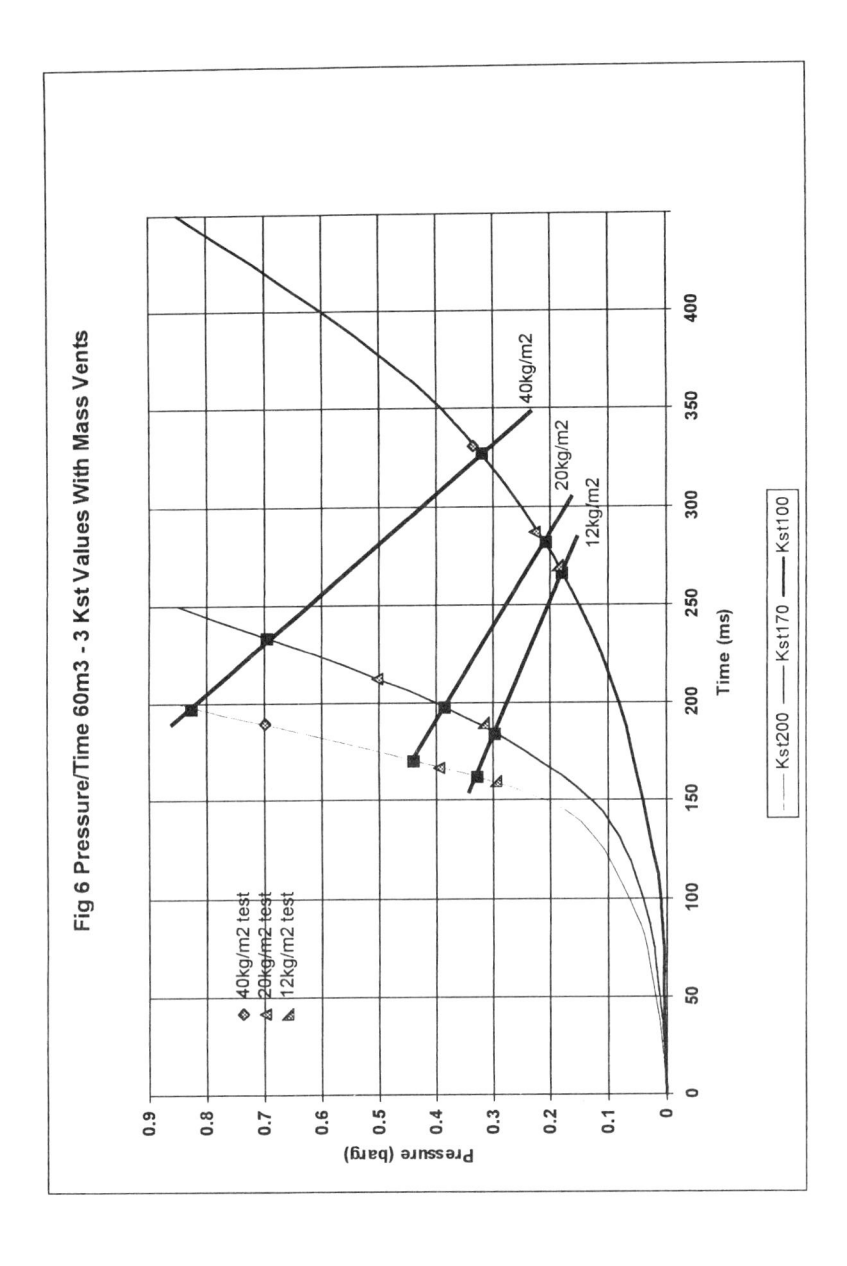

Fig 6 Pressure/Time 60m3 - 3 Kst Values With Mass Vents

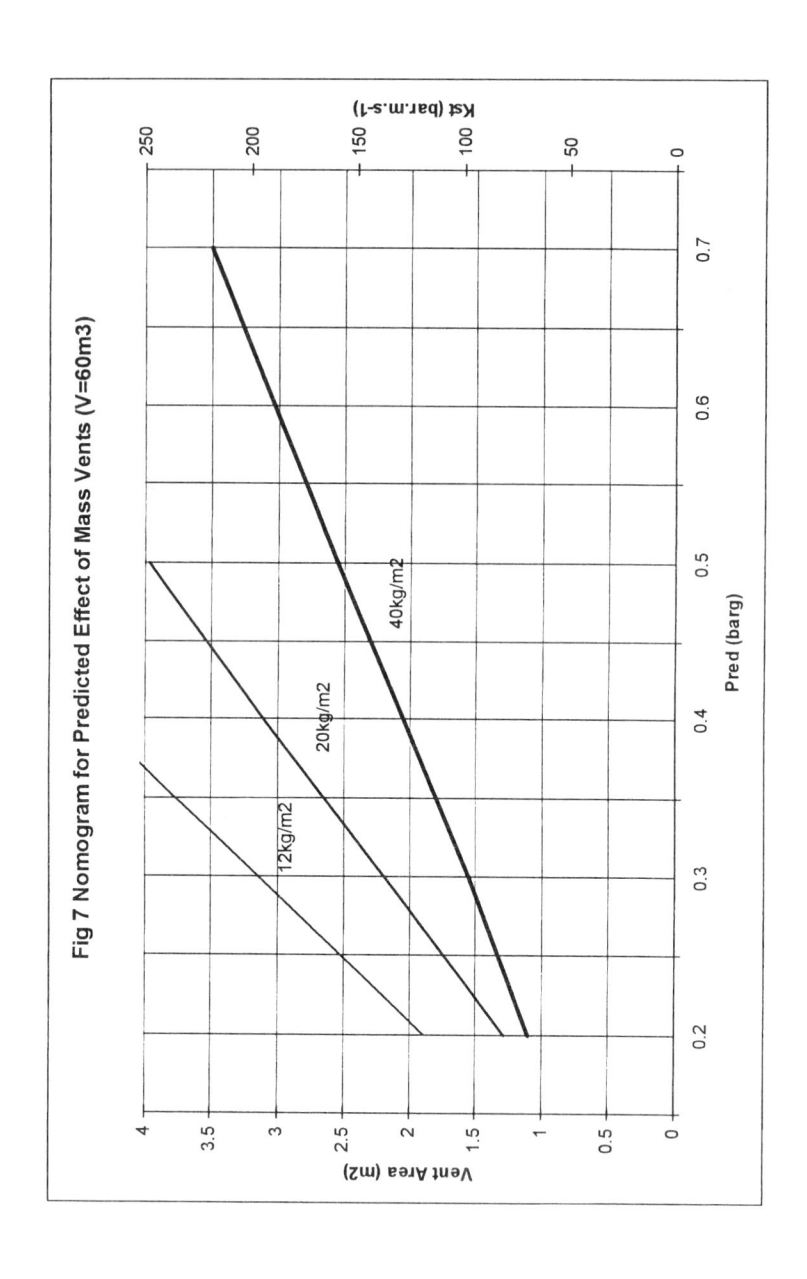

Fig 7 Nomogram for Predicted Effect of Mass Vents (V=60m3)

ICHEME SYMPOSIUM SERIES NO. 144

Fig 8 Nomograms for Vents With Mass in Comparison With VDI 3673 (Pred=0.2 barg)

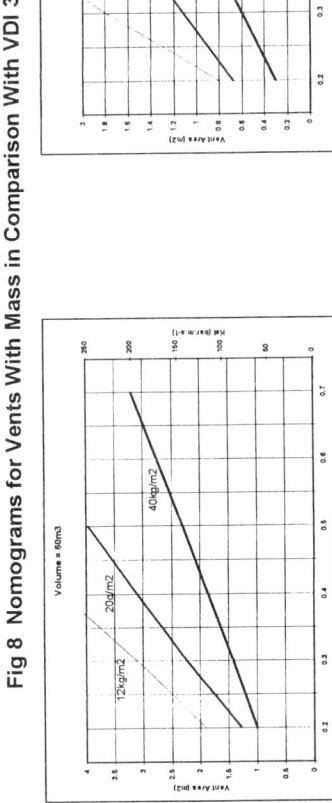

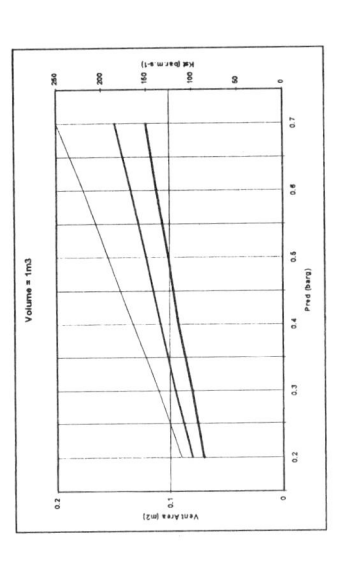

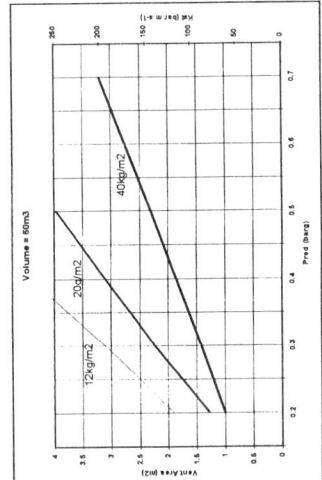

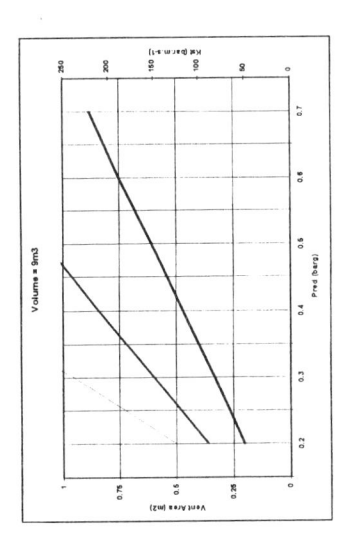

Fig 9 Reducing Effect of Mass on Large Volumes

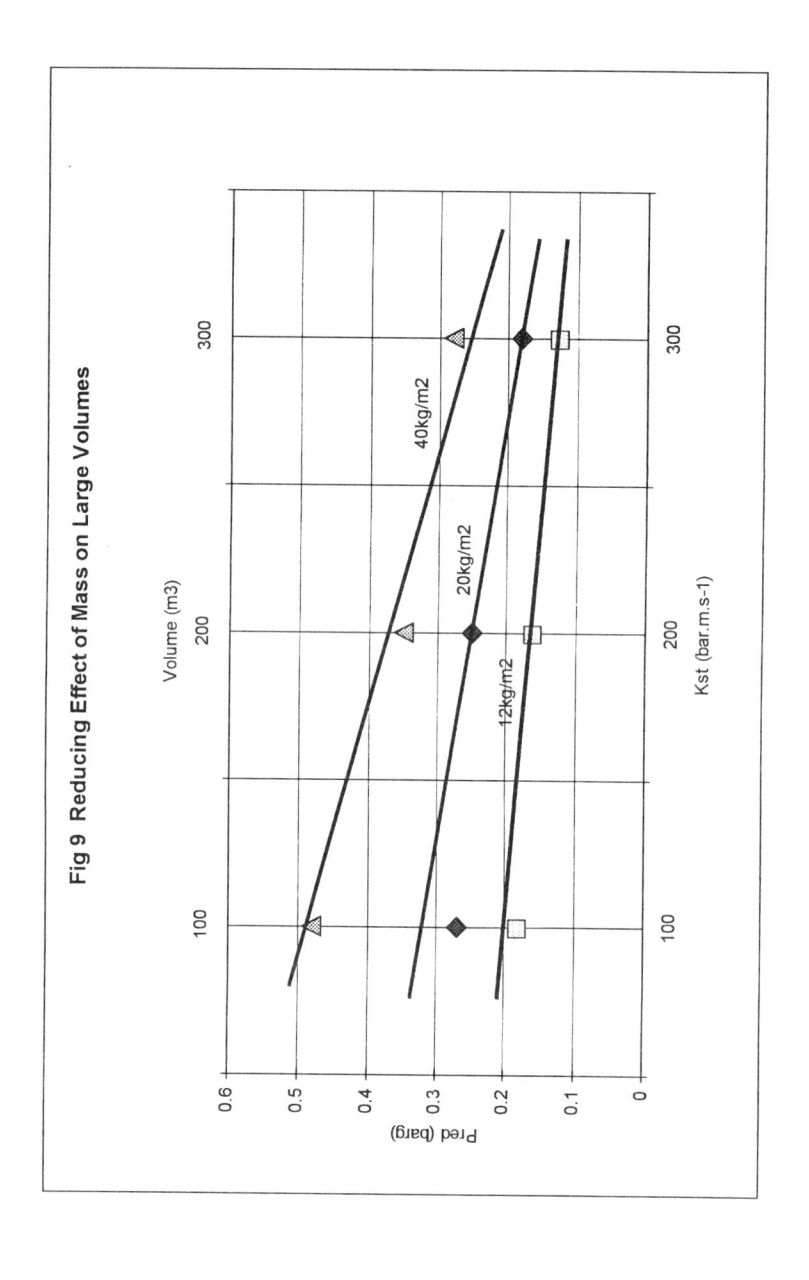

ICHEME SYMPOSIUM SERIES NO. 144

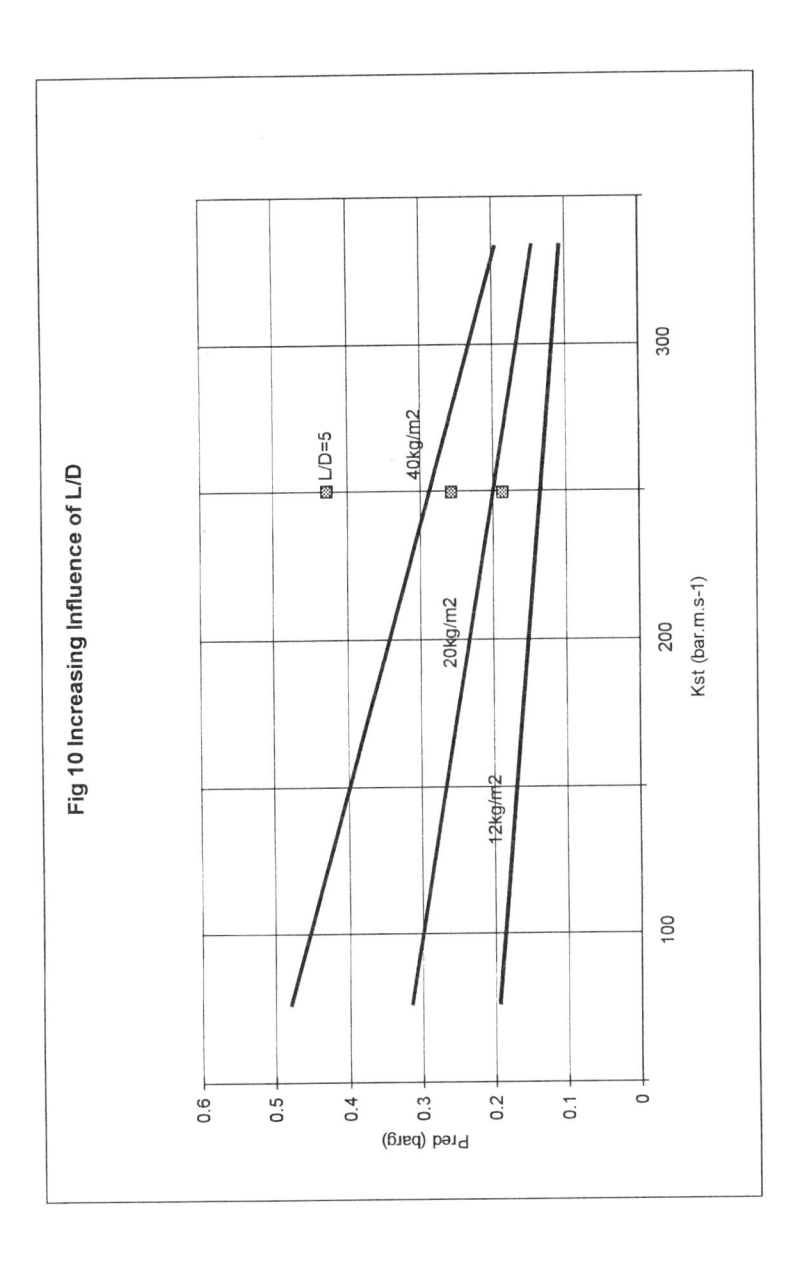

Fig 10 Increasing Influence of L/D

VOC ABATEMENT AND VENT COLLECTION SYSTEMS
"A structured approach to safe design" or "Do the safety risks outweigh the environmental benefit"

Peter J Hunt BSc (Hons) CEng MIChemE
Eutech, Belasis Hall, Billingham, Cleveland, TS23 4YS

An increasing number of environmental improvement initiatives are being implemented to meet environmental emissions limits for Volatile Organic Compounds (VOCs), using "end of pipe" abatement techniques. To achieve these limits Vent Collection Systems are typically required to collect the pipe vents and feed them to the abatement unit. Where vents contain flammable or reactive vapours, complex safety and operational problems can be encountered in terms of fire, deflagration, detonation or chemical reaction of mixing vent streams.

This paper will cover a number of deflagration / detonation incidents from vent collection systems.

This paper will also address initial consideration of VOC losses and whether they can be eliminated/minimised which is key to the selection of the Best Available Technique (BAT) for abatement, as well as achieving the design of a safe vent collection system. The paper will then cover a structured methodology to achieve a "Basis of Safety" for vent collection systems.

To illustrate the application of this approach to vent collection systems the following case studies will be used:-

(a) Multiproduct Batch Plant, where the chemical inter reactivity of mixing vent streams was evaluated to develop a safe design.
(b) Carbon Monoxide Vent collection system on a Titanium Dioxide plant waste gas stream where the approach was applied to develop a comprehensive basis of safety for operation.
(c) Monomer production plant where following a number of explosions in the vent collection system, the basis of safety was re-evaluated and as a result, inerting was used to reduce oxygen concentration to avoid formation of a flammable mixture in the header.

In all cases the installation of a vent collection system to achieve environmental compliance can result in significant safety risks. These risks must be evaluated to ensure the continued safe operation of sites and to consequently meet the environmental improvement objectives.

Keywords: VOC, Vent Collection, Deflagration, Detonation, Safety, Risk, Reactivity, Fire

© 1998 Copyright Eutech

INTRODUCTION

Air Quality Standards in Europe have been increasingly tightened over the years. Some of the most important industrial emissions are Volatile Organic Compounds (VOCs).

VOCs emissions result in photochemical ozone creation in the lower atmosphere which have both human health effects and can lead to damage to crops and vegetation. VOCs are classified according to their Photochemical Ozone Creation Potential (POCP) referenced to a standard of 100 for ethylene.

In 1993 the UK Government set out how it expects its obligations to be met under the United Nations European Committee on the Environment (UNECE) VOCs protocol to reduce its emissions by 30% (based on 1988 levels) by 1999. A 65% reduction is forecast for the chemical sector based on the application of environmental improvement programs. Also under review is the proposed VOC Solvent Emissions Directive which aims to cut VOC emission by 67% by 2007, compared to 1990 levels. VOCs are produced/consumed in a wide range of industry sectors including power, food, chemical, petrochemical, finechemicals, pharmaceutical, manufacturing, electronic and automotive. For large companies to small and medium sized enterprises/ (SME)to achieve environmental compliance for Volatile Organic Components (VOC) emissions as part of their improvement program, end of pipe abatement can often be the only practical option. This would typically follow a Best Available Techniques Not Entailing Excessive Cost (BATNEEC) and Best Practicable Environmental (BPEO) option assessment.

In many cases, to meet the EA IPC Guidance Note emission limits for VOCs, vents are collected and fed to an abatement unit.

Techniques for VOC abatement can be broadly characterised as :

Recovery and Re-use Techniques
- Absorption
- Adsorption
- Condensation

Destruction Techniques
- Thermal oxidation
- Catalytic oxidation
- Biological
- Flares

Where vents contain flammable or reactive vapours there are complex safety and operational implications from the deflagration, detonation and chemical reactivity risks. These hazards are a particular problem on VOC abatement vent collection systems which feed thermal oxidisers or where other ignition sources are present. Activated Carbon VOC abatement systems are also known sources of ignition from hot spots being formed by high heat of absorption, Reference 14.

This paper will outline some of the risks related to vent collection systems, and detail a structured methodology to achieve 'A Basis of Safety' which is applicable to both batch and continuous processes. This approach will be illustrated by a number of case studies.

BACKGROUND

In order to collect process vent materials from process vessels (including reactors and storage tanks) a vent collection system is typically required.

On many installations the vent collection has been considered as just another pipe system. Hazards have not been assessed in the Hazard Studies, no basis of safety detailed and the system not managed / operated as a main plant item. The interactions with the plant normal pressure relief system have also not been considered.

However vent collection systems present significant risk such as :-

- The vent can contain flammable or reactive components
- Ignition of flammable mixture within the vent header can lead to deflagration or detonation hazards
- Transmittal of fire / explosion to other areas of plant via vent system
- Vent collection systems also expensive to install and operate

INCIDENTS

There are many examples of hazardous events in vent collection systems involving deflagration / detonation. A great deal of the early work in the 1960's was in the American oil industry where environmental legislation led to vapours being collected from ship offloading facilities and fed to incineration systems. This, plus the increasing environmental pressures on the process industries, has resulted in a number of incidents, including:-

- Reference 24 A Waste Gas incinerator near Houston experienced a pressure wave in the suction vent pipe resulting in extensive damage. A well designed system was overcome by an "unforeseen" combination of failures. Through a combination of automatic and operator responses to shut off the waste gas feed, a fuel rich stream was suddenly introduced to the incinerator. The flame front generated a pressure wave which blew apart the flame arrestor, piping etc. The damage could have been minimised with an in line detonation arrestor.

- Reference 22 This discusses an incident which resulted in the destruction of a large fluid hydroformer, and 63 tanks. Investigators were unable to trace the propagation of detonation through the piping system but the velocity was estimated at 1000 - 2000 ms^{-1} with pressures up to 28 barg.

- Reference 35 In a vent system connected to an incinerator, a flash back occurred. The inline flame arrestor failed with a subsequent fire in the dump tank.

- Reference 39 A waste gas incinerator experienced a flash back with a pressure wave in the suction vent gas system. This resulted in extensive damage to the vent system fan, valves, arrestor and piping.

- Reference 31, 32, 33 An incident in August 1991 highlighted the risk of vent collection systems. Terminal Pty Limited, Coode Island, Melbourne operation on "Site A" Compound with 45 tanks with a total capacity of 45,000 m^3. The tanks were connected by a vapour recovery system. An explosion occurred in Tank 80, causing the tank wall and roof to be propelled approximately 20m in the air. The subsequent fire propagated through the vent collection system. Figure 1 shows the site before the incident and Figure 2 after the event which gives an indication of the scale of destruction. Ignitions in tanks are not uncommon with a recent incident at Shell Rotterdam Reference 37 being yet another example

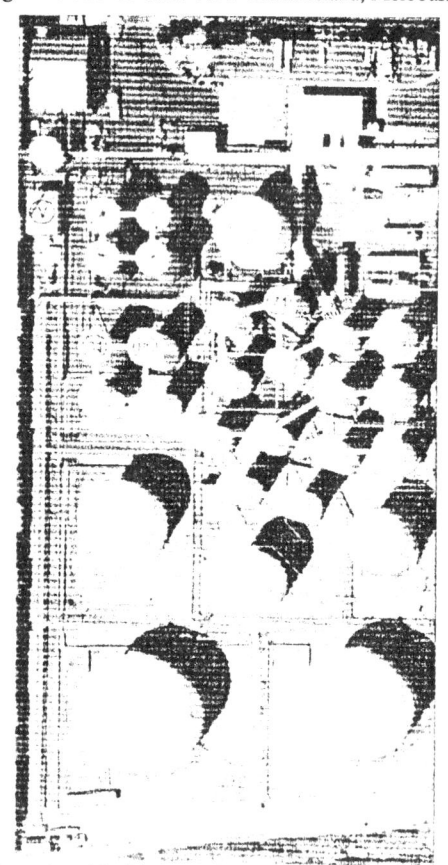

Figure 1: Site 'A' Plan View Coode Island, Melbourne

(Reproduced by kind permission of Terminals Pty Ltd)

Figure 2. Coode Island - Explosion Damage

(Reproduced by kind permission of Terminals Pty Ltd)

- Reference 29 details a number of incidents.

The number of vent collection system problems and incidents are on the increase with environmental pressures resulting in more end of line abatement systems being installed.

Where vent collection systems feed thermal oxidiser systems, a risk a they also present due to unburnt gas accumulation and ignition, which can be transmitted back to the plant.

VOC ABATEMENT

The need for vent collection systems, as described previously, can arise from the need for VOC abatement. A typical approach to VOC abatement is detailed in Figure 3.

Vent Collection can lead to significant risk, and where possible, attention should be given to eliminate, minimise and locally abate "VOC emissions" before a vent collection system is considered. In undertaking this, low/no solvent technologies should be considered to reduce or eliminate the requirement to use VOCs. Vent collection and end of line abatement, also represent a significant investment in capital, and ongoing operational costs. They have also been the cause of reduced plant reliability due to blockage, pressure control problems etc.

Before embarking on a VOC abatement project, whether it is based on source reduction or end of line abatement, it is imperative that vital preliminary data on vent emissions is collected in a systematic and structured fashion, Reference 33. This can be a time consuming exercise, especially for large sites with a significant number of emission points. It can also be equally complex for a Batch Process due to variation in cycle times, process operations and products. Establishing accurate emission data is essential to assess minimum, maximum and normal emissions at each step of a Batch cycle.

The approach normally consists of developing a model of emissions, to establish an accurate mass balance of VOC emission from all sources. Any errors, or inaccurate assumptions, at this point can lead to unsafe design, inappropriate selection of abatement techniques and high capital / operating costs.

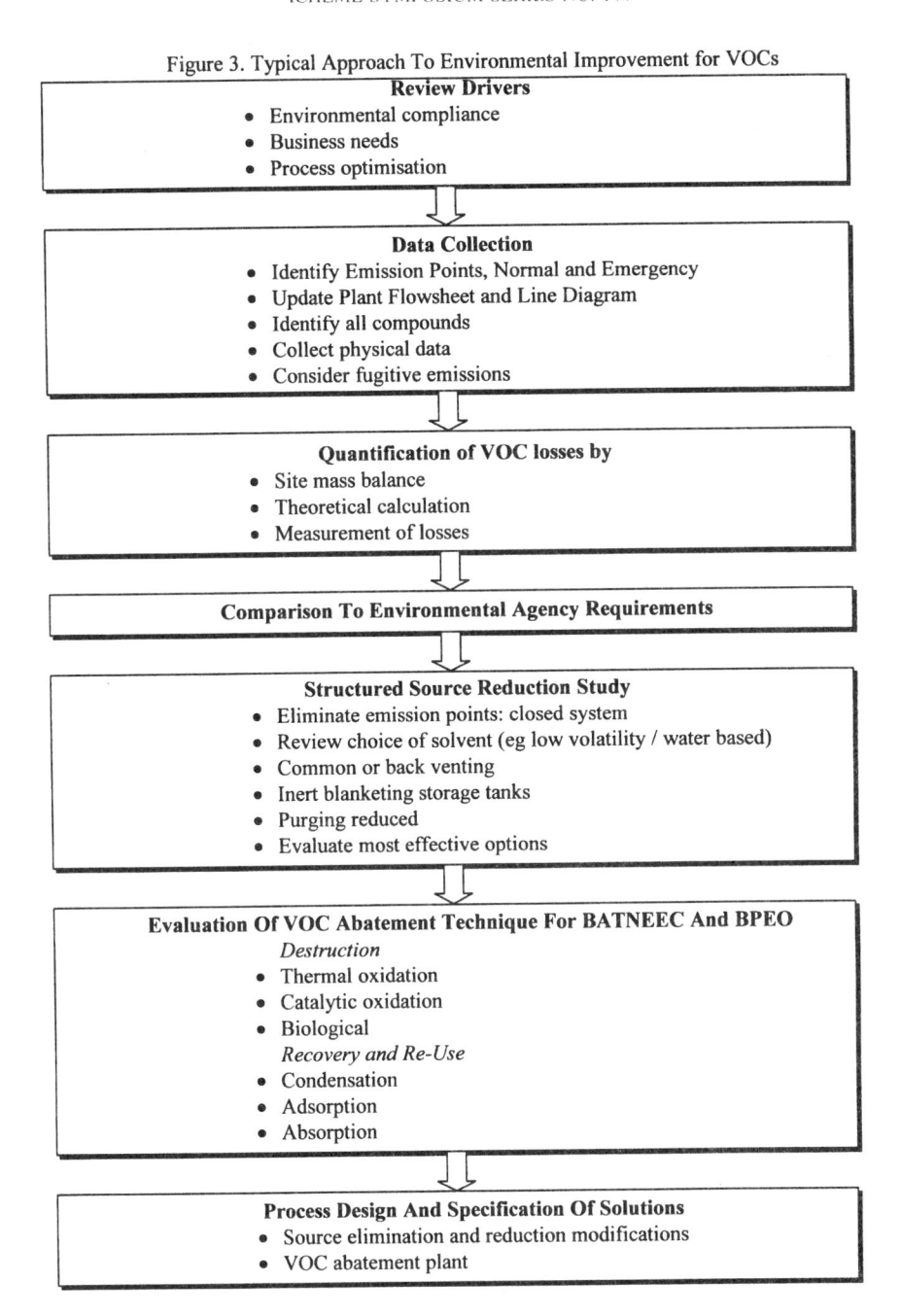

Figure 3. Typical Approach To Environmental Improvement for VOCs

Review Drivers
- Environmental compliance
- Business needs
- Process optimisation

⬇

Data Collection
- Identify Emission Points, Normal and Emergency
- Update Plant Flowsheet and Line Diagram
- Identify all compounds
- Collect physical data
- Consider fugitive emissions

⬇

Quantification of VOC losses by
- Site mass balance
- Theoretical calculation
- Measurement of losses

⬇

Comparison To Environmental Agency Requirements

⬇

Structured Source Reduction Study
- Eliminate emission points: closed system
- Review choice of solvent (eg low volatility / water based)
- Common or back venting
- Inert blanketing storage tanks
- Purging reduced
- Evaluate most effective options

⬇

Evaluation Of VOC Abatement Technique For BATNEEC And BPEO
Destruction
- Thermal oxidation
- Catalytic oxidation
- Biological

Recovery and Re-Use
- Condensation
- Adsorption
- Absorption

⬇

Process Design And Specification Of Solutions
- Source elimination and reduction modifications
- VOC abatement plant

VENT COLLECTION SYSTEM METHODOLOGY TO DEVELOP A BASIS OF SAFETY

In order to achieve safe design of a vent collection system, a structured methodology is required.

As a result of significant experience in the safe design of Vent Collection systems over a range of projects, Eutech has established a structured methodology as defined in Figures 4 and 5.

This will provide the information to develop a "Basis of Safety" and then assess the risk with a full Hazard Study program being be applied to any proposed design.

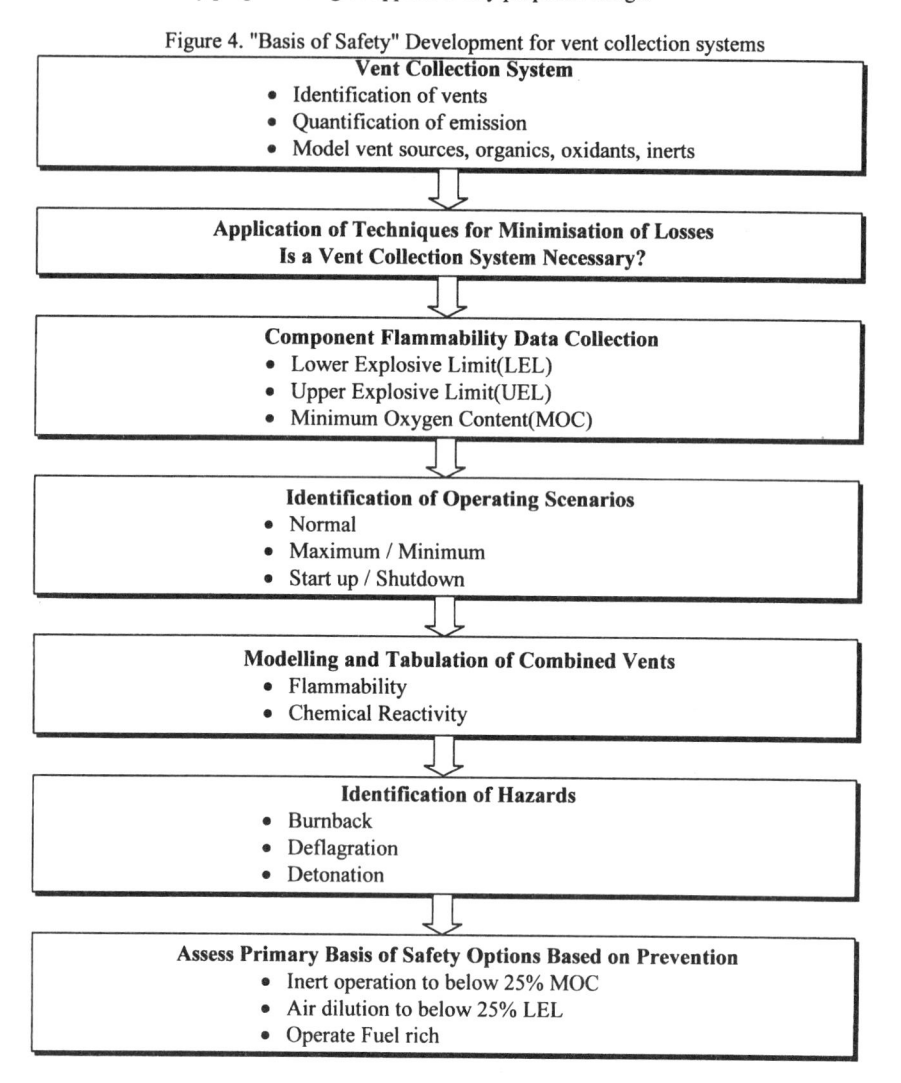

Figure 4. "Basis of Safety" Development for vent collection systems

Vent Collection System
- Identification of vents
- Quantification of emission
- Model vent sources, organics, oxidants, inerts

⇩

Application of Techniques for Minimisation of Losses
Is a Vent Collection System Necessary?

⇩

Component Flammability Data Collection
- Lower Explosive Limit (LEL)
- Upper Explosive Limit (UEL)
- Minimum Oxygen Content (MOC)

⇩

Identification of Operating Scenarios
- Normal
- Maximum / Minimum
- Start up / Shutdown

⇩

Modelling and Tabulation of Combined Vents
- Flammability
- Chemical Reactivity

⇩

Identification of Hazards
- Burnback
- Deflagration
- Detonation

⇩

Assess Primary Basis of Safety Options Based on Prevention
- Inert operation to below 25% MOC
- Air dilution to below 25% LEL
- Operate Fuel rich

In order to maintain and/or monitor the vent stream, an oxygen or flammable gas analyser(to monitor concentration as % LEL) is typically installed. Having established a proposed primary 'Basis of Safety' for the vent collection system (eg inerting, air dilution) it is necessary to carry out a risk assessment to establish whether the level of risk is acceptable versus the subsequent consequences

Figure 5. Vent Collection System Risk Assessment

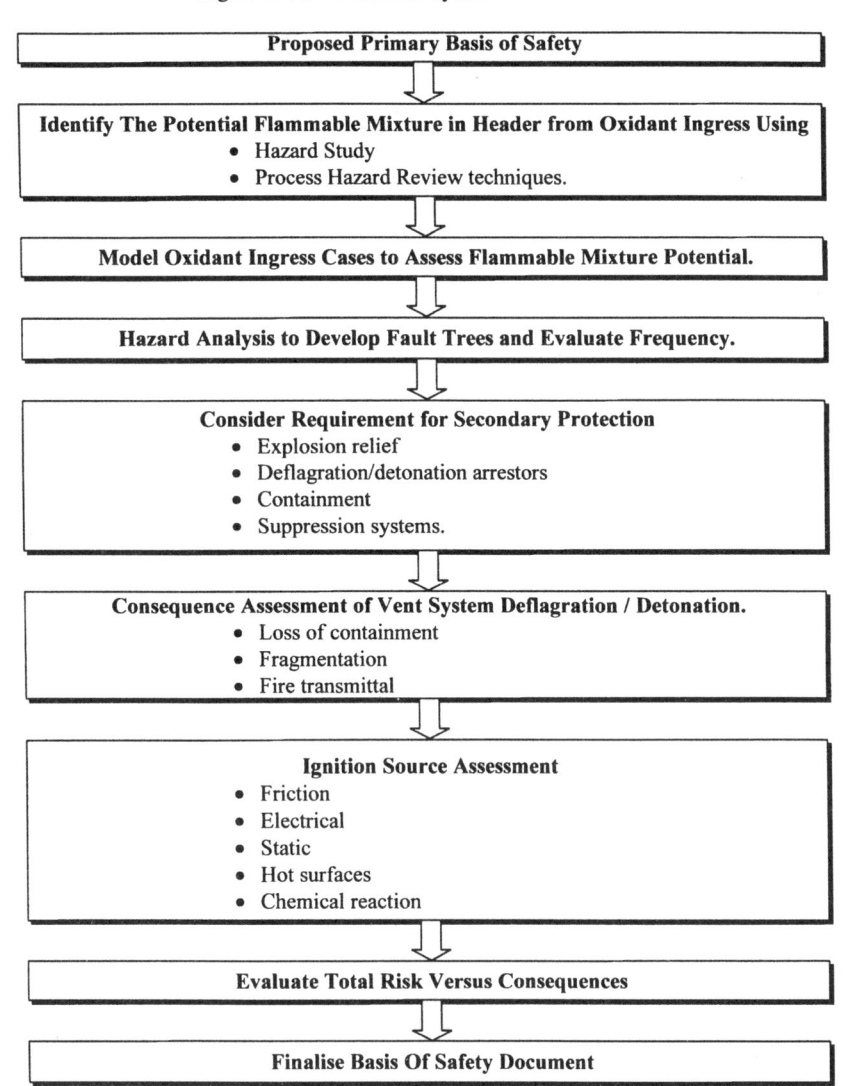

Proposed Primary Basis of Safety

⇩

Identify The Potential Flammable Mixture in Header from Oxidant Ingress Using
- Hazard Study
- Process Hazard Review techniques.

⇩

Model Oxidant Ingress Cases to Assess Flammable Mixture Potential.

⇩

Hazard Analysis to Develop Fault Trees and Evaluate Frequency.

⇩

Consider Requirement for Secondary Protection
- Explosion relief
- Deflagration/detonation arrestors
- Containment
- Suppression systems.

⇩

Consequence Assessment of Vent System Deflagration / Detonation.
- Loss of containment
- Fragmentation
- Fire transmittal

⇩

Ignition Source Assessment
- Friction
- Electrical
- Static
- Hot surfaces
- Chemical reaction

⇩

Evaluate Total Risk Versus Consequences

⇩

Finalise Basis Of Safety Document

As an outcome of this approach, it is also necessary to develop operating and maintenance procedures, taking into account identified hazards for safe commissioning and operation of the Vent Collection System. The final outcome should include a fully documented design, mass balance and quantified risk assessment.

For operation of the vent collection system, an owner of the process should be identified who is responsible for safe operation, design changes and maintenance.

DEFLAGRATION / DETONATION CONSEQUENCES

Ignition of a flammable mixture in a vent collection system can result in burn back from the point of ignition and run up to deflagration / detonation.

Deflagration

Deflagration is defined as a combustion wave, propagating at a velocity less than the speed of sound, (as measured at the flame front), which propagates via a process of heat transfer.

The consequences of ignition of a flammable mixture in a vent collection system, can result in deflagration with a pressure ratio up to 10 times initial pressure and maximum propagation velocities of typically 10 - 300 m/s. See Figure 6.

Deflagration to Detonation Transition

Following deflagration it is possible to achieve Deflagration to Detonation Transition (DDT) which results from acceleration of a deflagration flame to detonation via combustion generated turbulent flow and compressive heating effects. During DDT, overdriven detonation peak pressures of up to 100 barg can be observed, assuming the initial start pressure is atmospheric.

Overdriven detonation pressures cannot easily be estimated as they are dependant on many factors such as pipe layout, surface roughness etc. For example from a range of tests carried out by IMI AMAL assessing run up distances in 2" nominal diameter piping (Figure 6), the results indicated peak pressure from overdriven detonations in a pipe, were in the region of 70 to 80 barg for propane air and ethylene/air mixtures with a duration from microseconds to milliseconds. From IMI AMAL tests in 6" nominal diameter piping pressures up to 150 barg have been recorded. References 24 and 40 quote overdriven detonation pressures of up to 100 barg based on initial atmospheric pressure. However the pressures are only very short lived pulses, applying momentary stress on the walls and, hence, unlikely to lead to failure. These pressures are also supported by References 26 and 32.

Stable Detonation

This is the fully developed detonation wave, propagating at a constant velocity of typically 1600 - 1900 ms^{-1}. IMI AMAL test data, Figure 6, indicates typical stable detonation pressures of 20 - 50 barg and 20-30 barg for propane, based on a range of tests. Reference 24 advises on stable pressures of 18 - 30 barg and Reference 19 20 - 24 barg.

ICHEME SYMPOSIUM SERIES NO. 144

Figure 6. Experimental Effect of Run up length on Detonation Pressure
(IMI AMAL specific test results)

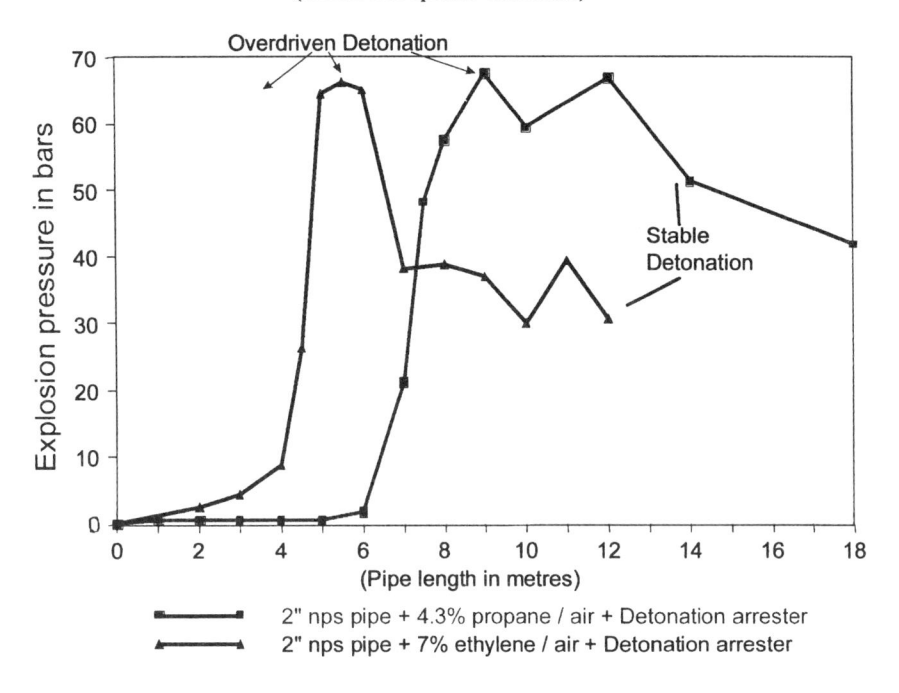

■——■ 2" nps pipe + 4.3% propane / air + Detonation arrester
▲——▲ 2" nps pipe + 7% ethylene / air + Detonation arrester

(Data reproduced by the kind permission of IMI AMAL)

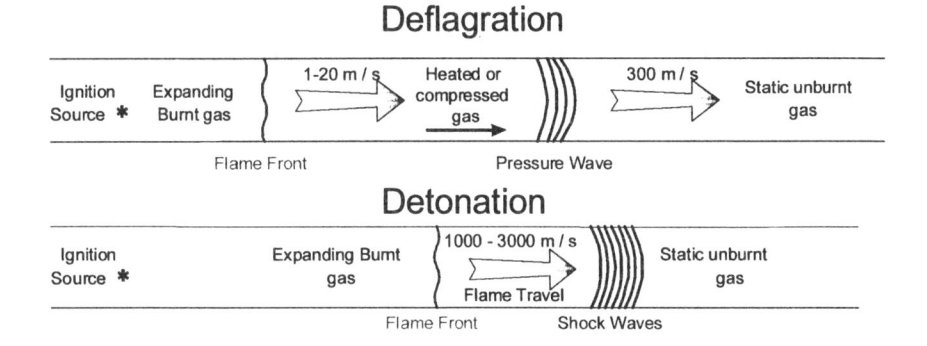

Deflagration

| Ignition Source ✱ | Expanding Burnt gas | 1-20 m/s Heated or compressed gas | 300 m/s | Static unburnt gas |

Flame Front — Pressure Wave

Detonation

| Ignition Source ✱ | Expanding Burnt gas | 1000 - 3000 m/s Flame Travel | | Static unburnt gas |

Flame Front — Shock Waves

Conclusions

The available literature data for the ignition of a flammable mixture in a vent system indicates that there is the potential for deflagration and acceleration to detonation. In a vent collection system the presence of obstructions, bends and flanges promote turbulence, and therefore provide increased acceleration.

Table 1. Typical Deflagration - Detonation Phenomena

	Deflagration	Overdriven Detonation	Stable Detonation
Maximum Pressure	~10 barg	~150 barg	~20 - 50 barg
Timescale	Millisecs	≤1 Millisecs	>1 Millisecs

At present, it is virtually impossible to predict the potential for detonation and run up distances. Typically, a worst case design basis needs to be assumed or experimental work must be carried out on the likely vent mixtures to establish a design basis and potential for loss of containment. Investigation into the phenomena of detonation in pipework is ongoing at the Department of Physics, University of Wales.

DEFLAGRATION / DETONATION PROTECTION

In many cases, it is not practical to eliminate ignition sources and prevent all sources of oxygen ingress to the header, hence options for deflagration/detonation protection need to be assessed. The options available to protect the vent collection systems in the event of a deflagration/detonation include:-

Containment

This method would require constructing the whole plant to withstand deflagration, overdriven detonation and stable detonation pressures as defined in the previous section. Although this approach can be used it can be expensive and often not practical except in small diameter pipes which can contain to pressures > 100barg

Deflagration Protection

This would arrest propagation of a flame in its incipient stages, (ie during subsonic flow) and can be achieved in a number of different ways.

<u>Passive Flame Arrestor.</u> Installed to quench the flame before Deflagration - Detonation Transition (DDT) can occur. Flame arrestors utilise a property known as Maximum Experimental Safe Gap (MESG) which is the largest gap through which a flame will not transmit when tested in accordance with test standards. Arrestors are designed with cells to be smaller than the MESG, such that as the flame front travels through each cell there is a transfer of energy between the

flame front and the cell walls. This heat transfer through the boundary layer to the cell wall results in cooling of the burning gases to below the autoignition temperature of the unburnt gas ahead of the flame front (34, 36). Current UK guidelines for testing of deflagration and detonation arrestors are covered in BS 7244, and the draft European standard Pr EN12874. The US Coast Guard (USCG) standard covers detonation arrestors only.

Design of Flame arrestor elements is based on the gas groupings in BS7244 1990 as follows:

$$\begin{array}{ll} \text{Group IIA} & \text{MESG} >= 0.9\text{mm} \\ \text{Group IIB} & \text{MESG} = 0.5 \text{ to } 0.9\text{mm} \\ \text{Group IIC} & \text{MESG} <0.5\text{mm} \end{array}$$

Group IIA comprises the majority of hydrocarbon gas, Group IIB more reactive gases/vapours such as ethylene, whilst Group IIC contains the most reactive gases such as hydrogen and acetylene. For a single gas/vapour the MESG can easily be found. However for mixtures, it is more difficult unless mixture data is available otherwise the component with the smallest MESG has to be considered. The concept of endurance burning is also an important issue in the use of flame arrestors. Under certain conditions, a flammable mixture in a header could ignite, burn back and then form a stable flame on the arrestor. Under this condition the arrestor element can gradually heat up to the Auto ignition temperature (AIT) of the inlet gas/vapour -this is the "Safe burning time" which can vary from 2 hours for small units to 15 mins for larger, although burning tests of greater than 15 mins are considered optimistic.

Explosion Venting. The explosion vent is a weak membrane in the pipework to relieve pressure and discharge the flame to atmosphere.

Vents should be placed at intervals less than the predicted run up distance to detonation. Great care is needed on location of a vent due to the flame and pressure which is vented. Explosion vents will reduce the effects of pressure, but will not stop the flame continuing past the vent. The main problem is identifying a safe location into which relieve the vented products/flame.

Explosion Suppression. Explosion Suppression involves the detection and extinguishing of the flame in its incipient stages by rapid injection of chemical suppressants (eg Kidde or Fike type system) and arresting the propagation of the flame front. The distance between detection and suppression must be less than the run up to detonation or a combination of detectors, suppressant and slam shut valves would be required. These have been used on vent collection systems. (Reference 25)

Hydraulic Arrestor. This is based on a liquid seal to act as an arrestor. It has the advantage of not being affected by blockage. The disadvantages are increased pressure drop, operational problems, and reliability. In addition extensive instrumentation is required including level measurement to maintain water level, Gas flow-to ensure design flow is not exceeded and Temperature to detect if burnback deflagration/detonation has occurred

Explosion Isolation. Explosion isolation involves the activation of mechanical valve or chemical barrier to arrest the propagation of flame in a pipe. Valves or chemical barriers are effective when located near potential ignition sources and present no restriction to flow.

Conclusions on Deflagration Protection. For deflagration protection it is necessary to have identified the point of DDT, pressures etc, as covered in the previous section. If however ,it is not possible to quantify this or locate the arrestor close to the ignition point, it is assumed that run up to detonation could occur and detonation rather than deflagration protection is required. The requirement for effective detonation protection is to arrest the propagation of a detonation, limiting the potentially destructive force of the pressure shock waves and transmittal eg a flame front to the main plant.

Detonation Protection

Passive Detonation Arrestor. These are passive bi or uni–directional arresting devices used to quench or destroy the transverse structure of the detonation flame front and are inherently safe. The problem with an arrestor is that it acts as a filter and is subject to blockage.

Active Detonation Arrestor. These are systems that detect the propagating flame front and activate rapid response valves and suppressors to prevent the propagation of a flame. There are high integrity trip systems designed specifically for the duty.

Suppressors and valves are located in strategic positions and supplemented with vents. These systems should be considered where blockage is a problem and have been installed on a number of vent systems (25, 28).

Conclusion. Passive detonation arrestors present an inherently safe simple solution with no moving parts when compared to an active system. This is on the basis that acceptable plant on-line time is possible due to potential blockage problems and that particulate build up has no effect on the integrity of the arrestors.

OTHER ISSUES ASSOCIATED WITH THE DESIGN OF VENT COLLECTION SYSTEMS

In the design of vent collection systems, other factors need to be taken into account including:-

a) Liquid condensation/freezing: Liquid collection in the vent pipe work can require knock-out pots to prevent carry over to the abatement system. Lagging and/or trace heating may also be required to prevent condensation in the vent pipe work. (Reference 35)

b) Fouling: Potential for blockage from solids or liquid build up in the vent, deflagration/detonation arrestors needs to be assessed in terms of operation of the plant and is a particular problem for Group II arrestors with their smaller and longer quenching cells. In cases of fouling problems, a parallel arrestor may be required so one arrestor is in service while the standby one is cleaned (see Figure 7 and 8 which shows a parallel arrestor installation.) An interlocked valving system must be used to ensure the header cannot be isolated which could result in over pressure of the plant.

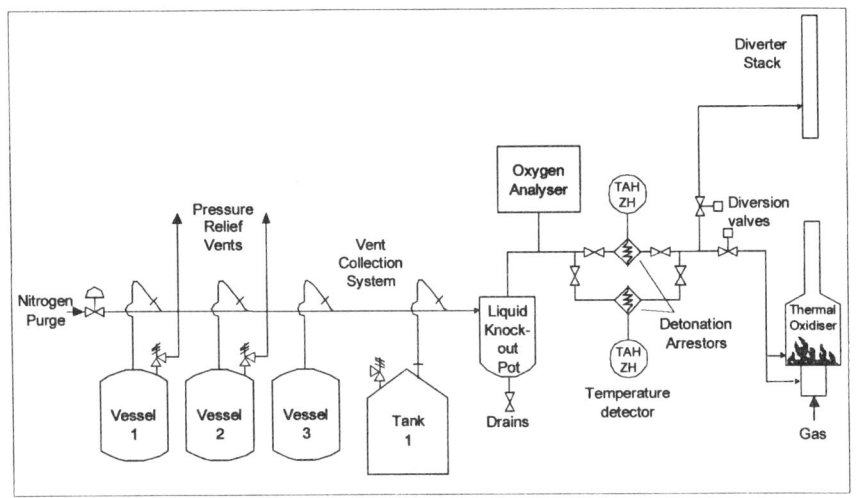

Figure 7 Inerted Vent Collection System with Detonation arrestors

Figure 8 Detonation arrestor with installed standby

c) Pressure Relief: VOC Vent collection systems are typically designed for normal vent losses. Pressure relief for relief as a result of fire, thermal expansion, reaction should be a totally separate system which could be individual relief vents or vents to a relief header. Any relief system design should follow a structured process from initial identification of relief cases to consideration of final vent discharge and disposal eg flaring / absorption etc.

d) Pressure Drop: Operation of vent collection systems are usually such that they do not affect the main process. A vent system can operate at slight positive pressure which has the advantage of preventing air ingress or slight negative pressure where a fan is often required to overcome pressure drop in the vent system with a pressure control loop. Design and control of such a system can be complex and require modelling to assess operating scenarios. The reduced MESG for more reactive gases/vapours ie Group II B/IIC can result in high pressure drops, for arresters

e) Divert Stack: Vent collection systems typically require a stack to divert flow into during startup / shutdown conditions when the vent stream may be in the flammable range; during fault conditions or in cases of abatement plant malfunction. The risks need to be considered as part of the risk assessment and discussions held with the Environment Agency to ensure the effect on the environment is understood. Diversion is often automated and linked to the plant control system (IPS).

CASE STUDIES

The structured approach to Vent Collection System design has been applied by Eutech to a number of studies.

Case study 1: Vent Collection System Design for a Multiproduct Batch Plant

At their factory at Seaton Carew, Oxford Chemicals manufacture a range of over 400 flavour and fragrance intermediates. Chemicals are processed in any of twenty-two reaction vessels, which range in size from 20 to 2500 litres, and are constructed from QVF glass-lined steel or stainless steel. Expansion and future development plans to incorporate a biofilter stimulated a review of process venting. An internal study quickly determined that the final vent treatment systems were adequate. The common vent collection system, however, was identified as an area of concern, as it had developed over the years with some undesirable interconnections between vessels that handle chemicals that can react violently if mixed.

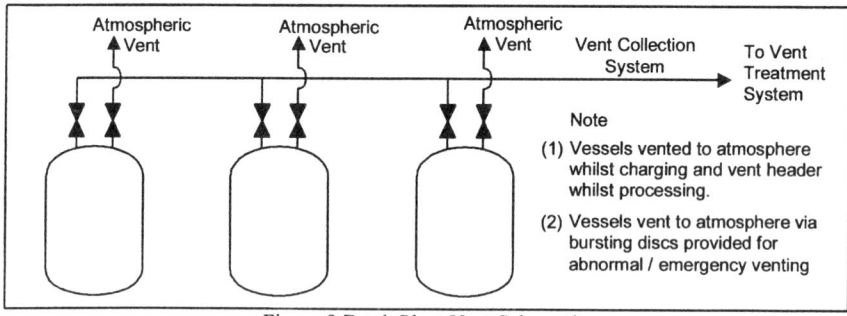

Figure 9 Batch Plant Vent Schematic

Liaising closely with Oxford Chemicals production and technical staff, data was collected on potential hazards in two areas - chemical reactions and potential flammable mixtures. By applying a reaction matrix it showed that no dangerous reactions would be introduced; action was needed, however, to protect against the possibilities of fire or explosion.

This conclusion was backed by detailed flammability studies, using knowledge-based analysis of the chemicals and mixtures involved.

Eutech proposed two possible control measures, inert gas purging or air dilution. Inert purging required the oxygen concentration in the header to be diluted with nitrogen to a design concentration of 2 per cent, with an action level of 5 per cent, representing an alarm point at which, if the oxygen level continued to rise, the header fan and vessel agitators would shutdown. This was based on operating at 25% of the minimum oxygen concentration (MOC) of organic chemicals.

Air dilution of the vent vapours, the other option, provided for the concentration of flammable vapours to be 25 per cent of the lower explosive limit of the worst case chemical. An additional advantage of dilution is that at very low concentrations, the risk of chemical reaction is insignificant. The design included proposals for sizing the new header, on air and nitrogen flows, and on elimination of ignition sources.

Oxford Chemicals selected air dilution with a flammable gas analyser for the replacement vent collection system which is now operating. This incorporates a large fan to draw air through the system continuously, an approach which has the added advantages of preventing leaks, and avoiding back pressure which could expand and stress the vessels.

Case study 2: CO/COS Vent Collection System, Tioxide

Tioxide's ICON Titanium Dioxide plant at its Greatham site produces Carbon Monoxide (CO) and Carbonyl Sulphide (COS) as waste gas from the process. In order to comply with agreed consent levels ,abatement options were evaluated by Tioxide .

The selected option was a thermal oxidiser with down stream cooling and absorption Figure 10. This raised concern about the potential for flammable mixture formation in the vent collection system being ignited by the oxidiser and burn back into the plant. Eutech working with Tioxide developed a Basis of Safety for the vent collection system.

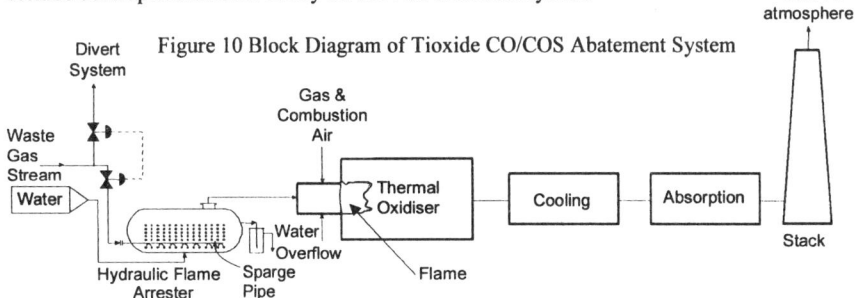

Figure 10 Block Diagram of Tioxide CO/COS Abatement System

The vent components consisted of N_2, CO, COS, H_2O, Cl_2, CO_2. The hazard identified was the potential for CO/COS air mixture to be ignited by the Thermal Oxidiser which could run up to Deflagration/Detonation. The methodology as detailed in "Vent Collection System Design Methodology" was applied as defined in Figure 4 & 5.

The vent flows were quantified and modelled on a spreadsheet for all potential operating conditions, eg peak rates, start up, shut down etc, with Le Chatelier's Equation used to assess the mixture flammability limits and these were plotted on a Flammability diagram Figure 11. The flammability data was collected for all components as well as data on deflagration/detonation potential for CO.

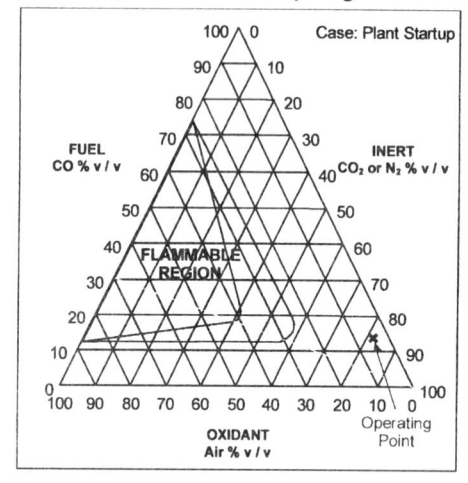

Figure 11 Flammability Diagram

A hazard assessment, lead by a Eutech Hazard Study Leader, was carried out with a plant team to assess potential for air ingress for normal, abnormal, startup / shutdown situations. Cases were eliminated that would not lead to a significant air / oxidant ingress which might approach the MOC, with the remaining cases being evaluated to quantify the frequency of a flammable mixture being established.

The study also considered ignition sources from the thermal oxidiser, vent fan and static from the GRP pipework.

The Basis of Safety for the Vent Collection system was established as operating at less than 25% MOC, with a slight positive pressure to minimise potential air ingress. Due to the potential for particulates in the waste gas, a hydraulic arrestor was selected as opposed to a standard detonation arrestor and designed to prevent burn back as a result of ignition of the vent gas steam. A hydraulic flame arrestor design is based on the velocity of gas through the sparge pipe, where the flame propagation will be stopped at the water surface because the water layer between rising bubbles prevents ignition transfer and flash back through the flame arrestor. Increasing the gas flow rate above the maximum gas flow will cause ignition transfer between bubbles and flash back.

The quantified risk assessment established the likely frequency of a flammable mixture being formed and defined recommendations to reduce this to achieve an acceptable level of risk.

Case study 3: Monomer Plant Vent Collection System Design

Following an explosion incident, as a result of liquid carry over to the oxidiser and a second incident of deflagration/detonation in the vent system, a comprehensive study was undertaken to achieve safe design and operation. The study assessed the two hazards considered ie liquid carry over to the thermal oxidiser and burn-back to deflagration/detonation.

A comprehensive study was carried out to accurately quantify vent losses from the plant to establish an accurate mass balance. Early in the study, considerable design effort was focused on elimination and minimisation of vent losses at source. This had the benefit of reducing flammable potential and blockage problems from monomers. The methodology, as defined in Figure 4 and 5, was followed with inert operation to give O_2 at less than 2% vol ie <25% MOC selected as the basis of safety. The vent system also had a divert stack system which would act as a safe location to divert vent flow on start up or shutdown. Consideration was given to the divert stack design and location to ensure dispersion of the vent gases such that acceptable flammability of less than 25% LEL and odour criteria were met at the site boundary. Dispersion modelling was undertaken to evaluate this.

The study addressed, in a structured way, the potential for air ingress from normal and abnormal cases. These were modelled on a spreadsheet and used to assess deviations, with Le Chateliers Law being applied to quantify mixture flammabilities. As part of the hazard analysis (HAZAN), fault trees were developed to establish the frequency of a flammable mixture being formed and subsequent ignition and deflagration/detonation potential.

A significant part of the study focused on the potential for run up to detonation potential, detonation pressures and consequences. Due to the complexity of the pipework and line length, the detonation was considered a risk.

In order to evaluate the consequences following detonation, detailed piping stress calculations were carried out on the proposed vent pipe; these indicated that the 4" Sch 40 had a failure pressure >200 barg and 6" >160 barg. The study also addressed the requirements for Bi-directional detonation arrestors and their optimum location. In order to overcome the risk of a stable flame being formed on the arrestor and subsequent burnback, temperature probes were to be installed on the relevant face. These would be used to trip feeds to the divert stack system and initiate a Nitrogen quench flow. It was also necessary to establish the system response time to ensure burn-through could not occur. Endurance burn requirements are defined in BS 7244 and USCG; AMAL advised that for the proposed arrestors, 15 minutes should be taken as the action point. The Quantified Risk Assessment was finalised to take into account risks versus consequence.

CONCLUSIONS

From the paper it can be seen that installation of vent collection systems can present significant risks in terms of safe operation of a plant or process. Elimination or minimisation of emissions at source should be the first priority and vent collection systems with end of line abatement should be avoided where possible. The safety risks include the potential for ignition in a vent collection system leading to deflagration/detonation and subsequent destruction. They are also costly to install and operate and can lead to reduced plant reliability.

To achieve safe design of a vent collection system, it is vital to have accurate data on vent sources as errors in this can lead to incorrect assessment of flammable potential and safe vent collection system design.

In summary, a structured approach is required to achieve safe design of a vent collection system and we should ask

'Do the Safety Risks outweigh the Environmental Benefits'?.

ACKNOWLEDGEMENTS

I would like to thank Oxford Chemicals and Tioxide for their kind permission to publish details of their study.

I would also like to thank Terminal Pty for the information provided on the Coode Island incident.

DEFINITIONS

BATNEEC. Best Available Technologies Not Entailing Excessive Cost.

BPEO. Best Practical Environmental Option.

Flame Arrestor. A device that permits gas flow, but prevents flame propagation beyond it in a flow system. The most common type is the dry flame arrestor. It consists of a matrix of small diameter holes or channels that permits the flow of gas, but quenches flame that propagates into it Other kinds of arrestors are the liquid seal, and high speed shut off valve types.

Deflagration. A flame front that propagates by the transfer of heat and mass to the unburned gas ahead of the flame front in a flammable gas mixture. Flame speeds can range from less than 1 m/s to (based on the unburned gas temperature) greater than 350 m/s (supersonic) for very high pressure, turbulent flames. Peak overpressures can range from a very small fraction to as much as twenty times the initial pressure.

Detonations (Gaseous). A flame front that propagates by shock wave-compression ignition in a flammable gas mixture. Flame speeds are supersonic (based on the unburned gas temperature) with Mach numbers ranging from 5 to 15. The pressure of a stable detonation usually ranges from about 20 to 30 times the initial pressure but can achieve compression ratios in excess of 100 at the moment of transition from deflagration to detonation.

Run-Up System. The flow system - pipes, bends, valves and any other flow devices - that a flame front travels through from the point of ignition to the flame arrestor.

End-of-Line Flame Arrestor. This type of flame arrestor is used where the potential ignition source is located outside of the vessel or flow system that is being protected. One end of the unit is open to the atmosphere directly or through a vent valve, cowl or a short length of open ended straight pipe. It is also referred to as a vent flame arrestor.

In line Flame Arrestor. A flame arrestor that is installed within a run-up system that does not vent to the atmosphere directly. The flow system between the potential ignition source and the flame arrestor is made through lengths of pipe that exceed end-of-line limitations and/or contain bends, tees, valves, or any other flow restricting or turbulence generating fittings.

Quenching Diameter. The largest diameter of a tube which will just quench a flame front in a particular fuel/air mixture. If the diameter is increased any further, the flame front can propagate in the tube without being quenched.

Maximum Experimental Safe Gap (MESG). The maximum gap between equatorial flanges in a spherical volume that will just prevent flame transmission from the vessel to the flammable gas mixture surrounding it.

Minimum Oxygen Concentration (MOC). This is defined as the lowest concentration of oxygen which will just support the combustion of fuel.

Burning Velocity. The fundamental burning velocity is the velocity with which flame moves, normal to its surface through the adjacent unburned gas.

Lower Explosive Limit (LEL). The minimum concentration of gas or vapour in air below which the propagation of a flame will not occur in the presence of an ignition source.

Upper Explosion Limit (UEL). The maximum concentration of a gas or vapour in air above which the propagation of a flame will not occur in the presence of an ignition source.

GUIDES AND STANDARDS

Pr EN12874: Draft Flame Arrestors – Specifications, Operational Requirements and Test Procedures.

HS(G)158 Flame Arrestors, HSE.

BS 7244 1990 "Flame Arrestors for General Use".

BS 5345: Part 1 1989"Selection, installation and maintenance of electrical apparatus for use in potentially explosive atmospheres".

Environmental Agency IPC Guidance Note S2 4.02 BAT in the Fine and Speciality Chemical Sector 1998(Draft for consultation).

HMIP Technical Guidance Note(A2) Pollution Abatement Technology for the Reduction of Solvent Vapour Emissions (HMSO), January 1997,ISBN 0-11-310115-5.

REFERENCES

1. Parker, R.V. "Safety Systems for a Continuous Petrochemical process under Vacuum". Plant/Operation Progress (Vol 3 No 2).

2. Bollinger, L.E., Fang, M.C., Edse, R. "Experimental Measurements and Theoretical Detonation Induction Distances". ARS Journal May 1961.

3. Lapp, K. "Safeguards cut tank explosion risk during gas flaring" Technology Aug 14 1989 Oil and Gas Journal.

4. Lewis, B. Von Elbe, G, "Combustion Flames and explosions of gases" Academic Press Inc, 1987.

5. Kirby, G.N. "Explosion Pressure Shock Resistance", Chemical Engineering Progress November 1985.

6. Abbott, J.A. "Prevention of Fires and Explosions in Dryers - A User Guide" IChemE P29-29, P36-39.

7. Lapp, K. "Flame arrestor failures illustrate needed design changes" P75, Oil and Gas Journal.

8. Hansel, J.G. "Predicting and Controlling Flammability of Multiple Fuel and Multiple Inert Mixtures" Plant/Operation Progress (Vol 11, No 4) October 1992

9. Reed, R.D. "Design and Operation of Flare Systems" Chemical Engineering Progress (Vol 64, No 6) June 1968.

10. Kilby, J.L. "Flare System Explosions" Chemical Engineering Progress (Vol 64, No 6) June 1968.

11. Chatrathi, K. "Deflagration Protection of Pipes" Plant/Operation Progress April 1992.

12. Martin, A.M. "Control Odours from CPI Facilities", Chemical Engineering Progress December 1992.

13 Craven, A.D. "The development of Detonation over pressures in pipelines" IChemE Symposium Series 25 (1968).

14 Loss Prevention Bulletin 105 P-15 'Activated Charcoal Filter Fire'.

15 Johnson, O.W. "An Oil Industry Viewpoint on Flame Arrestors in Pipelines" Plant/Operation Progress (Vol 12, No 2) April 1983.

16 Phillips, H. "Performance Requirements of Flame Arrestors in Practical Applications" IChemE Symposium Series No 97.

17 Broschka, G.L. "A Study of Flame Arrestors in Piping Systems", Plant/Operation Progress (Vol 2, No 1) January 1983.

18 Howard, W.B. "Use Precautions in Selection, Installation and Operation of Flame Arrestors" Chemical Engineering Progress, April 1992.

19 Roussakis, N. "A Comprehensive Test Method for Inline Flame Arrestors" Plant/Operation Progress (Vol 10, No 2), April 1991.

20 Watson, P.B. "Flame Arrestors" Conference March 1977.

21 Watson, P.B. "Flame Arrestors and the Chemical Tanker" Loss Prevention Bulletin 079.

22 Flessner, M.F. "Control of Gas Detonations in Pipes", US Coast Guard, Washington DC.

23 Thomas, G. "On Practical difficulties encountered when testing flame and detonation arrestors to BS 7244", Department Physics, University of Wales, Trans IChemE August 1993.

24 Guidelines for Engineering Design for Process Safety CCPS AIChE 1993.

25 Cooper, S.P., Moore, R.E., Kidde Gravinar Ltd
Capp, B. - HSE "Investigation into the use of active detonation arresters for solvent and waste gas recovery process".
I Chem E Symposium Series 130.

26 Clark, D.G., Sylvester, R.W., DuPont Engineering Limited
"Ensure Process Vent Collection System Safety" January 1996, Chemical Engineering Progress.

27 Burgoyne Consultant's 'Safe Design' letter The Chemical Engineer, 25 January 1996.

28 Fenwal Safety Systems 'Detection of in pipe combustion using optical flame sensors and pressure Detectors 1/8/97.

29   Nichols, F.P. "Design of Vent Collection and Destruction Systems", International Symposium on Runaway Reactions, Pressure Relief Design and Effluent Handling ,11-13 March 1998.

30   Coode Island, The Chemical Engineer, 31 October 1991, Page 17.

31   Coode Island, The Chemical Engineer, 12 September 1991, Page 14.

32   Reddie, P.A. 'Storage Terminal Safety Issues", Emergency Prevention and Management Conference, Signapore 1994.

33   Henton, J.E. 'Chemical Industry Case Study End of Pipe Treatment is Expensive', 7th Annual Conference on Volatile Organic Compounds, 9 - 10 March 1998.

34   Abrahamsen, A.R. "The Use of Flame Arrestors in Incineration and Flare Systems", IMIAMAL.

35   HSE Private Communication.

36   Long, D. "Flame Arrestors for Deflagation and Detonations The European View Point", Port Technology International.

37   The Chemical Engineer 'Fatal Explosion rocks shell' 11 June 1998, P3.
38   Leite, O.C. "Operating Thermal Incinerators Safety", Chemical Engineering June 1998.

39   Anderson, A.E., Dowell A.M., Mynaugh J.B. "Flashback from Waste Gas Incineration into Air supply Piping" Plant Operations Progress April 1992.

40   Reidewald, F. "Explosive Mixture" The Chemical Engineer,9 November 1995.

41   Lapp, K., Werneburg H. "Detonation Flame Arrestor Qualifying Application Parameter For Explosion Prevention in Vapour Handling Systems" Process Safety Progress (Vol 14,No2) April 1995.

THE DANGERS OF GRATING FLOORS: DISPERSION AND EXPLOSION

A E Holdo
The Department of Mechanical & Aeronautical Engineering, The University of Hertfordshire,
Dr G Munday
ISAA Consultants, 89, Chandos Avenue LONDON N20 9EG
D B Spalding FRS
Concentration, Heat and Momentum Limited. Bakery House, Wimbledon, SW19 5AU

> This paper describes a series of numerical simulations which explore the potential dangers of grating floors in partially enclosed process operations which involve flammable fluids. The circumstances which are being investigated involve the spill of a liquid or two-phase fluid below the grating which provides the source of vapour which will mix with air to form a potentially explosive vapour cloud.
> The only realistic technique available for the analysis of this complex situation is Computational Fluid Dynamics (CFD). However, there are limitations to the use of this technique and this paper examines these very carefully.
> The work described in this paper provides lessons to be learnt about possible hazardous situations and about the pitfalls and difficulties which may be met in the employment of highly sophisticated techniques for their analysis.
>
> Keywords: Safety, gratings, flow, dispersion, combustion, C.F.D.

INTRODUCTION

The evaluation of safety of operations which handle flammable materials has traditionally been factored into components, for example, release of flammable material, dispersion of that material, ignition leading to combustion whereby the loss is sustained by damage. However, even within this formalised structure, there are so many variations of possible circumstances that much of the understanding of safety issues follows from the analysis of experience rather than the synthesis of potential scenarios. The purpose of this paper is to demonstrate, using a specific instance, how observations from evidence of eyewitnesses to an accident involving fire/explosion of a release of hydrocarbon can lead to a new view of potential dangers. The approach is, of course, not a novel one since all accident investigations attempt to conclude their inquiries with a number of lessons which the industry should learn but the lesson in this case is new and it is hoped that the way in which the analysis of the experience is developed provides guidelines to the use of a combination of simple concepts of the basic phenomena involved and the subsequent use sophisticated tools to simulated their behaviour.

The accident under consideration is the fire on the offshore platform Piper Alpha on July 6th 1988. The evidence relevant to the present paper involves two areas which were of real concern to the Inquiry held subsequent to the accident by Lord Cullen. The first of these areas relates to the evidence of the only witness to observe the commencement of the explosion itself and involves the description of a precursor flame which spread across the end of the platform just before the witness felt the blast from the explosion. The other area of evidence relates to the absence of alarms from gas sensors which were located in the body of one of the process modules on the platform.

The Inquiry held by Lord Cullen culminated in a report, now known as the Cullen Report, which reached certain conclusions as to how the accident occurred. Subsequent litigation lead to a further examination of the evidence and the judgement at the end of this hearing was in general agreement with these conclusions. It is not our intention to suggest that the wrong conclusions concerning causation were reached but to show how the evidence can be attributed to an alternative explanation which in turn can provide useful lessons which were not identified in the Cullen Report.

EVIDENCE

The Piper Alpha platform is shown diagrammatically in Figure 1 as both plan (1a) and the east elevation (1b). It consists of a tubular steel structure standing on the sea bed and protruding some 68 feet above sea level. The platform was used for drilling wells and as a production platform. Oil production was handled in a number of modules on the 84ft level and the 107ft level. Eyewitness evidence lead the Inquiry to believe that the fire which eventually destroyed the platform was initiated by an explosion in one of two process modules, either Module B or Module C. These modules, are identified in the diagram together with Module A, which enclosed the terminations of the risers from the many production wells, and Module D, which contained the control room, various maintenance workshops and the power unit for the platform.

Module B was employed mainly in the processing of the well fluids which involved the separation of water and gas from the crude oil, the metering of the oil and pumping of the oil into the main oil line to the shore facilities. Module C dealt in the main with processing of the gas stream which to a large extent involved compression in two electrically powered reciprocating compressors and three centrifugal compressors powered by gas turbines. Normally, some of this gas was exported, some was used as fuel and some was used to assist the flow of oil from the wells. In addition certain components of the gas stream were condensed and exported from the platform to shore by injecting them as liquid condensate into the main oil line.

At the time of the explosion around 10 in the evening the wind was blowing from the southwest as shown by the wind arrow.

The inquiry concluded that condensate vapour leaked into Module C and drifted to form a flammable cloud at the east end of the module where the centrifugal compressors were located. When the cloud ignited flame accelerations lead to an explosion which breached fire walls between the modules and caused oil leakage in Module B. The fire in Module B rapidly developed in size and intensity and the rest is history. The main evidence for this interpretation of causation came from eyewitnesses in the control room who heard and dealt with the alarms from flammable gas sensors placed in and around the compressor enclosures at the east end of Module C. These witnesses testified to the absence of any alarms from the gas sensors in Module B. The conclusion that the release of the explosive source of gas was in Module C and its accumulation occurred at the east end of the module would appear to be obvious.

However, the evidence of the Master of a ship located a short distance from the west face of the platform as indicated in Figure 1 (at point X) placed some doubt on this interpretation. The Master was watching the west face of the platform as part of his duties regarding the positioning of his ship alongside the platform. He observed a blue flame which emanated from a location on the west face of the platform near to the boundary between Module B and Module C (the point marked Y). This flame spread across the face of the platform at the same level as the process modules downwind to the edge of the platform beyond module D and upwind to the radiation screen adjacent to Module A (as shown by the pair of arrows on the plan). The flame then faded away and the Master then, almost immediately, felt/heard the blast of the explosion. It is difficult to explain how the flammable vapours moved from a source within Module C to the west face and then drifted both with and against the wind to form a flammable environment through which the flame could spread before the explosion. A simpler, more obvious, explanation was possible if the source of flammable material was in Module B near the west face so that gas could drift to both ends of the module, east and west. At the west end the gas would also drift downwind so that a flammable envelope could exist from the south edge of Module B to the north edge of the platform. At the east end the gas could pass out of the module and be drawn into the air ducts which provided ventilation for the centrifugal compressors thus causing the gas sensors to trigger alarms in the control room. The difficulty with this explanation is the absence of any alarms from the flammable gas detectors in Module B.

Figure 2 shows in a simplified way the type of construction of Module B and a hypothetical layout of the kind of equipment which such a module might contain. No attempt has been made in these preliminary evaluations of the effects of gratings to investigate yet again the way in which the accident occurred on Piper Alpha. The equipment is supported on skids and these rest on the steel plates which form the floor of the module. The base of the skids are made from steel channel sections and the equipment layout is such that the skids form islands across the floor of the module. To provide process operators with easy access across the module and to items of equipment a walkway is placed above these sections. This is made of steel grating which has an open mesh so that air can pass between the lower and upper spaces. The ends of the module, both below and above the grating are open to the ends of the module so that there is free passage for air to flow above and below the grating from one end of the module to the other. Drains are placed in the floor of the module so that oil spillage does not normally accumulate under the grating.

Figures 2a and 2b show the elevation and plan of the hypothetical module and Figure 3a shows a perspective view which will be used to show the results of the evaluation which is described later in the paper.

CAUSATION

A possible series of events was considered to see if it could explain what happened and fit in with the evidence of eyewitnesses. The sequence is described in this section but as a preamble it should be stated that there is insufficient evidence due to the lack of physical evidence and the absence of survivors to offer this explanation as any more than a basis for concerns for future safety of operations designed on the same basis. It is important to note, in connection with our thesis concerning how issues of safety involving speculation can be usefully developed, that at this stage a description is developed which both assists in a explanation which is consistent with the evidence but also has a sound basis in the science which governs the behaviour of the phenomena which are involved.

It is supposed that there is a spill of crude oil below the grating which contains a sufficient quantity of the light hydrocarbons to produce sufficient quantities of vapour. For some reason the normal flow of oil down the drains is blocked and the oil accumulates and spreads across the floor of the module. As it spreads the oil forms quite a large surface for the escape of gases. The difference in ambient pressure between the west and east faces of the module produced by the action of the wind passing the platform provides a driving force which produces air flow in the space above the grating, past the equipment in the module, and below the grating, past the islands formed by the equipment skid bases.

The two flow regimes, below and above the grating will differ due to the differences in flow resistance and the size of the spaces between the walls, floor and ceiling and the items of equipment. If air velocities in the body of the module are of the order of ½ metre per second then simple considerations of Bernouilli's equation suggest that velocities of the order of 1/10th of a metre per second might be expected below the grating. In these circumstances it is quite possible that the flow above the grating could be turbulent whereas the flow below might be laminar. These contrasting conditions would lead to poor dispersion above the source of flammable gas (the spreading pool of oil below the grating) and good mixing in the body of the module where any gas detectors would be located. The presence of the grating would thus produce more than just a partial barrier to the flow of the flammable gases since it would also create conditions which would lead to low dispersion of vapours from the source to the grating and rapid dispersion of any gas which passed through the grating into the body of the module.

It is possible to envisage a steady-state flow regime developing in which the vapours coming off the oil are drawn through below the grating by the pressure differential across the platform and flow into the atmosphere at the leeward face. Any vapours passing through the grating are rapidly mixed in the body of the module so that concentrations never reach alarm levels near the various gas detectors distributed through the module. At the same time, if the wind strikes the platform at an angle recirculation will occur within the module spaces so that flammable vapours will also be dispersed through the opening below the grating on the windward face of the platform. In this way it might be possible to have flammable regions of gas/air mixtures developing at both ends of the module without the initiation of any of the alarms in the body of the module.

Ignition of these gases would lead to the propagation of a flame which might be through a well mixed gas/air cloud as a premixed flame or might be over the surface of a poorly mixed cloud as a diffusion flame. The precise nature of the developments of the combustion thereafter would be very complex as the expansion process associated with the flames would produce gas movements through the module spaces and across the grating boundary.

PROBLEM DEFINITION

Two questions now arise;

"Is this explanation of causation realistic?" and

"Under what circumstances do the conditions leading to such an event exist?".

Both these questions can only be answered by a scientific analysis - with the problem still to be defined in detail. This section of the paper looks at the proposed hypothesis for the causation described above and attempts to define the problem so that an approach can be made to providing answers to the two questions.

Module construction

The geometry of the module (height, width and length), the items of equipment (location, general shape and overall dimensions) installed in the module and the location of the piping (diameter, and locations where there is an appreciable density of pipework) connecting that equipment needs to be defined. The nature of the grating (thickness, size and spacing of web) and its location (height above the module floor and plan of areas of floor covered) must also be determined and the extent to which the space below the grating is blocked by the equipment skid bases must also be included in such definitions.

Source terms

The spill of oil is the first source term to consider. In the context of the hypothesis set out under causation it will be necessary to define the following properties of this source:

The flow rate of the spill;

The properties of the oil in connection with the flashing of the gaseous components

The extent to which the spill spreads to cover part of the floor of the module below the grating

The location of the spill or, alternatively, the location of the pool of oil which forms on the floor of the module as a result of the spill.

The flow of air through the module is the second source term which must be fixed in the definition of the problem. It cannot be defined as a variable since it is necessary to evaluate the way the relative cross sections of the above- and below-grating spaces of the module effect the process. Various levels of sophistication can be introduced at this stage. Should conditions be modelled in terms of global circumstances such as the wind interaction with the platform as a whole which would require the definition of the entire platform structure and the wind speed distribution and direction.? Or, at the next level, is it adequate to define just the wind speed and direction at one face of the module? Or, at the simplest level, is it sufficient to fix a pressure difference between the two faces of the module and analyse the flow through the module using this as the driving force?

In making a decision as to which of these approaches should be employed it is necessary to examine the objectives of the exercise. The simplest model will not allow us to examine the effects of recirculation within the module and will therefore not take into account the reverse flow from the windward face. On the other hand, the full model employing a complete description of the platform and the movement of the wind around it as well as the analysis of the flow through the modules would place a very heavy burden on computational resources. The practical approach will be to take the middle course and then explore the effects of downstream conditions at the leeward face of the module in conjunction with the variations in wind speed and direction on the windward face to allow for the platform as a whole.

Flow and dispersion

The problem must include the flow of two fluids, the air flowing through the module and the gases transferred from the surface of the oil. It is unlikely that it would be necessary to involve the flow of the spill of oil across the module floor.

The problem must include the diffusion and turbulent mixing of the gases in the air flow. It may be necessary to include the more complex behaviour of the diffusion of gases from a liquid surface into a stream of flowing air which would include the effects of concentrations gradients in the space directly above the oil on the source term for gas release from the oil.

Two additional aspects might be relevant to the problem and would require definition at some stage in the analysis. These aspects are related and are connected with the buoyancy forces which would exist as a result of the differences in density between the gases from the oil and air and the presence of heat sources. Thus density differences in the flow processes could be of importance and it might be necessary to introduces the process of heat transfer into the problem as well.

Combustion

Taking the analysis of causation to completion it would be necessary to include all those aspects associated with the fire and/or explosion which would follow on from the ignition of the gas cloud if its concentration reached flammable levels. Such an extension would involve the introduction of innumerable new definitions. Examples include the definition of all the flammable properties of the air gas mixture (such as upper and lower flammable limits, heat of combustion, laminar flame speed, turbulent gas flow effects on flame speed), the location of the source of ignition and the way in which the combustion process would manifest its behaviour outside the module confinement.

Problem solution

The range of processes involved in this problem definition requires the use of very sophisticated tools to provide even the simplest answers. The best solution would involve the complete solution of all the equations governing the behaviour of reacting gases. At the present time Computational Fluid Dynamics (CFD) provides a technique which approaches this goal most closely and in this paper the use of CFD techniques are considered as the only method for the analysis of the hypothesis in a manner which is both realistic and practical. We attempt to show in the following sections how both practicality and realism is achieved.

However, before tackling the analysis of the problem we refer back to the two questions posed at the beginning of this section. The first question asks for a judgement as to whether the hypothesis put forward for causation is realistic and to answer this we take what we have loosely termed the QUALITATIVE approach. To answer the second question regarding the conditions which lead to the events described in the hypothesis we apply a QUANTITATIVE approach.

'QUALITATIVE' ANALYSIS

To cover the topic adequately in this short paper we only describe the analysis of the non-reactive events before ignition occurs. In both the qualitative and quantitative approaches a Computational Fluid Dynamics (CFD) model of the module shown in Figure 2 was made. The CFD program used was Phoenics version 3.1 from CHAM Ltd. (1). The program solves the governing equations of heat and fluid flow and is able to model a large range of physical phenomena associated with fluid flow, heat transfer, gas dispersion, combustion processes and other chemical reactions. Furthermore the program also contains a selection of proven turbulence models.

To study the event in a question for a variety of conditions and to allow for a subsequent QUANTITATIVE approach, the initial model was set up as a schematic model of the module. Although all the equipment items were simplified and represented as hexahedra they were placed at the correct locations inside the module. Pipework was not included but as will be seen later modifications to the physical model of the module can be accomplished very easily as the analyses are developed and extended. The simplified model contained in total 4500 computation cells.

For this initial analysis, bearing in mind we are hoping to determine the plausibility of our hypothesis, the problem definition has been kept as simple as possible.

Source terms: The gas vaporisation rate was modelled from the assumed hydrocarbon spillage underneath the grating covering the entire floor space of the module. The gas was considered as a separate phase in the flow but having the same density as the air thus eliminating the buoyancy forces. Air flow through the module was based on the simple wind concept with the wind having horizontal velocity components along the axis of the module and perpendicular to it.

The spill of oil is the first source term to consider. In the context of the hypothesis set out under causation it will be necessary to define the following properties of this source:

Flow/dispersion : The turbulence model used was a generalized length scale model suitable for flows within complex geometries with many obstacles. As mentioned above buoyancy forces were neglected and the flashing of gaseous components from the oil was treated as a source term with a constant rate of production not influenced in any way by the flow behaviour. The grating was also modelled using imposed velocity conditions at the grating level.

Others: In addition, in this first approach, the solutions were based on steady state simulations rather than time dependent simulations. Thus the results are those that would have been reached after a sufficient period for them to attain time invariance.

Results

The results shown in Figure 3. There are three perspective views of the module in this figure each opened out to show its interior by the removal of the ceiling and the front wall. The windward face of the module is at the top of each figure and the direction of the wind is from top to bottom and from left to right. The shading on the horizontal plane indicates contours of constant gas concentration. The actual contours can be seen as faint white lines on the background. The shading identifies three levels of concentration. The darkest shade shows effectively zero gas concentration. The light grey shade shows a medium level of gas concentration and the white shows the highest level. In this 'qualitative' approach it is not necessary to define the numerical values of these levels.

Figure 3a shows the gas concentration levels just above the grating. Nearly the entire space in the module body is free from gas. A small patch about three quarters of the way from the air entrance shows concentrations at the medium level but the area extends over a very small part of the module. Figures 3b and 3c show the same concentration levels below the grating with the latter depicting the conditions close to the oil surface and the former a horizontal plane between the oil and the grating. In both these figures the maximum concentration of gas is observed to cover larger and larger areas as the oil surface is approached. At the lowest level the high concentration covers about 1/8th of the floor area and the medium concentrations nearly half the floor area.

Figures 4a and 4b are similar views as 3a and 3c but with twice the air velocity along the axes of the module. Even for this large change in the external conditions the gas concentration distributions at the two extreme levels do not change a great deal. The gas concentrations above the grating are still confined to quite a small location near the centre of the module while the corresponding concentrations of gas below the grating cover roughly the same area of the floor as in the lower velocity case. It would appear from this preliminary 'qualitative' analysis that a tentative conclusion can be drawn that the results of the analysis in terms of gas distribution are not very sensitive to wind speed.

Whilst the above results may not be quantitatively correct, they show that the vaporisation of hydrocarbons from a spill underneath a grating floor may cause high gas concentrations just underneath the grating floor. A further analysis in which the grating was removed shows quite different results which are illustrated in the three gas concentration contour diagrams in Figure 5. These diagrams have the same interpretation as those shown in Figures 3a, 3b and 3c with one exception. In Figure 5 the gas concentration levels associated with each shaded area are half those depicted in Figure 3. An examination of the results also shows that the maximum concentration in the simulation from which Figure 5 was drawn is half the maximum concentration for the simulation shown in Figure 3. This result shows that the gas cloud formed above the oil spill is much more extensive when the grating is not present and the gas concentration throughout the cloud is much more even. The halving of the maximum gas is consistent with the increase in the extended distribution of the gas in the module and is indicative of one of the shortcomings in the 'qualitative' approach adopted in this section. The limitation leading to this result is the fixed nature of the gas production rate chosen for one of the source terms. A more realistic gas source term would have allowed the source rate to increase as the gas concentration driving force between the oil surface and the bulk gas increased. This the expansion of the cloud with the absence of the grating would have enabled more gas to flash from the surface of the oil pool and a more realistic modelling of the flashing behaviour would have produced higher concentrations over a larger gas cloud.

The 'qualitative' approach has shown that there may be a real problem associated with the presence of gratings in process modules when flashing oil spills could occur. This indication is sufficient to encourage the modelling of these phenomena at a more detailed level in order to ascertain more quantitative results both in terms of temporal and spatial distribution. In other words, an attempt to answer the second question, "Under what circumstances do the conditions leading to such an event exist?" is warranted.

'QUANTITATIVE' ANALYSIS

In order to obtain quantitative results using a CFD package it is essential that not only the physical part of the modelling is a correct as possible, but also that the models of phenomena such as turbulence and its effects are at a realistic level. Furthermore, numerical diffusion must be kept to a minimum through the use of a high mesh density and/or the use of a higher order discretisation scheme. This requires a significant increase in the sophistication of the performance of CFD numerical routines and considerable additional computational effort. There is thus the justification for the preliminary work on smaller and less sophisticated models to identify the conditions that could be of special interest. If time dependent simulations are required as in the present case, the computational storage is also an issue, together with time required to analyse the data.

Modern CFD computational packages are very easy to run. The level of analysis we have described as 'qualitative' can be set up very simply and as long as solutions converge results at this level of sophistication can be obtained with comparatively little effort. The greatest burden for the user is the setting up of the system with the proper physical properties of the problem under investigation and the right boundary and initial conditions for a realistic solution. The sophistication of the user interfaces supplied with these packages provides for this ease of use but can lead a unwary user to produce erroneous conclusions when the quantitative aspects become important. It is at this level that expertise in the use of these systems and a thorough knowledge of their fundamentals becomes essential for the correct formulation of the input data for the calculation, the proper operation of the numerical computational system and the efficient interpretation of the very large quantity of numerical results which are produced.

The PHOENICS 3.1 package employed in the present calculations has a number of features which enable some of the difficulties identified to be accommodated without excessive cost or trouble. The CFD computational engine is complemented by a pre-processor and a post-processor which use virtual reality to assist in the entering of data for problem definition and for the interpretation of the results. The pre-processor known as the VR-Editor enables the user to construct a physical picture of his problem on the computer thus defining the problem in numerical terms for the subsequent CFD calculations. The domain for the calculation can be set up and a computer graphics picture constructed by setting up boundaries introducing obstacles and/or sources, selecting fluid properties and defining initial and boundary conditions.

A data file produced by the pre-processor becomes the input file for the CFD calculation. The calculation can be performed on the computer which was used for the pre-processor or may be exported to another more powerful machine to enable very large models to be run efficiently. Alternatively, the file can be sent to a Consultancy practice who have experience in these calculations to benefit from their expertise. The CFD computational package will produce an output data file which can then be dealt with in the post-processor known as the VR-Viewer, which provides an identical virtual reality view of the problem but within which the results can be displayed.

Thus the initiator of the problem can see his analysis through from the start when he is in control of the definition to the end when he can pick those results which best illustrate his conclusions. Initially, the computations can be handled by experts outside his organisation to ensure that the problem has been handled properly. However, as the user gains experience in the use of the system he can either start to do the calculations on his own machine by installing the CFD package. Or he can run his problems remotely on more powerful computers located elsewhere and maintain tight control of their development. This would be achieved by by initiating a larger proportion of the computational controls from his pre-processor but retaining the greater power of a computational machine run by a consultant organisation possessing better computational facilities.

The 'qualitative' results described in the earlier section were obtained using a complete PHOENICS system which included the VR-Editor and the VR-Viewer as well as the CFD package. A computer running on a Pentium 130 with 32 MBytes of memory performed quite adequately without excessive run times when the simple models with a fairly coarse mesh size was employed. The results for a 50 000 cell simulation show similar trends to the smaller models, however, grid independence may not have been reached and a 500 000 cell model is presently being constructed. This model will be used to explore the effects of a number of factors which have been identified in this paper as having an important bearing on the validity of a 'quantitative' analysis. The results of these calculations will be reported at the meeting.

CONCLUSIONS

The dangers associated with spills of oils which can produce gases has been identified as presenting a real risk to the safe operation of process equipment in enclosed buildings which possess void spaces below floors formed from gratings. The problem has been examined employing Computational Fluid Dynamic (CFD) methods and the authors of the paper have concluded that this approach is useful for preliminary studies but must be used with care when realistic quantitative results are desired.

REFERENCES

(1)   PHOENICS Version 3.1 user manuals (1998), CHAM Ltd, Bakery House, Wimbledon, SW19 5AU

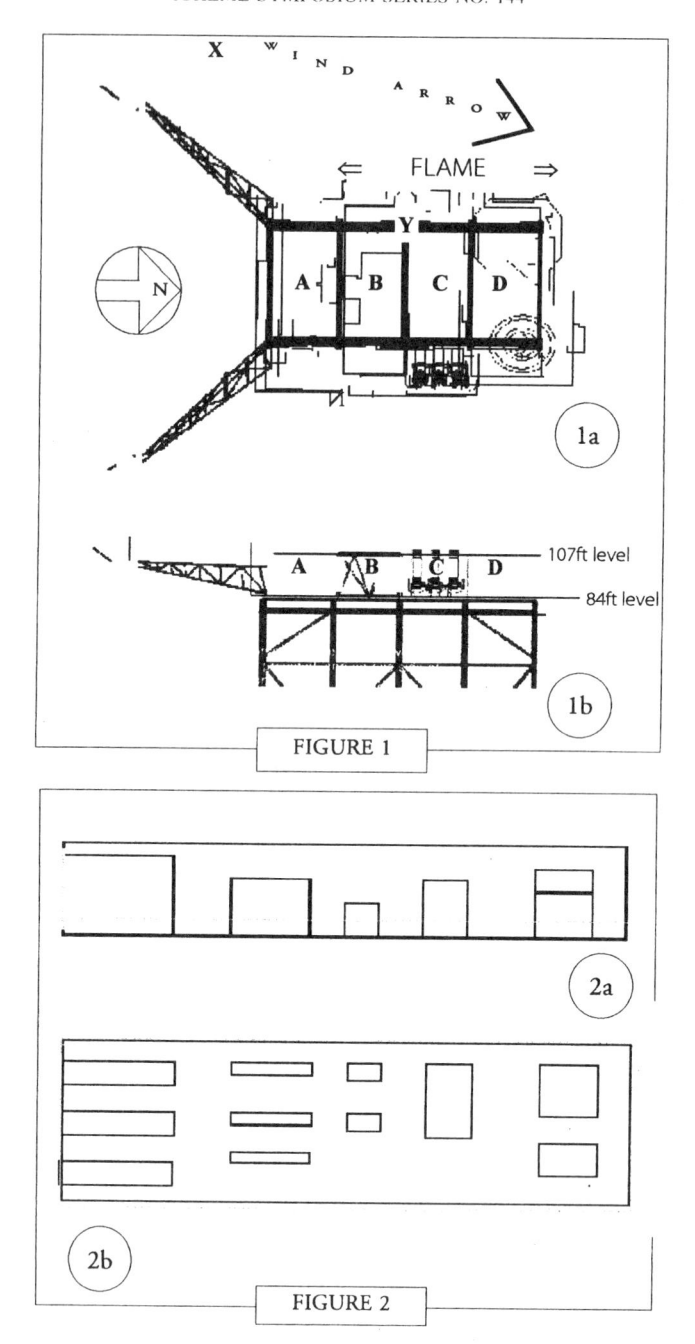

FIGURE 1

FIGURE 2

3a

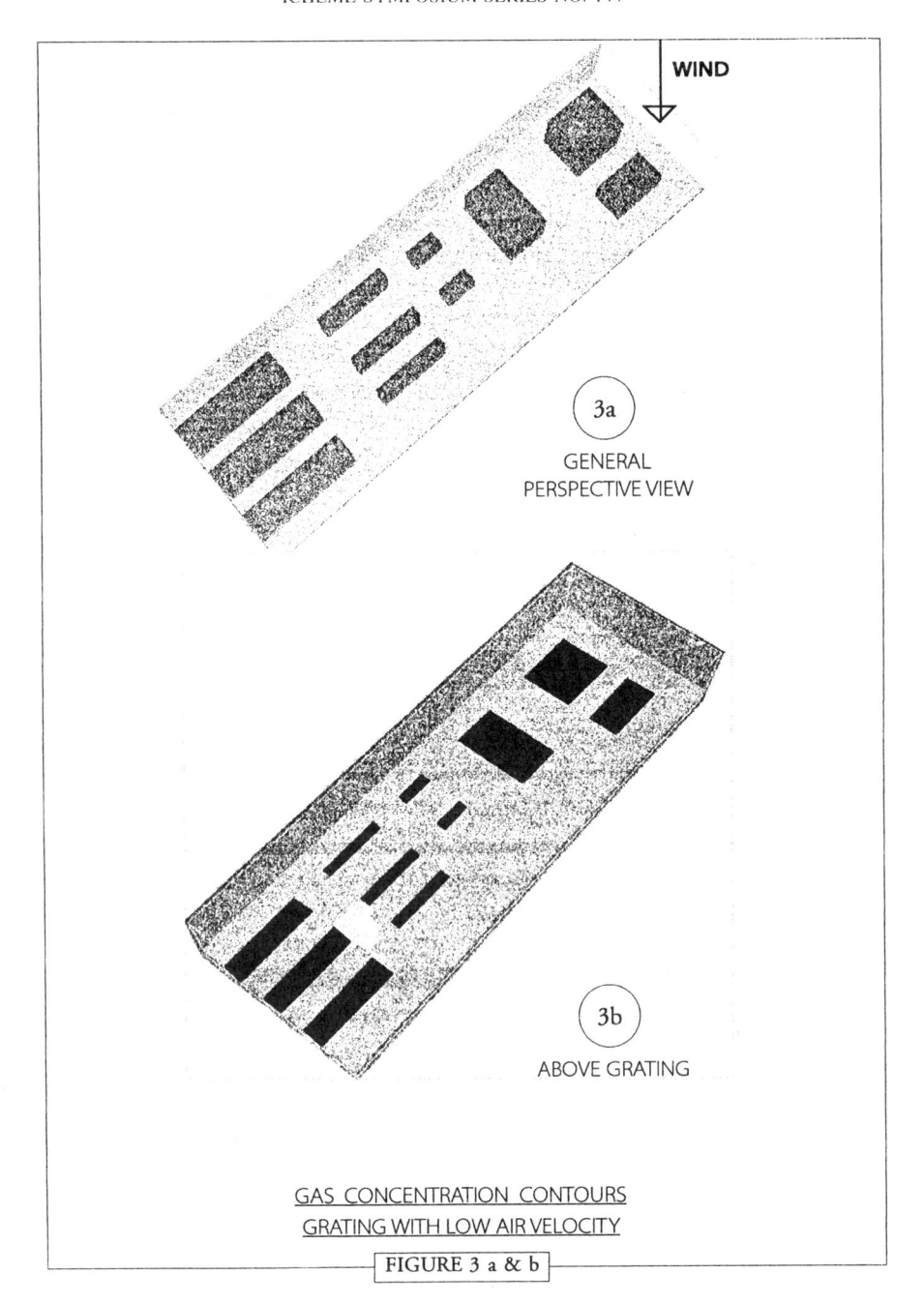

GENERAL PERSPECTIVE VIEW

3b

ABOVE GRATING

GAS CONCENTRATION CONTOURS
GRATING WITH LOW AIR VELOCITY

FIGURE 3 a & b

ICHEME SYMPOSIUM SERIES NO. 144

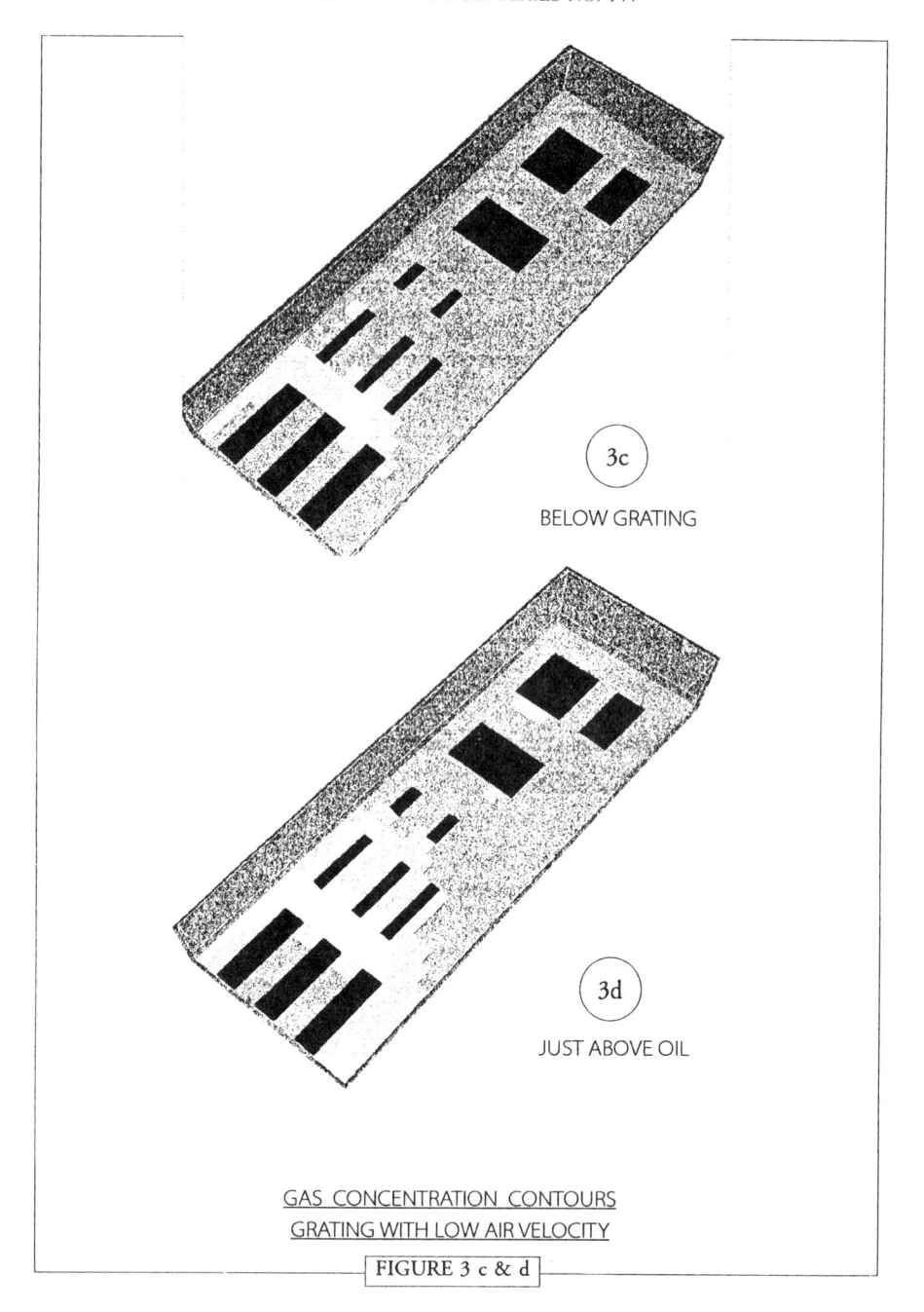

3c

BELOW GRATING

3d

JUST ABOVE OIL

GAS CONCENTRATION CONTOURS
GRATING WITH LOW AIR VELOCITY

FIGURE 3 c & d

4a
ABOVE GRATING

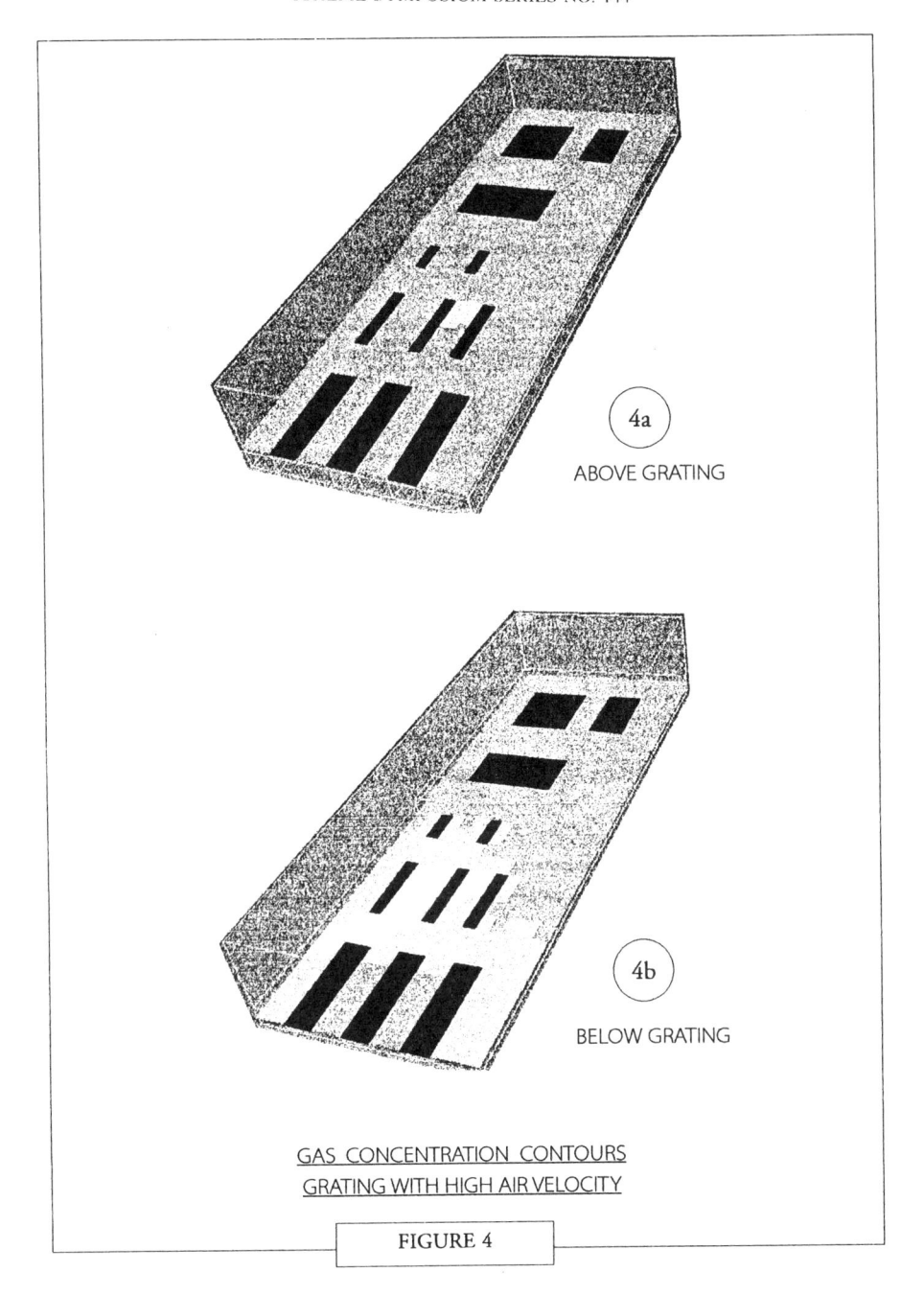

4b
BELOW GRATING

GAS CONCENTRATION CONTOURS
GRATING WITH HIGH AIR VELOCITY

FIGURE 4

5a ABOVE 'G'-LEVEL

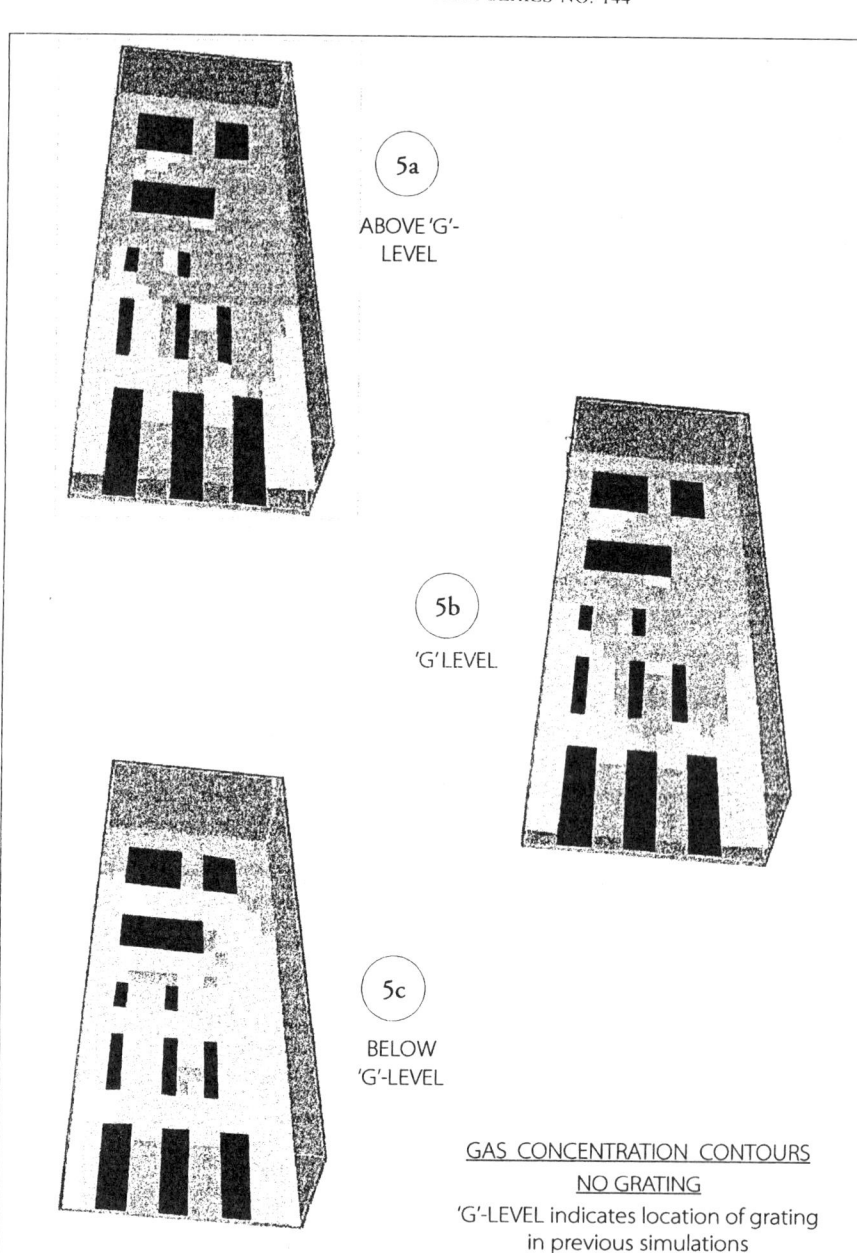

5b 'G' LEVEL

5c BELOW 'G'-LEVEL

GAS CONCENTRATION CONTOURS
NO GRATING
'G'-LEVEL indicates location of grating
in previous simulations

FIGURE 5

SUPPRESSION OF HIGH VIOLENCE DUST EXPLOSIONS USING NON-PRESSURISED SYSTEMS

Steve Cooper and Paul Cooke - Stuvex Safety Systems Limited

Explosion suppression is an accepted form of explosion protection against gas and dust explosions. Typically, systems have employed high pressure storage canisters, actuated using explosives as the means by which to inject chemical extinguishing agents. These systems can be costly to purchase and often require considerable maintenance. A new system has been developed that requires little maintenance and does not use pressurised canisters or high explosives. The system is actuated by a gas generator that is activated on explosion detection, propelling the chemical into the vessel to be protected. The technique and system development will be discussed. The test and research data compiled during ST1 and ST2 dust explosion testing will be presented and compared with pressurised systems. Future system developments, particularly with respect to ATEX compliance, will be addressed.

KEYWORDS: EXPLOSION, SUPPRESSION, HIGH VIOLENCE, NON PRESSURISED, GAS GENERATOR

INTRODUCTION

In the past, one of the most widely-used safety techniques to prevent the consequences of industrial gas and dust explosions was 'explosion venting'. However, the associated side effects are increasingly perceived as a problem in the modern processing industry. The (fire) damage, the refurbishing and start-up costs and reduction of productivity, associated with this method, can be considerable.

Furthermore, there is increasing resistance, on environmental and safety grounds, to the emission of large jets of flame, as well as burnt and unburnt product. Even if this only occurs occasionally.

One very good alternative to explosion venting is offered by the technique known as 'explosion suppression'. This method detects an explosion in its 'infancy' and extinguishes the flame in milliseconds. On detection of ignition, an extinguishing agent is released into the fireball, where the combustion process is interrupted by interaction between the flame and the extinguishant.

The main obstacle to the widespread acceptance of modern suppression technology is often the initial investment and the intensive maintenance is imposed on process productivity and efficiency.

OPERATING PRINCIPLE

If a cloud of gas or fine flammable powder, mixed in the right ratio with air, is ignited, a very rapid combustion occurs. The speed at which the flame spreads through the fuel cloud is dependent on a number of factors including the type of fuel, the geometry of the cloud and the initial conditions (pressure, temperature, turbulence).

The deflagration flame speed can range from less than 10 m/s to over 100 m/s. Compared with high-explosives, where the flame speed is measured in km/s, this is relatively slow. This offers the possibility of extinguishing the explosion before it assumes catastrophic proportions, provided that it is detected at an early stage and the extinguishing agent is injected quickly.

If the explosion is confined, say within a steel process vessel, the explosion can develop rapidly. Providing there is enough oxygen and fuel, the explosion can generate a maximum explosion pressure of <10 barg, in tens of milliseconds (figure 1). This pressure load would, normally, be well beyond the strength of vessels such as filters and silos. By using effective explosion suppression techniques the maximum explosion pressure can be reduced to a fraction of its expected intensity.

Although, of course, there are differences between one manufacturer of explosion suppression systems and another, until recently, the basic principles of virtually all such systems were identical.

All systems use a detector (usually pressure or optical), some suppression device and a control unit to perform power control and process interface functions.

Stuvex embarked on a project known as FLASH to improve on conventional techniques and to find less expensive ways of achieving installed protection and ongoing maintenance. The project was to look at all aspects of explosion suppression and to find some form of technical and commercial improvement.

DETECTION

The first area to be assessed was detection. Frequent calibration of detectors can often be the cause of high maintenance costs. It was decided that the detector must have fixed calibration that need not be checked frequently. Looking into the market place showed that conventional venting panels and vessel bursting discs have extremely good calibration properties but lacked the response time required to operate the system. With the help of one of the worlds leading burst disc manufacturers, Stuvex were able to design a fixed pressure sensor with close tolerances and good temperature characteristics (figure 2). The detector is able to withstand high negative pressures and is not affected by high mechanical shock or vibration.

The most important aspect however, is that once the detector is installed there is no need (or way for that matter) to check the calibration level. Maintenance is limited to checking for excessive powder accumulation or hard deposits on the PTFE lined sensing face.

On activation, the detector sensing membrane is permanently deformed and destroyed, providing conclusive evidence of over-pressure. The "spent" detector is then removed and replaced with a new, calibrated unit and the detection system is ready for further operation.

SUPPRESSORS

For the suppressors the improvements would need to be substantial. The FLASH project aim was to design a system eliminating pressurised vessels and high explosives.

The technique that was to be investigated uses techniques similar to those used in the inflation of vehicle air bags. The bags are inflated using rapid acting gas generators activated from a signal received from a central control unit. This technique is very similar to the way suppression systems are activated.

Again, specialist help was sought to design a gas generator that could produce clean, high pressure, gas - very quickly. A company at the leading edge of rocket motor development and gas generators for aerospace applications, produced the prototype gas generators. The initial discharge tests proved that the suppressant could be discharged from a non-pressurised canister, further and faster than conventional pressurised systems. The main problems were that the generators were incredibly expensive to produce and operated too quickly for the mechanics of the rest of the system.

The design was refined and the gas generator re-tested. Further discharge tests were very encouraging and the production costs reduced dramatically. The gas generator was sent for third party testing for classification as a pyrotechnic actuator, and was approved without the need for further design. Following this exercise Stuvex owned a range of gas generators with known properties (<5ms to pressurise the canisters) that can be stored and transported in a simple cardboard box without the need for special licenses.

With the suppressor canisters and gas generator designs finalised (figure 3) the discharge system was configured to accept sealed cartridges filled with powder suppressant (figure 4). The cartridges are produced in humidity controlled conditions thus ensuring that the contents are free from moisture. This had the added benefit that the suppressors can be easily refilled on site, possibly by the User, following an activation (figure 5) and that the cartridge can be checked for leakages and even checked weighed if necessary.

The discharge system would also need a dispersion nozzle. Many tests were conducted to achieve a discharge profile that would provide the correct concentration within the powder cloud, combined with forward throw. Many different configurations were tested. Telescopic systems provided the most hygienic option. Stuvex designed a two piece system comprising a stub pipe, which is welded to the vessel, and a nozzle which is retained in a retracted position until the system is activated (figure 6). The stub pipe aperture is sealed using a red silicone cap which is also securely retained.

The results were, that Stuvex now had a range of nozzles that could disperse suppressants in different ways and concentrations, that would enable the system to tackle odd shaped vessels and ducts.

TESTING

Test vessels of 4.0 m^3 and 9.4m^3 were used and were configured in accordance with ISO standard 6184 for determination of efficacy of explosion protection systems. The test vessel configuration is shown in figure 7. In the case of the 9.4m^3 vessel, three pressure transducers located at the top, middle and bottom of the vessel were used to record test data. Four dust discharge canisters, filled with a specified quantity of combustible dust, were equally distributed around the vessel and pressurised to 20 bar. The canisters, on activation by the sequence control unit, simultaneously discharge the pressurised dust into the vessel, generating a near homogeneous dust cloud within the test vessel. Ignition was by two 5kJ igniters at the geometric centre of the test vessel. The test vessel system was calibrated to the desired K$_{st}$ by adjusting the ignition delay time (t$_v$), between activation of dust canisters to the point of ignition.

Investigation showed that a large percentage of the suppressant was left inside the suppressor despite the large gas generators installed. New discharge tests were conducted which conclusively showed that, in some instances, smaller gas generators proved more efficient in dispersing the suppressant than large units.

Modifications were made to the discharge system and the tests recommenced. The unsuppressed explosions provided a new insight into the characteristics of the system which could be confirmed and developed in suppression trials.

Test	Gas Generator	Suppressor Type	Detection Pressure P$_a$ (mbarg)	Measured Suppressed Explosion Pressure (barg)
1	600 600 600	14 14 14	50	0.27
2	600 600 600	14 14 14	100	1.2
3	600 325 600	14 7 14	50	0.25
4	600 325 325	14 7 7	100	0.27

Table 1 - Suppressed explosion results

A summary of the many trials is shown in table 1. Here the system is tested against a standard test configuration with the same number and size of suppressors. The detector setting (P$_a$), the suppressant quantity and the gas generator type are altered to compare performance.

In test 1 with P_a=50 mbarg, a reduced explosion pressure of 0.38 barg was measured. This was achieved using a system with large gas generators on each suppressor. Test 2 proved that doubling the detection pressure (P_a=100 mbarg) caused the system to fail (1.23 barg achieved). On test 3 the P_a was set to 50 mbarg but the central suppressor was reduced in both suppressant capacity and gas generator size. Here an improvement of reduced explosion pressure was measured at 0.21 barg. This was corroborated by test 4 when two of the three suppressors were reduced in both size and gas generator volume, and the P_a set to 100 mbarg. A reduced explosion pressure of 0.28 barg was measured.

In addition, the contribution of propelling gas to the overall reduced explosion pressure was seen to be lower than pressurised systems, owing to the small volume of gas produced by the gas generator.

Having established the most efficient gas generator/suppressant fill combinations, tests were conducted using materials with a higher K_{st} to establish the limits of applicability. At low detection pressures the system suppressed easily. But as shown in the table above the efficiency decreased as the detection pressure increased to the point where a threshold value was achieved. For instance with a detection pressure of 70 mbarg a K_{st}130 yielded a P_{red}<0.27 barg. A pressure time graph for this test is shown in figure 8. The identical system however, struggled to suppress once the detection pressure reached 100mbarg. These tests are vital in determining the limits of applicability.

Having achieved successful results at K_{st}100 and 130 the same system combination with the same detection pressure was tested against K_{st}200 resulting in P_{red}<1.0 barg. A pressure - time trace in figure 9 clearly shows (in comparison with the trace in figure 8) an oscillating pattern to the suppression. This is a clear sign that the suppressors are struggling to deal with the higher violence explosion. There were two important considerations. Firstly the detection pressure of 70 mbarg is probably too high for K_{st}200 with this system configuration. Secondly the dispersion characteristics were not correct.

The tests were repeated using detection pressures of 50 mbarg and four suppressors instead of three with a different gas generator combination. The tests were then successful but, more importantly, another set of limits had been defined relating to the maximum detection pressure, the minimum required suppressant concentration and propellant configuration.

CONTROL UNIT

In this electronic age, the improvements in control offered many possibilities. Apart from the normal coupling of detection and activation circuits, the new control unit (figure 9) features a 'black box recorder'. This is a real time memory that records faults and activations that can be analysed at a later date, even on power loss. The main control panel has a display showing information on alarms and faults, and a keypad to choose from a menu of different analytical operations.

The interrogation of the memory and the analysis of other data is password protected. The unit also provides data transmission via an RS 232 interface, or via fibre optics. The control system is DEN approved, and meets all the relevant European standards, including 'fail-safety', and 'EMC' (STS 032) approval.

OTHER FEATURES

Accessibility and ease of replacement by semi unskilled personnel have been uppermost in the minds of the designers. MIL-SPEC electrical plug and socket connectors allow the rapid changing of the detectors and gas generators. Systems can be easily tested using simulation units for both detectors and suppressors. Hinges on the suppressor stub pipe (figure 10) and beneath the gas generator (figure 5) allow easy access to both the nozzle and the powder cartridge without having to strip down and remove the complete suppressor assembly.

The success of the gas generating system spawned another product aptly titled MULTI-FLASH. In this case a central gas generating system is directly coupled to powder canisters via high pressure hoses (figure 11). On activation, the gas generator rapidly pressurises all of the canisters attached to the main housing and discharges the powder cartridges into the vessel at the same time. The major benefit is that the number and size of activation devices are reduced dramatically. This system has now been installed on a number of grain elevators.

Suppressors were also developed and tested for application on single pipes and ducts both as chemical barriers and advance inerting suppressors.

CONCLUSION

The FLASH project has utilised new and emerging technologies and brought the concept of suppression up to date. Many of the inherent problems such as high initial cost, expensive maintenance and false activations, have been addressed and in many cases solved.

Testing and refinement of the system will be an ongoing project for years to come with benefits to industry each time a new development is achieved. The system has been successfully tested against high violence explosions up to $K_{st}200$ and the latest work in progress seeks to take the system into the ST2 range of applicability.

The whole FLASH project has been focused on allowing the User to take more control and ownership of, what have often been considered to be, complex systems.

Fig 1 Explosion Pressure vs Time

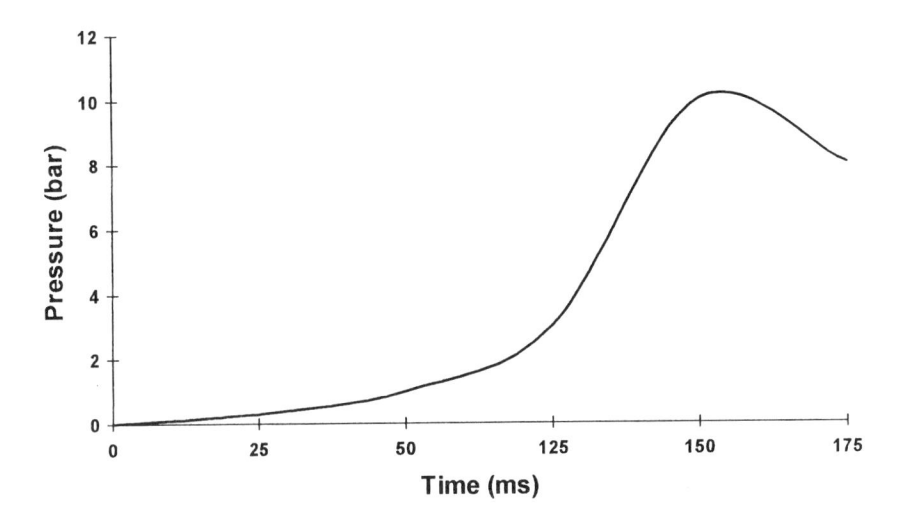

Fig 2 Static Pressure Detector

Fig 3 Explosion Suppressor

Fig 4 Powder Cartridges

Fig 5 Access to gas generator and powder cartridge

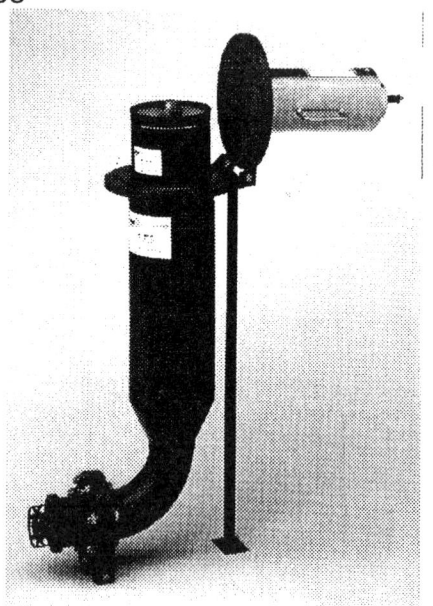

ICHEME SYMPOSIUM SERIES NO. 144

Fig 6 Dispersion Nozzle

Fig 7 Test vessel arrangement

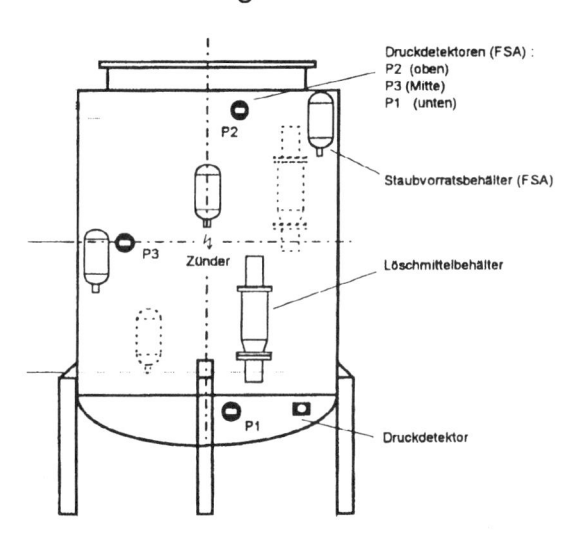

Druckdetektoren (FSA):
P2 (oben)
P3 (Mitte)
P1 (unten)

Staubvorratsbehälter (FSA)

Löschmittelbehälter

Zünder

Druckdetektor

Fig 8 Test with Kst130

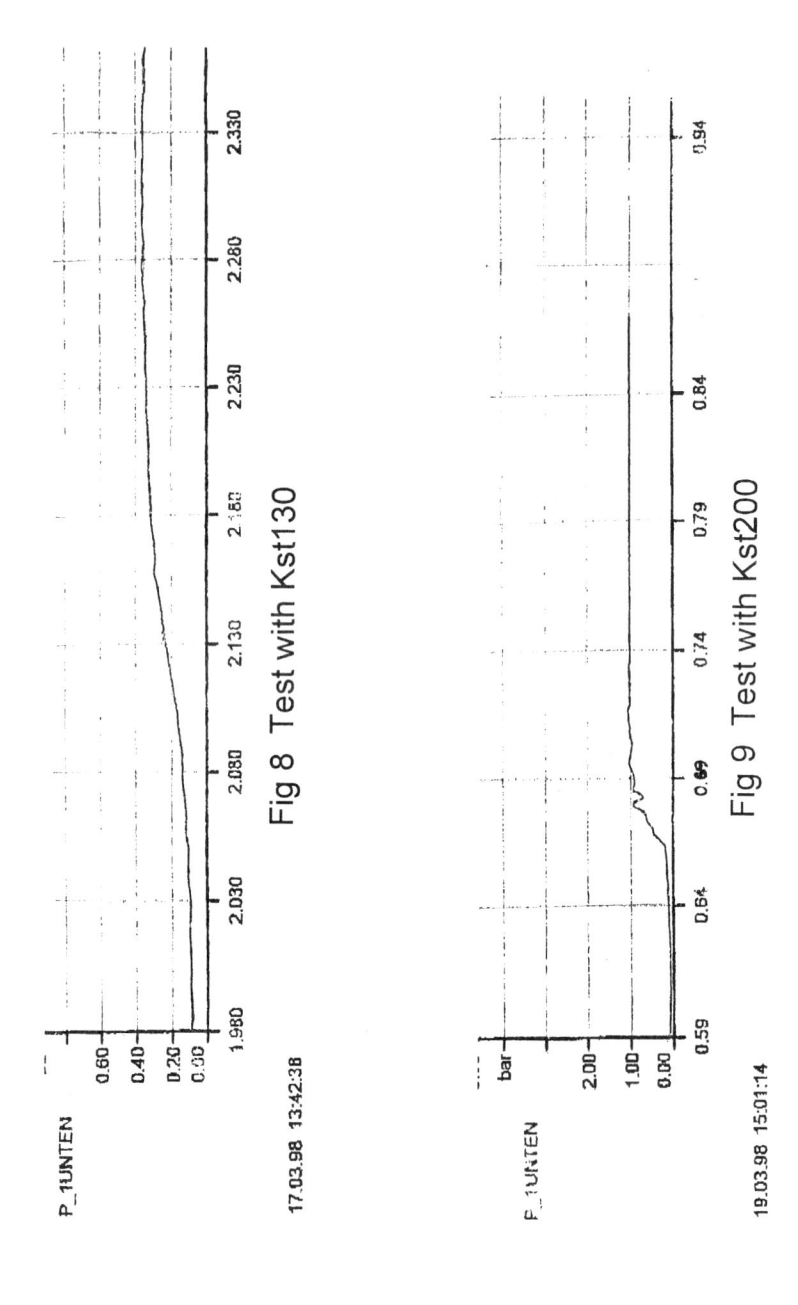

Fig 9 Test with Kst200

Fig 9 Control Unit

Fig 10 Hinged flanges

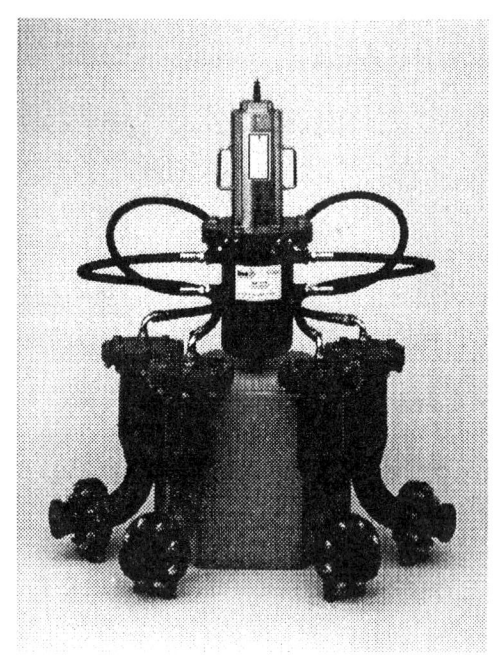

Fig 11 Multi-port discharge suppressor

MANAGING HAZARDS AND RISKS IN FINE CHEMICAL AND PEROXYGEN OPERATIONS

P.G. Lambert, J. Phillips and R.J. Ward,
Fine Organics, Laporte Organics, Seal Sands, Middlesbrough, TS2 1UB, U.K.

The Organic Specialities Division of Laporte operates in two distinct areas: the Peroxygen Group manufactures organic peroxides, special catalysts and uses active oxygen for chemical synthesis. The Fine Chemicals Group manufactures intermediates and finished products for the pharmaceutical, agrochemical, food products, and associated speciality industries. There are considerable variations in the chemistry employed by both groups and the range of chemicals, reactions, hazards and their potential consequences also vary. This paper describes one part of the divisional management strategy relating to loss control. It uses, by way of illustration, three internal guidance procedures covering the risks and consequences of identified hazards in the manufacture of fine chemicals, organic peroxides and vapour phase explosion hazards.

Keywords: Loss Control, Chemical Reaction Hazards, Vapour Phase Explosions.

STRUCTURE

Laporte's five business divisions currently operate about 90 sites worldwide. The Corporate & Group Safety, Health and Environment (SHE) departments devise overall strategy, monitor company performance as well as providing training. They also create the tools for line managers to direct the safety, health, hazards & environment programmes in the areas under their control. A strong commitment therefore exists to line responsibility and authority in safety, health, hazards, and environment issues. A co-ordinator in each division supports line management, - collating and disseminating information for the Group or for the individual businesses.

The Organics Division manufactures at ten sites world-wide, with the Peroxygen Group having manufacturing operations in Germany (Peroxid Chemie), the USA (Aztec Peroxides), Australia and Brazil. The Group also has joint venture companies in Spain and South Africa, as well as storage facilities in Thailand, The Netherlands and France. The speciality Fine Chemicals Group has operations in the U.K. (Fine Organics), Canada (Raylo) and Italy (Laporte Organics Francis) and a new joint venture in India.

These businesses operate with different cultures and regulatory bodies, and face differing hazards, levels of operational complexity as well as rapid change. This has resulted in a strong divisional approach to the management of hazards and risks founded on the strategy of the company's Group SHE Department.

DIVISIONAL SAFETY, HEALTH, HAZARDS AND ENVIRONMENT PROGRAMME

The divisional Safety, Health, Hazards and Environment Programme was based on the model originally outlined by the International Loss Control Institute (I.L.C.I., now owned by D.N.V.) and the well established approach of POLICY, STANDARDS, PROCEDURES, AUDIT.

The Chief Executive has endorsed the Laporte Group's safety and environment policies. In addition, individual sites or businesses have their own policy, reflecting the needs of the business. It is therefore not necessary, or appropriate, for the division to have its own policy. However, it was appropriate to build on and interpret the company policy (or statements of intent) as a basis for the overall Programme.

To execute these *statements of intent* effectively, a number of agreed *objectives* were required based on sound *strategy*. To support the strategy, *standards* were also needed which would sustain our collective corporate memory and provide the tools to manage the risks and hazards in our businesses. The final two elements of the Programme are an *audit and monitoring plan* and the *identification of responsibilities* of site and divisional management.

The disciplines of loss control and high standards were given senior management support through a "profit through safety" initiative, introduced by the divisional Chairman.

The eleven statements of intent include several "givens" - such as compliance with local and national regulations, line management responsibility and the provision of training, procedures and overall suitability of manufacturing plant. In addition, we confirmed our stance on risk management and our commitment to operating an integrated S.H.H.E. programme at each site that could include other regulated activities, such as cGMP and quality.

Hazard, Consequence, and Risk Management.

Many of the key definitions adopted by our Programme will be familiar. *Hazards* are physical situations with the potential for harm. *Consequences* are the outcomes or potential outcomes from events. *Risk* is the likelihood of a specific undesired event occurring in a specified period and is reported as a *frequency* of an event occurring in a unit of time or as the *probability* of an event occurring during a specific operation. *Risk management* is the term applied to the whole process of risk identification, estimation, reduction and control.

Risk Management has at least five phases and, dependent on the particular process or project, work may be necessary in each phase. Normally, these stages are *Hazard Identification, Risk Estimation, Risk Evaluation, Implementation and Monitoring* and *Auditing*.

Many sources of hazards are associated with chemical processing. In simple terms, they can be defined as:
- general safety considerations ....working at heights, with machinery etc.,
- specialist operations.......electricity, welding, radiation sources, etc.
- health hazards,
- environmental concerns.....emissions to air, waste, land, etc.,
- reactivity of chemicals,
- interaction of chemicals and the plant.....flammability, dusts etc.

To establish control and minimise loss, all of these issues need to be addressed. The large potential overlap between these general categories makes it critical to define the responsibilities and authority for the management of hazards.

By their nature, projects vary in complexity and in the degree of hazard involved. Potential environmental hazards and risks, for example, will dominate some projects; others may be primarily concerned with health risks associated with certain chemicals or with the risk of fire and explosion. A generic hazard management strategy has to accommodate these variations and provide the tools that line managers need, whatever circumstances apply. In Laporte's Organics Division, our approach is to consider the management of the risk as an integral part of the whole process or project, particularly since a number of activities may occur in parallel.

Our integrated hazards management strategy is known as the *Umbrella* approach. It provides protection to the company, its people and assets, and covers the range of hazards that we need to manage. To provide a comprehensive, fast and cost effective approach, we needed to develop guidelines and tools for the rapid assessment of hazards. Some of these tools are in the early stages of development, whilst others have been used effectively for many years.

The Umbrella Structure

The Umbrella structure is outlined below in figure 1. All its sections and subsections can be seen as ribs or supports to the shield provided by the umbrella. Failure in any one area lowers our overall protection.

A large number of Standards, Codes of Practice and Guidelines have been developed or adopted in Laporte. The terms of reference for these Guides, Codes and Standards are available to each site, and management is expected to apply them when required.

HAZARDS MANAGEMENT STRATEGY

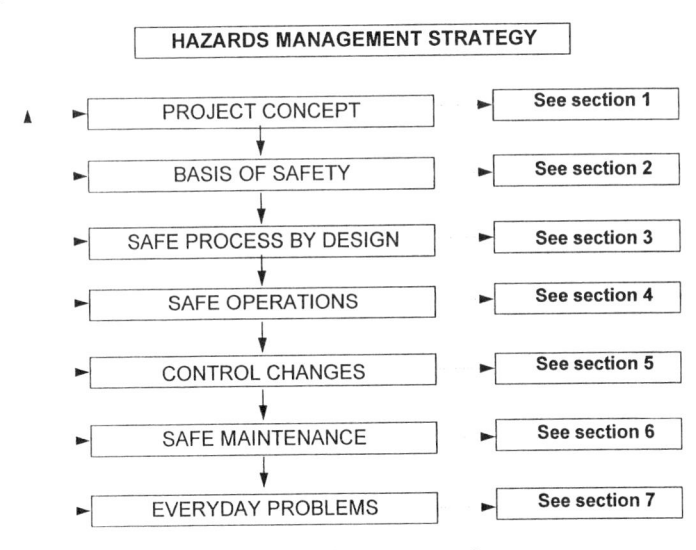

- PROJECT CONCEPT — See section 1
- BASIS OF SAFETY — See section 2
- SAFE PROCESS BY DESIGN — See section 3
- SAFE OPERATIONS — See section 4
- CONTROL CHANGES — See section 5
- SAFE MAINTENANCE — See section 6
- EVERYDAY PROBLEMS — See section 7

Figure 1 The Hazards Management Strategy

- A STANDARD is a technical specification for defining a system (plant, store, container, operating facility etc.) which is fit for its intended process.
- A GUIDELINE explains what, from the point of view of Laporte, is current good or best practice
- A CODE OF PRACTICE explains what, from the point of view of Laporte, is reasonably practicable.

A critical element is that each *process* or *project* should have a *basis of safety*. This is defined as the principle and methodology used to protect the business, people and assets from a known hazard. Examples are:

- Prevention (by avoiding the use),
- Minimise (by limiting the inventory),
- Render harmless (by treatment),
- Intrinsically safe (as a specific hazard does not exist),
- Emergency venting (to a safe location),
- Process control (via hardware, software, or human intervention),
- Elimination of ignition sources (by procedure, and equipment),
- Containment (by design).

Risk Criteria

The absolutely safe condition involves the lowest amount of energy and is therefore impossible to achieve in technological businesses. Processing chemicals necessarily increases the energy in a system and we therefore needed a definition of safety that recognises that it is not an absolute state. In this context, safety has been defined as a situation where the risk is no higher than the risk limit. At first sight, this definition seems to be of only partial use as it also raises other questions. However, this is the crux of the issue. It is difficult to obtain agreement on the risk limit, which is dependent on a large number of factors (such as perception, voluntary or involuntary, individual or societal). The risk limit, or tolerable risk, varies with the potential consequences. There are numerous examples of this principle in our everyday lives, for example whether we smoke, drive, climb mountains or simply eat certain types of food.

The Organics division uses consequence criteria to classify processes according to a basic scheme. This classification includes three classes which have been defined as having tolerable risk criteria, based on international guidance and best practice such as the Health and Safety Executive A.L.A.R.P. (as low as reasonably practical) methodology as shown in figure 2 below. The frequency of an event is defined as the demand rate of the system multiplied by the fractional dead time and is set at once in **X** years.

Class A includes processes where the simultaneous failure of the control and preventative system as well as the protective system would result in a major incident with a high probability of fatality, major environmental harm or considerable loss in terms of production, assets or business loss. Reliability is an essential characteristic of systems that may have these severe consequences.

Class B includes processes with a more limited potential for harm to the plant, environment or the business and where the probability of fatality is low - even if there was a simultaneous failure of the control/preventative system and the processes protective system.

Class C includes processes where simultaneous failure would mainly represent an inconvenience, such as a contained spillage, loss of minor equipment, or the momentary loss of production. This Class of process also includes systems that are adequately backed up by separate safety controls.

Reliability requirements for class A are ten times greater than for class B which in turn have a one hundred times greater requirement than class C.

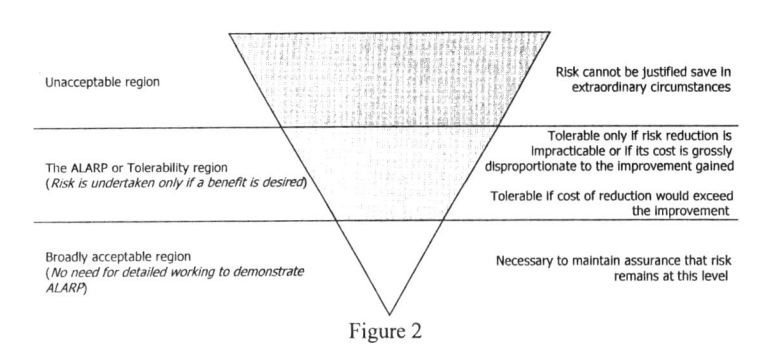

Figure 2

This basic system is currently under review. The aim is to have four classes, in line with the IEC 1508 (1) and DIN V 19 250 standards as in figure 3 below. The top tier reliability figures are likely to remain unaltered, and the fourth class is likely to fit between the current B and C classes.

The Risk Parameters are defined below:

Damage
- D1 slight injury
- D2 serious, irreversible injury of one or several persons
- Death of one person
- **D3** **death of several persons**
- D4 catastrophe, very many casualties

Duration of Stay in the Hazard Area
- S1 rare to more frequent stay
- **S2** **frequent to permanent stay**

Hazard Prevention
- H1 possible under certain circumstances
- H2 not possible

Probability of the Undesired Event
- P1 Extremely Low Probability
- **P2** **Low Probability**
- P3 Relatively High Probability

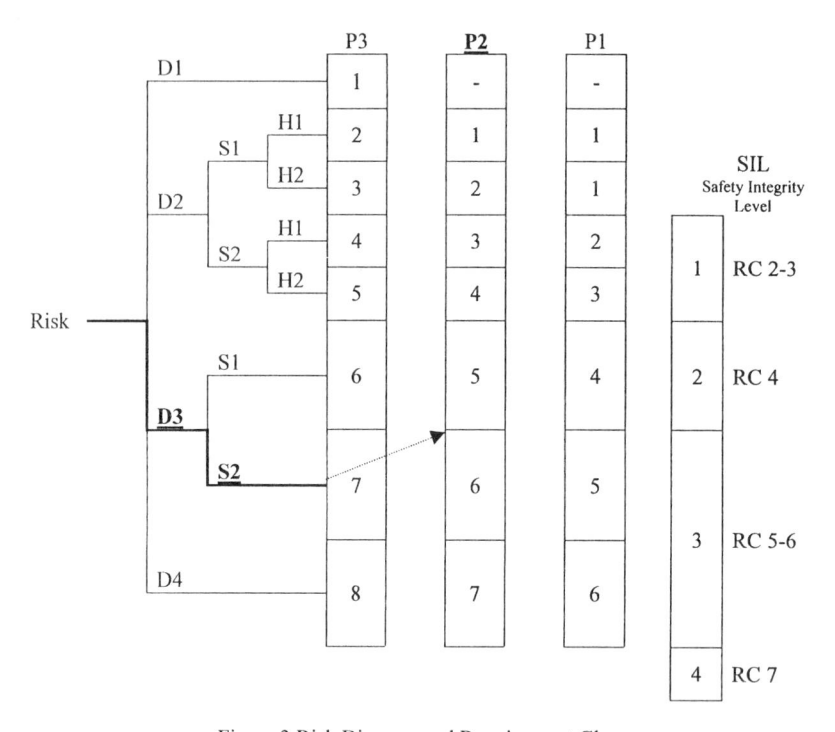

Figure 3 Risk Diagram and Requirement Classes

The purpose of the Standards, Codes and Guidance Notes is, at least in part, to help management to design, operate and maintain their processes with the *safety reliability* as defined above, whilst maintaining the *availability* of the plant to perform to the businesses' requirements.

SPECIFIC GUIDANCE NOTES

The remainder of this paper concentrates on three specific standards for the safe design of processes.

All three are important elements in the Umbrella system. They are (like all of the standards), based on the IDENTIFY, ASSESS, CONTROL, MONITOR principle. All the Standards or Guides are performance-based unless a prescriptive method is better suited, perhaps because there are only a limited number of ways to control a hazard.

We will examine the:
- handling and use of flammable gases and liquids.
- design of processes for organic peroxide manufacture.
- determination of chemical reaction hazards.

It is important to remember that these can rarely be taken as stand alone problems.

The Handling and Use of Flammable Gases and Liquids

Perhaps the most common hazard on the manufacturing plant for both fine chemicals and organic peroxide manufacture is the risk of fire and explosion associated with flammable gases and vapours.

The risk depends upon two factors:

- *the physical properties of the material*
- *the nature of the process*

To maintain a high standard at all our sites, a method of analysing and rating the explosive risk from a wide range of materials and processing operations has been developed by three Laporte personnel. The method was published in 1995.

Briefly, all of the common flammable solvents and gases have been categorised according to three physical properties, namely,

- *flammability rating* (the ratio of the flammable range divided by the lower explosive limit), *given as 0, 1, or 2*
- *minimum ignition energy rating* *given as 0, 1, or 2*
- *conductivity rating* *given as 0, 1, or 2*

Each material therefore has a rating that comprises three numbers. For example, Cyclohexane is coded 1,1,2 and hydrogen sulphide is coded 2,2,1.

This system provides five categories for assessing the explosive risks associated with materials using the coding outlined above:

- Very low risk   materials coded 0,0,0
- Low risk   materials coded 0,0,1 / 0,1,0 / 0,1,1
- Standard risk   materials coded 1,0,0 / 1,1,0
- High risk   materials coded 1,1,1 / 1,1,2 / 2,0,0
- Very high risk   materials coded 2,1,* / 2,2,* where * is 0, 1 or 2

The range of operations occurring on the plant were also assessed and categorised as:

- Very low risk,
- Low risk,
- Medium risk,
- High risk,
- Very high risk

Each operation has its own risk rating. For example, centrifuge filtration is very high, and off gas scrubbing with hydrogen peroxide is medium. Examples are shown in figure 4 below.

	Operation	Rating
EXTRUSION	Open	**High**
	Closed	**Medium**
BULK STORAGE **And** **TRANSFER**	Storage tank	**Very Low**
	Tanker transfer to Storage	**Very Low**
	Transfer to closed operations	**Very Low**
	Transfer to open operations (see also Level 2)	**Low**
	Tank transfer to tanker	**Very Low**
VACUUM SYSTEMS, **RELIEF SYSTEMS &** **DUCTS (downstream)**	Centrifugal	**High**
	Ejectors	**Medium**
	Open	**Medium**
	Closed	**Low**
DUST COLLECTION	Open	**High**
	Closed	**Medium**

Figure 4 Unit Operation Risk Rating

Knowing both the material and the operation involved enables the site management to use a matrix, rating the material risk vs. the operational risk to establish the level of control required to maintain safe operation.

There are four operational levels of control and the option of a fifth requiring input from a specialist as shown in figure 5 below.

The four operational levels of control become more sophisticated as the overall risk increases. For example, Level 1 requires attention to a number of issues including earthing, anti-static protection and flow rates. Level 4 includes the use of continuous on-line oxygen monitors.

The transfer of hexane by pump to a closed vessel provides an example of the rating system. Hexane has a material rating of *1.1.2 (High Risk)* and the process has a *very low risk* process rating. This implies an Operational Standard of Level 1.

This system has been used throughout the division (and other parts of the company) for about three years and, whilst some sites have fine-tuned the system to their own needs, the overall guidance has operated well.

PROCESS RATING

MATERIAL RATING	Very Low Risk (level)	Low Risk (level)	Medium Risk (level)	High Risk (level)	Very High Risk (level)
Very Low Risk	1	1	1	1	1
Low Risk	1	1	1	2	2
Standard Risk	1	1	2	3	4
High Risk	1	2(*)	2	3	4
Very High Risk	2	3	3	4	5(⁺)
Flammable Gases	2	3	4	5(⁺)	5(⁺)

(*) In general, Level 5 represents an unacceptable level of risk and expert advice should be sought.

(⁺) Recommending inerting, for example, when crystallising from low conductivity liquids appears harsh, but localised areas with high charge densities can develop which can discharge to the vessel walls.

Figure 5 The matrix of material risk *vs.* operational risk

The Manufacture of Organic Peroxides

In very simple terms, organic peroxides are manufactured because they are unstable! There are seven basic types of organic peroxide (O.P.). The particular O.P. used as a catalyst or promoter depends on the physical characteristics and properties involved.

If the O.P. is stable under particular conditions then it will be unable to perform its function which, in turn, presents problems for the manufacture of such materials. Not only are the products unstable, but the manufacturing processes are often highly exothermic and use energetic, corrosive and toxic raw materials. The sequence of additions, reaction temperatures, quantities, agitation and processing methodologies are also often critical.

With seven sites manufacturing O.P.'s world-wide, it was important everyone understood the hazards and risks associated with these materials and that a common standard was in place for implementation by line management.

A Laporte standard was devised based on the knowledge and experience of the operations, safety, hazards and engineering staff. Internally, the standard is known as *OPTIMIST* (Organic Peroxide Testing Information Minimum Acceptable Safety Techniques).

The early versions of OPTIMIST were based on an *open* flowchart and checklist system that was found to be inappropriate, as the number of new products, was limited when compared with fine chemicals. A new *closed* system is based on the same principles, but results in a more focused approach.

Processes are categorised according to the energy in the system and the ease with which materials or mixtures that present a major hazard can be made. Each category has certain reliability and control requirements according to it risk and consequence. Once categorisation has been completed, process changes, engineering improvements or procedural changes can be initiated to change the category to a safer one, if that is possible or is required.

> *Category 1* processes are those that would result in a detonable reaction mixture if the quantities or sequence of addition were incorrect.
>
> *Category 2* processes are those with a product or intermediate with detonable properties.
>
> *Category 3* processes are those where a rapid thermal runaway could not be vented.
>
> *Category 4* processes are those where a rapid thermal runaway is possible but could be vented.
>
> *Category 5* processes are those where no rapid thermal runaway possibility exists.

An additional category exists where the reaction mixture or a product has detonable properties and the initiating stimuli are small. These sensitive systems cannot be operated safely and are not allowed; thus, the category is an "X".

A chart lists recommendations for each process category and has sections on:

- Detonable properties,
- Runaway reactions and kinetics,
- Fire and vapour phase explosions,
- Decomposition,
- Toxicology,
- Environment, and
- Pressure burst

A review of the available information about the process, the operating plant, the preventative and protective systems and a comparison with the required standard will show any areas for priority improvement.

Category 2 processes are the highest category processes that are allowed to be operated in our production plants. They correspond to the class A processes described earlier and are required to be operated with a very high levels of safety reliability and availability.

An example of the first part of a study is outlined in figure 6 below.

1 Condensed phase explosion

1.1 Formation of a detonable mixture if Hydrogen peroxide is present
Data: Triangular diagram

#	Identification	Assessment	HR	Control Measure	Remark
1	Wrong sequence of addition	If H_2O_2 is added before tert.-Butanol: detonable mixture possible.			
2	If order of addition is correct, the H_2O_2 excess has to be 4 times the required quantity to reach the detonable area				

1.2 Formation of an autodetonable mixture (e.g.: undiluted ketoperoxide)

#	Identification	Assessment	HR	Control Measure	Remark
3	Omission of diluent	N/A No diluent used			
4	Wrong raw material	Use of MIBK instead of TBA causes serious detonation hazard			
5	Wrong order	N/A			
6	Wrong quantities	N/A			

Figure 6 Example of an OPTIMIST study.

To check the OPTIMIST requirements against the reliability criteria, three Laporte sites performed independent quantitative risk assessments (QRA) on the safety shutdown system (SSS) in place for their Category 2 processes. The three sites were chosen because they had attempted to reach the reliability goal by slightly different means, mainly for geographical and historical reasons.

One SSS was based on a double PLC system, another on a single PLC with a hard-wired backup and the third system used a double PLC system with a hard-wired backup. The results varied by a factor of about 3 between the systems, but all three sites were within the criteria for Class A processes. This was a very acceptable result given the nature of QRA studies and proved that the standards were compatible.

The Manufacture of Fine Chemicals

The range of chemistry and products in the Fine Chemicals business means that a different approach is required for the identification and control of hazards. A closed classification system cannot be operated.

For the last eight years, an internal standard has operated which has met most of the company's requirements and which categorised processes primarily on the energy content of reactive mixtures and products. At a higher level, the rate of gas generation and pressurisation in an enclosed system was also taken into account.

With an increased knowledge base and with the availability of new techniques, the old method is being replaced with a standard called *COMPASS* (COMputerised Process Assessment Safety System). The aim is to easily categorise processes so that the appropriate safety techniques can be incorporated into operational design in a safe, but efficient manner.

COMPASS will be used to assess the hazards involved in a chemical process and is based upon the physical and chemical properties of the process components, along with historical data on similar processes.

COMPASS is essentially broken down into six sections

1. Basic Data
2. Desk Screening
3. Reaction Data
4. Thermal Screening
5. Undesired Reactions
6. Consequences

In the prototype version of the system, the first two sections listed above have been completed with test data being entered for around 50 chemicals. The remaining sections will be coded and tested over the course of the next two months. The system is being written in Microsoft Visual Basic 5.0 with the underlying data being stored in Microsoft Access tables. All database activity is controlled using SQL constructs.

A key feature of the system is that it should be flexible enough to allow use in several countries, using different languages. With this in mind, all on-screen text labels are stored in a system database and loaded when the system is run (they are not 'hard-coded'). This allows efficient translation to be carried out.

The programme can be accessed at four levels. For everyday use, the data input screens are supported by a complete help system, giving basic information and guidance. A computer-based tutorial system is included to allow the user to gain additional knowledge and experts can access a high-level knowledge system to carry out specialised assessments.

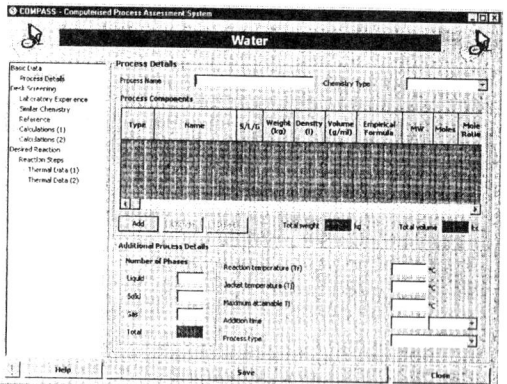

Figure 7. A sample screen, taken from the current prototype.

The database allows rapid access to the properties of a wide range of solvents and reagents. A variety of calculators has been built into the system that operate automatically (such as a molecular weight calculator, and an automatic oxygen balance calculation when the empirical formula is entered).

For desk screening, hyperlinks to standard texts such as Bretherick, the I. Chem. E. Accident Database and CHETAH® can be set up. Interrogation of the Corporate Memory is also required.

The system requests data on a wide range of variables and processes this information. Constant checks ensure that the data are self-consistent. From the entered data, the thermal properties of the chemical reaction are studied in order to generate the following key data:

- Heat of reaction,
- Rate of heat generation,
- Accumulation, which is a function of the reaction rate,
- The minimum temperature that a decomposition could be initiated,
- The maximum achievable temperature, and,
- The maximum temperature achievable for technical reasons.

The characterisation is based on the concepts of the desired reaction, undesired reaction, thermal runaway and the time to maximum rate established by the team at Ciba Geigy as well as other experts in the field.

The levels and reliability of the preventative and protective equipment for each process are set by internal standards. The preventative and protective systems increase on a scale from Class 1 to Class 5. No specific protection is necessary for class 1 but primary, secondary, tertiary and special measures are needed as the categories increase. These measures include duplication of temperature probes, automatic quench, secondary cooling, dual alarms and active or passive protection.

Reactions are classified according to the relative values of the reaction boiling point, the maximum temperature achievable in the runaway reaction and the temperature at which the time to maximum rate of reaction (and therefore heat evolution) under adiabatic conditions (TMR_{ad}) is 24 hours.

Class 1 reactions are those where the boiling point of the reaction mass is below the temperature where the TMR_{ad} = 24 hrs (hereafter called T_t) and the maximum theoretical heat release cannot raise the temperature of the reaction mass to its boiling point.

Class 2 reactions are those where the boiling point lies above T_t, but the maximum theoretical heat release cannot raise the temperature of the reaction mass to T_t.

Class 3 reactions are those where the maximum theoretical heat release can raise the temperature of the reaction mass to above its boiling point, but not to T_t.

Class 4 reactions are those where the maximum theoretical heat release can raise the temperature of the reaction mass to above its boiling point and also above T_t.

Class 5 reactions are those where the maximum theoretical heat release can raise the temperature of the reaction mass to above T_t, but not to its boiling point.

Neither Class 1 nor Class 2 reactions are safety critical. Simple control of temperature is all that is needed.

Class 3 reactions have a higher level of criticality in that the boiling point can be reached, with obvious consequences for the condenser capacity and release of V.O.C.'s.

Additional safeguards are required for Class 4 reactions where the loss of solvent may result in a temperature rise that could trigger the decomposition reaction. Here dual independent temperature probes may be required as well as indication of the performance of the condensers. Interruption of flow rates on loss of cooling will also be indicated.

Class 5 reactions need to be subject to the most rigorous analysis. Class 4 requirements will be needed together with automatic interruption of reagent addition on loss of cooling or loss of agitation. Quenching the reaction mass may also be required.

The original assessment system has been validated against the risk criteria using a QRA study. To date, only one process has been evaluated but the results were within the consequence criteria.

CONCLUSION

Risk management covers the whole business process and has been the subject of numerous books. In this paper, our aim has been to cover only one critical part of the risk management process, and to show the approach of Laporte and more specifically, its Organics division.

The integration of hazard management and the careful generic treatment of key problems have resulted in common high Standards, Guides and Codes of Practice at all manufacturing sites. A high level of awareness of the hazards and the tolerable residual risk results in cost effective controls or *PROFIT THROUGH SAFETY.*

ACKNOWLEDGEMENTS

The authors wish to acknowledge the contribution of colleagues within Laporte including, Dr F. Diem, Dr R. Band, Dr P. Bekk, Dr R. Owen and the late Dr N. F. Scilly all of whom wrote or contributed to specific standards.

REFERENCES

1. The International Electrochemical Commission IEC 1508
2. German national DIN standards.
3. H. L. Walmsley, 1992 Journal of Electrostatics, 27(1/2)
4. Scilly N. F., Owen R., Wilberforce J. K., 1995, A.C.S. forum, New Orleans.
5. Stöessel F.et al, 1997 Organic Process Research and Development, 1(6): 428, and references therein.
6. Lambert, P.G., 1993, Chemical reaction hazards. IBC symposium, London

UNDERSTANDING VINYL ACETATE POLYMERIZATION ACCIDENTS

J.L. Gustin and F. Laganier
Rhône-Poulenc Industrialisation - 24, avenue Jean Jaurès 69153 Décines - FRANCE

Vinyl acetate is processed to produce polymers and copolymers used in water based paints, adhesives, paper coatings or non-woven binders and various applications at moderate temperatures. The polymerization processes used include solution, suspension and emulsion processes. Many incidents involving the runaway polymerization of Vinyl Acetate Monomers (VAM) are known. The incidents happened either in batch or semi-batch polymerization processes in connection with wrong catalyst introduction.
In processes where the polymerization catalyst is dissolved in the monomers, the catalyst premix has polymerized violently in the premix vessel. In polymerization processes where the vinyl acetate monomer conversion ratio is not 100 %, storages of recycled monomers containing no polymerization inhibitor and possibly some traces of polymerization catalyst have exploded due to the VAM violent bulk polymerization.
In this paper, a review of polymerization incidents is given. Radical chain polymerization kinetics are used to explain some accident features such as polymerization isothermal induction periods. Experimental results on bulk VAM polymerization obtained in DSC, Dewar flask and VSP are given. The accidental polymerization of VAM in a vessel is simulated using the WIN SIM software. The simulation is based on radical chain polymerization kinetics and thermochemical data on VAM and polymers and on specified initial and boundary conditions. Simulation of accidental polymerisations is useful to validate accident scenarios and evaluate various cause hypotheses.

Introduction

Vinyl acetate (VAM) is processed to produce polymers and copolymers used in water based paints adhesives, paper coatings or non-woven binders and various applications at moderate temperatures.
The polymerization processes used include semi-batch or continuous, solution, suspension and emulsion processes. The bulk polymerization of vinyl acetate is not used in industrial operations due to the violent reaction obtained. However the unwanted bulk polymerization of vinyl acetate is involved in many industrial accidents where the polymerization occurred in storage vessels containing fresh or recycled monomers or in premix vessels where a polymerization initiator was dissolved in vinyl acetate monomer, previous to its use in the polymerization process.
Therefore, the study of vinyl acetate bulk polymerization is of great interest from a process safety point of view, even if it is not a commercial wide-spread process.
The bulk polymerization of vinyl acetate in storage vessels occurs spontaneously under constant temperature conditions, due to a chemical acceleration phenomenon related to the free radical nature of vinyl acetate chain polymerization.

The incidents in polymerization processes happen either in semi-batch or continuous polymerization processes due to wrong catalyst introduction, absence of agitation or inadequate priming operation.

The consequences of the unwanted or uncontrolled vinyl acetate polymerization depend on the various process conditions. The bulk polymerization of vinyl acetate is extremely violent and may generate a pressure surge to above 40 bar, a pressure exceeding most storage vessels pressure resistance.

The uncontrolled polymerization of vinyl acetate in the presence of light solvents in polymerization reactors may also result in a high pressure surge due to the solvent high vapour pressure under the runaway reaction final temperature. If the runaway polymerization causes the vessel rupture, this may occur with severe mechanical and blast effects.

The vessel inventory consisting of polymers would be released in the neighbourhood. In most incidents involving the bulk polymerization of vinyl acetate, the release did not ignite. This is probably due to the high conversion ratio reached. The presence of a flammable solvent or a lower conversion ratio of the monomers may allow the release ignition.

Therefore vinyl acetate runaway polymerization incidents are very serious and they occur with a high frequency. The polymerization of vinyl acetate is probably the second most frequent cause of runaway reaction accidents in the chemical industry, after the phenol + formaldehyde runaway reaction.

Therefore it is of interest to describe the circumstances where a runaway polymerization can occur and to explain the causes of accidents and the factors influencing their occurrence. As an example, the bulk polymerization of vinyl acetate initiated by a peroxide was studied on an experimental and theoretical point of view based on radical chain polymerization kinetics and on simulation using the Win Sim software.

Thermal stability of vinyl acetate monomer.
Runaway reaction hazard in storage vessels

Thermal stability of commercial monomers

Vinyl acetate is a reactive monomer which may undergo a free radical chain polymerization phenomenon under constant temperature conditions. The chain polymerization of monomers may be initiated by radical initiators i.e. inorganic or organic peroxides, azobisisobutyronitrile (AIBN), others ...

In monomer storage vessels, there is no such initiator in normal process conditions. However, there is a slow thermal production of free radicals in the bulk liquid monomer even under ambient temperature. The thermally produced radicals may accumulate and further initiate the free radical chain polymerization of the monomers in the storage vessel. To provide enough thermal stability, some polymerization inhibitor is added to the commercial product, which is a radical scavenger.

The polymerization inhibitor used for the stabilization of vinyl acetate monomers is hydroquinone (H.Q.). Rhône-Poulenc is a leading supplier for hydroquinone and other polymerization inhibitors.

Two grades of vinyl acetate monomers are currently supplied :
- the low hydroquinone grade containing 3 - 7 ppm H.Q. to be used within two months of delivery
- the high hydroquinone grade containing 12 - 17 ppm H.Q. for storage up to four months before use.

Compared to other vinylic or acrylic monomers i.e. acrylic acid (AA), methacrylic acid (MAA) or methyl methacrylate (MMA), the polymerization inhibitor concentration in VAM is very low owing to the low thermal activity of VAM in storage conditions.

Influence of oxygen

Many questions arise concerning the recommended storage conditions of vinyl acetate.

It is sometimes recommended to store vinyl acetate monomer under ambient air atmosphere to enhance the polymerization inhibitor efficiency as it is known that phenolic inhibitors need oxygen to be active. For example, air is bubbled in acrylic acid storage vessels to allow the polymerization inhibitor, hydroquinone mono methyl ether (MeHQ) to be active (1).

However, contrary to acrylic acid which is stored at a temperature below 25°C and has a flash point of 49°C (2), above the storage temperature, vinyl acetate monomer is a highly flammable liquid with a flash point of -7°C (2). Under ambient temperature conditions, the storage vessel gas phase under air atmosphere would be flammable for VAM. Therefore, storage of VAM under air atmosphere should be avoided.

It was further pointed out by A. Nicholson (3) that the optimum stability of methacrylic acid in the presence of HQ or MeHQ as polymerization inhibitor was obtained at much lower equilibrium oxygen concentration than that provided by an air atmosphere.

The most favourable oxygen concentration was shown to depend on the inhibitor considered and on its concentration in the monomer. A higher oxygen concentration than the optimum value was shown to decrease the monomer thermal stability due to unstable peroxide formation. Therefore, any monomer/inhibitor couple is specific and requires a special consideration for the choice of storage conditions.

In the case of vinyl acetate containing hydroquinone as an inhibitor, it was later pointed out byLeon B. Levy (4) (5) that the optimum thermal stability was obtained under dry nitrogen atmosphere without any oxygen present in the storage vessel gas phase.

It was shown that between 30°C and 120°C, the length of vinyl acetate polymerization isothermal induction period is about - 0.4 order in oxygen partial pressure. The detrimental effect of the presence of oxygen in the monomer gas phase, on the thermal stability of vinyl acetate containing hydroquinone as an inhibitor, is explained by the low thermal stability of the peroxides formed when VAM is stored in the presence of dissolved oxygen. The decomposition of accumulated peroxides near ambient temperature can induce an increased inhibitor consumption.

Therefore, for flammability and thermal stability reasons, it is recommended to store vinyl acetate under dry nitrogen blanket. The above conclusion holds for high temperature (100°C) and ambient temperature conditions. However, when stabilized with 3 - 5 ppm H.Q., both air saturated and oxygen-free VAM exhibit adequate thermal stability at normal transport and storage temperature (25°C - 50°C).

Influence of impurities

It was shown by Leon B. Levy that acetaldehyde impurity does not cause oxygen induced destabilization of vinyl acetate (5).

The presence of humidity in vinyl acetate would produce some degree of hydrolysis and alter the monomer quality.

Influence of the storage vessel wall material

As for other reactive monomers and especially in the case of polymerization accident inquiries (6) the influence of the storage vessels and transport containers wall material has been investigated.

The influence of carbon steel on the thermal stability of VAM containing 5 ppm HQ, at 48.9°C, blanketed with 10 % oxygen, was investigated by Leon B. Levy (5). It was found that untreated carbon steel covered with rust has a strong destabilizing influence on VAM compared to a glass vessel, whereas clean carbon steel would increase thermal stability.

The influence of carbon steel on vinyl acetate thermal stability can also be measured by monitoring of the inhibitor concentration as a function of time.

Two 2 litre samples of vinyl acetate containing approximately 20 ppm hydroquinone were submitted to aging at 20°C, one in a coloured glass vessel, the other in a 2 litre oxidized carbon steel vessel. Both vessels where closed under air atmosphere.

The following depletion of H.Q. concentration was observed by UV spectrometric determinations.

Table 1 : Vinyl acetate in 2 litre coloured glass vessel

Time (days)	H.Q. concentration (mg/kg)
0	20.1
5	19
18	18.4
41	17.8
55	18
67	17.6

Table 2 : Vinyl acetate in 2 litre oxidized carbon steel Vessel

Time (days)	H.Q. concentration (mg/kg)
0	21.7
1	20.8
7	19.2
12	19.1
22	17.6
48	16.5
61	16.5
76	15.9

In these experiments the hydroquinone consumption in carbon steel vessel was twice the H.Q. consumption in coloured glass vessel. Also enough inhibitor concentration would be present under ambient temperature, to prevent polymerization over a very long period of time in both containers.

However, considering the lower thermal stability of VAM in oxidized carbon steel vessels, it is recommended to design stainless steel storage vessels for VAM and to prevent existing carbon steel storage vessels from rust. This should be easier under dry nitrogen blanket.

Accidental polymerization of recovered vinyl acetate

The data found in the literature and additional data provided, show that commercial vinyl acetate should never polymerize in normal storage conditions. However polymerization accidents in storage vessels are known concerning VAM which is not exactly the commercial product. This is the case for vinyl acetate recovered from polymerization processes where the conversion ratio is not 100 %. An example of such a process is the manufacture of polyvinyl alcohol by polymerization of vinyl acetate in the presence of methanol initiated by azobisisobutyronitrile (AIBN), followed by alkaline hydrolysis of the polyvinyl ester.

The unreacted vinyl acetate is separated from the polymer and recycled to the polymerization step. The recovered vinyl acetate is free of polymerization inhibitor and may possibly contain some traces of polymerization initiator. A 35 m³ storage of recovered VAM happened to polymerize during the summer vacations in a southern country.

The long residence time and the warm temperature, possibly one month and 50°C, allowed the polymerization to occur. The polymerization was violent. The storage vessel roof was ejected to 100 m distance and polymers were spread in the neighbourhood.

The conditions where this accident occurred are fairly compatible with the stability data of Leon B. Levy for VAM with no H.Q., 10-20 % vol. O_2 and the presence of carbon steel wire.

Also it was shown that oxidized iron lowered the product thermal stability measured in DTA stainless steel closed cell under 5°C/min. temperature scan.

The polymerization exotherm onset temperature was shifted from 300°C to 230°C in the presence of oxidized carbon steel. The product polymerization exotherm was 1221 j/g close to the literature data of 1036 j/g (7).

A sample of recovered VAM showed polymerization isothermal induction periods at high temperature in DTA but such data cannot be extrapolated to ambient temperature where induction periods are too long to be measured using DTA machines.

Our comment on this incident is that one should take care of the thermal stability of non commercial or recovered vinyl acetate in particular if high storage temperature and long residence time are possible. The storage of uninhibited monomer should be avoided.

Polymerization incidents in premix vessels

It is a surprizing practice, from a process safety point of view, that concentrated solutions of polymerization initiators are used in premix to be injected in a continuous or semi-batch polymerization process. Premix vessel polymerization accidents are known for many reactive monomers. This type of incident has occurred also in vinyl acetate polymerization processes.

1.8 % wt dilauroylperoxide was dissolved in vinyl acetate in a premix to be injected in a VAM polymerization process. The premix was prepared in advance, to be used 8 hours later. Due to the hot summer temperature, the premix happened to polymerize. The reaction was extremely violent. A two phase flow of polymer and vapour was ejected in the vessel vent header. The release did not ignite. This incident has been investigated on an experimental and theoretical point of view to assess the conditions of occurrence and to evaluate the venting requirement of this scenario.

The relevant experimental study and simulation using Win Sim are given in the experimental and simulation sections of this paper.

Our conclusion on this issue is that the practice of dissolving the polymerization initiator in the reactive monomer in a premix, should be eliminated from polymerization processes for process safety reasons.

Incidents, accidents in polymerization processes

Runaway reaction incidents in polymerization processes are related to the reactive monomer accumulation in the polymerization reactor and the subsequent runaway polymerization of the accumulated quantity of unreacted monomer.

The monomer accumulation may have several causes.
- The polymerization initiator was not injected and could reach later the reaction vessel.

In a semi-batch solution polymerization process where VAM was polymerized in acetone as a solvent, a valve was closed on the polymerization initiator feed pipe. A large quantity of vinyl acetate was charged to the reactor when the polymerization catalyst could reach the reaction vessel through the vent line. A runaway polymerization occurred with a severe pressure surge which may have been caused by the use of a light solvent.

Other frequent causes of monomer accumulation are :
- Too low a process temperature which does not allow the monomer charged to react.
- No agitation. This process deviation allows the monomer and the polymerization catalyst to be in two separate liquid phases. The accumulated monomer may react violently if the agitation is restarted.
- The process is a batch process with respect to the reactive monomer and the polymerization initiator.

In a series of incidents on various VAM polymerization processes, the common factor seems to be the fact that the processes are batch processes with respect to the reactive monomer and the polymerization initiator. This made the reaction mixture temperature control more difficult and the process sensitive to loss of cooling. In a batch process the heat of reaction is released over a short period of time, producing a high heat flux and making the reaction mixture temperature control critical. Batch processes are better replaced by semi-batch operations where the controlling reactant and the initiator are introduced over a period of time long enough to adjust the heat flux produced to the reactor available cooling capacity. This advantage is well explained for vinyl acetate polymerization by H.U. Moritz (8).

- Incidents in priming procedures

The initiation of polymerization in semi-batch or continuous processes is frequently made using a priming procedure where a limited charge of controlling reactant is introduced in the reaction vessel to check the reaction initiation through the detection of the reaction exotherm.
On detection of the reaction exotherm, a continuous feed of controlling reactant to the reaction vessel is established. If the continuous feed of reactant is established too early or too late with respect to the priming charge reaction, an "unexpected" faster reaction is obtained due to a larger amount of unreacted monomer in the reaction mixture, if the continuous feed is established too early or due to the priming reaction extinction and monomer accumulation, if the continuous feed is established too late. See also ref. (8).

Experimental investigation of the bulk polymerization of vinyl acetate initiated by a peroxide.

Cause of the isothermal induction period phenomenon

As pointed out earlier, this bulk polymerization of VAM occurs during the storage of a premix where the polymerization initiator has been dissolved in the monomer. The very violent polymerization occurs after an isothermal induction period caused by the chemical acceleration of the radical chain polymerization. This phenomenon can be accounted for as follows.

The kinetic behaviour of radical chain polymerisations has been presented by Flory (9) And will be described in the simulation section of this paper.
In this theory, the rate of monomer consumption is referred as the rate of polymerization.

$$R_P = -\frac{d[M]}{dt} = k_P[M][P]$$

where [M] is the monomer concentration, [P] is the polymer radical concentration
k_P is the propagation rate constant.

According to the theory of Flory, the rate of polymerization is controlled by the initiator concentration (9) (10).

$$R_P = -\frac{d[M]}{dt} = k_P[M]\left(k_d f[I]/k_t\right)^{1/2}$$

Where k_d is the initiator decomposition rate constant, k_t the termination rate constant, [I] the initiator concentration, f the fraction of initiator radicals successfully reacting with the monomer.
Consequently, the rate of polymerization is an Arrhenius function with first order with respect to the monomer concentration and half order with respect to the initiator concentration which controls the population of polymer radicals.

Considering the maximum rate of polymerization, the higher the initiator concentration, the faster the maximum rate of polymerization. This explains the very violent polymerization observed in premix vessel polymerization accidents, where very high initiator concentration are present.

In storage vessels of monomer where some polymerization inhibitor is present, the polymerization induction period is the time necessary for the thermally - generated radicals R° or M° to consume the polymerization inhibitor and produce enough radical concentration to initiate the chain polymerization reaction.
Radicals are produced by the polymerization initiator if any, or by the monomers themselves. Consequently, the isothermal induction period is influenced by the inhibitor and polymerization catalyst concentrations in the monomer.
In the case of commercial monomer storage, there should be no polymerization catalyst present, and the induction period is the time necessary for the monomer thermally generated radicals to consume the inhibitor concentration (11).

$$\tau = \frac{[Z]}{k_0 \, e^{-E/RT} \cdot [M]} \qquad \tau = \frac{m \, [Z]}{k_0 \, e^{-E/RT} \cdot [M]}$$

depending on whether one or more monomer radicals react with each molecule of inhibitor Z. Consequently, the induction period is an Arrhenius function of the inhibitor concentration with an activation energy characteristic of the monomers.

$$\tau = K[Z] e^{E/RT}$$

For storage vessels of commercial monomer, the slope of the Arrhenius plot of the isothermal induction period is a characteristic of the monomer.
This may not be the case if a polymerization catalyst is present and if the inhibitor is eliminated first.

Then the isothermal induction period is :

$$\tau = \frac{m[Z]}{k_0 \cdot e^{-E/RT} \cdot [M] + k_{d0} \, e^{-E_d/RT}[I]}$$

In this case, the polymerization inhibitor is consumed by the monomer thermal generation of radicals and by the decomposition of the polymerization initiator.

Under moderate temperature where the polymerization initiator is active and the thermal generation of monomer radicals negligible compared to that of the initiator, the above equation reduces to :

$$\tau = \frac{m[Z]}{k_{d0} \cdot e^{-E_d/RT}[I]}$$

In this case, the polymerization isothermal induction period is an Arrhenius function of the inhibitor and polymerization initiator concentrations with an activation energy Ed characteristic of the polymerization initiator

$$\tau = K \, [Z][I]^{-1} e^{E_d/RT}$$

The slope of the Arrhenius plot of the induction period is a characteristic of the polymerization radical initiator. The Arrhenius plot of the polymerization induction period is a key piece of information to prevent runaway polymerization accidents. Also interesting is the experimental data on the polymerization exotherm, rate of reaction, pressure effect. This experimental information is presented in the following for the bulk polymerization of VAM containing 1.8 % wt dilauroylperoxide, concerning a premix vessel polymerization accident.

DTA determinations

VAM samples containing 1.4 % wt, 1.8 % wt, 2.2 % wt dilauroylperoxide were submitted to DTA test under 5°C/min. temperature scan conditions in stainless steel closed cell.

The samples exhibit two separate exotherms :
- the first asymmetric exotherm, $\Delta Q1$ = 639 - 676 J/g is due to the monomer initiated radical polymerization,
- the second exotherm of ΔQ = 296 - 469 J/g is due to the monomer high temperature thermal polymerization.

This is proved by the uncatalysed thermal polymerization of VAM giving up an exotherm of 1221 J/g above 290°C. The polymerization overall exotherm of 970 - 1221 J/g is of the order of magnitude of VAM heat of polymerization known in the literature (7). ΔH = -1036 J/g.

Polymerization induction period of vinyl acetate in the presence of 1.8 % wt dilauroylperoxide

The bulk polymerization of vinyl acetate containing 1.8 % wt of radical initiator dilauroylperoxide can be initiated by constant temperature exposure under different temperatures.

The polymerization induction period was measured using various experimental techniques : (12)
- DTA for high temperature short induction periods
- closed Dewar experiments and VSP experiments for low temperature long induction periods.

The isothermal induction periods are given for the exotherm onset detection since in some VSP experiments the test cell exploded due to the very fast pressure surge. Consequently, the maximum rate of reaction could not be observed. The Arrhenius plot of the polymerization induction periods obtained is given on Fig. 1.

VSP closed cell experiments

The runaway polymerization of vinyl acetate containing 1.8 % wt dilauroylperoxide was initiated on a 80 g sample under isothermal exposure conditions in a 116 cm³ Hastelloy C VSP closed cell. The isothermal exposure temperatures were 37°C and 45°C. In one experiment at 37°C the test cell did not rupture. The maximum heat-rate was 4967°C/min. The maximum rate of pressure rise was 2544 bar/min. i.e. 42.4 bar/s. The maximum rates were obtained for a temperature of 144-174°C. The final temperature was 268°C and the final pressure 44.5 bar. The heat-rates obtained are given on Fig. 2. The pressure corrected for the nitrogen pad as a function of the reciprocal temperature is given on Fig. 3 for the three experiments. The VAM vapour pressure is also represented (13).

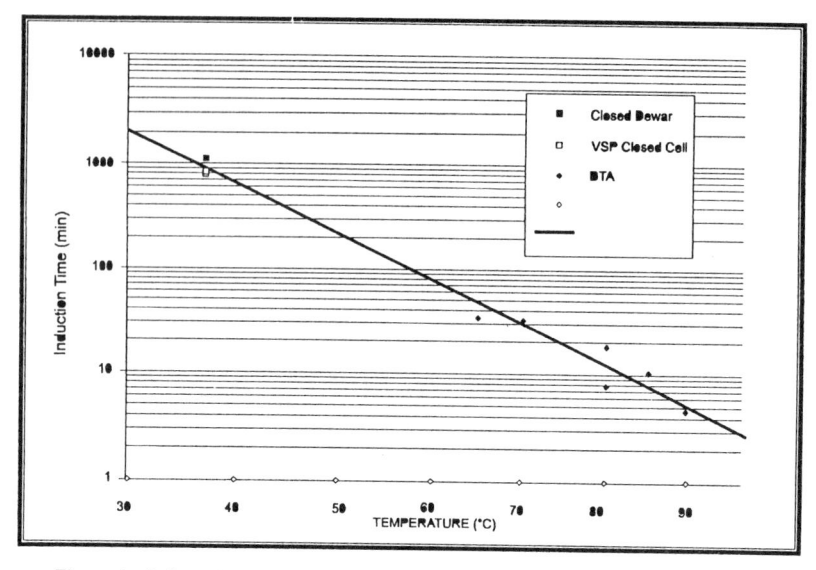

Figure 1: Polymerization of vinyl acetate monomers containing 1.8% w/w peroxide. Induction time vs Temperature.

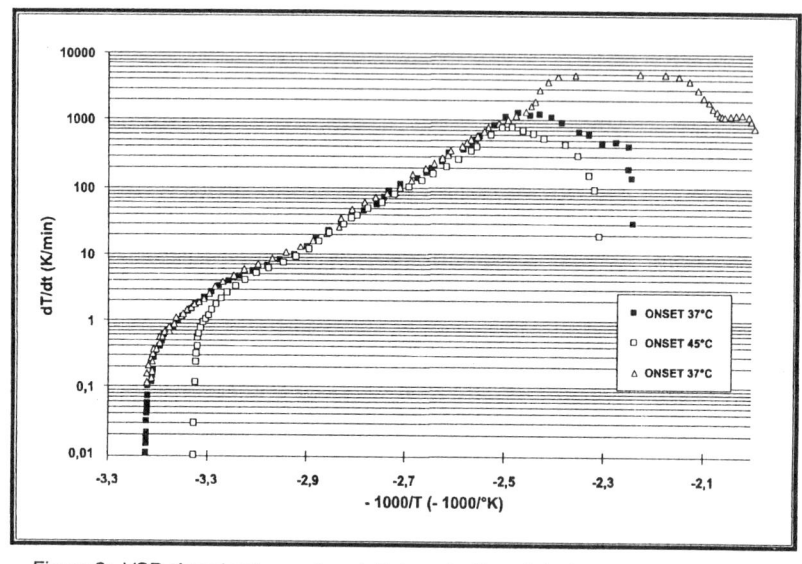

Figure 2: VSP closed cell experiment. Polymerization of vinyl acetate monomers with 1.8% w/w dilauroyl peroxide initiated by constant temperature exposures at 37°C and 45°C. Heat rate curves.

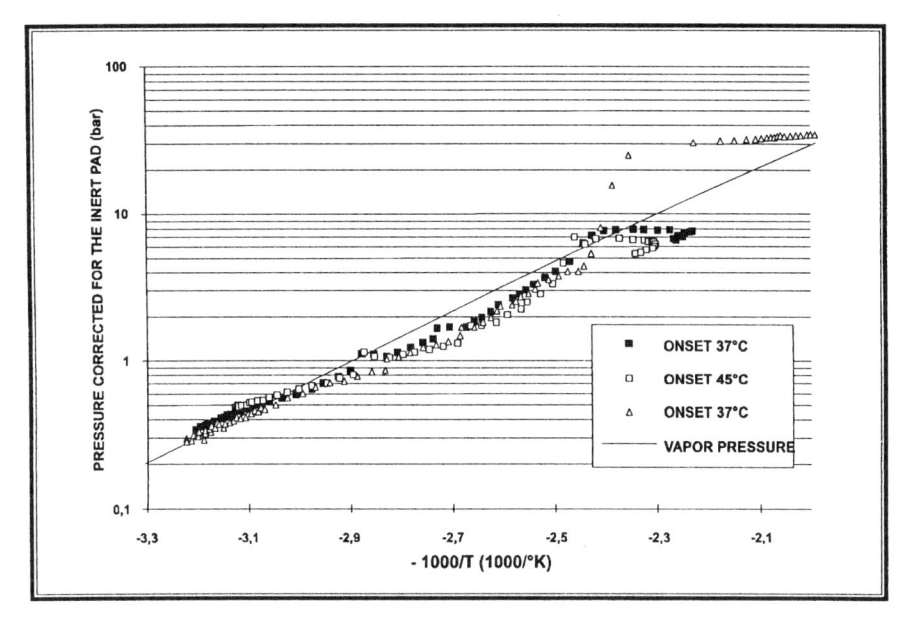

Figure 3 : VSP closed cell experiments. Polymerization of vinyl acetate monomers with 1.8 % w/w dilauroyl peroxide initiated under constant temperature exposures at 37°C and 45°C -
P vs T data - Curve represents vapour pressure of monomers.

Simulation of vinyl acetate runaway polymerization using Win Sim

In this section, a kinetic model is proposed to represent the radical chain polymerization of vinyl acetate. The relevant kinetic data are obtained from the literature and discussed. The software used for the simulation is described. The use of this software is exemplified by the prediction of the runaway polymerization of vinyl acetate in the presence of dilauroylperoxide as an initiator i.e. to represent a premix vessel polymerization incident.

Kinetic model

A reaction mixture containing the following stable components is considered : Polymerization initiator (I), Monomer (M), Polymer (D), Solvent (S). The following radical components are also present : Monomer radical (R*), Polymer radical (P).
The kinetic model includes the following reaction steps : Initiation, Propagation, Termination together with the relevant kinetic equations. The kinetic model chosen is the model used in the "Polymer Plus" software of Aspen Tech using Arrhenius type rate equations.

Reaction mechanism

The model reaction steps and rate equations are the following :

Initiator decomposition	$I \longrightarrow nR^*$	$R_d = k_d [I]$
Chain initiation	$R^* + M \longrightarrow P_1$	$R_i = k_i [R^*][M]$
Thermal generation of radicals	$M \longrightarrow P_1$	$R_{th} = k_{th} [M]$
Chain propagation	$M + P_n \longrightarrow P_{n+1}$	$R_p = k_p [M][P]$
Transfer to monomer	$M + P_n \longrightarrow D_n + P_1$	$R_{trm} = k_{trm} [M][P]$
Chain termination by disproportionation	$P_n + P_m \longrightarrow D_n + D_m$	$R_{td} = k_{td} [P]^2$
Chain termination by combination	$P_n + P_m \longrightarrow D_{n+m}$	$R_{tc} = k_{tc} [P]^2$

The following constituent balances must be satisfied :

Initiator $\dfrac{d[I]}{dt} = -R_d$ Monomer radicals $\dfrac{d[R]}{dt} = n \cdot f \cdot R_d - R_i$

Monomers $\dfrac{d[M]}{dt} = -R_i - R_{th} - R_p - R_{trm}$ Polymer radicals $\dfrac{d[P]}{dt} = R_i + R_{th} - 2R_{td} - 2R_{tc}$

Polymers $\dfrac{d[D]}{dt} = -R_{trm} + 2R_{td} + R_{tc}$

The quasi stationary state hypotheses is assumed and is written as follows :

$$\frac{d[R]}{dt} = 0 = f n R_d - R_i \qquad \frac{dP}{dt} = 0 = R_i + R_{th} - 2R_{td} - 2R_{tc}$$

$$[P] = \sqrt{\frac{R_i + R_{th}}{2(k_{tc} + k_{td})}}$$

Hence : $R_i = f \cdot n \cdot R_d$

Gel effect

The Gel effect or Trommsdorff effect is related to the reaction mixture increase of viscosity with increasing polymerization conversion ratio. The Gel effect may induce

an increase in the rate of polymerization due to slower termination reaction caused by increased viscosity.
The kinetic interpretation of the Gel effect is that given in the "Polymer Plus" software where the termination reaction rate constants are multiplied by a fraction X_p depending on the polymer mass fraction in the reaction mixture.

$$k_t = GF \cdot k_{t^o} \quad \text{where} \quad GF = \exp(-gf_1 X_P - gf_2 X^2_P - gf_3 X^3_P)$$

Dynamic model

The reaction mixture is assumed to be enclosed in a vessel under constant volume conditions.
The initial mass, composition, temperature, must be specified by the user.
In step one the model determines the reaction mixture thermodynamic state : vapour fraction, composition of liquid and vapour phases, vessel pressure, vessel filling ratio.
In step two, the boundary conditions are specified : vessel wall under constant temperature, reaction mixture and vessel under constant temperature, specified heat exchange through the vessel walls.
The model determines the reaction mixture temperature, pressure, composition in the vessel, as a function of time.

The following system of differential equations is integrated using a variable step Runge Kutta method :
- Enthalpy change = heat exchanged through the wall + heat of reaction released.
- Enthalpy change of the vessel wall = heat exchange with the reaction mixture + heat flux specified by the user.
- Change in constituents mole number = contribution of every reaction.

The polymer molecular weight is calculated assuming a constant total mass of the reaction mixture.
The following simplifying hypotheses are assumed : the liquid volume is constant during the polymerization reaction, the reaction mixture specific heat is constant during the polymerization reaction.
We know that both hypotheses are not true since the polymerization conversion ratio can be determined by measuring the reaction mixture volume and the liquid specific heat changes during the reaction. However the relevant experimental data is not available and no model to account for these properties change was introduced. The liquid volume and liquid specific heat are calculated for the initial conditions together with the reaction mixture thermodynamic properties.

Kinetic data obtained in the literature

Polymerization initiator decomposition

The reaction is : $I \longrightarrow nR^{\bullet}$

The initiator decomposition rate is : $-\dfrac{d[I]}{dt} = k_d [I]^{nd} = k_d \exp(-E_d/RT) \cdot [I]^{nd}$

The initiator considered is dilauroylperoxide (LPO). The following decomposition kinetics are available for dilauroylperoxide.

1 - Paper of Korbar and Malavasic (14)

Order n_d	E_d (kJ/mol)	T = 65°C		T = 75°C		T = 85°C	
		K_d (s⁻¹)	T½ (h)	K_d (s⁻¹)	T½ (h)	K_d (s⁻¹)	T½ (h)
1.1	135.7	0.22.10⁻⁴	8.8	0.88.10⁻⁴	2.20	3.27.10⁻⁴	0.60

2 - Data of manufacturer mentioned by Korbar and Malavasic (14)

Order n_d	E_d (kJ/mol)	T = 65°C		T = 75°C		T = 85°C	
		K_d (s⁻¹)	T½ (h)	K_d (s⁻¹)	T½ (h)	K_d (s⁻¹)	T½ (h)
-	127.3	0.28.10⁻⁴	6.8	1.04.10⁻⁴	1.90	3.56.10⁻⁴	0.50

3 - Data of Warson (15) used in Polymer Plus

Order n_d	E_d (kJ/mol)	$K_d°$ (s⁻¹)	Half-life = 60 s		Half-life = 1 h		Half-life = 10 h	
			T (°C)	T½ (h)	T (°C)	T½ (h)	T (°C)	T½ (h)
1	127.4	1.44.10¹⁵	116		80		62	

4 - Other data

Half life periods are given in various works. A half-life period of 10 hours at 62°C is mentioned in (16). A half-life of 1 min. at 115°C is mentioned by SRI. The following data was measured :

T½ (h)	42	12	3.2	1
T (°C)	50	60	70	80

Thermal generation of radicals by the monomer

The reaction is :   M ⟶ M•
The monomer radical generation rate is : $R_{th} = k_{th} [M] = k_{th}° \exp(-E_{th}/RT) [M]$
No kinetic data is available on this reaction rate in the literature, due to the many factors influencing the thermal generation of radicals by the monomer.

Chain propagation

The reaction is : $M + P_n \longrightarrow P_{n+1}$
The heat of polymerization is basically the heat of propagation reaction.
The data available on this reaction is :

1 - The data of Flory (17)

ΔHr (kJ/mol)	kp° (mol-l-s)	Ep (kJ/mol)
-	24.10⁷	30.5

2 - The data of the Encyclopedia of Polymer Science and Engineering

ΔHr (kJ/mol)	kp° (mol-l-s)	Ep (kJ/mol)
87.5 à 89.1	3.2.10⁷	26.2

Termination reaction

The chain termination can occur following two reactions :
by combination $P_n + P_m \longrightarrow D_{n+m}$ by disproportionation $P_n + P_m \longrightarrow D_n + D_m$
The data from the literature suggests that the combination reaction is negligible compared to the disproportionation reaction. The data on the disproportionation reaction are :

1 - The data of Flory (17)

K_{td} at 60°C	$K_{td}°$ (mol-l-s)	E_{td} (kJ/mol)
$7.4.10^7$	$2.1.10^{11}$	21.8

2 - The data of the Encyclopedia of Polymer Science and Engineering

K_{td} at 60°C	$K_{td}°$ (mol-l-s)	E_{td} (kJ/mol)
$2.9.10^7$	$3.7.10^9$	13.3

Transfer to monomer

There is a scarcity of data on the transfer to monomer reaction. Encyclopedia of Polymer Science and Engineering gives data on the ratio $C_m = k_{tm} / k_p$:
   $C_m = 1.10^4$ at T = 0°C $C_m = 3.10^4$ at T = 70°C
C_m may influence only the polymer mean molecular weight.

Gel effect

Limited data is available concerning the gel effect of vinyl Acetate polymerization. The factor values of Polymer Plus are assumed :

gf1	0.44
gf2	6.36
gf3	0.17

Use of the model

The model described above is used for the simulation of vinyl acetate bulk polymerization but could be used also for any other polymerization with or without a solvent provided that the thermodynamic properties are available through UNIPHY.
The program run provides various information displays.
Main screen : pressure, temperature, temperature gradient and liquid/gas phase compositions as a function of time.
Secondary displays : kinetic descriptions, initial conditions, vessel characteristics, alarm conditions.

Application to the simulation of the bulk polymerization of vinyl acetate

The Software is used to simulate the bulk polymerization of vinyl acetate in the presence of 1.8 % dilauroylperoxide under isothermal temperatures of 37°C and 47°C in a VSP test cell.

The test conditions are the following :
   Test cell volume :   115 cm³
   Charge :   81.7 g
   Dilauroylperoxide : 1.8 % w/w

Initial pressure : 1 bar abs.

The following kinetic parameters are used :

	k°	E (kJ/mole)
Initiator decomposition	$1.44 \cdot 10^{15}$	127.4
Thermal generation of radicals	0	0
Propagation	$3.2 \cdot 10^{7}$	26.2
Transfer to monomer	0	0
Termination by combination	0	0
Disproportionation	$3.7 \cdot 10^{9}$	13.3

Initiator efficiency factor : f = 0.6
Enthalpy of polymerization Δhr = 88 kJ/mole.

A first simulation was made without gel effect.
A progressive polymerization was obtained which did not represent the runaway polymerization obtained in the VSP, because the rate of polymerization decreased as a function of time. If a gel effect is introduced using the Polymer Plus parameters an induction period phenomenon is obtained. For an isothermal exposure at 47°C, an isothermal induction period of 90 min. is obtained. The gel effect is found to occur for a conversion ratio of approximately 50 %. By tuning the kinetic parameters and the heat losses from the test cell, the simulation can be adjusted to represent the experimental data.

Conclusion

The prediction of the runaway polymerization of vinyl acetate based on simulation using a kinetic model and kinetic data from the literature provides an estimate of the polymerization isothermal induction period.
In the model used, the only cause of chemical acceleration is the gel effect obtained for a conversion ratio of 50 %. The radical chain reaction is not found to speed up the reaction under isothermal conditions and the presence of a polymerization inhibitor is not taken into account. Much effort will be devoted to better define the initial and boundary conditions to represent experimental conditions.
The simulation is specific for a premix vessel containing a high concentration of polymerization initiator.
For storage vessels containing monomer, the polymerization induction period near ambient temperature is controlled by the inhibitor consumption as explained in this paper.
The authors hope that this work will be useful to Process Safety in polymerization processes. We thank our colleagues from the Process Safety Laboratory for the experimental work and the Rhône-Poulenc Company for allowing this research to be done.

Literature

(1) Leon B. Levy, "Inhibitor-oxygen interactions in Acrylic acid stabilization". Plant/Operations Progress, Vol. 6, n° 4, 188-189, October 1987.

(2) "Les mélanges explosifs", INRS, 1980.

(3) A. Nicholson, "The effect of O_2 concentration on Methacrylic Acid Stability", Plant/Operations Progress, Vol. 10, n° 3, 171-183, July 1991.

(4) Leon B. Levy, "The effect of oxygen on vinyl acetate and Acrylic monomer stabilization". Process Safety Progress, Vol. 12, n° 1, 47-48, January 1993.

(5) Leon B. Levy, L. Hinojosa, "Effect of oxygen on vinyl acetate polymerization", J. Applied Polymer Science, Vol. 45, 1537-1544, 1992.

(6) J. J. Kurland and D. R. Bryant, "Shipboard Polymerization of Acrylic Acid", Plant/Operations Progress, Vol. 6, n° 4, 203-207, October 1987.

(7) Encyclopedia of polymer science and Engineering 2nd Edition, Vol. 17, 504.

(8) H. U. Moritz, "Reaktionskalorimetrie und Sicherheitstechnische Aspekte von Polyreaktionen" in "Sichere Handhabung chemischer Reaktionen", Praxis der Sicherheitstechnik, Vol. 3, 115-173, Dechema, 1995.

(9) Ham G.E., "Kinetics and Mechanisms of Polymerisations", Marcel Dekker Inc. New York, 1967.

(10) J.L. Gustin , "Thermal stability screening and Reaction calorimetry. Application to Runaway Reaction Hazard assessment and Process Safety management". J. Loss Prev. Process Ind., Vol. 6, n° 5, 275-291, 1993.

(11) J. Brandrupt, E. H. Immergut, Polymer Handbook 3rd ed. II - p. 371, John Wiley editor.

(12) J.L. Gustin, "Calorimetry for Emergency Relief System design" - "Safety of Chemical batch reactors and storage tanks". Eurocourse on reliability and risk analysis, Vol. 1, 311-354, 1991. Kluwer Academic Publishers, Dordrecht, The Netherlands, for the Commission of the European Communities.

(13) Encyclopedia of polymer science and Engineering 2nd edition, Vol. 17, p. 394.

(14) A. Korbar, T. Malavasic, "Influence of different initiators on methyl methacrylate polymerization, studied by differential scanning calorimetry", J. of Thermal Analysis, Vol. 44, p. 1357-1365, 1995.

(15) H. Warson, Per-compounds and per-salts in polymer processes, Solihull Chemical Services, 1980.

(16) Encyclopedia of polymer science and Engineering, Vol. 11, p. 4.

(17) P.J. Flory, Principles of polymer science, Cornell University Press, 1958.

A METHODOLOGICAL APPROACH TO PROCESS INTENSIFICATION

M.Wood and A.Green
BHR Group Limited, The Fluid Engineering Centre, Cranfield, Bedfordshire, MK43 OAJ

> Process Intensification (PI) is a design philosophy where process plant is designed to match the fundamental requirements of the chemical process and meet business needs. The benefits of applying PI include smaller, inherently safe plant; reduced energy requirement; improved product quality; lower capital cost. This paper describes a methodology that assesses the feasibility for applying PI to a chemical process. Application of the methodology is demonstrated on the design of a continuous, intensified reactor to replace a semi-batch stirred tank reactor. The resulting conceptual PI plant has an inventory three orders of magnitude smaller, eliminates runaway potential, and provides significant economic benefits.
>
> Keywords: Process intensification, static mixer, compact heat exchanger, continuous process

INTRODUCTION

It is likely that the chemical plant of the future will be far smaller than that of today (1). This can be achieved only by a step change in the plant technology used, rather than incremental improvements of existing plant items. The philosophy of size reduction has been in existence for several years under the name of Process Intensification (PI). Smaller equipment can result in reduced capital cost and reduced operating costs, whilst giving improved product quality. Just as important, according to Kletz (2), is that smaller often means inherently safer. Despite these benefits, uptake of PI appears to be low. There are many possible reasons for this. Standard process design and development has stressed the use of batch reactors (3), often with limited available knowledge of reaction kinetics. Lack of awareness of novel technology has to be overcome, from new graduates right through to top level management. Conservatism within the chemical industry may also result in unwillingness to take the risk with novel technology. The challenge of increasing the use of PI lies in promoting a different approach to process development, which should assist in overcoming these barriers.

Current procedures for applying PI technology also need to be considered, as this can tend to be done with an equipment driven approach. Organisations that have developed novel technology will look for applications where a chemical process can be run in their particular equipment. The equipment driven approach can be summed up through the opinion that PI is currently a solution looking for a problem. This situation needs to be reversed so problems look to PI for solutions, known as the process driven approach. Equipment should be chosen to match the process and allow it to run at its optimal rate, resulting in the consideration of a range of intensified equipment where normally only conventional plant would be used. The methodology set out in this paper uses a process driven approach to assess the feasibility for applying PI. It should be stressed that this methodology is not about forcing PI upon situations where it is not really required, but it aims to find the best solution for running a process. Improved understanding of the process as a result of the methodological approach can lead to benefits even if it is shown that full PI is not feasible.

PROCESS INTENSIFICATION (PI)

PI has been categorised as follows by Hannon and King (4):
a) Equipment - reducing the size of a unit operation. Full PI uses novel equipment to reduce size by 2-3 orders of magnitude, improving the process safety. Intensification can also apply to reducing the size of a conventional unit through more efficient operation.
b) Physical - combining two or more operations in one unit. Examples include pumps as mixers and reactive distillation. Compact reactor-heat exchangers, described by Edge *et al* (5), are an example of both equipment and physical intensification.
c) Chemical - improving the reaction scheme. Using different reagents or catalysts can improve yield or speed reactions up. Fast reactions are preferable for PI as they require shorter residence times and lead to smaller equipment.
d) Plant - Size reduction of the entire plant and integration of utilities to save energy and space.

PI can be applied across the whole flowsheet, but for the purpose of this work the focus is on the reactor. Any changes or improvements made here will affect the entire plant. The reacting inventory is often the most dangerous on the plant, as shown by Barton and Nolan (6) in a study of thermal runaway incidents. Reducing this inventory through the use of PI would be a major aspect in improving the safety of the process. Although PI has many benefits, there can be some potential drawbacks, such as lack of flexibility. PI equipment usually has to be tailored to a particular reaction scheme, whereas stirred tanks can run a number of chemical process, increasing plant occupancy and hence perceived value for money. To improve flexibility, a standard framework of feed pipes can be envisaged with interchangeable intensified reactor units to suit different reaction schemes. Not every reaction scheme can be intensified.

The 'S' curve (fig.1) is used to demonstrate how process performance is linked to plant performance. Factors under consideration for plant performance might be the mixing or mass transfer rate or heat transfer capability, while process performance might be yield of desired product, energy efficiency or product quality. Plant performance can be illustrated through the mixing sensitivity of some reaction schemes. For reactions to run at their inherent kinetic rate, the mixing has to be faster than the rate of reaction. If this is not the case, the reaction will be running slower than is theoretically possible, increasing residence time and reducing the process performance as by-products have more opportunity to form. Ideally operation would be close to the top of the S-Curve without moving too far to the right, which would entail excessive costs. PI can be the only means of improving plant performance enough to move up the S-Curve. The interaction between PI and chemistry is also shown. Improving chemistry (for example with a more selective catalyst) can push performance up to a higher S-Curve, but benefits will be lost if operation is lower down this S-Curve.

Removing reactor mechanical limitations to allow reactions to run at their inherent kinetic rate can be achieved by utilising a range of PI technology (7). In-line devices such as static mixers, ejectors (fig.2) and rotor stator mixers have proved to be effective as mixers and reactors, with good plug flow characteristics and mixing intensities up to three orders of magnitude greater than stirred tanks. Exploiting intensified force fields is another approach to intensification. Ramshaw (8) has shown that centrifugal fields can be used for separations, reactions, heat and mass transfer. The centrifugal field within a rotating disk reactor (fig.3) creates thinner, unstable liquid films, improving mass and heat transfer. Ultrasonic, electrostatic and magnetic fields can also be used for process intensification (9).

EXISTING PROCESS DEVELOPMENT AND SAFETY METHODOLOGIES

Hazop (10) is a well established safety methodology which is applied once the plant design is reasonably detailed, giving limited opportunity to intensify or redesign the plant for inherent safety. To gain the maximum benefits, it is necessary to consider safety and PI as early as possible in process development. This requires engineers being involved with development chemists to ensure the right chemical characteristics are being looked for. Several methodologies have been published exploring the inherent safety of a chemical process route (11, 12, 13). These include options to consider novel, intensified technology as a means of achieving inherent safety, though it will be necessary to follow a dedicated PI methodology to determine what this intensified plant might look like.

There can be apparent conflicts between PI and inherent safety methodologies, particularly for fast reactions which are most favourable for reactor PI. Slow reactions are preferred in conventional stirred tanks, particularly for exothermic reactions where the rate of heat generation will be limited. This enables the relatively poor heat removal capability of the stirred tank to cope. Fast, exothermic reactions could be considered as less inherently safe, or even completely undesirable from a conventional plant point of view. Hence, both the chemistry and plant need to be considered together to get a full grasp of inherent safety, as intensified plant can open up new, safe operating windows.

THE PI METHODOLOGY

The methodology sets out structured procedures to follow for considering PI during process development. The overall methodology, known as the framework, consists of a number of protocols detailing the information needed to ensure the potential for PI is fully examined. Figure 4 shows the framework which is formatted to apply to situations where an existing chemical plant is to be replaced or upgraded. Each of the methodological steps is described below.

a) Business Drivers

Determine why it is desirable to change the plant. This step is phrased 'business drivers' as these reasons are normally of an economic nature. Safety, health and environmental concerns are increasingly becoming important factors. Even so, these relate back to business issues as it is preferable to achieve these requirements in the most cost effective manner, or by ensuring costly incidents do not occur. Another major business driver may be to have a higher and more efficient production rate. These drivers are required to set targets for the plant design to meet.

b) Knowledge Elicitation

An understanding of the whole process is required which is gained through the knowledge elicitation stage. The approach is split into separate chemistry and plant audits, though there will be interaction between the two.

The Chemistry Audit examines the whole reaction scheme. The potential to use different solvents, catalysts or operating conditions should be considered. Ideal operating conditions and those conditions that promote byproduct formation should be determined, such as temperature of operation or residence time. Check if the chemical reaction rate is inhibited in any way. Some knowledge of the kinetics and thermodynamics of the reaction is essential.

The Plant Audit examines what the existing plant currently does. The audit should include all physical aspects of the reactor, including mixing and heat transfer capabilities, feed rate and position of feed addition. It is necessary to have a fundamental understanding of the reactor to determine where and how the reaction occurs. If the intention is to run a new chemical reaction scheme in existing equipment, as is the case in many fine and speciality chemicals processes, the equipment should be audited as if it were already running the new process.

c) Examine PI Blockers

Blockers are those properties or conditions of a process which may prevent the application of PI. Many are process blockers to do with the nature of the chemicals themselves, such as the presence of solids. PI equipment often has narrow channels, which large solids would not pass through. Fine solids can be handled. There may be some business blockers which relate to practical problems of running PI plants, such as flexibility or continuous operation versus batch production. Batch production is preferred in some sectors of the chemicals industry, such as pharmaceutical manufacture where there is a requirement for batch identification. Consider whether any identified blockers can be prevented or worked around.

d) Identify Rate Limiting Steps

Rate limiting steps are conditions preventing the overall process running at a faster rate. These may be mechanical limitations such as low heat transfer area, poor mixing or limited supply of feedstock to the reactor from an upstream operation. Chemical rate limiting steps, for example slow kinetics or mass transfer into a solid reactant, may occur. Rate limiting steps and blockers are considered in parallel as there can be common elements, such as slow reactions which are both a PI blocker and a rate limiting step. PI should aim to remove or improve rate limiting steps.

e) Assess PI Viability

The potential for intensifying a process is determined by pulling together the results of the audits, blockers and rate limiting steps into a mid-methodology assessment. This will ensure all the required information has been gathered and properly considered. Even if it is determined that full PI is not possible, it is worth continuing with the methodology as improvements to the conventional plant could be found that partially intensify it.

f) Drivers

Business and process drivers are required to set targets for the plant design to meet. The business drivers identified at the start of the methodology, which are the economic reasons why it is desirable to intensify the process, should be reviewed to keep a clear idea of the overall aims of the project. Process drivers are those characteristics of the chemical reaction scheme that determine the required operating conditions within, and performance of, reactor equipment to allow the process to run at its most efficient rate. A process driver example is the rate of heat release from a reaction determining the heat transfer capability required of the equipment.

g) Initial Concepts

Throughout the methodology, ideas or concepts will occur on how to intensify the process, which will tend to be equipment driven concepts for applying familiar equipment. These ideas should be documented for discussion in the proper manner at the appropriate methodological stage. Accepting an initial concept early on could introduce bias into the rest of the methodology, preventing further, possibly superior, plant concepts being suggested.

h) Generate Design Concepts

A creative problem solving session should be held in which plant concepts are suggested for meeting the process and business drivers. Include the initial concepts in this session. A database of available PI equipment and their capabilities would be useful here so that no possibilities are overlooked, but concepts should not be restricted to plant items already known about. The success of the concepts generation stage depends on thinking laterally to come up with possibly novel solutions to a problem.

i) Select Best Concept

All the concepts suggested must be analysed to study how each of them matches the business and process drivers. There may be factors which limit or rule out the use of a particular piece of equipment, such as it not being available in the required material of construction for corrosion resistance purposes. The best concept must now be chosen. Some economic analysis may be required if there is more than one feasible choice.

j) Laboratory Scale PI Protocols

It will be necessary to prove that the selected concept will work with actual process chemicals. PI laboratory protocols are being designed to demonstrate the performance of continuous, intensified operation without the need for a pilot plant. This will allow the quantification of any potential benefits of intensification, such as improvement in product quality, shorter reactor residence time and lower reacting inventory.

k) Compare With Conventional Plant

List the strong and weak points of the existing and conceptual plant. Showing that the conventional plant is not fully suitable for a process, due to mechanical rate limiting features, could be just as important as showing the benefits achievable by PI when trying to justify its use.

l) Final Choice of Plant

The person or team responsible for making the ultimate choice of plant equipment should have an open mind to the use of PI. This final decision process involves factors currently outside the scope of this methodology, such as the risk of using novel equipment, legislation and lead time to commissioning of plant. A high risk factor and long lead time to commissioning may rule out the use of PI, even if significant financial and operability benefits have been shown to exist.

PI CASE STUDY

The methodological approach will now be illustrated by a feasibility study recently carried out on a fine chemicals nitration process, which generated an intensified plant concept. The process has a multiple sequence of additions of which only the nitration step was initially considered for intensification. It soon became obvious that the whole process could be intensified.

a) Business Drivers

A runaway reaction and explosion occurred in another stirred tank reactor on the production site, emphasising the need for safer equipment. Production needs to be increased. The plant should be relatively cheap to build. Some knowledge of PI does exist within the company and there is a general feeling that continuous, intensified operation is the way forward.

b) Knowledge Elicitation

Chemistry Audit. All of the process steps consist of blending, reaction and heat transfer operations. Reactions in every step are almost instantaneous and some are very exothermic. All reactants are single phase liquid. Solids can exist in the initial stages, though controlling temperature prevents solids formation. The last process stages involve crystallisation, but crystal sizes are small. Byproduct formation for the nitration step at full scale operation (taking 18 hours) is far higher than that in laboratory production tests (taking 4 hours). This shows a PI plant with short residence time could significantly reduce byproduct formation.

Plant Audit. A 13,000 litre glass lined stirred tank (fig.5a) with cooling jacket and coil is currently used. The large reacting inventory is a major safety concern. For the exothermic nitration step, reactant feed is literally dribbled into the reactor over a period of 18 hours to allow the removal of all the heat of reaction. If feed rate was increased for any reason, there is large potential for runaway reaction conditions to occur. Total batch time for all reaction stages is 30 hours. Low heat and mass transport from the reaction zone at the feed pipe exit could promote byproduct formation.

c) PI Blockers

No particular PI blockers exist. Any solid formation can be controlled. Corrosiveness may become an important issue as a glass-lined stirred tank reactor is currently used to resist the operating conditions. Manufacturing intensified equipment in corrosion resistant materials will increase the cost several times over, but the equipment will still be relatively cheap due to its small size.

d) Rate Limiting Steps

As reaction kinetics are fast, the rate limiting steps are all mechanical. Poor mixing in the stirred tank, which restricts heat transport from the reaction zone, then low heat transfer from the vessel combine to cause the very long feed addition and batch time.

e) Assess PI Viability

The process is suitable for PI due to the lack of blockers and fast, single phase liquid reactions.

f) Drivers

Business drivers are improved safety and productivity at low capital cost. Process Drivers are fast kinetics and high heat release, meaning a plant has to deliver intensive mixing and heat transfer.

g) Initial Concepts

Concepts suggested during the project were based upon previous experience, using an equipment driven approach. These included a heat exchanger loop on the existing reactor, which would improve heat removal and reduce batch time, and a compact reactor-heat exchanger.

h) Generate Design Concepts

For the nitration reaction it is desirable to rapidly mix the reactants and then remove the heat as quickly as possible. From these process drivers, a number of concepts were generated in addition to the initial concepts. These possibilities include utilization of existing PI equipment and some more novel solutions involving new arrangements of existing equipment.

i) Select Best Concept

The concept eventually chosen to achieve the drivers is a static mixer followed immediately by a plate and frame heat exchanger (fig.5b). The reaction will take place in the static mixer, with the adiabatic temperature rise limited to an acceptable level by the presence of inert components from upstream stages. Byproducts formation should be significantly reduced due to the short residence time within the reactor. A similar concept is used for the other reaction stages. There are some novel features of the overall plant design that would not have resulted from an equipment driven approach.

j) Laboratory Scale PI Protocols

A requirement before this project can move into the detailed design phase is demonstration of continuous operation as proof of concept. Experimental procedures have been devised to do this.

k) Compare with Conventional Plant

Figure 5 is an approximate scale drawing of the existing reactor and the conceptual intensified nitration reactor, demonstrating the immense size difference. The PI plant will consist of five reactors, but even so, total inventory is three orders of magnitude smaller than the existing stirred tank. Although a full comparison with the conventional plant cannot be completed until the PI protocols are done, a preliminary economic comparison has been made. The product quality achievable, which would be determined by the PI protocols, is important as it could remove the need for a downstream purification stage with all its associated costs. Major points for comparison are:

	Current	PI	Comments
Production	15 tpa	50 tpa	Increased annual sales value of £2 million, based on continuous operation for two weeks per quarter.
Reacting Inventory	13,000 litres	0.2 litres	Full PI plant inventory (including inter-reactor piping) is approximately 15 litres.
Heat transfer	Poor	Good	Current process feed addition is limited by the poor heat transfer. PI reactor runs stoichiometrically.
Operating safety	High runaway potential	Minimal runaway potential	Runaway conditions should not occur in PI reactor, even if cooling fails, as it is designed to operate adiabatically with cooling after each reaction stage.
Capital cost	£100,000 for new reactor	£40,000 for plant	Cost of control system and other associated items will be evaluated in the next design stage.
Plant layout			PI reactors could literally be bolted to a wall and not require building space as the stirred tank does.
Nitration time	18 hours	0.25s for reaction	Total PI nitration time for reacting and cooling is 3 seconds.
Residence time	30 hours batch time	1 minute	Substantially shorter overall residence time limits the opportunity for byproduct formation

| Complexity | 4 plant elements | 11 plant elements | PI plant is more complex with six static mixers and five heat exchangers required to replace the vessel, impeller, cooling coil and cooling jacket. |

Other points under consideration include the filtration stage at the end of the process. Currently this is done batchwise. In order to get the maximum benefit out of the PI process, filtration should be continuous. The cost of installing a continuous filtration system will be examined at the next stage of this project. The alternative is using holding tanks to store product until there is enough to operate the batch filtration step. This would still allow the benefits of improved product quality and safer operation of the PI plant to be achieved. Manual intervention and labour required on the PI plant will be greatly reduced compared to the existing plant.

l) Final Choice of Equipment

The company is reviewing market demand for the product and looking into how the plant can be made in such a way that it can be reconfigured for other products, before deciding whether to replace the existing plant or not. PI laboratory protocols will be followed to fully determine the benefits the PI plant would produce before proceeding onto the production of a more detailed design.

CASE STUDY SUMMARY

Application of the PI methodology has been demonstrated on the conceptual design of an intensified plant for a nitration process, with the methodology acting as a checklist to ensure no important aspects were overlooked. The reasons why the existing stirred tank reactor has a long operation time and high byproducts yield have been identified, showing substantial improvements can be made. Following the methodology generated a PI concept with novel aspects that would not have resulted from an equipment driven approach. Comparison and selection procedures have yet to be completed.

CONCLUSIONS

The PI methodology presented in this paper operates as a decision route for assessing the feasibility for intensifying a chemical process. Consequences of applying PI include smaller, inherently safer plant that is cheaper to build and operate. The methodology is tailored for application to existing chemical processes, though it can also be applied to completely new processes. The methodology is not about forcing PI upon situations, but choosing the best possible plant design to achieve the business targets. Ultimately, integration of this PI methodology with inherent safety methodologies has the potential to produce large financial and safety benefits through enabling effective use of PI.

The case study applies the methodology to an existing fine chemicals process, showing there to be substantial benefits achievable through the adoption of PI. The conceptual PI plant has a reacting inventory five orders of magnitude smaller, and total inventory three orders of magnitude smaller than the existing reactor. Capital cost is less than half the price of a new glass-lined stirred tank reactor. Benefits to the company of applying PI will involve safer process operation, improved product quality and increased productivity. A successful application of this individual plant design would give impetus to modernising the whole site, making it a cleaner, safer and more efficient place.

Future work will focus on the laboratory protocols section of the methodology. This involves further development of experimental equipment and procedures to demonstrate intensified, continuous operation. This is a vital part in proving the success of a PI concept and will allow determination of the benefits achievable, without the need for building a continuous pilot plant.

Awareness of PI still has to be raised in some sectors of the chemical industry, though there are signs that many firms are looking towards innovation as a means of gaining a competitive edge and meeting legislation. A change in the way process development is traditionally done will be required for innovation to be properly adopted. This PI methodology provides a mechanism to promote such a change by encouraging PI to be considered where it may normally be overlooked.

REFERENCES

1. IChemE, 1998, Future Life, The Chemical Engineer, Issue 653: 16-17.
2. Kletz T., 1985, Make Plants Inherently Safe, Hydrocarbon Processing, Sept 1985: 172-180
3. Borland J., 1996, The Effect of Process Development on Manufacturing Economics, in 'Batch Processing III, Professional Process Development & Design', IChemE Symposium Papers No.5, 4th December: 1.1-1.3
4. Hannon J. and King R., Increased Yield and Energy Efficiency through Process Intensification, BHR Group Limited, unpublished.
5. Edge A., Pearce I. and Phillips C., 1997, Compact Heat Exchangers as Reactors for Process Intensification (PI), in 'Process Intensification in Practice' (Ed.J.Semel), Proceedings of the Second International Conference on Process Intensification, Antwerp, Oct 1997: 175-189
6. Barton J. and Nolan P., 1989, Incidents in the Chemical Industry due to Thermal-runaway Reactions, in 'Hazards X: Process Safety in Fine and Speciality Chemical Plants', IChemE Symp Ser No.115: 3-18
7. Green A., 1998, Process Intensification: The Key to Survival in Global Markets?, Chemistry & Industry, 2 March: 168-172
8. Ramshaw C., 1993, The Opportunities for Exploiting Centrifugal Fields, Heat Recovery Systems & CHP, 13 no.6; 493-513
9. Akay G., Mackley M. and Ramshaw C., 1997 Process Intensification: Opportunities for Process and Product Innovation, IChemE Jubilee Research Event: 597-606.
10. Kletz T., 1983, Hazop and Hazan - Notes on the Identification and Assessment of Hazards, IChemE
11. Edwards D. and Lawrence D., 1993, Assessing the Inherent Safety of Chemical Process Routes: Is There a Relation Between Plant Costs and Inherent Safety? Trans IChemE, 71 Part B: 252-258
12. Schabel J., 1997, The INSET Toolkit Stages I and II - Route Selection and Optimisation in 'Inherent SHE - The Cost-Effective Route to Improved Safety, Health and Environmental Performance', London, 16-17 June
13. Mansfield D., 1997, INSET Toolkit Stages III and IV - Process Front End and Detailed Design, in 'Inherent SHE - The cost-effective route to improved Safety, Health and Environmental Performance', London, 16-17 June

ACKNOWLEDGEMENTS

This research is part of the Postgraduate Training Partnership (PTP) Scheme between BHR Group and Cranfield University. The Scheme is managed by the Teaching Company Directorate and sponsored by the Department of Trade and Industry and the Engineering and Physical Sciences Research Council.

Figure 1. The S-Curve of Process Performance

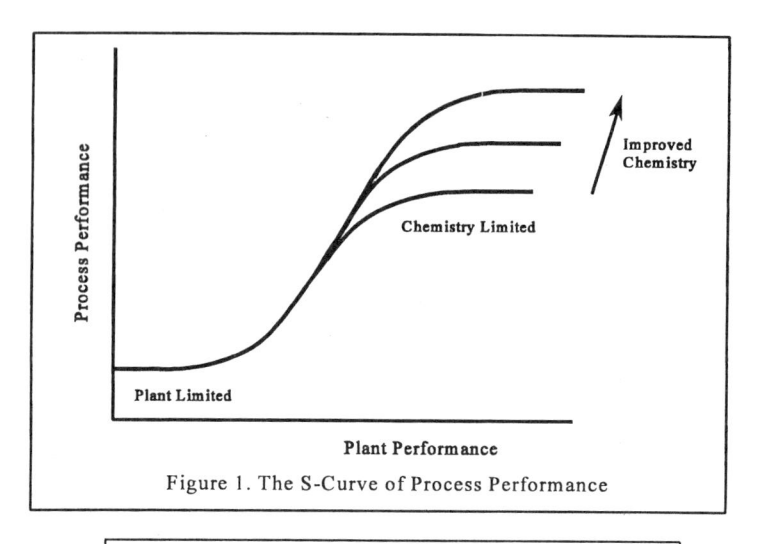

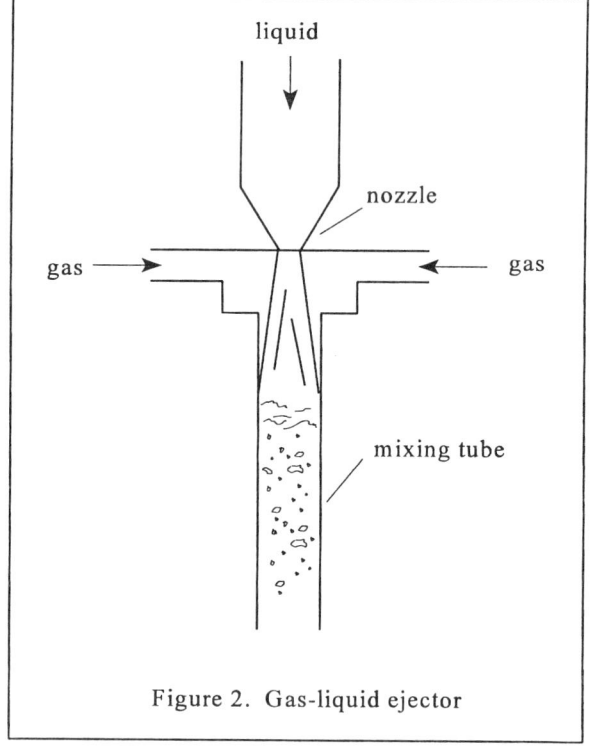

Figure 2. Gas-liquid ejector

Figure 3. Spinning Disk Reactor

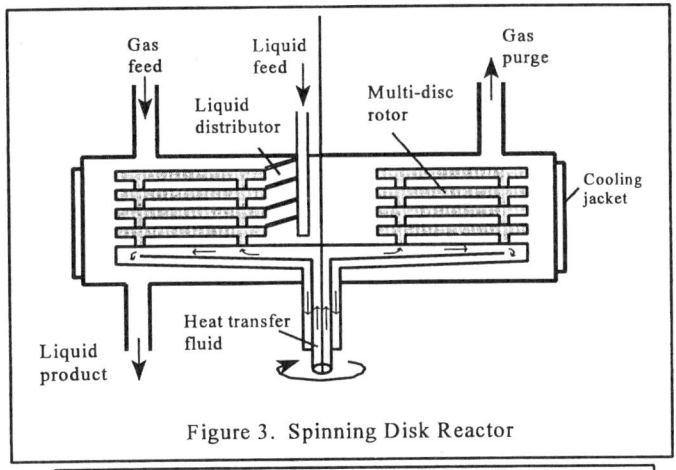

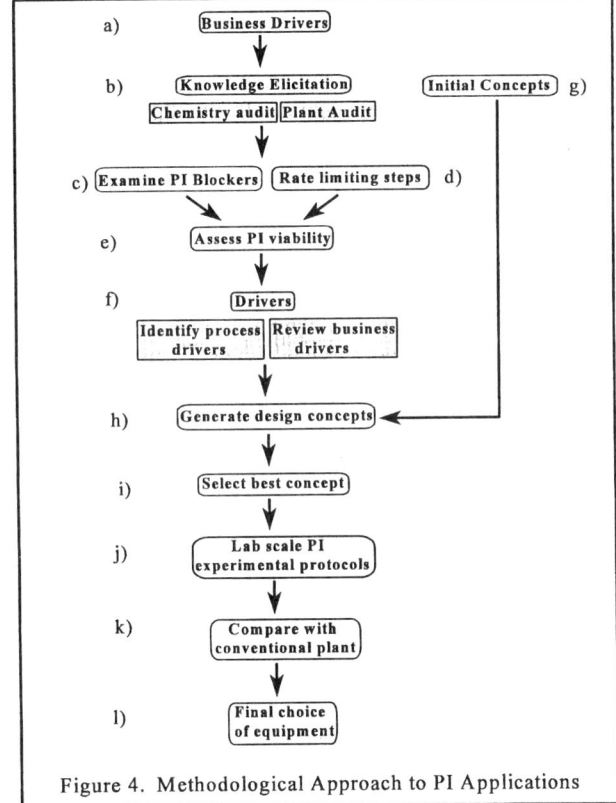

Figure 4. Methodological Approach to PI Applications

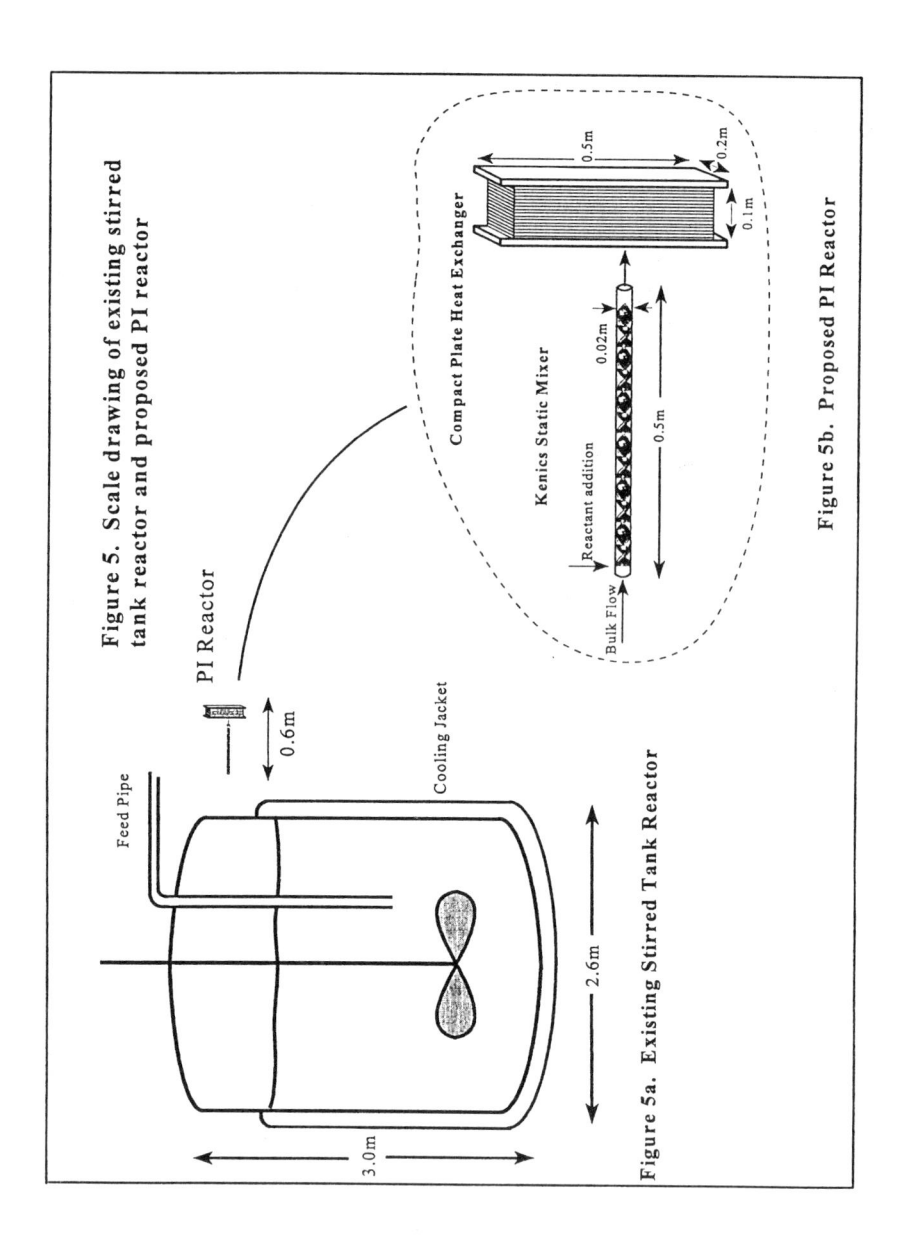

Figure 5. Scale drawing of existing stirred tank reactor and proposed PI reactor

Figure 5a. Existing Stirred Tank Reactor

Figure 5b. Proposed PI Reactor

CRITERIA FOR AUTOIGNITION OF COMBUSTIBLE FLUIDS IN INSULATION MATERIALS

J. Brindley[3], J F. Griffiths[1], N Hafiz[2], A C. McIntosh[2] and J Zhang[1]
School of Chemistry[1], Department of Fuel and Energy[2] and School of Mathematics[3],
The University, Leeds LS2 9JT, UK

> Criteria have been investigated for the conditions at which a "lagging fire" may occur when a flammable liquid penetrates and is dispersed within insulation material surrounding a hot pipe. The conditions at which a Frank-Kamenetskii thermal ignition criterion should be replaced by one derived from the heat release rate versus fluid evaporation rate were deduced. These conditions were related to the decreasing enthalpy of vaporisation of the fluid. Practical investigations were based on formal "cube test" methods for thermal ignition. The theory was tested against the behaviour of n-$C_{16}H_{34}$, n-$C_{18}H_{38}$ and n-$C_{20}H_{42}$, which represent alkanes of mid-range volatility, and also with reference to squalane ($C_{30}H_{62}$), which is representative of highly involatile alkanes.
>
> Keywords: lagging fires, spontaneous ignition, ignition criteria

INTRODUCTION

The practical motivation for this paper relates to an understanding and interpretation of the conditions at which spontaneous ignition may occur when a flammable liquid penetrates an insulation material surrounding a hot pipe, the so-called "lagging fire". The problem has received relatively little attention, either through theory or through systematic experimental investigation (Lindner and Seibring (1), Bowes and Langford (2), Gugan (3), Bowes (4), Britten (5), McIntosh et al (6)). The basic mechanism for the combustion instability that leads to a lagging fire is similar to that for thermal ignition (Frank-Kamenetskii (7), Gray and Lee (8)), namely encompassing the imbalance between the heat release rate with its dependence on temperature and the heat transport rate within the system or away from its surface by conduction and convection. Supplementary to these are the consequences of movement of the liquid or vapour that permeate the insulation. The fluid may become dispersed over an extremely high internal surface area, and ignition then may result from the exothermic oxidation of the fluid in contact with air within the hot, porous structure. For most processes oxygen is essential for exothermic reaction to occur, and its diffusion or limitation of access, as a result of displacement from the voids of the insulation as the liquid vaporises, may also play a part.

The present study is a combination of experimental measurement and theoretical interpretation. The experiments were designed to distinguish between the behaviour of relatively involatile and highly volatile hydrocarbons when they are dispersed in hot microporous insulation. The practical problem, which relates mainly to asymmetric heating of lagging material which surrounds hot pipe and is exposed to ambient temperature on its external surface, is complicated to interpret experimentally and theoretically (Thomas and Bowes, (9)), and it is important to address these conditions in further developments. The present approach builds on the normal procedures involving "cube tests" in a uniformly heated oven, which are formally adopted in EEC and ASTM tests to determine critical ignition criteria (e.g EC test L2951/ 86, auto-flammability).

Typically, insulation materials are made from glassfibre, mineral wool or amorphous silicate compounds. These materials have different thermal insulation properties and densities, and may present significantly different surface/volume ratios within their structure. All of these factors may have important implications for fire risk. The present experiments and modelling were based on microporous, silicate-based insulation which has an open cellular structure with average cell dimension of less than 0.1 μm.

Since many of the liquids that are susceptible to problems of combustion in hot lagging are hydrocarbon based (such as heat transfer fluids or diesel fuel), the alkanes, n-hexadecane (n-$C_{16}H_{34}$, B Pt 560 K, 287 °C), n-octadecane (n-$C_{18}H_{38}$, B Pt 590 K, 317 °C) and n-eicosane (n-$C_{20}H_{42}$, B Pt 616 K, 343 °C) were investigated as representative volatile hydrocarbons over the temperature range 450 - 500 K, (177 - 227 °C). By contrast, 2,6,10,15,19,23-hexamethyl-tetracosane (squalane, $C_{30}H_{62}$, B Pt 696 K, 423 °C), was studied as representative of the highly involatile liquid alkanes. The relevance of these choices is shown in a chromatogram obtained by capillary column gas chromatography of diesel fuel doped with squalane (Fig. 1).

The purpose of Figure 1 is to demonstrate the boiling range of the components of a typical diesel fuel, from about 470 K (~ 200°C) for the C_{11} alkanes to about 670 K (~ 400°C) for the C_{24} alkanes. Alkanes are not the only components of diesel fuel, but they are usually predominant. The alkanes are conspicuous as a regular series throughout the chromatogram. Markers are inserted at C_{17} and C_{18}, which refer to the respective isoprenoids appearing as the second peak of each doublet. The carbon number of each of the other n-alkanes can be deduced from these markers. The abscissa is a measure of the retention time on the chromatography column and, since a non-selective capillary column (BP1, Scientific Glass Eng.) was used for the separation, the boiling point of each component is directly proportional to the retention time. Squalane ($C_{30}H_{62}$) is retained longer on the column than the diesel fuel components, indicating that it has a higher boiling point, which was determined experimentally from the retention time (see below).

The reactivity of the alkanes investigated was regarded to be sufficiently similar that differences in their combustion characteristics could be attributed to their respective volatilities. To quantify the behaviour we have determined the critical oven temperature at which thermal runaway is able to occur when each of these hydrocarbons is present in a 5 cm cube of the insulation, thermostatted in an oven. The critical temperature is governed by the respective enthalpy of vaporisation (ΔH_v) and, by implication, the normal boiling point.

The theoretical foundation presented here is derived from thermal ignition theory. The classical, Frank-Kamenetskii criterion for thermal ignition, which is related to the imbalance of the heat release and heat loss rates by thermal conduction, represents one extreme of behaviour. It would appear that, to date, virtually all experimental and theoretical studies of the lagging fire problem have been confined to this limit (Brindley et al (10). However, as the volatility of the fluid becomes greater, a criterion based on the imbalance between the rates of heat release and evaporation of the fluid becomes more appropriate. The foundations for this new criterion, the transition from the classical thermal ignition to the evaporative criterion, and the relationship between theory and experiment are the subjects of the present paper.

THEORETICAL MODEL AND SAFETY CRITERIA

In this section we develop the theoretical background and the limiting criteria for safe operation, and we discuss the magnitudes of the parameters that may be used to compare theory with experiment. The theoretical model has been cast in such a way as to retain analytical tractability. This means that the representation of certain features which would be appropriate from a both physical and chemical points of view, has to be restricted to simplified forms (Brindley et al (11)). For example, the overall chemistry is assumed to be a first order exothermic reaction, as in thermal ignition theory (Bowes (2), Frank-Kamenetskii (7), Gray and Lee (8)). However, there are concerns that it is too simplistic a representation of hydrocarbon combustion, even in a condensed phase (Brindley et al (11)). The implications are addressed in the Discussion.

Model and Assumptions

The analytical treatment refers to the geometries which may be represented by a single characteristic dimension, namely the infinite cylinder, the infinite slab and the sphere (Bowes (2), Frank-Kamenetskii (7), Gray and Lee (8)). These are represented respectively in the terms that define thermal and mass diffusion in the conservation equations by the shape factor $j = 0$, 1 and 2. The following assumptions are made.

(i) Reactant is adsorbed in the liquid state on the pore surface within the insulation matrix. It may react in this condensed state or it may evaporate, the vapour being assumed to be inert. There is no inhomogeneity of the surface sites.
(ii) Exothermic oxidation occurs by reaction between gaseous oxygen in the pores and the condensed reactant, with exothermicity Q. In order to link with classical thermal ignition theory, this is interpreted as a single step reaction which is first order with respect to the gas-phase oxygen concentration and the condensed fluid density. The rate constant has an exponential dependence on temperature, with activation energy E.
(iii) Vaporisation is related to the condensed fluid density, and its temperature dependence is controlled by the enthalpy of vaporisation (ΔH_v), expressed in an Arrhenius-like exponential form ($E_v = \Delta H_v$)
(iv) Conductive heat transfer (coefficient λ) and oxygen diffusion (coefficient D_o) are assumed to occur in accordance with Fourier's and Fick's laws respectively. Convective heat transport of gaseous components is ignored, as are pressure gradients within the matrix.
(v) There is no resistance to heat transport at the surface $B_i = \infty$).
(vi) A known amount of fluid is soaked uniformly within the block of insulation initially and there is air present throughout the porous structure, as measured previously (Brindley et al (11)).

The model variables, each expressed as a function of spatial coordinate r and time t, are the matrix temperature T(r,t), the molar density of condensed reactant X(r,t) and the concentration of oxygen Z(r,t). The respective conservation equations are thus

$$\rho c \frac{\partial T}{\partial t} = Q A X Z e^{\left(\frac{-E}{RT}\right)} - Q_V F X \exp\left(\frac{-E_v}{RT}\right) + \lambda \left(\frac{\partial^2 T}{\partial r^2} + \frac{j \partial T}{r \partial r}\right) \quad (1)$$

chemical heat release rate — endothermic vaporisation rate — thermal conduction rate

$$\frac{\partial X}{\partial t} = \underbrace{-AXZ\exp\left(\frac{-E}{RT}\right)}_{\text{chemical consumption rate}} - \underbrace{FX\exp\left(\frac{-E_V}{RT}\right)}_{\text{evaporative loss rate}} \qquad (2)$$

$$\frac{\partial Z}{\partial t} = \underbrace{-\nu AXZ\exp\left(\frac{-E}{RT}\right)}_{\text{oxygen consumption rate}} + \underbrace{D_0\left(\frac{\partial^2 Z}{\partial r^2} + \frac{j\partial Z}{r\partial r}\right)}_{\text{oxygen diffusion rate}} \qquad (3)$$

With relatively light loadings of fluids, the volumetric heat capacity (ρc) may be assumed constant and represented by that of the inert matrix. The oxygen consumption rate in eq. (3) includes a stoichiometric coefficient ν for the relative numbers of moles of oxygen consumed per mole of fuel reacted.

Non-dimensionalisation of Equations and Reduced Parameters

To facilitate analytical interpretations and to generalise the predicted behaviour, it is appropriate to non-dimensionalise equations (1) - (3). The characteristic time is based on the Fourier time t_F ($= l^2\rho c/\lambda$). Temperature is non-dimensionalised with respect to the temperature coefficient for the chemical reaction (E/R) and given as u (Gray and Wake (12)). By contrast to the more common Frank-Kamenetski parameter ($\theta = E\Delta T/RT_a^2$) (Frank-Kamenetskii (7), Gray and Lee (8)), or $n = T/T_a$ (Aris (13)) this means that T_a (or u_a) can be used as an independent control parameter to explore the properties of the system. The species are non-dimensionalised with respect to the initial molar densities of the uniformly distributed fluid (X_0) and the initial oxygen concentration within the insulation block (Z_0), which is taken to be the same as the external oxygen concentration. Thus the non-dimensionalised terms and reduced parameters are

$$\tau = \frac{t}{t_F}, \qquad u = \frac{RT}{E}, \qquad x = \frac{X}{X_0}, \qquad z = \frac{Z}{Z_0}, \qquad \beta = \frac{E_V}{E},$$

$$q = \frac{QX_0 R}{\rho c E}, \qquad q_V = \frac{\Delta H_V X_0 R}{\rho c E}, \qquad a_1 = AZ_0 t_F, \qquad a_2 = \frac{\nu X_0 a_1}{Z_0}, \qquad f = Ft_F.$$

The dimensionless exothermicity q is equal to the dimensionless adiabatic temperature excess following complete consumption of the reactant in the absence of any vaporisation. Similarly q_v is equal to the dimensionless temperature decrease owing to complete vaporisation of the reactant with no reaction. The Lewis number, which represents the ratio of oxygen diffusion to thermal diffusion, is given by

$$L_0 = \frac{D_0 \rho c}{\lambda}$$

Equations (1) - (3) may then be represented in the non-dimensionalised and reduced forms

$$\frac{\partial u}{\partial \tau} = qa_1 xz \exp\left(\frac{-1}{u}\right) - q_V fx \exp\left(\frac{-\beta}{u}\right) + \frac{\partial^2 u}{\partial \xi^2} + \frac{j\partial u}{\xi \partial \xi} \tag{4}$$

$$\frac{\partial x}{\partial \tau} = -a_1 xz \exp\left(\frac{-1}{u}\right) - fx \exp\left(\frac{-\beta}{u}\right) \tag{5}$$

$$\frac{\partial z}{\partial \tau} = -a_2 xzv \exp\left(\frac{-1}{u}\right) + L_0\left(\frac{\partial^2 z}{\partial \xi^2} + \frac{j\partial z}{\xi \partial \xi}\right) \tag{6}$$

Stationary State Solutions and Limiting Conditions

This system of equations must be reduced to one variable for the purpose of analytical interpretation. Following thermal ignition theory, it is necessary to assume that there is always sufficient oxygen and fuel present in subcritical conditions so that both $\partial x/\partial t$ and $\partial z/\partial t$ are assumed to be zero. The behaviour of the system is then governed solely by the energy conservation equation (4). We have shown elsewhere that $\partial z/\partial t = 0$ is a robust criterion but, in practice, $\partial x/\partial t = 0$ cannot always be sustained (Brindley et al (11)). Nevertheless, as shown later, the assumption $\partial x/\partial t = 0$ leads to a critical criterion that errs on the side of safe operation.

From eq (4) there are two limiting physical conditions for a lagging fire to develop. The first is that of a fluid which is sufficiently involatile at the critical temperature that its behaviour resembles thermal ignition under Frank-Kamenetskii conditions (Frank-Kamenetskii (7)). Most examples of lagging fires that have been studied so far in the laboratory appear to approach this extreme (Bowes and Langford (2), Britton (5)). This condition may be represented as a stationary state solution to eq (4) in the limit of no evaporation (McIntosh et al (14)). That is,

$$\frac{\partial u}{\partial \tau} = qa_1 xz \exp\left(\frac{-1}{u}\right) + \frac{\partial^2 u}{\partial \xi^2} + \frac{j\partial u}{\xi \partial \xi} \tag{7}$$

Invoking the Frank-Kamenetskii exponential approximation (Frank-Kamenetskii (7)) and assuming constant x and z, equation (7) may be solved in one dimension to give the classical solution

$$\delta_{cr} = \frac{qa_1 xz}{u_{FK}^2} \exp\left(\frac{-1}{u_{FK}}\right) = \frac{QAEX_0 Z_0 l^2}{\lambda RT_{FK}^2} \exp\left(\frac{-E}{RT_{FK}}\right) \tag{8}$$

The Frank-Kamenetskii parameter, δ_{cr}, takes the values 0.878, 2.00 and 3.32 at the critical dimensionless ambient temperature u_{FK} or, in dimensional form, T_{FK}, for the shape parameters j = 0, 1 and 2. An equivalent harmonic mean square radius (R_o) can be defined for any given stellate shape (Boddington et al (15)), which, for a cube, yields j = 3.28 and $\delta_{cr}(R_o) = 3.665$ at Bi = ∞.

The second limiting condition arises when the volatility of the fluid is sufficiently high that the propensity for thermal runaway is controlled entirely by an interaction between the chemical heat release rate and the endothermic evaporation rate (McIntosh et al (14)). This condition can also be interpreted from a stationary state solution of eq (4) involving only the first and second terms on the right hand side. Thus, if x and z are assumed constant, a dimensionless critical temperature (u_∞) can be derived, in the analytical form

$$u_\infty = \frac{1-\beta}{\ln(qa_1 z / q_v f)} \tag{9}$$

For $u < u_\infty$, the system is net endothermic and is not subject to any temperature rise. This represents one extreme for an inherently safe system. Hitherto, the relevance of evaporation as a controlling factor and the conditions at which it becomes important seem not to have been addressed in a formal way.

There is a switch from u_{FK} to u_∞ as the more appropriate criterion for safe operation as the volatility of the fluid increases, and this occurs at a dimensionless ambient temperature for the transition, u_a^*. Numerical predictions of the dependence of u_{FK}, u_∞ and u_a^* on the parameters of the combustion system are discussed elsewhere (Brindley *et al* (14)). However, the conditions that are appropriate for a switch from u_{FK} to u_∞ are given by a transcendental relationship between the principal quantities, derived as follows from an equality of equations (8) and (9):

$$\frac{(1-\beta)^2}{\ln\left(\frac{q^* a_1^* z^*}{q_v^* f^*}\right)} \frac{1}{(q_v^* f^*)^{(1/(1-\beta))}} \frac{1}{x} (q^* a_1^* z^*)^{(\beta/(1-\beta))} = \frac{1}{\delta_{cr}} \tag{10}$$

where * signifies the values taken at the crossing point.

Although eq. (10) may be modified to make a specific parameter the subject of the formula, the dimensionless enthalpy of vaporisation at which the transition occurs (q_v^*), which would be of greatest interest in the present work, cannot be expressed in an explicit analytical form. The major difficulty arises because β is a function of q_v^*. The dependence of u_∞ on q_v, and that of its fully dimensional counterpart ΔH_v, and the relationship to u_{FK} are discussed further below.

Thermochemical and Kinetic Parameters and Comparison with Experiment

Fluids that are encountered in lagging fires are often hydrocarbon based, so the -CH2- moiety may be regarded to be a representative part of the chemical structure. Its complete oxidation is given by

$$-CH_2- + 3/2\ O_2 = CO_2 + H_2O \tag{11}$$

Whilst the overall heat of combustion (Q) is 4.4×10^7 J kg^{-1} of the reactant, most of this energy is released in the late stage of reaction. This is preceded by many competitive and consecutive, free radical chain propagation and branching processes, which lead to complex mixtures of partially oxidised or degraded intermediates prior to the formation of the final products (Griffiths (16)). The onset of spontaneous ignition of hydrocarbons is known to evolve at low temperatures (T < 700 K, ~ 430 °C)) through organic peroxides and other partially oxygenated intermediates. Thus the exothermicity associated with the preliminary stage should be used to represent Q in determining the critical criteria for spontaneous ignition and not the overall heat of combustion. The partial oxidation of -CH$_2$- might be take either of the forms

$$-CH_2- + 1/2\ O_2 = CO + H_2 \tag{12}$$

or $\quad -CH_2- + 1/2\ O_2 = CH_2O$. $\hfill (13)$

for which the exothermicity of reaction is approximately 6.5×10^6 J kg^{-1} in each case. This is a relatively small proportion of the overall heat of combustion. In the present calculations Q is taken to be 2.5×10^6 J kg^{-1}, in order to allow for the still lower exothermicity associated with peroxide formation, which cannot be interpreted directly from -CH$_2$- oxidation.

The activation energy associated with peroxide forming processes may be expected to be in the range 150 - 200 kJ mol^{-1} (Snee and Griffiths (17)). The representation of the chemistry as a single step, first order reaction requires the pre-exponential factor (A or a_1) to be obtained empirically. This parameter was derived in the present work by matching the critical temperature obtained for squalane at $\delta_{cr} = 3.665$. Kinetic and thermal parameters are summarised in Table 1.

Vaporisation Characteristics

The fluid vaporisation rate, expressed in a form related to the Clausius - Clapeyron equation for the temperature dependence of the vapour pressure of liquids, means that the enthalpy of vaporisation determines not only the endothermicity of the vaporisation process but also the temperature coefficient for the rate at which the vapour is generated. The temperature coefficient is expressed as an activation energy for the vaporisation process (E_v) in Table 1.

The enthalpies of vaporisation of the n-alkanes studied in this work fall in the range 55 - 70 kJ mol^{-1}. From the normal boiling point of squalane obtained from Fig. 1, ΔH_v for squalane is about 85 kJ mol^{-1}. The measured boiling point, 696 ($\pm$ 1) K (or 423 °C), seems more reasonable than that quoted in the Merck Index (18), 623 K (or 350 °C), which (when combined with other vapour pressure data) leads to the unreasonably high value $\Delta H_v = 135$ kJ mol^{-1}. The empirical factor F, representing the vaporisation rate coefficient, relates to the surface area within the porous structure as well as the molecular behaviour of the liquid itself, and so this was derived from comparisons with the combustion measurements made by McIntosh et al (6).

Table 1: Physical and chemical parameters related to the comparisons between theory and experiment

Exothermicity Q / J kg^{-1}	2.34×10^6
Enthalpy of vaporisation Q$_v$ / J kg^{-1}	$(2.0 \pm 0.1) \times 10^5$
Pre-exponential factor (reaction) A / kg^{-1} m^3 s^{-1}	1.5×10^{13}
Pre-exponential factor (vaporisation) F / s^{-1}	5.8×10^3
Activation energy (reaction) E / J mol^{-1}	1.5×10^5
Activation energy (vaporisation) E$_v$ / J mol^{-1}	$(5.0 - 8.0) \times 10^4$
Oxygen diffusion coefficient D$_0$ / m^2 s^{-1}	1.0×10^{-5}
stoichiometry coefficient ν	0.3

EXPERIMENTAL STUDIES

Experiments were performed by dispersing the liquid (10 cm^3) within a 5 cm cube of an amorphous silicate (micropore, matrix density 360 kg m^{-3}). An even distribution was ensured by dissolving the liquid in diethyl ether (50 cm^{-3}) and allowing this solution to soak through the block. The solvent was then allowed to evaporate completely at laboratory temperature before an experiment was started. A thermocouple (T_1/T_2, 0.1 mm dia wire) was located at the centre of the cube with a reference junction located external to the surface, so that the difference between the centre and the ambient (oven) temperature could be recorded as a function of time using a recorder.

For the purpose of mass loss measurements, a cube was suspended by wire from the arm of a torsion balance so that it was located, via a small entry port, at the centre of a recirculating air oven (void volume 0.25 m^3). Mass changes were recorded manually as the oven was heated up from laboratory temperature and stabilised at its pre-set value.

Oxygen was detected by continuous sampling to a mass spectrometer (RGA 10, magnetic sector instrument giving unit mass separation to 100 a.m.u.) via a silica microprobe (50 μm i.d.) inserted to the centre of the block and connected to the mass spectrometer by a heated stainless steel tube (0.5 mm i.d.). The pumping rate through the probe and the pressure drop to the mass spectrometer operating pressure (~ 10^{-6} mbar) were controlled by a two-stage pumping system with the intermediate pressure at the first stage being controlled by a needle valve, at approximately 4 mbar. Most of the pressure drop in the sampling line took place at the microprobe tip, so that quenching of the gaseous fraction occurred virtually instantaneously. Extensive tests were performed to optimise the mass spectrometer response whilst minimising the perturbation of the system by continuous sampling. The oxygen concentration was determined from the intensity of the mass peak at m/e = 32 relative to the background peak in the absence of oxygen, without further correction for other contributing species. Corrections were made for variations of the sampling rate as a consequence of temperature changes at the probe, which affected the viscosity of the gaseous sample. The mass spectrometer measurements were made by a repetitive scanning of the m/e range 29 - 33 over a period of one minute. This ensured that the maximum intensity of the m/e = 32 peak could be obtained repeatedly at regular intervals throughout the duration of an experiment (< 3 hours). The statistical variation in the peak height from successive scans was ± 5%. Temperature measurements were made simultaneously with the determination of oxygen concentrations, as described above. A different oven was used for the measurement of oxygen concentrations than that used for mass loss measurements, and it was necessary to use a different microporous block in these separate experiments.

EXPERIMENTAL RESULTS

Studies of Squalane Combustion

The critical oven temperature at which spontaneous ignition of squalane (10 cm^3) in a 5 cm cube was found to be 478 ± 1 K, 205 ± 1 °C (Table 2). Temperature change and oxygen concentration measurements during the combustion of squalane within the insulation at a supercritical oven temperature of 485 K are shown in Fig. 2. The average mass loss rate of liquid squalane from prior to onset of the hot stage of ignition was approximately 0.3 g per hour which is consistent with its vapour pressure at the oven temperature (4 - 5 mbar over the range 480 - 490 K, ca. 210 - 220 °C).

Ignition took place within the cube, at 485 K (7 K above criticality), at about 90 minutes after the experiment was begun (Fig. 2). There was an initial phase during which the block responded to the ramping and stabilisation of the oven temperature. This was recorded as an initial negative temperature difference resulting from the slower response of the centre of the cube with respect to the rising oven temperature. An equality between the centre temperature and the external temperature was reached after about 50 minutes. There then followed a period during which a further temperature rise at the centre occurred during which ΔT reached 50 K over an interval of about 10 minutes. In the subsequent development following an inflexion in the record, the centre temperature rose to nearly 500 K above that of the oven.

There was no depletion of oxygen at the centre of the insulation material throughout virtually the whole of the heating period and its concentration was still about half of the initial value even when the centre temperature had risen by more than 50 K above that at the edge. Complete consumption of oxygen at the centre of the block occurred just prior to the attainment of the maximum temperature. Undoubtedly reaction within the block had become oxygen diffusion controlled by this stage. A rapid recovery of the oxygen to its initial concentration occurred once the maximum temperature had been reached. It is possible that the inflexion of the thermal record in Fig. 2 may be associated with oxygen diffusion control but complex chemistry may also be contributory, such as the burn-out of carbonaceous residues once the centre temperature has become sufficiently high for this to take place.

The Combustion of Evaporation of Higher Volatility Alkanes

The behaviour of the alkanes n-$C_{16}H_{34}$, n-$C_{18}H_{38}$ and n-$C_{20}H_{42}$, and that of squalane ($C_{30}H_{62}$) is summarised in Table 2. The dimensionless critical temperature (u_{cr}) is derived from an activation energy of 150 kJ mol^{-1}. As indicated in the comments in Table 2, whereas n-eicosane (n-$C_{20}H_{42}$) showed a clear distinction between a limited extent of self-heating and fully-developed high temperature combustion ($\Delta T \sim 500$ K), n-hexadecane (n-$C_{16}H_{34}$) was capable of exhibiting only a very small temperature rise at the centre of the cube accompanying its evaporation at all oven temperatures up to its normal boiling point. This is illustrated in Fig. 3 at an oven temperature which is close to the boiling point of n-hexadecane (560 K).

Table 2: Combustion of selected alkanes in hot insulation experiments

Alkane	B Pt / K	T_{crit} / K	u_{cr}	Comment
2,6,10,15,19,23-hexamethyltetracosane, $C_{30}H_{62}$	696	478 ± 1	0.0265	no significant evaporation
n-eicosane, n-$C_{20}H_{42}$	616	505 ± 2	0.0280	clearly defined criticality
n-octadecane, n-$C_{18}H_{38}$	590	523 - 528	0.0291	marginal criticality
n-hexadecane, n-$C_{16}H_{34}$	560	-	∞	marginal self-heating only, no ignition

The behaviour of n-octadecane (n-$C_{18}H_{38}$) was intermediate between the lower and higher molecular mass alkanes. It exhibited a quite strong "parametric sensitivity" over the oven temperature range 513 - 528 K (Fig. 4). However, even at much higher oven temperatures the centre temperature excess did not exceed 100 K, as shown in Fig. 3. Mass loss measurement during the combustion of n-octadecane at 528 K shows that virtually all of the fluid had evaporated before significant extents of self-heating occurred (Fig. 4). Clearly a limitation on the residual fuel restricts the rate and extent of heat release during thermal runaway.

DISCUSSION

One practical consideration that emerges from this work is the experimental observation that, in general, alkanes may become susceptible to spontaneous ignition in insulation materials at relative molecular mass in the range 218 - 246. Other classes of organic compounds may have similar reactivity, especially if they have long alkyl side-chains. However, their volatility (reflected in their boiling points) may be quite different, and this must be taken into account when considering the hazard implications. For this reason it is important to formalise the criteria for criticality and ignition.

Relationship Between Theory and Experiment

Whereas the theory has been developed here on the basis of spherical symmetry, the experiments were performed in cubes of material. From both theoretical and experimental points of view there is no qualitative distinction between spontaneous ignition which occurs in cubes (or other axisymmetric shapes) and in spherically symmetric systems, and the criteria for criticality in a cube can be interpreted in a quantitative way from that in a sphere (Boddington *et al* (15), Beever and Griffiths (19), Jones and Puignou (20)). Moreover, at marginally subcritical conditions, spontaneous combustion evolves from the centre of the cube with virtually spherical symmetry. It is only when the temperature begins to rise in regions close to the surfaces that a distortion evolves as a consequence of the reduced thermal gradient along the axis into the corners relative to that perpendicular to the faces of the cube (Beever and Griffiths (21)).

Using the criteria of Boddington *et al* (15), we have derived the dimensionless critical temperature (u_{cr}) from the experimental result in terms of the Frank-Kamenetskii equivalent sphere (Table 2), so that comparisons may be made with the theoretical values of u_{FK} and u_∞ as the enthalpy of vaporisation increases (Fig. 5). Both of the theoretical criteria err to the safe side with respect to the critical conditions for the volatile alkanes in the range C_{16} - C_{20}, and the crossing point of u_∞ with u_{FK} coincides well with the departure of criticality of the more volatile fluids from the u_{FK} line. The more satisfactory theoretical criterion for safe operation then becomes the condition for u_∞. However, u_∞ exists even for fluids of sufficiently low enthalpy of vaporisation (or normal boiling point) that the fluid evaporates before exothermic reaction can develop. This discrepancy arises from the analytical constraint set by the assumption that $\partial x/\partial t = 0$, which signifies that there is no reactant loss. Whilst the chemical oxidation rate may be very slow (so that the condition $\partial z/\partial t = 0$ remains valid), in reality the evaporation rate becomes too high for the reactant density to remain even approximately constant. According to the data in Fig. 5, it would be extremely useful to establish a theoretical relationship which could give a prediction of the asymptotic limit for criticality that is demonstrated experimentally by n-$C_{16}H_{34}$.

Scaling Relationships and Assessment of Hazards Associated with Asymmetrically Heated Systems

When tests are performed in uniformly heated systems, and it is found that the classical ignition criterion (u_{FK}) is appropriate to a particular fuel, the conditions for safe operation may be interpreted from the rules established by Boddington et al (15) for a wide range of shapes at any size. These rules have been validated by experiment (Beever and Griffiths (19), Jones and Puignou (20), Egeiban et al (22)). Appropriate kinetic and thermal parameters are required for the chemical system, which can be derived from a series of experiment in cubes of different size if the parameters are not already known (Bowes (4), Beever and Griffiths (19)), Jones and Puignou (20)). Whether or not u_{FK} or u_z is the most appropriate criterion is determined by u_a^*, which is derived through equation (10). When u_z is relevant, the conditions for safe operation rest entirely on the relationship between the exothermic oxidation rate and the endothermic evaporation rate.

The simplest analytical form for u_z, equation (9), contains the empirical, dimensionless evaporation coefficient, f, which also relates to the surface area, porosity and transport of the vapour from the inert matrix (Brindley et al (10)). The properties of insulation and of the fluid can be separated in a more sophisticated formulation if the prediction of criticality in a different type of insulation material is to be made (Brindley et al (11)). However, this analysis is amenable only to numerical interpretation as yet because a fourth equation must be included to represent the vapour of the fluid and its transport through the material. The physical representation is of an interconnected pore structure with the possibility of multilayer build-up of liquid on the pore surface. The vapour density adjacent to the surface is at equilibrium with the liquid at the local surface temperature. The vapour density at the centre of the pore is governed by the diffusion gradient between it and the pore surface, such that evaporation and condensation are admitted, and also between the interconnecting pores and the external environment (Brindley et al (11)).

The conditions that relate to the classical imbalance between the heat release and heat loss rate in an asymmetrically heated system have been addressed by Thomas and Bowes (9). When evaporative loss becomes a controlling factor of spontaneous ignition in an asymmetrically heated system, an analytical criterion cannot be defined in a simple form, as is the case for uniform heating. Difficulty arises because the there is always a propensity for evaporation to take place at the hot surface, which may result in re-condensation in a cooler region of the insulation material. Thus the location of the seat of ignition may have a spatial dependence governed lay the thermal gradient and the diffusion characteristics throughout the system. These complicated but extremely interesting interactions are under further investigation.

Chemical Features of the Theoretical Model

The high non-linear nature of the temperature excursion in the experiments performed on squalane during ignition (Fig. 2) is not characteristic of that exhibited either theoretically or experimentally in a classical thermal ignition (Griffiths and Scott (23)). It is reminiscent of the branching chain - thermal interactions that occur during hydrocarbon oxidation both in gas and liquid phases (Griffiths and Scott (23)). Consequently the representation of the combustion chemistry in reactions of these kinds as a single step first order exothermic reaction may not be adequate. Also there is a potential difficulty with regard to reaction of the condensed phase expressed as a conventional concentration dependence. Whilst it is not practical in the present context to address a detailed kinetic model there has been an extensive development of reduced kinetic models to interpret the spontaneous ignition of hydrocarbons (Griffiths (16)). One possibility is to regard the chemistry to be represented as a quadratic branching process with allowance for reaction at the surface in the reduced form

$$v = kxZ\{(Z_0 - Z) + \zeta\}. \tag{14}$$

The term x represents the effective reaction density of the adsorbed liquid which is related to the Langmuir adsorption isotherm $x = X/(X_m + X)$ where X_m represents the monolayer coverage and X is the local density. Until the monolayer is exposed the oxidation rate depends only on the number of adsorption sites occupied by the liquid and not on the remaining adsorbed liquid density. Z_0 and Z represent the gaseous oxygen density initially and at time t respectively. The small parameter ζ is an initiating term ($\zeta \ll (Z_0 - Z)$) and is essential to permit the autocatalytic reaction to begin. The inclusion of ζ represents a seeding of the system with a small amount of the degenerate branching agent rather than it being generated in an initiation reaction (Gray and Scott (24)). The expression for quadratic autocatalysis has been shown to be appropriate for gas-phase oxidation where there is a large excess of fuel. The acceleration in the reaction rate is then independent of the fuel concentration but is a function of the oxygen concentration (Griffiths and Phillips (25)). The properties of this kinetic representation for reaction of hydrocarbons in hot lagging have been discussed elsewhere (Brindley et al (11)).

Anomalous Behaviour Observed in Laboratory Experiments

Our experiments are similar in design to those performed by Britton (5) in 1991. However, most of the fluids studied by him (amines, olamines and glycols) were reactive at fairly low temperatures and they had comparatively high boiling points, so the competition between reactivity and volatility was not put to rigorous test in that particular study. The corresponding behaviour in the present work is shown by squalane (Fig. 2). The longest ignition delay that we were able to reproduce for squalane in a uniformly heated 5 cm cube was about 7 hours. This time is entirely consistent with that which would be expected from the imbalance of the heat release and heat transport rate by thermal conduction at marginally supercritical conditions (Brindley et al (11)).

In some of Britton's experiments ignition delay times were recorded which exceeded 10 hours, and for the glycols these times extended to several days. In our own studies on a series of normal alcohols (McIntosh et al (6)), we also recorded ignition delays that exceeded 24 hours. Delays of this duration are incompatible with the thermal relaxation times for a 5 cm cube of any insulation material. Moreover, mass loss measurements involving the alcohols showed that there was a residual fraction of alcohol retained within the insulation matrix long after the normal evaporative process should have reached completion.

The substances involved in these abnormal cases tend to be polar in character. We believe that exceptionally strong adsorption forces may have been involved in the adhesion to the surface, which then controlled the development and eventual occurrence of ignition. Hydrogen bonding of the substrate to the silicate based insulation is one example of how adhesion may have been brought about. This implies that in certain circumstances, the normal expectation that evaporation will reduce the potential hazard of a fluid in hot insulation is not appropriate. The enhanced risk of fires developing for reasons connected with this anomalous must be taken into account. These aspects of the lagging fire problem require further investigation.

ACKNOWLEDGEMENT

The authors gratefully acknowledge financial support of this project from EPSRC (GR/J27431). Thanks are due also to B. Frere for the chromatography shown in Figure 1 and the boiling point determination of squalane.

REFERENCES

1. Lindner, H. and Seibring, H. 196,, Chemie-Ing. Tech., 39:667 - 671
2. Bowes, P.C. and Langford, B. 1968, Chem. and Proc. Eng., 49: 108 - 116
3. Gugan, K., 1974, I. Chem. E. Symp., Ser., 39: 28 - 43
4. Bowes, P.C., 1984, Self-heating: evaluating and controlling the hazard, Her Majesty's Stationery Office, London
5. Britton, L.G., 1990, Plant / Operations Prog., AIChE, 10: 27 - 37
6. McIntosh, A.C., Bains, M., Crocombe, W. and Griffiths, J.F., 1994, Combust. Flame, 99: 541 - 550
7. Frank-Kamenetskii, D.A., 1969, Diffusion and Heat Transfer in Chemical Kinetics (trans. J.P. Appleton) Plenum Press, New York
8. Gray, P. and Lee, P.R., 1968, Oxidation and Combustion Reviews, vol 2, ed C.F.H. Tipper, Elsevier, Amsterdam, 1 - 94
9. Thomas, P.H. and Bowes, P.C., 1961, Trans. Faraday Soc., 57:2007 - 2017
10. Brindley, J., Griffiths, J.F., Hafiz, N., McIntosh, A.C. and Zhang, J., 1998, AIChE Journal, 44: 1027 - 1037
11. Brindley, J., Griffiths, J.F., McIntosh, A C. and Zhang, J., 1998, Twenty-Seventh Symposium (International) on Combustion, The Combustion Institute, in press
12. Gray B.F and Wake G.C., 1988, Combust. Flame, 71: 101 - 104
13. Aris, R., 1975, The Mathematical Theory of Diffusion and Reaction in Permeable Catalysts, Clarendon Press, Oxford
14. McIntosh, A.C., Truscott, J.E., Brindley, J., Griffiths, J.F. and Hafiz, N., 1996, J. Chem. Soc. Faraday Trans., 92: 2965 - 2969
15. Boddington, T., Gray, P. and Harvey, D.I., 1971, Phil. Trans. R. Soc. Lond., A270: 467 - 506
16. Griffiths, J.F., 1995, Prog. Energy Combust. Sci., 21:25 - 107
17. Snee, T.J. and Griffiths J.F., 1989, Combust. Flame, 75: 381 - 395
18. The Merck Index, 11th Edition, 1989, Merck and Co. Inc., Rahway, N.J.
19. Beever, P.F. and Griffiths, J.F., 1989, International Symposium on Runaway Reactions, CCPS, AIChE, New York, 1 - 20
20. Jones, J.C. and Puignou, A., 1998, Trans. Inst. Chem. Eng. Part B: Process Safety and Environmental Protection. 76: 14 - 18
21. Beever, P.F. and Griffiths, J.F., unpublished results
22. Egeiban, O.M., Mullins, J.R., Scott, S.K. and Griffiths, J.F., 1982, Nineteenth Symposium (International) on Combustion, The Combustion Institute, Pittsburgh, 825 - 833
23. Griffiths, J.F. and Scott, S.K., 1987, Prog. Energy Combust. Sci., 13, 161 - 191
24. Gray, P. and Scott, S.K., 1990, Chemical Oscillations and Instabilities, Clarendon Press, Oxford
25. Griffiths, J.F. and Phillips, C. H., 1989, J. Chem. Soc. Faraday Trans. 1, 85: 3471 - 3479

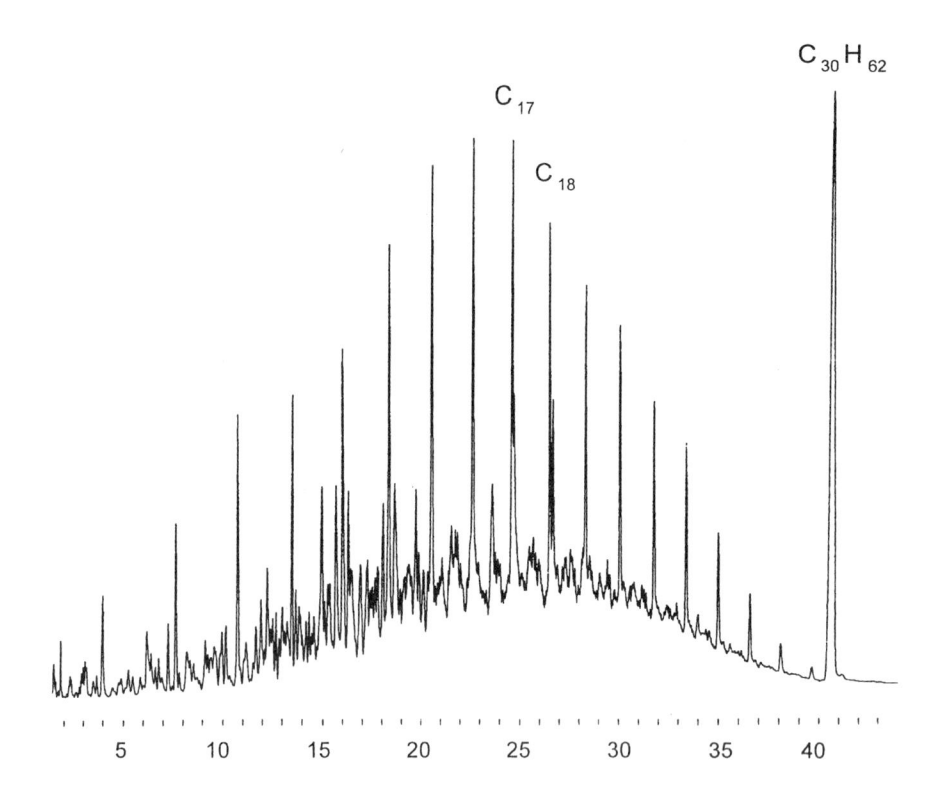

Figure 1   A chromatogram of a typical diesel fuel doped with squalane ($C_{30}H_{62}$),. The sample was separated on a 12 m BP1 capillary column temperature programmed from 50 to 250 °C for 40 minutes at 5 °C per minute. The ordinate represents the peak intensity, which is proportional to the amount of each component . The abscissa is marked with time of elution, in minutes, which is proportional to the boiling point of each component. The most prominent peaks are the series of n-alkanes, n-$C_{17}H_{36}$ and n-$C_{18}H_{38}$ being identified from the corresponding isoprenoids which appear as the second peak of each doublet.

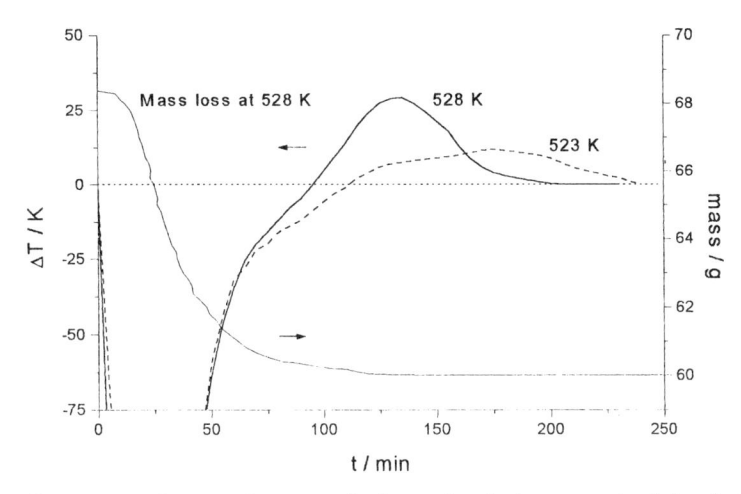

Figure 4 Temperature change at the centre of a 5 cm cube of microporous insulation during the combustion of n-C$_{18}$H$_{38}$ at oven temperatures of 523 and 528 K (250 and 258 °C). The measured mass loss of fluid at 528 K is also shown.

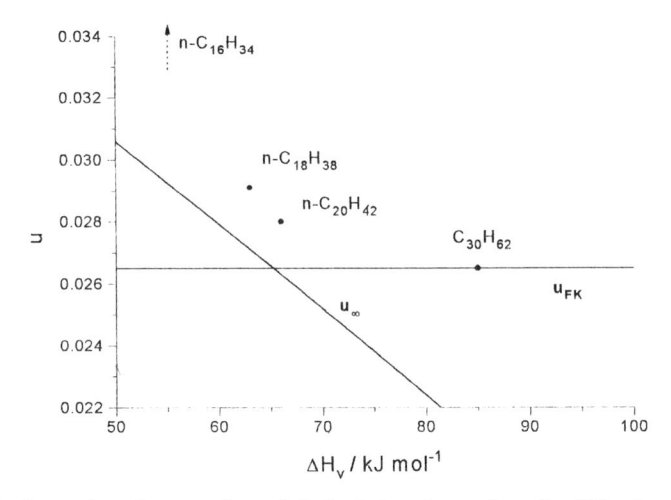

Figure 5 Comparisons between the analytical solutions for u$_{FK}$ (equation (8)) and u$_\infty$ (equation (9)), as a function of enthalpy of vaporisation, and the experimentally measured critical conditions for the n-alkanes, expressed in non-dimensional form as u$_{cr}$. The crossing point of the theoretical lines represents u$_a$*.

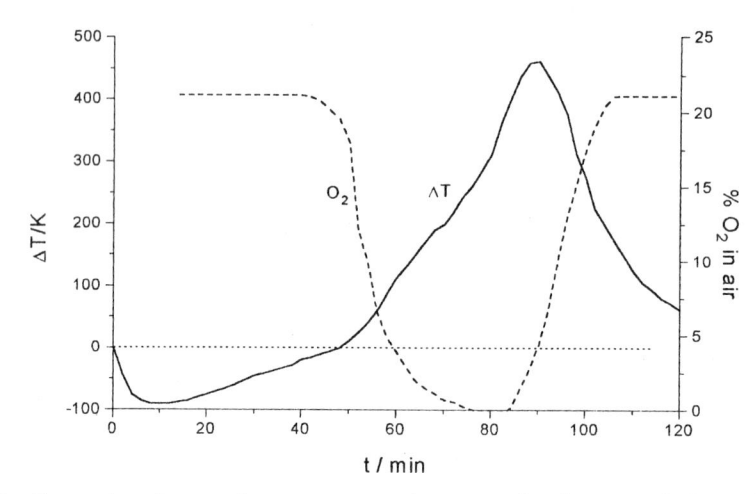

Figure 2 Temperature change and oxygen concentration measured at the centre of a 5 cm cube of microporous insulation during the course of the development of spontaneous ignition of squalane at an oven temperature of 485 K (212 °C).

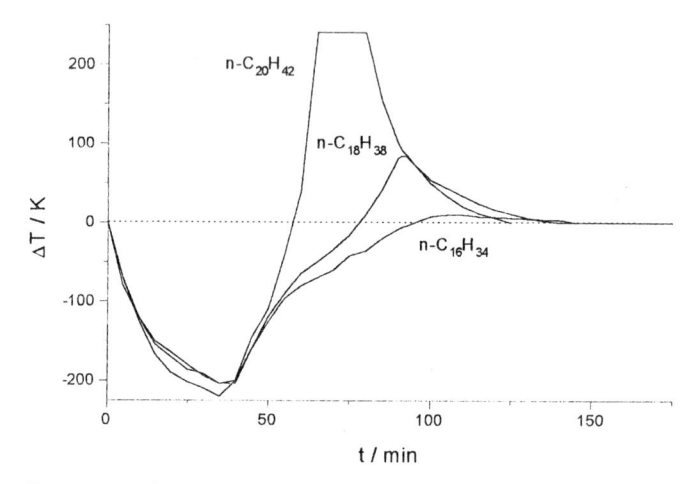

Figure 3 Temperature change at the centre of a 5 cm cube of microporous insulation, during the combustion of n-$C_{16}H_{34}$, n-$C_{18}H_{38}$ and n-$C_{20}H_{42}$ at an oven temperature of 558 K (285 °C).

ASSESSMENT OF THE THERMAL AND TOXIC EFFECTS OF CHEMICAL AND PESTICIDE POOL FIRES BASED ON EXPERIMENTAL DATA OBTAINED USING THE TEWARSON APPARATUS

Christian Costa, Guy Treand and Jean-Louis Gustin
Rhône-Poulenc Industrialisation, 24 avenue Jean-Jaurès – 69153 Décines - France

> The Tewarson apparatus is a combustion calorimeter developed by Factory Mutual Research Co. USA, in the 1970s. A modified and computerised version of this calorimeter is used at the Rhône-Poulenc Decines Centre to study the combustion of plastics, fabrics, chemicals and pesticides on 30 grams samples in a 0.1 metre diameter glass dish.
> The combustion of up to 100 products has been studied in this experimental set-up and the following thermal data obtained : mass of product burnt, experimental heat of combustion, combustion efficiency, burning mass flux, ratios of convection and radiant heat, flame height, flame temperature.
> The on-line analysis of combustion gases provides the following chemical data : production of CO_2, CO, HCN, NO_2, NO, SO_2, HCl, HF, HBr, chemical yield for the combustion of carbon, nitrogen, sulphur, chlorine, fluorine, bromine.
> The thermal data obtained is an input to the POOL 2.0 Computer code to estimate the thermal effect of chemical pool-fires.
> The combustion chemical data obtained is an input to atmospheric dispersion codes to estimate the toxic effect of chemical pool fires.
> The correlation of experimental data obtained using the Tewarson apparatus, based on the sample chemical formulae helps provide the missing combustion data. As an example, a correlation is given for the combustion characteristics of chlorinated organic chemicals.
>
> Keywords : Pool fires, combustion data, Tewarson apparatus, combustion efficiency, chemical yield.

Introduction

The assessment of fire hazards in chemical and pesticide storages and warehouses is based on both the determination of the material combustion thermal data and the identification of the toxic emission from combustion gases.

The combustion thermal data is the input data required in fire simulation softwares to estimate the consequences of industrial fires. This combustion thermal data includes the determination of the mass of product burnt, the experimental heat of combustion, the combustion efficiency, the burning mass flux, the ratios of convection and radiant heat, the flame height and flame temperature. (1) (2).

The simulation of large industrial fires provides information on the thermal effect of the accidental fire on adjacent equipment and on the protection needed to prevent the fire from spreading.

Large industrial fires are also source-terms for modelling atmospheric dispersion of volatile toxic combustion products. The input data to atmospheric dispersion models are combustion chemical data including the production of combustion gases Cl_2, CO, HCN, NO_2, NO, SO_2, HCl, HBr, depending on the burning material composition. Also necessary is the determination of the chemical yield for the combustion of the chemical elements present in the burning material formula : Carbon, Nitrogen, Sulphur, Chlorine, Fluorine, Bromine if any (3).

The thermal and chemical data characterizing the combustion of chemicals and pesticides can only be obtained using a bench-scale apparatus, due to the great number of experiments to be performed on a wide range of products.

Such a combustion calorimeter was developed by A. TEWARSON at Factory Mutual Research Corporation (USA) in the 1970s (4). A modified and computerized version of the Tewarson apparatus was built at the Rhone-Poulenc Decines Centre to study the combustion of plastics, fabrics, chemicals and pesticides on 30 grams samples. This new experimental set-up is described in the following section with special attention to the improvement of the original design.

Description of the modified tewarson combustion calorimeter

The principle of the modified Tewarson combustion calorimeter built in Decines is shown on figure 1.

The experimental set-up may be divided in three sections :

- The lower part of the apparatus is the combustion chamber, section A on figure 1.
  The combustion chamber consists of a standing cylindrical quartz tube 0.160 metre in diameter and 0.490 metre high. In this combustion chamber, a 30 grams sample in a 0.1 metre diameter glass dish is placed on the plate of a balance to measure the sample weight during combustion experiments.
  An external heat-flux is applied to the combustion chamber by eight infra-red heaters in an air flushed jacket, allowing the sample to be heated to a temperature where its vapours or fumes can be ignited by an ignition source. The maximum external heat-flux applied to the combustion chamber is 30 kW/m^2. The ignition of the sample by an electric spark was preferred to the original pilot-flame ignition source, to avoid additional heat input and combustion gases to the experiments.

The ignition of the sample is obtained without external heat-flux applied if the sample is flammable under ambient temperature. If not, an external heat-flux is applied to raise the sample temperature until ignition is obtained.

If the sample combustion is self-sustained, the sample is allowed to burn without external heat-flux applied. If not, an external heat-flux is applied to allow the sample combustion, in which case the radiant heat-flux cannot be measured. Preliminary experiments are necessary to choose the most suitable operating conditions.

A permanent air flow of 5 m^3/h is blown to the combustion chamber bottom through a glass sphere bed to obtain homogeneous inlet gas composition and regular air stream.

The inlet air is under flow control and its oxygen concentration may be varied by adding oxygen or nitrogen to the air flow. The combustion air composition is measured by a continuous oxygen analyzer during experiments.

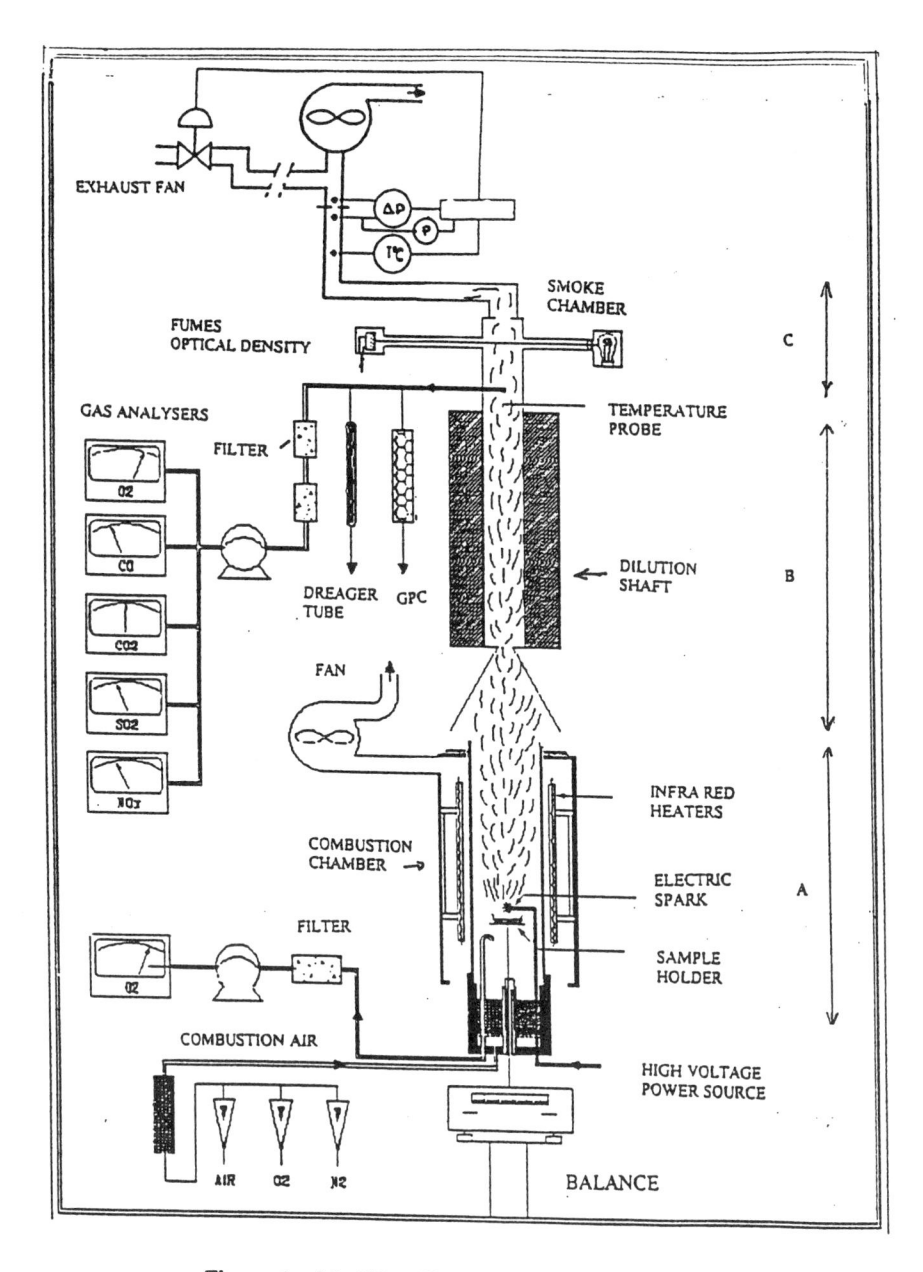

Figure 1 : Modified Tewarson apparatus

A calibration of the external heat input to the sample is achieved by replacing the sample holder by a heat-flux meter. The heat-flux received by the sample is measured as a function of the power input to the infra-red heaters. This calibration allows the compensation of the infra-red heater aging by an increased electric power supply to the external heating device.

- The intermediate part of the apparatus is the dilution shaft, section B on figure 1.

  The dilution shaft is a standing Teflon cylinder 0.1 metre in diameter and 0.6 metre high intended to dilute and mix the flow of combustion gases and smokes with air. Teflon was preferred to stainless steel, to avoid soot deposits which could absorb contaminants such as HCl, HCN, dioxines, etc . The dilution air inlet flow to the dilution shaft is controlled by the exhaust fan mass flow rate of 70 kg/h.

  The aim of the combustion products dilution is to avoid losses of volatile combustion products by condensation or leaks, while limiting the dilution ratio to keep the oxygen concentration analysis accurate and controlling heat losses from combustion products. Mixing of combustion products with dilution air is achieved by the convergent nozzle at the dilution shaft bottom.

  At the top of the dilution shaft, the smoke temperature is measured by a thermocouple and the gas flow is sampled for continuous on-line analysis.

  The on-line gas analysis of the diluted smoke includes the determination of O_2, CO, CO_2, SO_2 and NOx. This is achieved after passing the gas sample over a filter and a desiccant. Other analysis are performed after absorption of the gas on a Draeger tube followed by mass spectrometer analyses or after absorption on resins followed by gas chromatography.

- The upper part of the Tewarson apparatus is the smoke chamber, section C on figure 1.

  This device, intended to measure the optical density of smoke, is equipped with an external photoelectric system to measure the optical density in a standing stainless steel cylinder 0.1metre in diameter and 0.3 metre high, extending the combustion shaft. The smoke chamber exit is connected to the exhaust fan inlet. The fan flow control principle is shown on fig. 1. The fan volumetric flow rate is adjusted taking into account the smoke temperature to obtain a constant mass flow rate of exhaust gases.

- Data acquisition and processing system

  This section includes a CHESSEL recorded / converter and a graphic data treatment allowing display of the variation of the measured parameters as a function of time. A test result sheet is produced, giving the most important test characteristics and results.

Experimental results

To date, up to 100 products have been studied using the modified Tewarson apparatus. An overview of the results obtained on well known chemicals and pesticides is given in tables where the following data is listed under the product current name :

- Gross chemical formula
- Molecular weight (kg/kmole)
- External heat flux applied (kW/m^2)
- Initial sample mass (g)
- Mass fraction of product burnt (%)
- Net calorific value (kJ/kg)
- Heat of combustion per kg of product burnt (kJ/kg)
- Heat of combustion per kg of sample (kJ/kg)
- Combustion thermal efficiency (%)
  i.e. The ratio of the combustion heat measured on 1 kg initial sample to the net calorific value of this 1 kg sample.
- Average combustion mass flux ($g/m^2.s$)
- Maximum combustion mass flux ($g/m^2.s$)
- Ratio of convection heat to combustion heat (%)
- Ratio of radiant heat to combustion heat (%)
- Production of CO_2, CO, HCN, NO_2, NO, SO_2, HCl (g of gas / kg of sample)
- Chemical yield for the conversion of carbon into CO_2, CO and HCN (%)
  i.e. The ratio of the carbon contained in the CO_2 CO and HCN produced by 1 kg sample to the carbon present in that 1 kg sample.
- Chemical yield for the conversion of nitrogen into NO2, NO and HCN (%)
  i.e. The ratio of the nitrogen contained in the NO2, NO and HCN produced by 1 kg initial sample to the nitrogen present in that 1 kg sample.
- Chemical yield for the conversion of chlorine into HCl (%)
  i.e. The ratio of the chlorine contained in the HCl produced by 1 kg sample to the chlorine present in that 1 kg sample
- Maximum flame height (cm)
- Maximum flame temperature (°C)
- Specific extinction area (m^2/kg)
  i.e. The surface darkened by 1 kg of fuel, deduced from the measured optical density.

The detailed calculation methods for the combustion heat, convection and radiant heat, opacity of smokes and other parameters are given in reference (5) and are not reproduced in the present paper. The total combustion heat was deduced from the oxygen depletion during combustion according to Thornton (6) (7). The convection heat is the heat carried out in the gas plume over the combustion chamber. The radiant heat is deduced from the total combustion heat, the convection heat and an estimation of the heat-losses obtained from calibration experiments.

The combustion data for some chemicals and solvent is given in table 1.

The combustion data for some pesticides is given in table 2.

Table 1 : Combustion characteristics of chemicals and solvents, obtained in the modified Tewarson apparatus

Parameters	units	ACETONE	ACETO NITRILE	ISOPROPYL ACETATE	CYCLOHEXANE
Gross chemical formula		C_3H_6O	C_2H_3N	$C_5H_{10}O_2$	C_6H_{12}
Molecular weight	kg/kmol	58	41	102	84
External heat-flux applied	kW/m^2	0	0	0	0
Initial sample mass	g	22	33	22	21
Mass fraction of product burnt	%	100	100	100	100
Net calorific value	kJ/kg	28 560	29 580	25 995	43 388
Heat of combustion /kg product burnt	kJ/kg	27 500	29 000	24 220	42 570
Combustion thermal efficiency	%	96	97	93	98
Mean combustion mass-flux	g/m^2.s	12	11	12	14
Maximum combustion mass-flux	g/m^2.s	14	12	14	18
Ratio of convection heat	%	56	55	53	47
Ratio of radiant heat	%	16	22	20	21
Production of CO_2 / kg sample	g/kg	2 188	2 040	2 004	2 834
Production of CO / kg sample	g/kg	17	25	22	29
Production of HCN / kg sample	g/kg	-	0.2	-	-
Chemical yield for carbon	%	97	97	94	92
Production of NO_2 / kg sample	g/kg	-	1	-	-
Production of NO / kg sample	g/kg	-	19	-	-
Production of HCN / kg sample	g/kg	-	0.2	-	-
Chemical yield for N_2	%	-	3	-	-
Production of SO_2 / kg sample	g/kg	-	-	-	-
Chemical yield for sulphur	%	-	-	-	-
Production of HCl / kg sample	g/kg	-	-	-	-
Chemical yield for Cl_2	%	-	-	-	-
Flame height	cm	30	30	30	40
Flame maximum temperature	°C	816	865	796	812
Specific extinction area	m^2/kg	0	35	0	0

Table 1 (continued) : Combustion characteristics of chemicals and solvents, obtained in the modified Tewarson apparatus

Parameters	units	ETHYL ALCOHOL	DIISOPROPYL ETHER	HEXANE	METHYL ALCOHOL	TETRA HYDROFURAN
Gross chemical formula		C_2H_6O	$C_6H_{14}O$	C_6H_{14}	CH_4O	C_4H_8O
Molecular weight	kg/kmol	46	102	86	32	72
External heat-flux applied	kW/m^2	0	0	0	0	0
Initial sample mass	g	23	23	20	20	22
Mass fraction of product burnt	%	100	100	100	100	99.8
Net calorific value	kJ/kg	26 810	36 194	44 740	19 892	32 200
Heat of combustion /kg product burnt	kJ/kg	26 450	35 010	42 210	18 800	30 320
Combustion thermal efficiency	**%**	**99**	**96**	**96**	**96**	**94**
Mean combustion mass-flux	$g/m^2.s$	9	17	15	8	14
Maximum combustion mass-flux	$g/m^2.s$	10	21	19	9	17
Ratio of convection heat	%	54	47	52	59	53
Ratio of radiant heat	%	20	25	22	14	18
Production of CO_2 / kg sample	g/kg	1 843	2 378	2 868	1 330	2 290
Production of CO / kg sample	g/kg	0	21	33	0	21
Production of HCN / kg sample	g/kg	-	-	-	-	-
Chemical yield for carbon	**%**	**96**	**93**	**95**	**97**	**95**
Production of NO_2 / kg sample	g/kg	-	-	-	-	-
Production of NO / kg sample	g/kg	-	-	-	-	-
Production of HCN / kg sample	g/kg	-	-	-	-	-
Chemical yield for N_2	%	-	-	-	-	-
Production of SO_2 / kg sample	g/kg	-	-	-	-	-
Chemical yield for sulphur	%	-	-	-	-	-
Production of HCl / kg sample	g/kg	-	-	-	-	-
Chemical yield for Cl_2	%	-	-	-	-	-
Flame height	cm	25	> 40	> 40	15	40
Flame maximum temperature	°C	843	768	782	780	811
Specific extinction area	m^2/kg	0	0	0	0	0

Note : n.m. = not measured

Table 1 (continued) : Combustion characteristics of chemicals and solvents, obtained in the modified Tewarson apparatus

Parameters	units	XYLENE	TOLUENE	PROPYLENE GLYCOL	PHENOL	DIETHYL OXALATE
Gross chemical formula		C_8H_{10}	C_7H_8	$C_3H_8O_2$	C_6H_6O	$C_6H_{10}O_4$
Molecular weight	kg/kmol	106	92	76	94	146
External heat-flux applied	kW/m^2	0	0	0	0	0
Initial sample mass	g	30	30	28	30	30
Mass fraction of product burnt	%	100	100	100	100	100
Net calorific value	kJ/kg	40 900	40 550	21 630	31 000	18 600
Heat of combustion /kg product burnt	kJ/kg	27 100	29 055	21 310	27 625	17 700
Combustion thermal efficiency	%	66	71	99	89	95
Mean combustion mass-flux	g/m^2.s	18	35	6	12	9
Maximum combustion mass-flux	g/m^2.s	25.8	46	7	15	11
Ratio of convection heat	%	37	35	62	43	57
Ratio of radiant heat	%	29	27	28	33	25
Production of CO_2 / kg sample	g/kg	2 170	2 184	1 664	2 625	1 819
Production of CO / kg sample	g/kg	73	76	21	57	18
Production of HCN / kg sample	g/kg	-	-	-	-	-
Chemical yield for carbon	%	69	69	98	96	100
Production of NO_2 / kg sample	g/kg	-	-	-	-	-
Production of NO / kg sample	g/kg	-	-	-	-	-
Production of HCN / kg sample	g/kg	-	-	-	-	-
Chemical yield for N_2	%	-	-	-	-	-
Production of SO_2 / kg sample	g/kg	-	-	-	-	-
Chemical yield for sulphur	%	-	-	-	-	-
Production of HCl / kg sample	g/kg	-	-	-	-	-
Chemical yield for Cl_2	%	-	-	-	-	-
Flame height	cm	35	40	16	32	24
Flame maximum temperature	°C	630	715	826	758	850
Specific extinction area	m^2/kg	1 738	1 600	0	1 000	0

Table 2 : Combustion characteristics of various pesticides, obtained in the modified Tewarson apparatus

Parameters	units	DIURON	ISO PROTURON	ACID 2,4 D	MANCOZEBE
Gross chemical formula		$C_9H_{10}ON_2Cl_2$	$C_{12}H_{18}ON_2$	$C_8H_6O_3Cl_2$	$(C_4H_6N_2S_4Mn)_1 Zn_{0.1}$
Molecular weight	kg/kmol	233	206	221	269
External heat-flux applied	kW/m^2	25	25	25	25
Initial sample mass	g	30	30	15	30
Mass fraction of product burnt	%	99.7	100	100	60.9
Net calorific value	kJ/kg	20 300	32 800	(11 540)	14 000
Heat of combustion /kg product burnt	kJ/kg	10 240	23 870	4 500	15 590
Heat of combustion / kg sample		10 210	23 870	4 500	9 500
Combustion thermal efficiency	**%**	**51**	**72**	**(39)**	**69**
Mean combustion mass-flux	g/m^2.s	26	28	24	5
Maximum combustion mass-flux	g/m^2.s	45	50	n.m.	16
Ratio of convection heat	%	38	30	63	44
Ratio of radiant heat	%	n.m.	n.m.	n.m.	n.m.
Production of CO_2 / kg sample	g/kg	760	1 704	504	498
Production of CO / kg sample	g/kg	80	56	74	0
Production of HCN / kg sample	g/kg	12	7	-	0
Chemical yield for carbon	**%**	**53**	**70**	**39**	**76**
Production of NO_2 / kg sample	g/kg	2	1	-	1
Production of NO / kg sample	g/kg	3.5	8	-	7
Production of HCN / kg sample	g/kg	12	7	-	1
Chemical yield for N_2	%	7	7	-	4
Production of SO_2 / kg sample	g/kg	-	-	-	535
Chemical yield for sulphur	%	-	-	-	56
Production of HCl / kg sample	g/kg	144	-	121	-
Chemical yield for Cl_2	%	46	-	37	-
Flame height	cm	n.m.	n.m.	n.m.	n.m.
Flame maximum temperature	°C	n.m.	n.m.	n.m.	n.m.
Specific extinction area	m^2/kg	1 600	1 160	1 640	0

Note : n.m. = not measured

The net calorific value of 2,4 D acid was not available and was deduced from the combustion chemical yield for the conversion of carbon.

Combustion properties as a function of time

In addition to integral combustion data deduced from experiments, dynamic and time dependent properties are also obtained from combustion experiments performed in the modified Tewarson apparatus.

As an example, the combustion heat-flux during the combustion of a 30 g sample of cyclohexane is given as a function of time on figure 2. In this figure, the combustion heat per unit time or combustion thermal power is estimated using the oxygen consumption deduced from the oxygen depletion measured in the fumes. The reference liquid sample area is that of the glass dish.

The convection heat-flux measured in the same experiment is given as a function of time on figure 3. The convection heat was deduced from the diluted fumes mass-flow and temperature. The reference area for the estimation of the convection heat-flux is that of the glass dish.

The radiant heat-flux is deduced from heat-flux measurements and calibration experiments using the infra-red external heating device. The radiant heat-flux during the combustion of a cyclohexane sample is given as a function of time on figure 4. The reference area for the estimation of the radiant heat-flux is that of the glass dish.

Figures 2-3-4 show that the combustion heat-flux is not the sum of the convection heat-flux and radiant heat-flux. The missing heat is the experimental set-up heat-losses or conduction heat, mainly due to the combustion chamber quartz tube heat capacity.

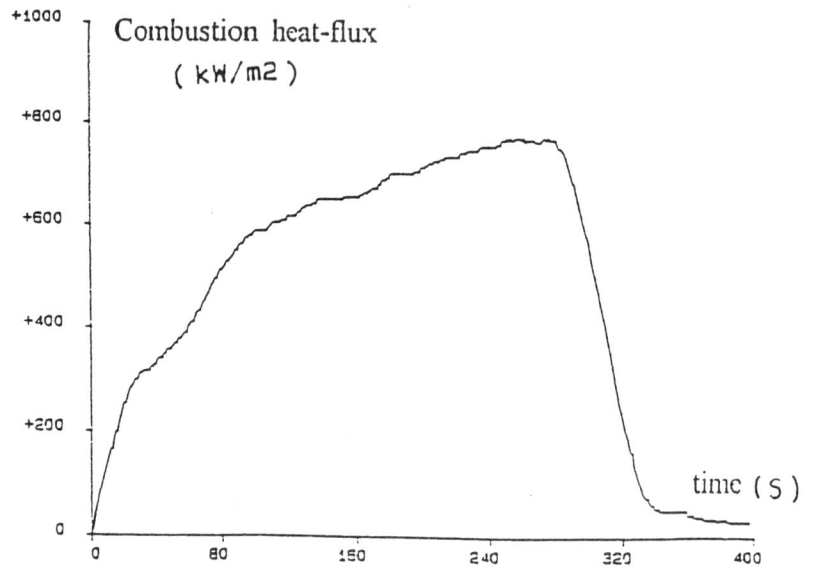

Figure 2 : Combustion heat-flux as a function of time measured during the combustion of a 30 g sample of cyclohexane in the modified Tewarson apparatus.

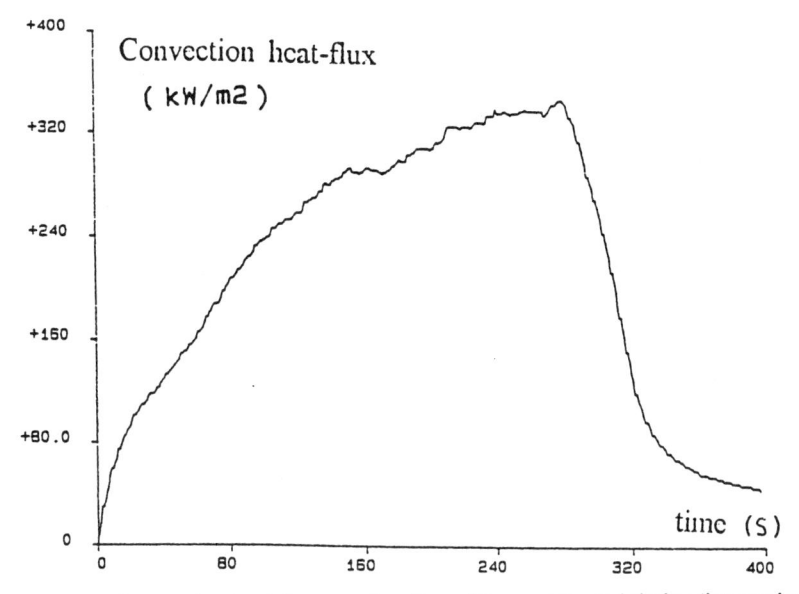

Figure 3 : Convection heat-flux as a function of time measured during the combustion of a 30 g sample of cyclohexane in the modified Tewarson apparatus.

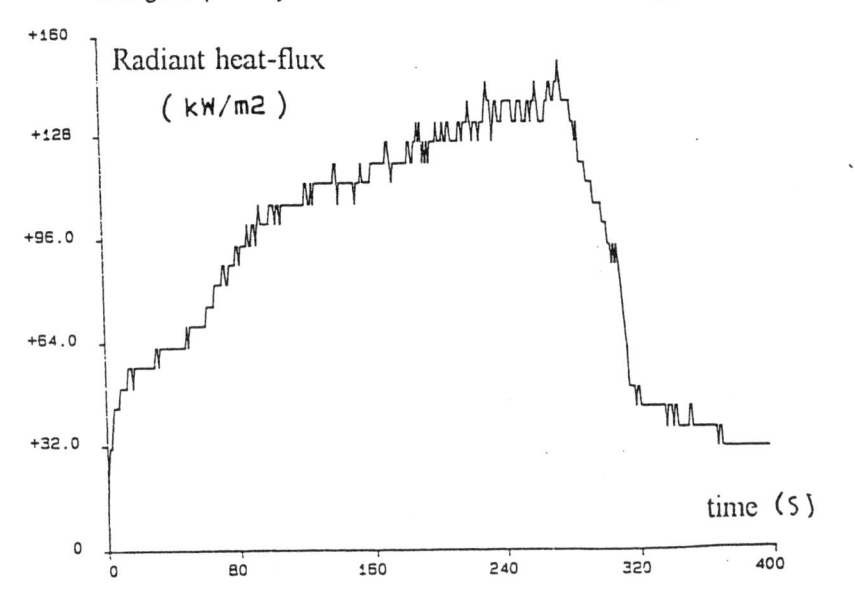

Figure 4 : Radiant heat-flux as a function of time measured during the combustion of a 30 g sample of cyclohexane in the modified Tewarson apparatus.

Influence of the chemical formula on the combustion thermal data

The experimental results obtained on the chemicals and pesticides studied in the modified Tewarson apparatus, show that the combustion thermal data is influenced by the chemical formula. For the most useful combustion characteristics used as input data for the simulation of large industrial fires i.e. the combustion thermal efficiency, the chemical yield for the conversion of carbon, the ratio of radiant heat, which were defined in the previous section, recommended specific values can be deduced from the experimental results obtained on the different types of chemicals. A summary of the thermal data suggested is given in table 3 for aliphatic derivatives, in table 4 for C, H, O, N aromatic and unsaturated cyclic compounds and in table 5 for miscellaneous organic compounds including chlorinated and fluorinated organic compounds.

The combustion thermal data in table 3, 4 and 5 may be used as a first estimate in fire simulation softwares such as the POOL 2 program in the absence of specific experimental data.

Table 3 : Suggested values of thermal data for aliphatic derivatives

Chemical compound	Combustion thermal efficiency (%)	Chemical yield for carbon (%)	Ratio of radiant heat (%)
Alkanes	97	93	22
Alcohols	98	97	22
Esters, Ethers, Ketones	96	95	23
Nitriles	98	98	26
Amines	91	82	12
C, H, O, N compounds	95	90	22
C, H, O, S compounds	95	91	/

Table 4 : Suggested thermal data for aromatic and unsaturated cyclic compounds

Chemical compound	Combustion thermal efficiency (%)	Chemical yield for carbon (%)	Ratio of radiant heat (%)
C, H, O, N compounds	70	70	30
C, H, O, N aniline derivatives	75	75	40
C, H, O, N, nitro derivatives	40	45	40

Table 5 : Suggested thermal data for miscellaneous organic compounds

Chemical compound	Combustion thermal efficiency (%)	Chemical yield for carbon (%)	Ratio of radiant heat (%)
Monochlorinated compounds with C, H, N, Cl atoms	55	51	35
Polychlorinated compounds with C, H, O, Cl atoms	50	50	40
C, H, N cyclic compounds	90	95	40
Fluorinated aromatic compounds	55	55	40

Influence of the fuel chemical formula on the toxic emission of fires

The combustion chemical data obtained in the modified Tewarson apparatus, on a great number of chemical compounds can be used to estimate the chemical yield for the conversion of Nitrogen, Sulphur, Chlorine, Fluorine, during combustion. This allows the prediction of missing data for chemical compounds which have not yet been studied in combustion experiments. As an example the chemical yield for the conversion of Chlorine into HCl during the combustion of chlorinated organic compounds was studied on a selection of 25 chlorinated organic derivatives. A summary of the results obtained is given in table 6 for chlorinated and poly-chlorinated derivatives. The data in table 6 can be used as a first estimate of the chemical yield for the conversion of chlorine into HCl, to replace missing data on specific chemical compounds for which no experimental data is available. The chemical yield for the conversion of hetero atoms is an input data to atmospheric dispersion models.

Table 6 : Suggested values of chemical yield for the conversion of chlorine into HCl during the combustion of chlorinated organic compounds

Type of chlorinated compound. Number and position of chlorine atoms	Chemical yield for the conversion of Chlorine into HCl during combustion
One Cl atom per chain or ring	80
Two Cl atoms per chain or ring	40
3 Cl atoms per chain or ring	20
4 Cl atoms per chain or ring	0
5 or 6 Cl atoms per chain or ring	0
3 Cl atoms on the same carbon (- C Cl$_3$)	100 %
Amine hydrochloride (RNH$_2$. HCl)	95 %

The definition of the chemical yield for the conversion of chlorine in table 6 is given in chapter 3 above.

Conclusion

The modified Tewarson apparatus described in this paper, is a useful experimental set-up to study the combustion of chemicals and pesticides. To date, this apparatus has been used to investigate more than 100 different chemicals and pesticides. The thermal and chemical data obtained was used as input data for computer simulation of large industrial fires and for atmospheric dispersion calculations to evaluate the toxic impact of industrial fires.

The modified Tewarson apparatus is also a valuable tool to compare the combustion behaviour of plastics, resins and fabrics while determining the nature of the combustion products.

The modified Tewarson apparatus was used as a bench scale apparatus in the STEP European Program to study fires and their consequences, in the MISTRAL 1 program. The partners associated with this European program are :

CEA / IPSN	(France)
CISI	(France)
CNRS / LCRS	(France)
ENEA / AEAS	(Italy)
ISSEP	(Belgium)
Rhône-Poulenc Industrialisation	(France)
University of Aveiro	(Portugal)
University of Poitiers	(France)

other partners who joined this project in the course of the program are :

EDF / CLI	(France)
INERIS	(France)

It was shown in the course of this European project that the modified Tewarson apparatus was a key item to investigate the consequences of large chemical fires on a wide range of chemicals and pesticides.

Literature

[1] G. Mangialavori, F. Rubino, "Experimental tests on large hydrocarbon pool fires".
7th Int. Symposium on Loss Prevention and Safety promotion in the Process Industries, Taormia, Italy, 4-8 may 1992 - Paper n° 83.

[2] S. Ditali, A. Rovati, F. Rubino, "Experimental Model to assess Thermal radiations from hydrocarbon pool fires". ibid
Paper n° 13

[3] L. Smith - Hansen, "Toxic hazards from pesticide warehouse fires".
8 th Int. Symposium on Loss Prevention and Safety promotion in the Process Industries
Antwerp, Belgium, june 6-9, 1995, I, 265-276

[4] A. Tewarson, F. Tamanini, "Research and Development for a Laboratory-scale flammability test Method for cellular plastics". Final report FMRC serial n° 22524 RC 76 - T 64 (1976)

[5] C.Costa, "Step European program. Study of fires and of their consequences. Mistral 1 program, small scale studies", 1994.

[6] W.M. Thornton, Philos. Mag. 33, 196 (1917)

[7] C. Hugget, "Estimation of the rate of heat release by means of oxygen consumption measurements".
Fire and Material, 4, n° 2, 61-65, 1980

TOP LEVEL RISK STUDY – A COST EFFECTIVE QUANTIFIED RISK ASSESSMENT

R.I. Facer, J.A.S. Ashurst, and K. A. Lee
EQE International Limited, 500 Longbarn Boulevard, Birchwood, Warrington, Cheshire WA2 0XF

> Probabilistic Safety Assessment (PSA) or Quantified Risk Assessment (QRA) is an internationally accepted approach to modelling the contributors to the risk from an industrial facility, plant or process. A full PSA/QRA is a time consuming and costly exercise. A Top Level Risk Study (TLRS) is a means of achieving the benefits of a QRA more quickly, for less cost, and in a form to allow much easier use and interpretation. A TLRS is based upon the combination of experience in modelling and assessing the reliability of complex systems. Risk and Safety Management are continuous processes, especially with respect to the design and implementation of loss prevention and loss control measures. These aspects generally involve a continued expenditure on the plant, and it is important that the expenditure is directed at the areas which will contribute the maximum return in the form of risk reduction. The TLRS forms a useful input to such Cost Benefit Analysis (CBA), and again this information contributes to efficient Risk and Safety Management.
>
> Keywords: Risk assessment, CBA, safety management, option studies, risk reduction.

INTRODUCTION

The Top Level Risk Study (TLRS) is a logical, structured, approach to deriving an indication of the risk from operation of an industrial facility, with identification of dominant weaknesses, and it facilitates modelling of alternative scenarios for safety improvements and Risk Management, utilising probabilistic techniques based upon demonstrated experience based assessments of system performance and reliability.

The TLRS is an Event Tree based assessment, with the top event reliabilities being assigned on the basis of judgement and the use of a set of experience based guidelines, instead of using the normal time consuming Fault Tree approach. It is accepted that this is a more approximate approach than a full PSA/QRA, however, the results are available in a considerably shorter timescale and, experience indicates, are not significantly different.

The information provided by a TLRS includes graphical representation of accident sequences in the form of Event Trees and a quantified analysis of the risk from operation. In addition, and perhaps just as important, the methodology also provides important analysis of both the Initiating Events and the system level Top Events included in the model to allow the most significant data items and assumptions to be most carefully reviewed.

Due to the approximate nature of the assigning of data, the model is developed in such a way as to facilitate numerous sensitivity studies to be performed on both the Initiating Event and system level Top Event data, in an efficient manner. This latter feature also enables ease of modification to the models in order to investigate the relative benefits of different proposed safety improvement and Risk Management scenarios.

EQE International, whilst working for the Government of Bulgaria, embarked on the development of a risk evaluation approach which utilised the completeness and coherence elements of a PSA/QRA, but that would develop useful results in a timely manner. The result is the Top Level Risk Study, which has been successfully applied to nuclear power and nuclear-chemical plants but can easily be adapted for other facilities and processes for the assessment of environmental, safety or business interruption aspects.

METHODOLOGY

The methodology developed for the TLRS is based on the PSA/QRA methodology, but modified to result in a tool which can be used to give timely and cost efficient information for both safety and risk management and audit purposes. A prime concern throughout the development process has been that the approach must remain logical and structured to be of use to Safety and Risk Managers and regulators.

The three key features involved in the construction of a TLRS are Operational Factors/Events, Protections and Consequences. Examples of these are shown in Figure 1. Subsections for each of the three features can be considered for the appropriate industry under review.

The key steps which make up the TLRS approach are as follows:

Understanding the Plant Design

The first, and one of the most important tasks in performing any risk study is to develop a thorough understanding of the plant or facility under investigation. This should cover the system and component design, including all support requirements and arrangements, normal and post fault duty, and any associated operator actions. In addition, it is important that operating regimes and principles, and the safety philosophy are understood.

Fault Schedule

The fault schedule development is similar to that which would be expected for a full QRA. Plant specific faults are identified and a comprehensive listing of all applicable Initiating Events compiled. Alternatively, it may be more appropriate to use a checklist/walkdown approach for identification of the potential hazards. However the listing is developed, it is then reviewed and faults may be removed on the basis of low potential consequence or acceptably low frequency, and others are bounded to give a listing of the faults which are to be carried through to the probabilistic analysis.

The TLRS is an Event Tree based approach which takes each of the identified and bounded Initiating Events and models the potential accident sequences which may follow, taking into account the success and failure of the safety systems and operator actions provided to mitigate against the consequences of the fault. The TLRS approach leads to a very fast run time which enables numerous Initiating Events to be included in the model. In order to model these sequences it is first necessary to identify what the requirements are to protect against each of the faults and the levels of protection required, e.g. fire detection, equipment protection or management control.

The study will address each of the Safety Functions appropriate for the facility, e.g. for a chemical plant such considerations are: inventory, Emergency Shutdown (ESD), pressure relief systems, etc., and the success criteria will be based on existing performance analysis. For process plant it may be more appropriate to adopt a checklist/walkdown approach to confirm the mitigating features which exist, e.g. fire detection and suppression systems. This latter approach is generally more appropriate when considering hazards, either internal or natural, and in practice the review of a process will involve a combination of both approaches.

System Dependencies

The Event Trees forming the model will generally only be detailed to a system level, and as such it is important that the interactions between the systems are thoroughly understood. This should include any support requirements. It may also be appropriate to investigate the relative locations of equipment in order that the potential vulnerabilities to environmental effects and hazards can be addressed.

The approach adopted is to use a dependency matrix which identifies all of the support systems required for each of the operating systems to function correctly. It is important that each support system is also identified as an operating system so that any second or higher order dependencies are identified. Any redundancy in supplying the support services should be identified.

The environmental dependencies are assessed in a similar manner by splitting the plant into separate regions, generally divided by some physical feature, and identifying which regions of the plant each system passes through, or is located in. This is performed for both operating and support systems.

Consequence Categories

To quantify the Event Trees used to model accident scenarios, it is necessary to define a set of consequence categories to which each sequence end state can be assigned. In general the consequence categories will be based on the functional degradation of the plant. The specific categories will depend on the specific facility under investigation, and may relate to safety issues, loss of toxic material, or business interruption and commercial impact.

The number of end states is dependent on many factors, including the specific facility, the level of detail of any transient analysis, and the purpose for which the risk study is being performed. It may be appropriate that only two consequence categories are defined, namely success or failure, however, it is generally found that a number of categories are better since they can be used to reflect the relative seriousness of sequences. In addition, they can assist in distinguishing between availability versus safety issues, and beyond design basis conditions for which some protection exists versus major accidents. For example four consequence categories were used in a TLRS for a nuclear facility, these being:

A. Equipment damage or operational delay. An availability issue, but of very limited hazard.
B. A contained release for which protection exists.
C. Minor release of inventory. A release for which some protection exists.
D. Major release of inventory. A major safety issue.

Event Tree Analysis

The possible accident sequences arising following each of the initiating events are modelled using the standard Event Tree approach of investigating the success and failure of systems provided to operate to protect against the Initiating Event, and assigning a consequence category to each end state. Due to the manner in which the data is assigned to the Top Events (see later), it is necessary that the Top Events are at a system level or, where not possible, at functional sub-system level. It is normal to treat the operator as a system for this purpose.

It is at this stage that the dependency and location matrices are used to ensure that systems are not modelled where they must be considered to have already been made unavailable due to failure of a support system or supply, and that the failure of a necessary support system is represented in a failure of the system.

Figure 2 shows an Event Tree model with Initiating Event A, Protection Factors 1, 2 and 3, and Consequence Categories A-D. The Consequence Categories are assigned once the effect of the failure of Protection Factors 1-3 have been studied and understood.

Setting Up the Model

A TLRS looks at the model under review as a series of systems, which are assigned unreliabilities as follows. The model can easily be adapted to represent an industrial facility or a management plan for a business.

The approach for modelling hazards is slightly different from that for investigating intrinsic plant faults, and it varies further for internal and external hazards.

For internal hazards, such as fire and flood, the possible locations of the hazard are identified and each treated as a separate initiating event. The different locations are defined as those with hazard barriers, or equivalent, separation. It should be noted that where there is significant distance between equipment that may be considered to be acceptably segregated.

For each hazard location, the impact on the plant and the success or failure of mitigating equipment, both detection and suppression, are investigated in a similar manner to that described above for intrinsic plant faults. The consequence categories assigned are generally the same as those used for the intrinsic plant faults.

For natural hazards of sufficiently greater magnitude, it is generally the case that they will result in some impact on the plant requiring it to be shutdown, e.g. a seismic event or extreme wind is generally assumed to result in a loss of off-site electrical supply. The approach to modelling the plant response to such hazards is therefore to identify what consequential Initiating Events could be considered to have occurred, and model these in a similar manner to that described for the intrinsic plant faults, but taking into account any potential impact on the mitigating equipment by the natural hazard. The impact on the plant equipment will be assessed using equipment fragility (probability of failure as a direct consequence of the hazard), either plant specific if available, or based on generic information. Again, the same consequence categories will be used for the sequence end states.

Fault Frequency and Top Event Data

The assigning of data to the Initiating Events follows the traditional approach based on available operational experience data, historical and world data for equivalent plant.

Assigning data to the Top Event is done using judgement and a set of guidelines. The guidelines used will be specific to the plant or facility being modelled, but will ensure that a consistent approach is adopted to the assigning of data. This is important to ensure that the results of the importance studies can be used to identify dominant weaknesses. For each facility there is one set of guidelines for the systems based Top Events and another for the operator based Top Events.

One of the factors on which the rules for the system based Top Events will be based is its function and any redundancy in the system to achieve this, e.g. the number of trains and any spare capacity. It will also be necessary to take account of the separation of any redundant items, system locations, i.e. the potential for the system to be affected by the initiating event or another safety system failure. The diversity between redundant equipment and system may also be represented in the event tree and taken into account.

The control philosophy of the system and the control logic and any associated instrumentation will have an effect on system reliability, for instance the number of signals required, or the majority voting logic may be taken into account.

Other important factors may be the level of maintenance and inspection, the design and manufacturing criteria, safety culture at the plant and the work control procedures, etc.

The rules applied to deriving the operator reliability are more simple and are generally based on a combination of the time available for the action, the indications available and the procedures governing the action. A band of data may be appropriate to allow for modelling of any administrative improvements, e.g. improved housekeeping, procedural development, training, etc.

The structure of the Top level Study is such that the effect of each of these rules and their application can be investigated by performing sensitivity studies on the model. This is discussed further below.

Evaluation of Risks

The purpose of the evaluation phase of the TLRS is to calculate the total predicted frequency of each of the pre-defined consequence categories. EQE use either a proprietary event tree code, or an in-house code which is tailored to provide extensive importance analyses and model manipulation capabilities.

The Outputs from the TLRS

The evaluation of the TLRS gives a numerical indication of the actual risk from each of the identified Initiating Events and the total frequency of any of the consequence categories. Although important and useful, this is by no means the extent of the information available from the TLRS. The importance analyses will give an indication as to the dominant contributors to the different consequence categories either as weaknesses in the protection (Top Events) or the major vulnerabilities of the design (Initiating Event). Top Events can also be grouped together for analysis. This information is important to identify which areas require special attention and should be the subject of loss prevention exercises. This information helps to prioritise improvement activities or can be used to justify that changes in some areas will make little impact upon either the risk or other consequences.

The model is easy to modify and manipulate which enables numerous sensitivity studies to be performed and therefore allows the effect of the data approximations to be investigated and provides some confidence in the results. It also makes it eminently suitable as a decision making tool for assessing the relative benefits of proposed modifications to the plant.

Figure 3 shows one of the outputs of the Event Trees analysis as a bar chart, detailing the frequency of risk for each of the consequence categories A-D.

The Advantages of the TLRS

The main advantage of the TLRS over a full PSA/QRA approach is the time and cost. The timescales enable safety improvements and Risk Management decisions and actions to be taken at an earlier time and therefore enables risk reduction in the short term. The low costs make the TLRS an affordable tool and available to all parties, including the designers, the safety managers, the operators, the regulators/licensers, risk managers, investors, insurers and brokers.

The TLRS does not compromise the requirement for a logical approach in reducing the timescales and as such proves very useful in removing any pre-conceived ideas as to which are the major contributors to the risk. The importance analyses provide great insight into this area.

The ease of changing data and the model facilitates numerous modification options to be investigated, and as such can form an invaluable tool in the execution of Cost Benefit Analysis or option studies. Although the data is derived on a approximate basis, the ease of changing the data also enables investigation into the sensitivity to the data used, and provides some confidence as the acceptability of the approach.

Figure 4 shows the results obtained from a sensitivity study, compared to the original TLRS analysis and the saving that can be made by carrying out specific modifications to the plant or business model..

The TLRS also forms the perfect basis from which a full scale PSA/QRA can be developed if that is the requirement. The TLRS has the advantage that it has the same structure as a PSA/QRA and as such the full PSA/QRA could be developed in a staged manner over a period of time, whilst maintaining a risk model which is available to be used throughout the full PSA/QRA development process.

The TLRS has been used in a comparison with a full approximately million pound PSA/QRA. The difference in the results was small (less than a factor 2) although the TLRS cost less than £50,000 for the equivalent study.

System Failure Probabilities

Empirical evidence suggests that the failure probability of a system, or groups of components, is dependent on a number of factors, i.e.

- the minimum level of redundancy within the system;
- the separation/segregation between similar components;
- the Control & Instrumentation logic;
- the maintenance practices followed; and
- the plant safety culture.

Furthermore, the evidence would suggest that the failure probability of a system cannot be reduced below a minimum value which is related to the level of redundancy provided within the system.

Given the above, the TLRS assumes that the unavailability of a system, or other groups of components, can be estimated on the following basis:

Basic System Failure Probability * Degradation Factors.

The basic system failure probability of a system, or a group of components is dependent on a number of factors, i.e.

- Commonality/separation/segregation.

This factor accounts for the affects on the system failure probability of the number of:
a) passive component failures required to fail the system;
b) separate locations of groups of similar components, i.e. the number of location in which failure needs to occur to fail all of the components; and

c) segregated areas of groups of similar components, i.e. the number of segregated areas in which failure needs to occur to fail all of the components.

- Control & Instrumentation logic.

This factor accounts for those aspects of the Control & Instrumentation, etc., that are not modelled explicitly within the TLRS. Factors considered are the minimum number of:

a) Control & Instrumentation sensors, transmission or voting logic failures required to fail the system; or
b) control actions which may prevent the system operating correctly.

- Maintenance practices.

This category addresses the impact of the maintenance practices that are followed, that is whether a regulatory approved maintenance, testing and inspection regime is followed or not.

- Safety Culture.

This category address the impact of the general attitude of the plant staff towards safety and equipment availability.

The appropriate degradation factor is determined by taking into account the safety culture, the minimum number of active component failures required to fail the system. It should be noted that, although where more than one aspect appears to apply, only the highest single degradation factor is applied for each of the above four categories.

Operator Actions

Following several faults, or fault sequences, the operator is required to perform certain actions in order to either maintain the continued operation of a safety system, or to prevent further deterioration. For a satisfactory assessment of the risk of operation, the likelihood that the operator fails to carry out the claimed actions must be estimated.

In the TLRS, the probability that the operator fails to perform a claimed action in good time is estimated depending on the time taken to take action. As an example of a particular application the categories can be split into:

<u>Operator action within an hour.</u> If it is necessary to initiate a safety system, or stop the fault from the control room or at a point local to plant; or to take action to prevent further deterioration;

Operator action in greater than one hour. In order to re-establish equipment starting procedures, or system which can be manually commissioned.

Operator action after 8 hours. Following common practice, an operator reliability of a further order of magnitude is claimed due the change of shift.

Operator action after 24 hours. Another order of magnitude is claimed due the multiple change in shifts and the availability of other advisory personnel.

It is assumed that all the claimed actions are effectively backed by the appropriate procedures. Furthermore, it is assumed that these procedures are reinforced by appropriate training.

TLRS IN SUMMARY

The benefits in time and cost when carrying out a TLRS are particularly of use in the identification of dominant weaknesses within the plant or business model, where the consequences may be significant either in terms of safety to workers, or loss of market share and business interruption. Follow-up sensitivity analysis can then be carried out to look at the ways in which the risk could be realistically reduced by modifying specific areas of weakness within the plant or model. This would enable the expenditure to be directed towards areas of the plant where the effect on risk reduction would be the most effective.

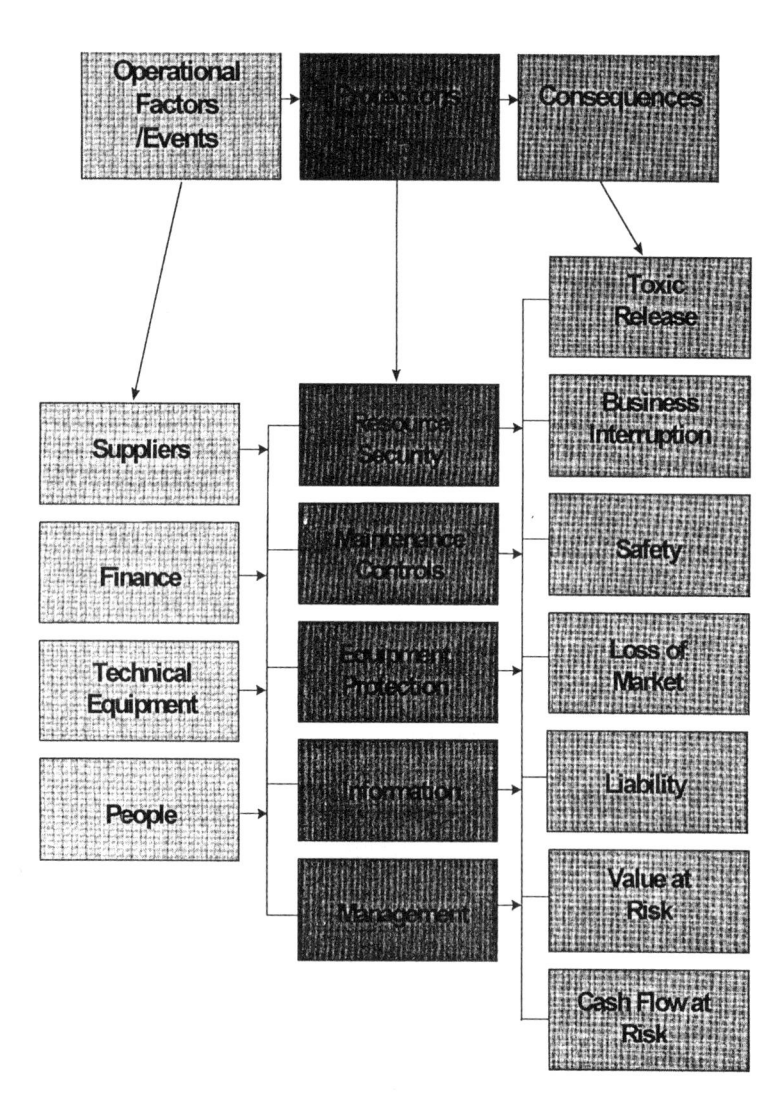

Figure 1: Construction of TLRS

Figure 2: Event Tree Layout

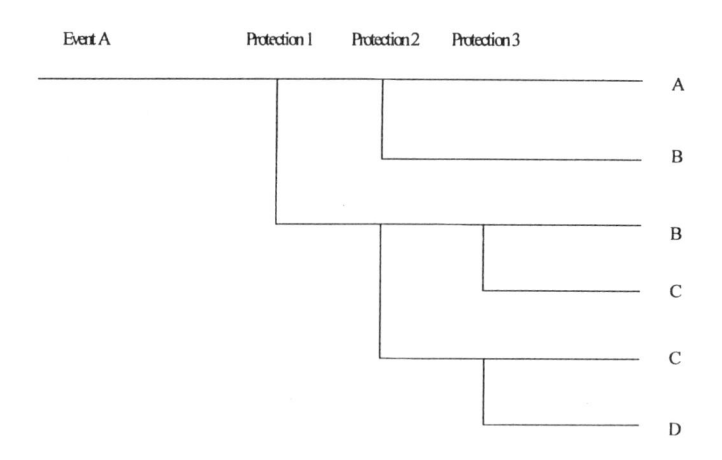

Figure 3: Quantification of Risk for Consequence Categories A-D

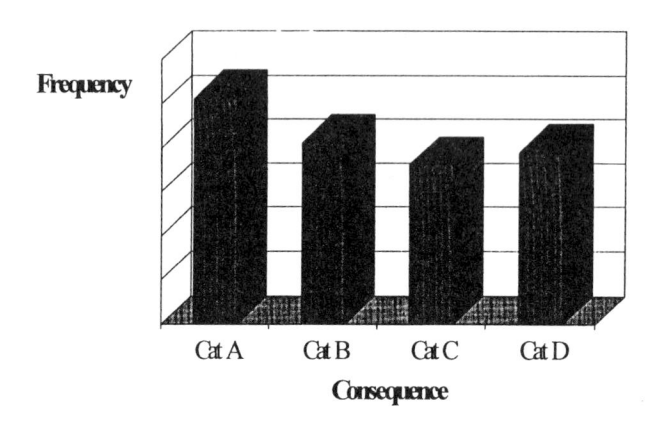

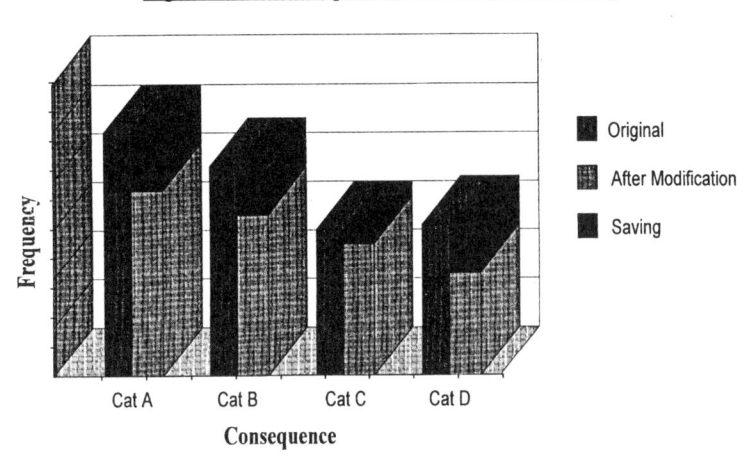

Figure 4: Cost of Improvement v Risk Cost Saving

APPLICATION OF CASE-BASED REASONING TO SAFETY EVALUATION OF PROCESS CONFIGURATION

A.-M. Heikkilä, T. Koiranen and M. Hurme
Helsinki University of Technology, Department of Chemical Technology,
P.O. Box 6100, FIN-02015 HUT, Finland

> An index based method for estimating the effect of process configuration on inherent safety has been presented. Process configuration means which operations are involved in the process and how they are connected together. An inherently safe process structure is not possible to define by explicit rules, but one has to rely on standards, recommendations and accident reports. When problem solving is based on experience it is possible to use case-based reasoning. Therefore CBR was employed for determining the value of safe process structure subindex. For this purpose a casebase of accident cases and recommended designs was created. A case can be retrieved on five levels of aggregation (from process to detail) from the casebase to ensure the relevancy of retrieved information in various phases of preliminary process design.
>
> Keywords: inherent safety, case-based reasoning, process configuration

INTRODUCTION

The most important process design decisions are made during the conceptual design phase when the process route is selected. Also the process safety - especially the inherent safety - is determined by the early design decisions. The essence of the inherent safety is to avoid hazards by proper design rather than to control them by added-on protective systems (Kletz, 1991). Thus inherent safety is related to the selection of chemicals, process conditions and operations which are used. Based on these factors Heikkilä et al. (1996) have developed a methodology and a safety index which allows safety comparison of process alternatives to be done in the conceptual design phase.

In the Inherent Safety Index the subindex of safe process structure describes the inherent safety of different process configurations. In this subindex the designed process structures are compared with known safe and unsafe process solutions. Experience based data from different process solutions can be found in safety standards, accident reports and design recommendations. While experience based data has no explicit rules, the safe process structure need to be evaluated by case-based reasoning. The advantage of case-based reasoning is its ability to present a more explicit representation of knowledge compared to rule-based reasoning, since it is based on detailed case histories rather than on their interpretation and recollection by an expert. Thus no heuristic rules which are based on generalizations are needed since case-based reasoning relies on analogies. This same approach is used mentally by practicing engineers to generate new process designs.

Extensive databases have been collected from accident reports (Anon, 1996). Also much data has been published as design recommendations of different process systems (Lees, 1996). From this data a database of good and bad designs can be collected. By using this database case-based reasoning can check, if a new design resembles known safe or unsafe design cases.

INHERENT SAFETY INDEX

In process synthesis an interactive rule-based system can be used for generating process alternatives (Hurme and Järveläinen, 1995). The comparison of generated alternatives is based on economics, safety and environmental considerations. For the safety comparison the Inherent Safety Index (ISI) is used (Heikkilä et al., 1996). ISI is formed of several subindices, which describe reactivity, flammability and toxicity of chemicals, inventory, process conditions, type of equipment and process structure.

Total Inherent Safety Index (Eq.1) is calculated for each process step separately as a sum of Chemical Inherent Safety Index (Eq.2) and Process Inherent Safety Index (Eq.3). More details of the method have been given by Heikkilä et al. (1996).

$$I_{TI} = I_{CI} + I_{PI} \tag{1}$$

$$I_{CI} = I_{RM, max} + I_{RS, max} + (I_{FL} + I_{EX} + I_{TOX})_{max} + I_{COR, max} + I_{INT, max} \tag{2}$$

$$I_{PI} = I_I + I_{T, max} + I_{p, max} + I_{EQ, max} + I_{ST, max} \tag{3}$$

The subindices for the heat of main and side reactions, I_{RM} and I_{RS}, are related to the maximum heat release from the process. Flammability (I_{FL}), explosiveness (I_{EX}), toxicity (I_{TOX}) and corrosiveness (I_{COR}) describe the hazardous properties of the chemical substances. Subindices I_{FL}, I_{EX} and I_{TOX} are summed for each substance separately, and the maximum sum is used in the calculations. Chemical interaction (I_{INT}) describes the reactivity between the substances present in the process.

The subindex for inventory (I_I) is related to the amount of process materials present in the process. Process temperature (I_T) and pressure (I_p) reflect the maximum temperature and pressure in the process. Equipment safety (I_{EQ}) describes the safety of individual process items such as a pump or a reactor. The estimation of the Safe Process Structure Subindex (I_{ST}) is discussed in more detail in this paper.

SAFE PROCESS STRUCTURE SUBINDEX

The safe process structure means which operations are involved in the process and how they are connected together. Therefore the Safe Process Structure Subindex describes the safety of the process from systems engineering point of view. It describes: how well certain unit operations or other process items work together, how they should be connected and controlled together. The index describes also how auxiliary systems such as cooling, heating or relief systems should be configured and connected to the main process.

The importance of this subindex is increasing as the processes are becoming more integrated through heat and mass-transfer networks.

The Process Structure Subindex does not describe the safety of process items as such or their interaction through nonprocess route (i.e. through layout), since this is described by the Equipment Safety Subindex (Heikkilä and Hurme, 1998).

EVALUATION OF SAFE PROCESS STRUCTURE

Many different alternative process configurations can be created for a process in the conceptual design phase. In choosing the most feasible alternative safety should be one of the major evaluation criterias. Therefore information on the safety features of alternative process structures are needed on preliminary process design (Heikkilä and Hurme, 1998).

Most of the subindices of ISI are quite straightforward to estimate since they are e.g. based on the physical and chemical properties of compounds present. The process structure subindex looks at the process from a systems engineering point of view. Therefore it is much more difficult to estimate. In fact there is no explicit way of estimating the safety of the process structure but one has to rely on experience based data which is documented as standards, design recommendations and accident reports.

SOURCES OF EXPERIENCE BASED SAFETY INFORMATION

Process solutions have shown their strong and weak, safe and unsafe points in operation practice. The knowledge of practising solutions consists of the details collected during the operation and maintenance. Practising solutions reveal for instance which unit operations are preferable for certain purposes and how the units can be connected safely together. Some of the information can also be found from design standards which have been created on the basis of the experience on the operation of existing process plants (Lees, 1996).

Another source of design information is the accident reports made after an accident. They give valuable information of the possible weaknesses that can occur in unit operations, while they are used for certain purposes. In the past many of the unit operations have shown their adverse characteristics. This information is mainly collected to accident reports and included to safety standards. Accident reports tell us for example:
* which process equipment configurations have unfavourable properties
* which type of chemicals do not suit to certain unit operations
* which unit operations/ configurations are risky
* when the connection of process units should be avoided.

The difficulty in utilizing accident reports lies in the lack of accident report standards. Reports vary a lot how they document the details of the accident itself, the path to the final event, the causes, and the consequences. Still the reports can tell much experience based information which can - and should be - utilized in designing new plants. In fact a major goal in improving the design of safe process plants should be to enhance the reuse of design experience. This is important since the same mistakes are done again and again (Kletz, 1991).

A more refined form of accident reports is an accident database, where all the reports are presented in a standardized format. Extensive databanks have already been collected from accident reports (Anon, 1996). This kind of standardized format allows easier retrieval of accident information also by computerized means.

STRUCTURE OF THE DATABASE

The basis for the estimation of safe process structure lies in the integration of the two types information sources: 1) recommendations and standards how the process should be designed, and 2) accident database which describes the negative cases from which one can learn. Therefore a casebase of good and bad design cases is needed. Both of these information sources should be readily available to the design engineer through the database. A design problem can be compared with the cases in this combined databank for instance by case-based reasoning.

In this approach accident cases and design recommendations are analysed level by level. In the database the knowledge of known processes is divided into categories of process, system, subsystem, equipment and detail (Fig. 1). Process is an independent processing unit (e.g. hydrogenation unit). System is an independent part of a process such as reactor or separation section. Subsystem is a functional part of a system such as a reactor heat recovery system or a column overhead system including their control systems. Equipment is an unit operation or an unit process such as a heat exchanger, a reactor or a distillation column. Detail is an item in a pipe or a piece of equipment (e.g. a tray in a column, a control valve in a pipe).

A search for cases in the databank can be made on these levels on the basis of the nature of the design problem. If a process is being designed from beginning the first search is made for a whole process. The search is then made for those systems, subsystems and equipment, which are informable for the design. On the basis of the retrieved information the designer can evaluate the right index value for the process structure of the section under review. The input data for a database search contains information on the process level and on the raw materials and products, reaction types and their details such as catalysts and phase of reaction. As output there is information about the unfavourable process configurations, recommended configurations and accident cases.

A plant is divided into inside and offsite battery limit areas. The configurations of ISBL and OSBL areas differ considerably. Generally the size of equipment, the amount of chemicals and also the spacings are larger in OSBL area. The safety of the process structure is also affected by these factors. Therefore this aspect is included also into the database.

The database does not always contain information which is directly related to the process under review. Therefore it is important to be able to use analogies. In general much of the design of new processes relies on analogies. For example most hydrogenation processes have similar features, most tanks of liquefied gases have similarities etc. For that reason information has been included into the database on the type of materials in incident (e.g. liquefied gas), the type of the reaction (e.g. oxidation), the thermal nature of the reactor (e.g. exothermic), the phase of the reaction and the type of catalyst.

To learn from the accident cases it is essential to indicate the type of incident which happened (e.g. explosion), the direct cause of the incident (e.g. static electricity), the reason why this could take place (e.g. filling through the gas phase) and finally - most important - the lesson how this can be avoided (e.g. fill the tank through the bottom).

CASE-BASED REASONING

When problem solving is based on experience which is difficult to define as explicit rules, it is possible to apply case-based reasoning (CBR). CBR uses directly solutions of old problems to solve new problems. The functional steps in CBR are (Gonzalez and Dankel, 1993):

1. New problem presentation.
2. Retrieval of the most similar cases from case-base.
3. Adaptation of the most similar solutions for generating a solution for a current problem.
4. Validation of the current solution.
5. Learning from the problem cases by adding the verified solution into the case-base.

A data table of a case-base can be divided into input and output sections. Input parameters are retrieval parameters and output parameters are design specification parameters. The problem is characterized as input data to the system. In the retrieval phase a set of retrieval parameter values of all cases in the case-base are compared to the input data. The most similar cases are then selected and ranked based on the comparison.

In the case of string data types suitability is simply:

$$X_i = C_{i,j} \Rightarrow Y_{i,j} = 1 \quad (4)$$
$$X_i \neq C_{i,j} \Rightarrow Y_{i,j} = 0 \quad (5)$$

where X_i is the input value of parameter i, $C_{i,j}$ is the value of parameter i of case j, and $Y_{i,j}$ is the suitability of a parameter i for the case j.

The quality of reasoning increases, if the importance of selection parameters can be altered. The user should determine the importances of selection parameters for the topic under study. Weighted suitability R_{ij} can be expressed:

$$R_{ij} = W_i Y_{i,j} \quad (6)$$

where W_i is weight factor of a selection parameter i evaluted by user. Overall suitability can be calculated for the case j based on the number of parameters N and parametric suitabilities R_{ij}:

$$S_j = \frac{\sum_{i=1}^{N} R_{i,j}}{N} \quad (7)$$

Case-based reasoning has earlier been used for instance for equipment design. Koiranen and Hurme (1997) have used case-based reasoning for fluid mixer design and for the selection of shell-and-tube heat exchangers. They have included an estimation of design quality for the case retrieval beside technical factors.

Chung and Jefferson (1997) have combined the IChemE Accident Database (Smith et al., 1997) with case-based reasoning to create an automatic data retrieval for designers' and operators' use. They intend to develop an intelligent system, which takes for example the term 'electrical equipment failure', works out all the related terms and retrieves the relevant information automatically. The method should be integrated with computer tools used by designers, operators and maintenance engineers so that appropriate accident reports can be automatically presented to the user. The employed IChemE database contains much information on accident causes. The aim of the system presented by Chung and Jefferson (1997) is to find all relevant causes of past accidents to improve processes, whereas our CBR system is intended for reasoning on the structure of a process and its favourable and unfavourable characteristics for preliminary process design purposes. The database used by Chung and Jefferson (1997) is an accident database, whereas our database contains also design recommendations. On the other hand our CBR system is intended specifically for the use of process designers, but the system of Chung and Jefferson (1997) is developed for wider use from chemical plant designers and operators to maintenance teams.

DESCRIPTION OF PROTOTYPE APPLICATION

Prototype CBR application has been implemented on MS -Excel spreadsheet. The program has been organized on several sheets. A database of cases was created which consists of accident cases collected from literature (e.g. Lees (1996) and Loss Prevention Bulletin) and of design recommendations. The application program includes retrieval functions which are used to retrieve the most suitable cases from the database.

Input and output parameters

The scope of a database search is defined by using categories of process, system, subsystem, equipment and detail as input parameters. This hierarchy is used for clarifying the process structure and for making the use of process analogies more feasible in reasoning. E.g. a condenser has certain safety characteristics undependent on the process it is located. Beside the process structures input parameters include the raw materials and products and some reaction details. The importance of the parameters may be evaluated by using weighting factors.

Output parameters contain the input parameters plus information on the safety characteristics of the process and information on accidents and their causes. Specific design recommendations are included in the output. On the accidents the output describes e.g. following information:

* what kind of incidents have happened
* what is the actual cause of the incident
* what are the contributing factors or circumstances of the incident
* how to improve the application for better safety

All stored cases are validated on the basis of the Safe Process Structure Subindex. The validations are given for every case and included in the output. Further information on the cases is given as appendices, which describe the case in more detailed.

Retrieval of cases

In this work the cases in the database are stored on their own MS-Excel worksheets. The stored cases are copied on a retrieval calculation data sheet during the retrieval phase. All retrieval parameters in this application are textual string parameters. Thus the comparison between casebase and input problem is simple. When the input value is equal with the case value, the distance is 1, otherwise the distance is 0. The weighted suitability of parameters is then calculated by Equation 6. The weighting factors are introduced by the user. Overall suitability is calculated by Equation 7. Cases are ranked according to their overall suitability and the five nearest cases are shown for the user on an output worksheet.

The retrieval of cases can be done in several steps. The first step is the evaluation of the process with the stored cases. This way can be seen, if the process is safer or unsafer than the alternative processes. The second step is the safety evaluation of specific process systems, subsystems or pieces of equipment. The database contains improvement recommendations to avoid the same accidents happening again. The evaluation of processes can be extended to detailed level. Also the equipment details or safety valves etc. can be checked on this level.

INHERENT SAFETY INDEX OF SAFE PROCESS STRUCTURE

All included processes and their subprocesses in the database are evaluated according to the Inherent Safety Index. Process structures are divided into six groups of scores from 0 to 5 according to the knowledge of their safety behaviour in operation.

The first group is the safest group with the score 0. It consists of recommended and standardized process and equipment solutions. The second group is based on sound engineering practice, which implies the use of well known and reliable process alternatives. In the third group there are processes which look neutral, or on which there is no safety data available. The fourth group includes configurations which are probably questionable on the basis of safety even accidents have not occured yet. The fifth and sixth groups contain process cases on which documented minor or major accident cases exist.

Table 1. Values of the Safe Process Structure Subindex I_{ST}

Safety level of process structure	Score of I_{ST}
Recommended (safety etc. standard)	0
Sound engineering practice	1
No data or neutral	2
Probably unsafe	3
Minor accidents	4
Major accidents	5

CASE STUDY

As a case study an acetic acid process is discussed. Acetic acid is produced by the liquid-phase methanol carbonylation. The reaction is carried out at 175 degrees of Celsius and 30 bar pressure. The process diagram is shown in Figure 2.

For the safety evaluation CBR database searches were done on two levels. First level was the acetic acid process as a whole. On the second level the reactor system was studied in more detail.

Reasoning on the acetic acid process alternatives

First the acetic acid process was studied as a whole to find out if the alternative processes have differences in the safety on the conceptual (i.e. process) level. The search (Table 2) found cases for carbonylation and oxidation processes (Table 3). It can be seen that there has been explosions and fires on both types of plants. The explosion in the carbonylation plant was due to static electricity in loading of a storage vessel. This type of explosions are not specific to carbonylation plants, but

Table 2. Input data of the search for the acetic acid process

INPUT DATA

Retrieval parameters	Active	Importance	Value
raw material	TRUE	9	methanol
product	TRUE	9	acetic acid
reaction type	FALSE		
termic type of reaction	FALSE		
phase of reaction	FALSE		
catalyst	FALSE		
ISBL / OSBL	TRUE	6	isbl
system	FALSE		
subsystem	FALSE		
equipment	FALSE		
detail	FALSE		

they are possible also in many other processes. The fires and explosions on the oxidation plants were related to the chemicals present in that process. They are more likely to happen in such a plant than somewhere else. Thus the carbonylation process can be considered safer than the oxidation process based on the information from this search.

Reasoning on the reactor heat transfer system

In the second phase searches were made on the system and subsystem level. This is needed for the design of the reactor and its heat transfer systems. Carbonylation of methanol is an exothermic reaction. Thus only the exothermal reactors were searched. The CBR search found two cases which are recommendations on the design of exothermic reactors with heat transfer systems. They are shown in Figures 3 and 4.

The case in Figure 3 represents a reactor with two different cooling systems. In the not recommended case (right) the cooling system presents a feedback loop between a reactor heat rise and the rise in the coolant temperature, which should be avoided. On the left is the recommended system, where the coolant temperature does not depend on the reactor temperature.

The case in Figure 4 shows a heat recovery system of a reactor. The not recommended case on the left shows the feed to an exothermic reactor being heated by the product. In this case the temperature rise in the reactor may lead to the temperature rise in feed. The recommended case on right is safer since the connection is broken because the heat transfer is done by generating and using medium pressure steam.

Table 3. Output data of the search for the acetic acid process

	1st Case	2st Case	3st Case
PROCESS:			
Raw material	methanol	butane	butane
Product	acetic acid	acetic acid	acetic acid
Reaction type	carbonylation	oxidation	oxidation
Thermic type of reaction	exo	exo	
Phase of reaction	liquid	liquid	liquid
Catalyst	Rh complex		
Isbl / Osbl	isbl	isbl	isbl
SYSTEM		reaction	reaction
SUBSYSTEM	intermediate storage	purging	feed
EQUIPMENT	tank	reactor	boiler
DETAIL	inlet pipe		
Incident:	explosion	fire	explosion
Cause 1:	static electricity	self-ignition of acetaldehyde	oxygen leak
Cause 2:	filling through vapor phase	methane ignited	
Recommendations:	fill through bottom		
Material:	acetic acid	acetaldehyde	butane/air
Nature of material:	organic acid	aldehyde	LPG
Safety Index (0-5)	4	4	5
Appendix:			App.1

Appendix 1: Explosions occured because pure oxygen entered a gas-fired boiler and mixed with the butane and steam used to form acetic acid. The first blast occured near a gas fired boiler and the second blast occured at a nearby reactor. (3 killed, 37 injured)

Score of the Safe Process Structure Subindex

From the reasoning on the process level we get score 2 (no data or neutral) for the carbonylation process, since the found case was not specific to this process. For oxidation process we get score 5, since a major accident has taken place.

For the recommended reactor system we can get scores 0 (recommended/standard) or 1 (sound engineering practice) depending how we value these recommendations.

The final score of the Safe Process Structure Subindex for the carbonylation process would be 2 based on this limited reasoning, since the final score of I_{ST} is chosen on the basis of the worst case. Of course in practice one should do the reasoning on all the systems and subsystems in the process.

This case study was given only to represent the principle of CBR in reasoning the value of the Safe Process Structure Subindex.

CONCLUSIONS

In this paper an index based method for estimating the effect of process configuration on inherent safety has been presented. Process configuration means which operations are involved in the process and how they are connected together. The Safe Process Structure Subindex describes the safety of the process configuration from system engineering point of view. Importance of this aspect is becoming more important since the processes are becoming more and more integrated.

Since an inherently safe process structure is not possible to define by explicit rules, information which is based on cases and engineering experience has to be used. This type of knowledge is presented as standards, engineering recommendations and accident reports.

When problem solving is based on information which is difficult to define as rules, it is possible to use case-based reasoning. The advantage of case-based reasoning is its ability to present a more explicit representation of knowledge compared to rule-based reasoning, since it is based on detailed case histories rather than their interpretation and recollection by an expert. Thus no generalizations are needed since CBR relies on analogies. This same approach is used mentally by practicing process engineers to generate new process designs.

For the estimation of the Safe Process Structure Subindex a casebase of good and bad cases was created from recommendations, standards and accident reports. The cases can be retrieved on five levels of aggregation (process, system, subsystem, equipment, detail) to ensure the relevancy of retrieved information for various phases of process design. The cases are valued by using the Safe Process Structure Subindex. The subindex has six values (0-5) representing recommended design, sound engineering practice, no data or neutral, probably unsafe, minor and major accident cases. The final score of the subindex is chosen on the basis of the worst case of different levels of the reasoning. The results can be used with other subindices for estimating the total inherent safety of process alternatives for the selection of process concept or details of the process configuration.

The results of the database search can be presented as reports which highlight possible danger points of the process. These reports should follow the process alternatives till the end of process design and even till the operation stage of the process.

REFERENCES

Anon. (1996), Accident Database Expands, *The Chemical Engineer*, 25 April, No 610, 5.

Chung, P.W.H. and Jefferson, M., Accident Databases - Indexing and Retrieval, *EC/EPSC seminar on "Lessons Learnt from Accidents"*, Linz, 16-17 October 1997.

Gonzalez, A.J. and Dankel, D.D., (1993), *The Engineering of Knowledge-based Systems: Theory and Practice*, Prentice-Hall International, Inc., New Jersey.

Heikkilä, A.-M., Hurme, M., Järveläinen, M., (1996), Safety Considerations in Process Synthesis, *Computers chem. Engng*, **20** Suppl. S115-S120.

Hurme, M. and Järveläinen, M., (1995), Combined Process Synthesis and Simulation System for Feasibility Studies, *Computers Chem. Engng,* **19**, Suppl., S663-S668.

Kletz, T. (1991), *Plant Design for Safety*, Hemisphere, New York, 11.

Koiranen, T. and Hurme, M., (1997), Case-Based Reasoning Application in Process Equipment Selection and Design, *SCAI'97 Sixth Scandinavian Conference on Artificial Intelligence*, G. Grahne (Ed.), IOS Press, Amsterdam, 273-274.

Lees, F.P., (1996), *Loss Prevention in the Process Industries: Hazard Identification, Assessment and Control*, Butterworth-Heinemann, Oxford, UK.

Smith, M.K., Bond, J. and Mellin, B., (1997), *The Accident Database: capturing corporate memory*, SHE Department, IChemE, Rugby.

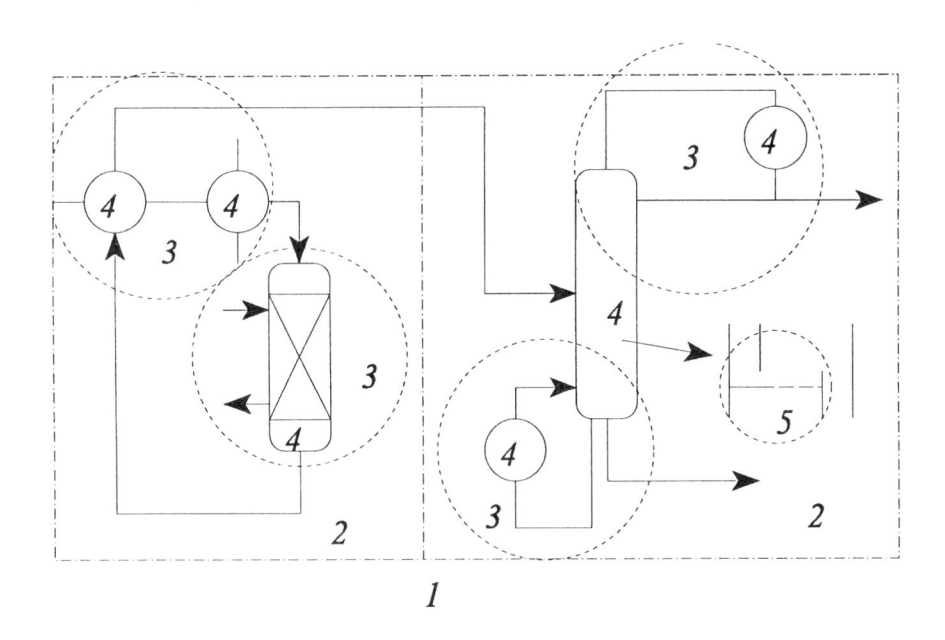

Figure 1. Example of the levels of the process as used in the CBR database. (1 = process, 2 = system, 3 = subsystem, 4 = equipment, 5 = detail)

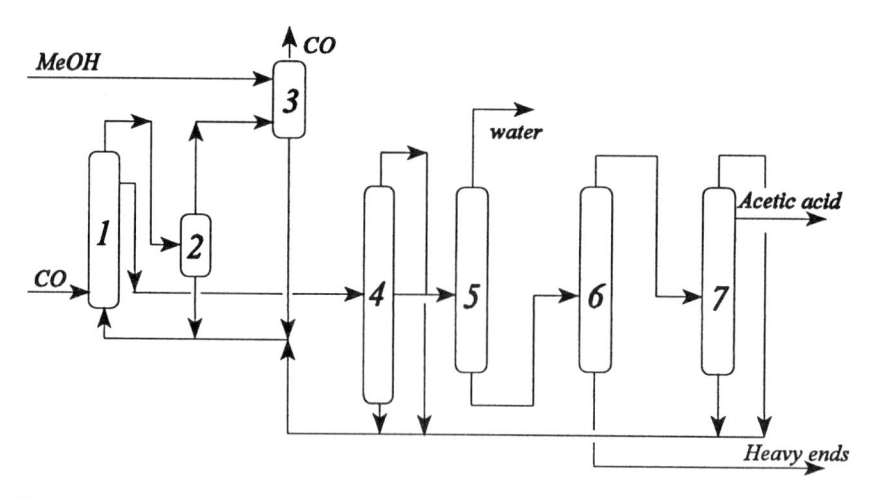

Figure 2. Flowsheet of the acetic acid process: 1) reactor, 2) separator, 3) scrubber, 4) light ends separator, 5) drying column, 6) product recovery, 7) product finishing

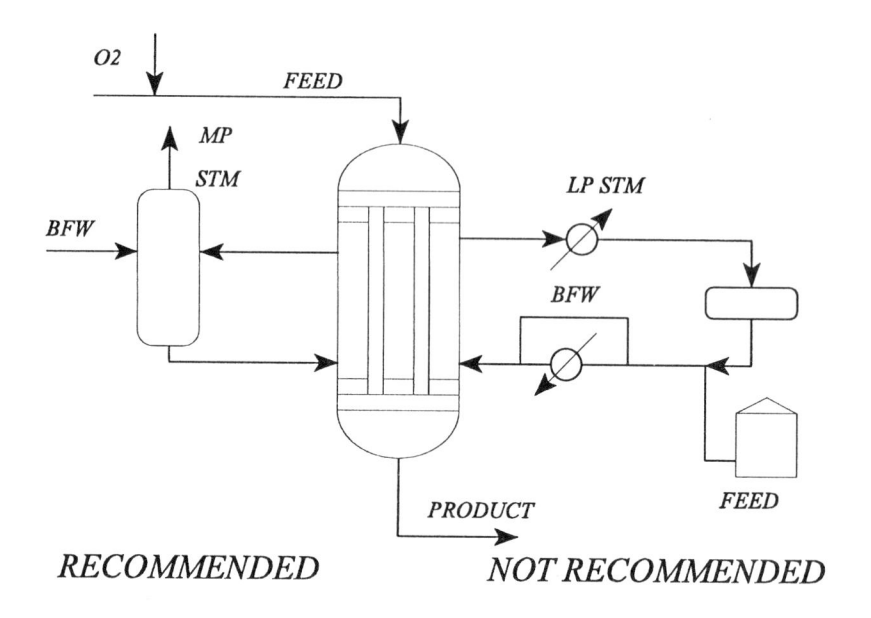

Figure 3. A recommendation to avoid the feedback loop between a reactor heat rise and a rise in coolant temperature.

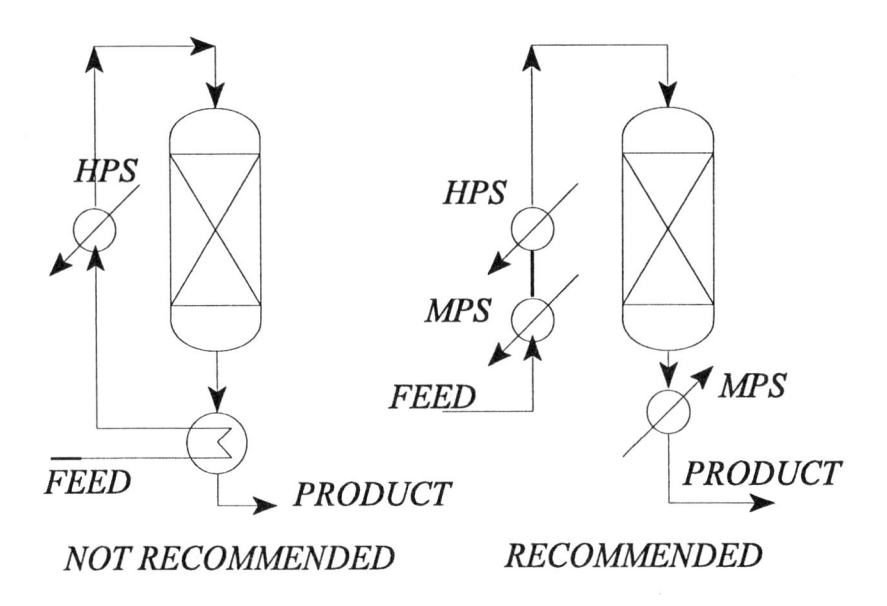

Figure 4. A recommendation for preheating the feed of an exothermic reactor.

INDEX METHOD FOR COST-EFFECTIVE ASSESSMENT OF RISK TO THE ENVIRONMENT FROM ACCIDENTAL RELEASES

A J Wilday[*], M W Ali[+] and Y Wu[^]
[*] Health and Safety Laboratory, Broad Lane, Sheffield, UK
[+] Universiti Teknologi Malaysia, [^] University of Sheffield

>The objective of this study was to develop a methodology by which the risk from major accident hazards to the environment can be assessed or quantified. An index method, the Environmental Risk Index (ERI), has been developed based on an existing method, the Environmental Hazard Index (EHI), and the Department of the Environment (DoE) definitions of events that would constitute a major accident to the environment. The method includes proposed criteria for risk tolerability. The method requires more extensive testing and revision to accommodate forthcoming revisions to the DoE criteria for major accidents to the environment.

>Keywords: Risk, environment, major hazard, risk index.

INTRODUCTION

Risk assessment is a useful technique for allocating resources for protection of human safety, product quality and the environment in such a way that the highest priority is given to the highest risk. Methodologies, e.g. HAZOP and HAZAN, are well-established for safety purposes. However, such methodologies are not yet well-developed for accidental release of chemicals into the environment.

The problem with risk assessment for environmental accidents is that the environment is extremely complex. The Department of the Environment (DoE) (1) has produced guidance on the type of events which would comprise a major accident to the environment and such events include both short-term and long-term effects to land, water, eco-systems, buildings and public access. See Table 1. This guidance is in the process of revision by the Department of Environment, Transport and the Regions (DETR) but serves to illustrate the wide diversity of possible accidental harm effects to the environment.

A full risk assessment for environmental accidents might include: identification of possible release events; estimation of the frequency of such events; development of an event tree for each release event to determine all possible types of environmental harm which could result; dispersion/persistence modelling to determine the area affected and duration; and eco-system modelling to determine whether and how long recovery would take. This would then allow assessment of whether the event would be a major accident under the DoE definitions. However, toxicity data linking concentration or dose with particular long-term or short-term harm effects are usually sparse and often non-existent. A full quantified risk assessment

Table 1
Summary of types of event that could constitute a major accident to the environment (1)

Criterion No.	Description
5.2	Permanent or long term damage to more than 10% or >0.5 hectares of National Nature Reserves, Sites of Special Scientific Interest (SSSIs), Marine Nature Reserve or an area protected by a limestone pavement order.
5.3	Permanent or long term damage to wider environment such as area of scarce (> 2 hectares affected), intermediate (>5 hectares) or unclassified (>10 hectares) habitats.
5.4	Effects on a significant part (>10 km or> 1 hectare) of freshwater and estuarine habitat which may include stream, river, canal, reservoir, lake, pond or estuary according to the National River Authority (NRA) classification scheme for more than 1 year.
5.5	Damage to aquifers and groundwater leading to precluding its use for public domestic or agricultural water supply or have significant adverse impact on the surface waters and biotic system its supports.
5.6	Permanent or long term damage to the marine environment. The area of concern is damage to about 2 hectares or adjacent to the coast an area of about 250 hectares of the open sea, or a casualty count of about 100 sea birds (excluding the commoner species of gull), or 500 sea birds of any species, or 5 sea mammals of any species found dead or unable to reproduce.
5.7	Death or inability to produce of 1% of any species.
5.8	Release of persistent toxic substances into the environment of 10% or more of the "top-tier" threshold quantity of a persistent dangerous substance.
6.2	Damage to a built heritage such as Grade 1 listed or a scheduled ancient monument or an area of archaeological importance.
6.3	Damage to recreational facilities such as Long Distance Route National Trail), Country Park.
7.2	Contamination of 10 hectares or more of land which, for one year or more, prevents the growing of crops or the grazing of domestic animals.
7.3	Contamination of water sources or supply such that the supply to 10,000 or more consumers is rendered unfit for human consumption.
7.4	Direct or indirect damage to a sewerage system or sewerage treatment works which results in a significant risk to public health.
7.5	Socio-economic effects which can result from a major accident, such as destruction of homes and industrial premises or loss of income from contaminated farmland of fisheries.

(QRA) for effects to the environment would be very difficult and time-consuming. Environmental QRA would be more difficult than QRA for public safety because of the very wide range of possible environmental consequences involved.

A relatively quick and cost-effective solution is provided by index methods which use a readily calculated index in place of the actual measures of harm to the environment. Such methods are well-accepted in other areas, for example the use of the Mond and Dow indices in the fire and explosion field. This paper will describe the development of a risk index method for the full range of major hazard accidents to the environment and its demonstration for a case study involving the accidental release of pesticide into a river.

Although published work carried out for DoE and the Health and Safety Executive (HSE) was used as an input, the work described in this paper was an independent academic study which led to Ali's PhD (2). The topic of study was embarked upon because a literature review revealed an absence of practical risk assessment methodologies for environmental accidents. The work is reported here in the hope that it will be useful to others working towards the development of such a methodology. This work has no status as a method approved or accepted by either regulators or industry.

EXISTING ENVIRONMENTAL RISK METHODS

Ecological risk assessment

Ecological risk assessment is a process for evaluating the likelihood of adverse ecological effects occurring as a result of exposure to environmentally active agents. It is intended for application to planned not accidental releases into the environment and is required by legislation in the USA. Methods are proposed by the USA EPA Risk Assessment Forum (3) and the US National Academy of Sciences (4). The methodology is very resource intensive and is at least as concerned with being able to measure the onset of environmental problems as with being able to predict it.

Environmental hazard index (EHI)

This is a hazard index method proposed as a result of a European project in which AEA Technology was a participant (5). It was intended to allow a practical assessment of risk against criteria, using data which are likely to be available.

Use of a generic ecosystem consisting of five trophic levels was proposed. The levels are shown below:

Phytoplankton	Primary producers
Zooplankton	Primary consumers
Benthos	Decomposers
Vertebrates	Secondary consumers
Higher vertebrates	Tertiary consumers

An Environmental Harm Index (EHI) was then proposed which quantifies the potential for damage from any accident to that generic ecosystem. EHI was developed only for releases into rivers. The simple version of EHI is given by the equation below:

$$\text{EHI} = \frac{PEC_{max} \, S_{max}}{\min LC_{50} \, S_{ref}} \qquad \text{Equation 1}$$

where PEC_{max} is the predicted maximum concentration of toxic material in the environment, S_{max} is the predicted distance to the dangerous concentration, $minLC_{50}$ is the concentration which would cause 50% fatalities of the most sensitive species in the generic ecosystem, and S_{ref} is the reference distance given for a river in the DoE Green Book (1). This definition of EHI may cause an overestimate of risks because the maximum concentration is used and no account is taken of the plume behaviour of the contaminant as it moves downstream in a river.

A more accurate version divides the river into several segments with distance downstream from the release point. The continuous decrease in maximum concentration over distance is estimated by stepwise calculation over j steps. Then:

$$\text{EHI} = \frac{\sum_{j=2}^{N} PEC_j \left(S_j - S_{j-1} \right)}{\min LC_{50} \, S_{ref}} \qquad \text{Equation 2}$$

The value of PEC as a function of distance can be obtained from river dispersion modelling software. EHI can be seen as a toxicity factor multiplied by a damage factor.

AEA Technology (5) also proposed risk tolerability criteria in terms of EHI values. See Figure 1. This is based on the tolerability framework in HSE guidance (6). An EHI of 1 represents an event which is just a major accident to the environment. This was given the same borderline tolerable frequency, 10^{-4} per year, as a major accident causing offsite human fatality. Also, historical data suggested that small environmental accidents, equivalent to an EHI of 0.01, are currently being tolerated at a frequency of 10^{-2} per year.

Comments on existing methods

The EHI method was seen to have the potential for further development in view of its simplicity and requirement for minimal toxicity data (although even the few data required may not be available). The present authors are aware that AEA Technology have developed the EHI method beyond the latest published version in reference (5). The present authors' comments on the EHI method, as given in reference (5), are:

a) The EHI method was originally developed for water-borne hazards only, in particular for releases into rivers. Analogous indices need to be developed for all other types of release which can contribute to major accident hazards to the environment.

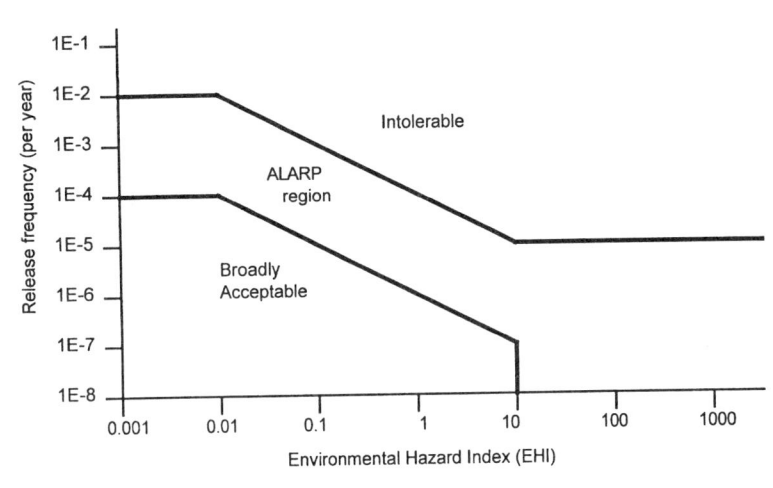

Figure 1
Tolerability criteria for the EHI method

[Graph: Release frequency (per year) on y-axis from 1E-8 to 1E-1, vs Environmental Hazard Index (EHI) on x-axis from 0.001 to 1000, showing regions labeled "Intolerable", "ALARP region", and "Broadly Acceptable"]

b) The calculation of EHI uses environmental concentration compared with the LC_{50}. However, it is well known that it is the dose (which is a combination of concentration and exposure time) and not the concentration which determines harm.

c) The EHI assumes that all chemicals causing environmental harm are non-persistent. The method needs further development to include the effects of persistent chemicals and the effects of bioaccumulation.

d) The proposed tolerability criteria in Figure 1 include horizontal sections in the lines defining tolerable and broadly acceptable risk. Standard societal risk criteria graphs do not include such horizontal sections.

e) Case studies (2, 5) indicate that very high values of EHI are possible from credible accidental releases into rivers. There is some measure of double-counting between the factors in EHI. A high environmental concentration (high toxicity factor) will tend to also result in a large distance being affected (high damage factor).

PROPOSED ENVIRONMENTAL RISK INDEX METHOD

The authors have attempted to develop a method which overcomes the limitations expressed above about EHI. This new method retains the use of an index but carries out the calculation for all the DoE major accident criteria (see Table 1). The method incorporates event probabilities which are the likelihoods that any given release would result in each DoE criterion. Because of the inclusion of these probabilities, the index is a risk index rather than a hazard index and is termed the Environmental Risk Index (ERI).

The ERI for a given release scenario combines event probabilities and a hazard index, the Environmental Severity Index (ESI) for all i of the DoE Green Book criteria :

$$ERI = {}^i\Sigma p_i ESI_i \qquad \text{Equation 3}$$

The method of calculation of values of ESI depends on whether the DoE criterion in question concerns a short or long term harm effect. In both cases, in order to remove double-counting of the factors comprising ESI, the geometric mean, rather than the product, is used. For acute, short-term criteria (e.g. 5.5, 5.7, 6.2 in Table 1)

$$ESI = \sqrt{\text{(toxicity factor)(damage factor)}} \qquad \text{Equation 4}$$

For long-term criteria (e.g. 5.2, 5.4, 7.2 in Table 1)

$$ESI = \sqrt[3]{\text{(toxicity factor)(damage factor)(recovery factor)}} \qquad \text{Equation 5}$$

Toxicity factor

The toxicity factor gives a measure of the level of toxicity in the environment caused by the particular release. As for the AEA Technology EHI method, if possible, toxicity data for the chemical released should be found for a number of species at different levels in the food chain which are representative of the eco-system as a whole. In practice, toxicity data are usually very difficult to find in the literature, and, if necessary, the data for whatever species found may have to be used.

For our proposed method, several possible equations can be used for toxicity factor, depending on the application. A toxicity factor in terms of concentration is given by Equation 2. However, there will be occasions when it is more appropriate to use a toxicity factor in terms of dose.

In an accident, exposure in a river will only last for a limited time as the contaminated water moves past any given point. Also the concentration may be high compared with the LC_{50}. If a persistent chemical is released to land, the exposure could be so long-term that concentration could be a better measure of risk than dose, assuming toxicity data were available for very long-term exposures. However, LC_{50} or LD_{50} (dose in mg/kg body weight giving a 50% chance of death) data are the measure of toxicity most likely to be found in the literature, and these are measured for short exposure times. This may not matter for long/high exposures because once an organism is dead it does not matter if the exposure lasts longer than the time required to kill it.

It is therefore proposed that dose should be used in cases when the exposure time is less than the measurement time for the LC_{50} (usually 96 hours). This will apply, for example, to short-term releases to flowing water or air, and to releases of non-persistent chemicals to any medium. Concentration effect should be used for exposures longer than the measurement time of the LC_{50} or LD_{50}. Concentration should therefore be used for release of persistent chemicals to land or relatively stagnant water such as lakes or ponds.

The toxic effects factor in terms of dose, to be used for relatively short-term exposures is :

$$\text{Toxicity factor (dose)} = \frac{\sum_{j=2}^{N} D_j \left(S_j - S_{j-1} \right)}{S_{total} \left(\text{Dose equivalent to } LC_{50} \text{ or } LD_{50} \right)} \qquad \text{Equation 6}$$

where :
- N = number of sections in the system
- j = section number of the system
- D = predicted average dose affecting the section ($C^n t$), where C is concentration, t is time and n is the exponent which best fits toxicity data for the particular chemical
- S = predicted distance (m), area (m^2) or volume (m^3) affected by the concentration of the section
- S_{total} = total distance (m), area (m^2) or volume (m^3) in the system

The authors consider that a maximum value should be set for the toxicity factor. If the level of toxicity in the environment is high enough to kill all the species present, it does not matter how much higher it is. If a probit equation were available, the maximum value of the toxicity factor would be the ratio of the dose giving 100% fatality to the dose giving 50% fatality. For inhalation of chlorine vapour, this ratio is approximately 30.

Toxicity factors have also been developed for toxic effects to humans (2, 7). There is a factor in terms of occupational exposure standard (OES) for DoE criteria involving contamination of land, preventing human access and a factor in terms of water quality standards for contamination of aquifers or other water supplies. For those criteria which are concerned with effects at a particular distance from the point of release, e.g. at a Site of Special Scientific Interest (SSSI) :

$$\text{Toxicity factor} = \frac{\text{concentration at specific distance/point}}{\text{min } LC_{50}} \qquad \text{Equation 7}$$

Damage Factor

The damage factor is the ratio of the magnitude of effect on the environment to the DoE criterion (see Table 1). The form of the damage factor is different for each criterion.

For example, for criterion 5.3, permanent or long-term damage to the wider environment, for scarce habitat :

$$\text{Damage factor} = \frac{\text{(area of scarce habitat affected)}}{2 \text{ hectares}} \qquad \text{Equation 8}$$

For criterion 5.4, damage to a river:

$$\text{Damage factor} = \frac{(\text{length of river affected})}{10 \text{ km}} \qquad \text{Equation 9}$$

For criterion 5.6, concerning seabirds killed:

$$\text{Damage factor} = \frac{(\text{number of seabirds of any species killed}}{500} \qquad \text{Equation 10}$$

The full set of damage factors for each DoE criterion are reported elsewhere (2, 7).

Recovery Factor

The recovery factor gives a measure of the time that the environment would take to recover from the release.

The recovery factor has to be based on a subjective judgement or estimation of the recovery time which is then used in equation 11. The authors made attempts to derive a recovery factor from such information as the half-life (a measure of persistence) and the octanol/water partition coefficient (a measure of bioaccumulation), but these attempts were unsuccessful.

$$\text{Recovery factor} = \frac{\text{estimated time for recovery}}{\text{reference recovery time}} \qquad \text{Equation 11}$$

where the reference recovery time is 5 years for aquatic habitat; 15 years for terrestrial habitat; 1 year for accidents which prevent access to crops, domestic animals and other foodstuffs; also 1 year for quality of water courses. These are quoted in the DoE criteria (1).

Risk Criteria

Although the AEA Technology EHI and the ERI proposed here are different, the EHI tolerability criteria (5), shown in Figure 1, can be used for both methods. This is because the tolerability criteria were developed independent of the EHI method. They were calibrated using a major accident to the environment (EHI=ERI=1) and an accident much less than a major accident EHI=ERI=0.01).

The present authors propose a modification to Figure 1. Most FN curves have no horizontal regions whereas the EHI criteria do. The horizontal section at low values of EHI would mean that no accident with any effect on the environment, however small, could be justified with a frequency greater than once in hundred years. This could probably not be achieved by industry. The horizontal section at high EHI means that there is no advantage, in terms of the criteria, of reducing the frequency of very severe accidents. This horizontal section makes it relatively easy to achieve the tolerability criteria in spite of the very high values of EHI which can be calculated. The authors' proposal to use the geometric mean, rather than the product, of the toxicity factor, damage factor and recovery factor would reduce this problem.

The authors consider that it would be preferable to retain a standard FN curves (without horizontal sections) at high ERI. An accident 10 times worse that a "standard" DoE major accident is nowhere near as severe as certain accidents which could be imagined and which could sterilise large areas of the countryside including important habitats. It is reasonable that such very catastrophic potential accidents should be reduced to an extremely low frequency. The authors therefore propose the tolerability criteria shown in figure 2 (on which the original EHI criteria are shown as dotted lines).

Figure 2
Revised tolerability criteria and Case Study results

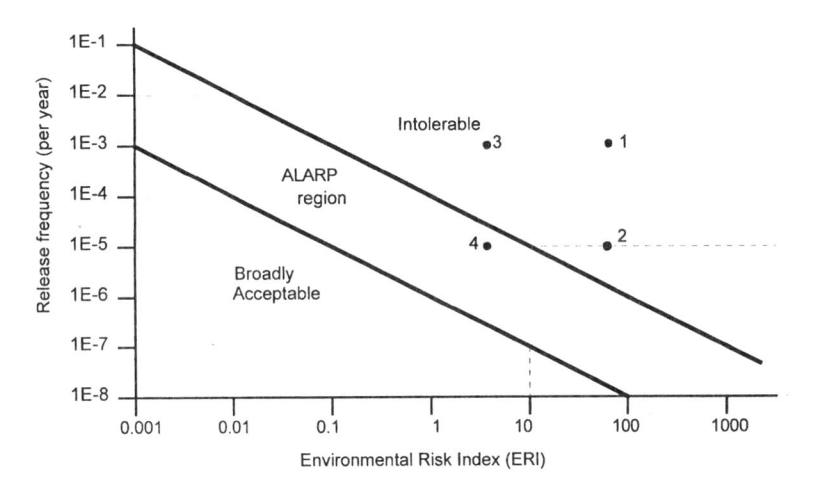

PROCEDURE FOR USING ERI METHOD

A flowchart for the use of the method is given in Figure 3. The procedure is based on that for the Mond Fire and Explosion Index method (8). An ERI is calculated for the current design. The frequency at which the accident scenario might occur can be estimated using standard quantified risk assessment methods (e.g. using historical failure rates or fault tree analysis). The ERI and frequency combination can be compared with the risk criteria in Figure 2. A sensitivity study can then be performed in which the effects of various mitigation measures can be evaluated. Some of these will change the ERI, some the frequency and some both. This can allow a judgement to be made about the best design so that the risk is tolerable and as low as reasonably practicable.

Figure 3
Flowchart for use of the ERI method

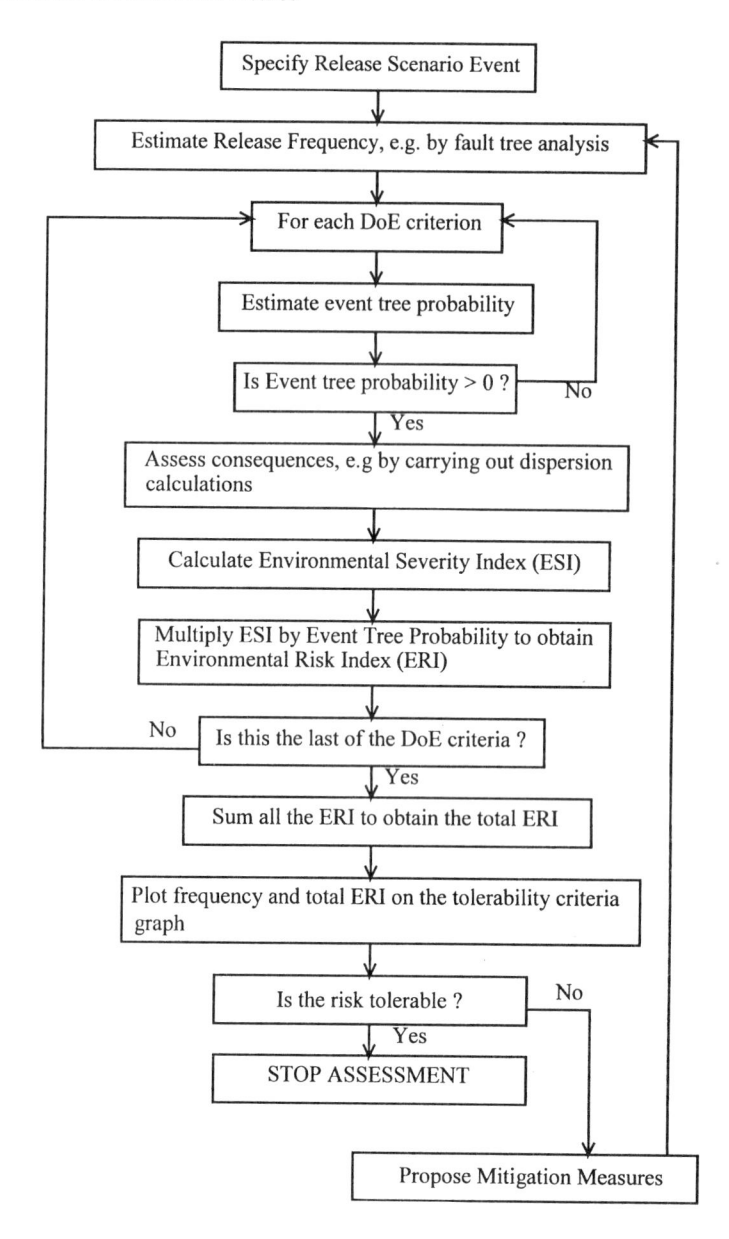

```
Specify Release Scenario Event
        ↓
Estimate Release Frequency, e.g. by fault tree analysis ←─────┐
        ↓                                                      │
  For each DoE criterion ←──┐                                  │
        ↓                    │                                  │
  Estimate event tree probability                               │
        ↓                                                      │
  Is Event tree probability > 0 ? ──No──┐                      │
        ↓ Yes                            │                      │
  Assess consequences, e.g by carrying out dispersion           │
  calculations                                                  │
        ↓                                                      │
  Calculate Environmental Severity Index (ESI)                  │
        ↓                                                      │
  Multiply ESI by Event Tree Probability to obtain              │
  Environmental Risk Index (ERI)                                │
        ↓                                                      │
  No ── Is this the last of the DoE criteria ? ─────────────────┘
        ↓ Yes
  Sum all the ERI to obtain the total ERI
        ↓
  Plot frequency and total ERI on the tolerability criteria
  graph
        ↓
  Is the risk tolerable ? ──No──┐
        ↓ Yes                    │
  STOP ASSESSMENT                │
                                 ↓
                        Propose Mitigation Measures
```

CASE STUDY

A case study was carried out involving a hypothetical release of pesticide from a storage tank into the River Don in Sheffield (2, 9). The pesticide was chosen to be one for which toxicity and other necessary data were readily available. The river was surveyed to obtain data to allow a river dispersion model to be run. A sketch plan of the river is shown in Figure 4. The ERI was calculated for a base event and a number of mitigation measures of which a selection are shown in Table 2. The values of ERI shown were calculated using the PRAIRIE river dispersion code (10) but other models were also used for comparison. The results are also shown on Figure 2.

Figure 4
Sketch plan of river used for case study

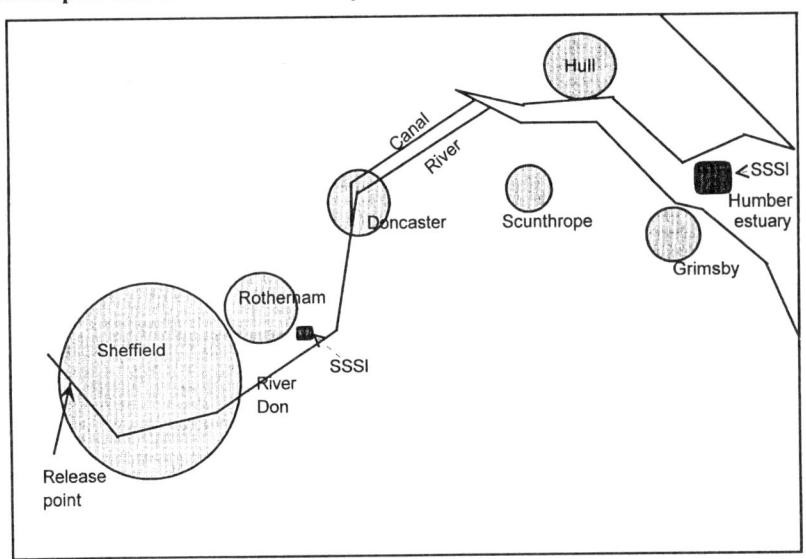

Table 2
Selected case study results

Case	Frequency (year^{-1})	ERI
1. Base case. Bunded tank with manual drain valve from bund to river to remove rainwater	10^{-3}	73
2. Hold tank used for rainwater before discharge to river	10^{-5}	73
3. Pesticide stored as 5% solution in water	10^{-3}	5
4. Both hold tank and storage as 5% solution in water	10^{-5}	5

The results of the hypothetical case study showed that the ERI obtained varied by about a factor of 2 depending on the dispersion model used. PRAIRIE gave the lowest values and a simple plug flow model gave the highest. It would be expected that the EHI Method would show similar sensitivity to the dispersion models used.

None of the dispersion models used could predict the behaviour of chemical after the River Don goes through the Humber Estuary and then into the sea because the models were intended for non-tidal rivers. The authors therefore estimated concentrations in the estuary and sea in order to demonstrate the use of the method. This also demonstrates that the method is usable when only very approximate consequence information is available.

In all cases, recovery times had to be estimated. This was done in a common sense way taking account of factors such as whether or not the entire population of species would be killed or whether some would be left to repopulate the area. A better quality of judgement could have been obtained if it had been performed by ecologists rather than by the authors.

It can be seen from Figure 2 that the ERI is in the intolerable region for the base case scenario. It is recommended that mitigation option 4 (using holding tank to contain the rainwater from the bund and storing pesticide as a 5% solution in water) be implemented. These clearly sensible precautions reduce the risk into the ALARP region.

DISCUSSION

Use of the proposed ERI method for the case study described above showed that it was reasonably quick and easy to use, once data on river hydrology had been obtained. Obtaining river data required several days effort, visiting different locations on the river in order to measure width, depth and flow rate. This was done at the height of summer in order to obtain a conservative low value for river flow. Better results might have been obtained using a boat (and a competent sailor!). The case study was chosen so that toxicity data and degradation rate data (used by dispersion models such as PRAIRIE) were available. Even so, LC_{50} data were only found for a very limited number of species which did not cover the full range in the generic ecosystem of the EHI method. Dispersion modelling was relatively quick once the data had been assembled and familiarity with the models had been achieved.

Testing of the proposed method has been limited to the case study described above. Further testing for a range of scenaria which impact on different DoE criteria is needed. Index methods such as the Mond Index were calibrated by using them in a large number of case studies and comparing results with other risk calculations and with experience of whether or not the required risk reduction measures were in place. Much more extensive testing of the proposed ERI method in this way would be needed to improve its robustness and increase confidence in its use.

The DoE criteria (1) for the types of event that would constitute a major accident to the environment, on which the proposed ERI method is based, are currently under revision. The ERI method would need modification following this revision in order to remain consistent with DETR guidance.

CONCLUSIONS

1. Quantified risk assessment (QRA) for major accidents to the environment is much more complex and time-consuming than QRA applied to hazards to humans. This is because of the extreme complexity of the environment and of ecological systems.

2. The use of hazard or risk index methods has the potential for allowing a simplified and more cost-effective risk assessment to be carried out.

3. An index method, the Environmental Risk Index (ERI) method, which includes risk criteria, has been developed from an existing Environmental Hazard Index (EHI) method and making use of DoE criteria for events that would constitute a major accident to the environment.

4. The proposed ERI method has been successfully applied to a semi-hypothetical case study. It was found to be reasonably quick and easy to use. The design option indicated by the method seemed sensible in the opinion of experienced engineers.

5. Any environmental risk method will be subject to difficulties in obtaining necessary data, particularly toxicity data.

6. The proposed ERI method would need considerably more testing for a range of scenaria to increase confidence in its use.

7. The proposed ERI method will require revision to make it consistent with the current revision of the DoE criteria for major accidents to the environment.

REFERENCES

1) UK Department of the Environment, "Interpretation of Major Accident to the Environment for the Purposes of the CIMAH Regulations - A Guidance Note" DoE, 1991.

2) M W Ali, "Development of Risk Assessment Framework for Major Accident Hazards to the Environment", PhD Thesis, University of Sheffield, 1997.

3) US Environment Protection Agency, "Framework for Ecological Risk Assessment", EPA/630/4-92/001, EPA Risk Assessment Forum, Washington D.C, 1992.

4) National Research Council, Committee of the Institutional Means for Assessment of Risks to Public Health, "Risk assessment in the Federal Government: Managing the Process", National Academy Press, Washington D.C., 1983.

5) AEA Technology, Ministry of Housing, Spatial Planning and Environment (VROM) & AGEL-CBI, "Environmental Risk Criteria for Accidents: A Discussion Document", 1995.

6) HSE, "The tolerability of risk from nuclear power stations", HSE Books, 1992.

7) M W Ali, A J Wilday & Y Wu, "Risk Assessment for Accidental Releases to the Environment from Major Hazard Chemical Plants", paper presented at International Meeting on Reliability & Risk Assessment, Rio de Janeiro, Brazil, December 1997

8) P Doran & T R Greig, "The Mond Index", 2nd edn., published under license from Imperial Chemical Industries PLC, 1993, by Mond Index Services, 40 Moss Lane, Cuddington, Northwich, Cheshire, UK, CW8 2PX

9) M W Ali, A J Wilday & Y Wu, "Development of Environmental Risk Index (ERI) Method for Assessing Risk from Accidental Releases to the Environment", IChemE Research Event, 1998

10) S Welsh, "Assessment and Management of Risks to the Environment", Trans IChemE, Part B, Vol 71, 3-14, 1993

ACKNOWLEDGEMENT

This paper is based on the PhD research programme of Mr M W Ali at Sheffield University, UK, under the supervision of Ms A J Wilday and Dr Y Wu and sponsored by the Universiti Teknologi Malaysia, Johor, Malaysia.

PLEASE NOTE

The views expressed in this paper are those of the authors and not necessarily those of the Health and Safety Laboratory nor the Health and Safety Executive

INCORPORATION OF BUILDING WAKE EFFECTS INTO QUANTIFIED RISK ASSESSMENTS

I.G. Lines, D.M. Deaves and R.C. Hall
WS Atkins Safety & Reliability, Woodcote Grove, Ashley Road, Epsom, Surrey KT18 5BW

Most risk assessments currently use flat terrain dispersion modelling without considering the effects of building wakes. Where such effects are considered, they are often assessed in an idealised manner, by taking a single building wake model and defining the dimension of an equivalent building, which may not be appropriate to dense releases. The use of computational fluid dynamics (CFD) has facilitated the investigation of such wake effects, enabling source conditions and multiple buildings to be considered. In addition, a simple passive dispersion wake model has been extended to cover dense releases. CFD results have been obtained for a set of realistic releases for a real site, and have been compared with the results of simple modelling. This has demonstrated the effects of incorporating wake models into risk assessments, and allowed the development of an optimum methodology for their inclusion.

Keywords: QRA, CFD, Gas dispersion, Building wakes

INTRODUCTION

Accidental releases of hazardous materials may affect surrounding areas if toxic or flammable vapours disperse in the atmosphere. Calculation of such dispersion, for inclusion within risk assessments, or Safety Cases, is straightforward for uniform unobstructed flat terrain. However, practical releases may be affected by the presence of adjacent buildings, generally resulting in enhanced dispersion. The effects of buildings on dispersion have been reviewed by Lines et al (1) and a substantial effort in the application and validation of CFD modelling to the problem has been presented by Hall (2). An investigation into the extent to which simple models could be used to calculate the dispersion of releases, particularly of dense gases, in building wakes has also been given by Lines and Deaves (3).

The effects of building wakes on the source term for the dispersion of released vapours can be assessed either using simple 'box' or 'zone' type models, or using more sophisticated methods, such as computational fluid dynamics (CFD). The simple approach has been reasonably well developed for single rectangular buildings, and this has been extended to cover dense gas effects. Although there are some models becoming available for the treatment of arrays of obstacles, these tend to be confined to regular arrays, and to focus upon the 'street canyon' effects on urban pollution. For non-regular groups of buildings, there may be some merit in the use of CFD either to determine the dispersion, or to ascertain the appropriate source term for input to a standard flat terrain dispersion model.

This paper considers the application of both simple and CFD modelling to chlorine releases from a real site. It compares the results, demonstrates how the simple modelling may be used most effectively, and shows the effects on risk calculation of including these building wake effects.

BASE CASE RISK ASSESSMENT

The example that has been chosen to act as the base case is a toxic gas risk assessment for a chlorine bulk storage site, as such sites are relatively common in the UK and can lead to significant risks to off-site populations at some distance from the plant. The particular site that has been chosen is a water treatment works in the North of England. The chlorine bulk store and off-loading bay are located within one of the main site buildings, which is approximately 38 x 30 m wide and 7m high. There are a few other buildings on site of similar dimensions. The site comprises the following main items of equipment which represent major hazards:

- 2 chlorine bulk storage vessels (stored under pressure at ambient temperature)
- a chlorine road tanker off-loading bay
- various sections of 25 mm diameter liquid chlorine pipework

Loss of containment failures involving any of these items of equipment will lead to the formation of a toxic gas cloud.

In order to quantify the risks associated with potential chlorine events, it is first necessary to define a set of representative events which cover all possible significant accidents that could occur. The release rate of chlorine and the frequency of each of these events then needs to be determined. A set of representative scenarios (from Carter, Deaves and Porter (4)), corresponding to a typical small chlorine installation has been used. There is a range of possible weather conditions that may occur at the site, and so each of the 40 events identified is considered in 4 representative weather conditions, namely D2.4, D4.3, D6.7 and F2.4, where the letter corresponds to the Pasquill stability category and the numbers correspond to the wind speed in m/s. The percentage frequencies of these four weather conditions are taken to be 17%, 20%, 45% and 18% respectively, based on the average data over 20 years from a nearby meteorological station; for ease of application, a uniform wind rose has been used.

The dispersion of chlorine vapour clouds has been assessed using the models in the latest version of HGSYSTEM (Version 3.0; Post (5)). Continuous releases have been modelled using the HEGADAS-S code, and instantaneous releases have been modelled using the HEGABOX followed by the HEGADAS-T codes. The risk calculations involve a summation of the risks from each event in each of the representative weather conditions. The risks have been calculated for a typical residential population, which is assumed to be present for 100% of the time, and which is outdoors for 10% of the time, except in F2.4 weather conditions, where 1% is assumed to be outdoors. The population is assumed to be indoors for the remainder of the time.

The risks to persons indoors are based on a calculation of the time-varying concentration inside the building, using an air exchange rate of 2 air changes per hour (ach) for all conditions except D6.7, where the higher wind speed implies a higher air exchange rate of 3 ach. The persons indoors are assumed to remain indoors for 10 minutes after the cloud has passed before evacuating to fresh air, but in no case does evacuation take place until at least 30 minutes has elapsed from the start of the release.

Six of the most significant individual events, contributing 63% of the overall total risk at 500 m and 67% of the risk at 1000 m, were then selected for use in the subsequent analysis in which wake effects are considered in some detail. The conditions for these six events are summarised in Table 1.

Scenario Number	Description	Release Rate (kg/s)	Duration* (min)	Frequency ($\times 10^{-6}$/year)	% of Total Risk 500m	1000m
1	**Storage Vessels** Liquid space	44	6.8	2	5.95	9.85
	Pipework: 25mm diameter to vessel					
2	Full	4.6	20	9	6.07	9.42
3	25	3.6	20	45	25.14	38.5
4	Flanges	1.4	20	55	11.73	0.08
	Pipework: 25mm diameter vessel outlet					
5	Flanges	1.1	20	40	6.91	0.05
	Pipework: 25mm diameter Road tanker coupling					
6	Full	7.4	20	6	7.64	8.65

*In Scenarios 2-6, it is assumed that the duration is limited by the correct operation of an isolation valve.

Table 1 Summary of representative scenarios

SIMPLE MODELLING

The program WEDGE is used to evaluate the effects of building wakes on each scenario. WEDGE contains a choice of two models, those of Fackrell (6) and Brighton (7), both of which calculate the wake dimensions and the average concentration in the wake region. The choice of model is dependent on the release conditions and building parameters of each scenario. Fackrell's model is recommended when the resultant release density is low or effectively 'passive', and Brighton's model is recommended when the density of the release gas is high or 'dense'. Values of mass flow rate (kg/s) that will result in a transition from using the Fackrell model to using the Brighton model in the WEDGE program are dependent upon building dimensions and wind speed. For a single building of the size considered, the values range from around 2kg/s at 2.4m/s to 40-50kg/s at 6.7m/s.

WEDGE calculates the wake dimensions and the average mixed concentration of the release within the wake. The released gas is assumed to have no source effects ie. the momentum of the gas is destroyed once it escapes from its primary containment or from the building. This excludes the cases where the jet momentum of the release takes the gas straight through the wake region, without being affected by the building wake. For all scenarios, it is assumed that the incident wind is normal to the face of the building. The results of WEDGE can be incorporated into a HEGADAS-S input file as a TRANSIT block. This TRANSIT block enables the HEGADAS-S model to start the dispersion modelling at the breakpoint, which in this case is the downstream edge of the wake, with cloud width, temperature and concentration that are determined by WEDGE.

The WEDGE results are based on the assumption that the mixing of gas and air occurs throughout the wake region, with an average uniform concentration that extends up to the wake boundary, which is dependent on the building dimensions. In some release cases, large building cross-sectional areas give rise to long wake lengths and the wake concentration emerging from the downwind end of the wake could be higher than the concentration predicted at the same position by the 'standard' dispersion model, possibly resulting in longer hazard ranges than those predicted by the 'standard' method. Alternatively, the increased mixing in the wake region may lead to lower concentrations and shorter hazard ranges.

The wake conditions were used as input to the dense gas dispersion model, and concentrations at downwind location compared for the with/without building wake calculations. The 'concentration ratio' thus defined was plotted against distance for varying wind conditions (Figure 1) and release rates (Figure 2). Comparison of results in this way allowed the selection of scenarios which were considered further in the CFD studies.

CFD MODELLING

As a result of the initial application of simple modelling, the following conditions were selected for analysis using CFD to model the complete site selected:

Case	Release rate (kg/s)	Wind speed (m/s)	Wind direction	Wind stability
1	3.6	2.4	SW	D
2	7.4	2.4	SW	D
3	3.6	4.3	SW	D
4	7.4	4.3	SW	D
5	3.6	2.4	NE	D
6	3.6	2.4	NW	D
7	3.6	2.4	SW	F

Table 2 Identification of CFD runs

A standard mesh was used for each case described in Table 2 and required a simple modification for the different wind directions. The mesh consisted of a site section which includes all the buildings and is fixed for all of the cases considered. An additional section is attached to the downwind side of the site section to capture the dispersion further away from the buildings. The width of the domain is taken to be approximately twice that of the plume predicted using preliminary HGSYSTEM calculations. Accordingly, a width of 400m is used for the domain. This dimension is also used for the sides of the site section so that the attached downwind section may easily be transferred from one side of the site to another. The downwind length of the attached section is 300m. The chlorine release location is offset 50 metres away from the centre of the site, so that the maximum distance from the release point to the downstream boundary is 550m and the minimum distance is 450m. The domain extends to a height of 80m above ground level.

For higher accuracy of the solution, grid resolution was considered to be important in the region close to the release point and in regions around the edges of buildings. A cell size of approximately 2.0m x 2.0m x 0.4m was chosen for the mesh at the release point, with an expansion ratio of about 1.2 for cells away from the release. A further level of grid refinement was considered necessary in the region close to the chlorine source. Fluid cell lengths were therefore halved up to about 10m away from the release in the plane of the jet and up to about 1.2m vertically from ground level. Hence, the smallest cell actually solved for was approximately 1.0m x 1.0m x 0.2m in size. In the vertical direction, 40 cells were used, and the final mesh comprised a total of 179,087 fluid cells.

The inlet velocity profiles were modelled using standard formulae for equilibrium atmospheric boundary layers. The standard k-ε turbulence model is used, and appropriate equilibrium profiles of k and ε applied at the inlet boundary.

The chlorine gas release is modelled as an area source located at the loading bay entrance. The gas is discharged horizontally and at right angles to the bay wall. It is assumed that the liquid has flashed to vapour and that air has been entrained by the time it has reached the loading bay entrance. Exit conditions are obtained by assuming adiabatic conditions prior to reaching the bay entrance. Turbulence levels for the jet are obtained using a turbulence intensity of 5% applied to the calculated efflux velocity, and an integral length scale of 0.1m is used for calculating ε.

Example CFD results for a 3.6kg/s release into a 2.4m/s wind speed are shown in Figures 3, 4 & 5 for SW and NE directions in D stability, and SW direction in F stability respectively. It is clear from comparing these results that the whole site is affecting the dispersion, and that the width of the dispersed cloud is strongly dependent on both wind direction and atmospheric stability.

COMPARISONS OF CFD AND SIMPLE MODELLING

Initial comparisons were undertaken between the CFD results and results of using the simple wake model for a single building (that from which the chlorine is released). From this comparison, the CFD results indicated:

- The width of the cloud in the very near field (on site between buildings) may be much greater than that predicted by the simple modelling.
- Channelling effects between buildings may be significant in the near field.
- Ground level concentrations immediately downwind of the near wake region may be substantially lower than those predicted by the simple model depending on the building orientation.

It was therefore considered that better agreement between the CFD and simple modelling results could be achieved if the simple modelling was based on the overall dimensions of all the site buildings, rather than just the chlorine building. It also appears to be important that the release location and building orientation with respect to the wind should be included within the analysis.

On the basis of these comparisons, the following minor changes were incorporated into the simple modelling:

- threshold on buoyancy parameter for dense gas spreading in wake reduced
- effect of release position (upwind/downwind) incorporated
- methodology for inclusion of all site buildings adopted.

This updated version of WEDGE was then used to model Runs 1-7, as identified in Table 2. In order to set the subsequent discussion in context, preliminary comparisons were undertaken between the simple modelling results of Runs 1, 5 and 6 with those for Run 8, which is for the same release and wind conditions, but without the buildings, and hence using HEGADAS only. The release is 3.6kg/s in D2.4 wind conditions, and Runs 1, 5 and 6 are for different wind directions. The results are presented in Table 3, which shows the area and hazard range for each of 3 threshold concentrations.

Run	Direction		Threshold concentration (ppm)		
			100	300	500
1	SW	Area (m^2)	1.2×10^5	3.8×10^4	1.9×10^4
		Hazard range (m)	579	240	134
5	NE	Area (m^2)	1.2×10^5	5.2×10^4	3.6×10^4
		Hazard range (m)	703	353	250
6	NW	Area (m^2)	1.2×10^5	4.6×10^4	3.0×10^4
		Hazard range (m)	692	330	220
8	No buildings	Area (m^2)	1.1×10^5	4.5×10^4	3.0×10^4
		Hazard range (m)	827	467	354

Table 3 Comparison of simple modelling for 3.6kg/s release in D2.4 wind conditions

The following observations can be drawn from these results:

- The predicted area of the 100ppm contour is increased by about 10% by the presence of buildings, although the hazard ranges are reduced slightly, by around 15% for Runs 5&6, and by around 30% for Run 1.

- The effects of buildings become more marked with increasing concentration threshold, which corresponds to regions closer to the buildings.

- The predicted area of the 500ppm contour is increased by the presence of the buildings, for Run 5 (by 20%), but reduced by nearly 40% for Run 1 and unchanged for Run 6. Hazard ranges for this concentration are reduced for all runs; by around 30-40% for Runs 5 and 6 and by more than 60% for Run 1.

From these observations it is clear that the greatest effects occur in the near field, and that the effects are enhanced when the release is blown through the building array, rather than away from it or around the side.

Figures 6 and 7 show the results of applying WEDGE/HEGADAS for a single building and for the whole site respectively. The change in width is evident, and it is clear that the wider effective plume of Figure 7 would give a better match to the CFD results of Figure 3.

A quantitative comparison of the hazard ranges and areas covered by various concentration contours has been undertaken for all the cases analysed. It was shown in the preliminary comparisons that there was generally little effect of the building wake on the 100ppm contour, with the greatest effects felt closer to the buildings. It was also found that, for all the CFD runs undertaken, the 100ppm contour extended beyond the end of the computational domain, thus rendering estimation of hazard range and contour area inaccurate. For these reasons, the comparisons presented below relate only to the 300 and 500ppm contours. In the tables which follow, the ratios given in the final columns are the CFD prediction divided by the WEDGE/HEGADAS (W/H) predictions.

Run		CFD	W/H	Ratio
1	Area (m^2)	2.4x10^4	3.8x10^4	0.63
	Hazard range (m)	240	240	1.00
2	Area (m^2)	*7.2x10^4	1.0x10^5	0.72
	Hazard range (m)	*450	457	1.00
3	Area (m^2)	1.1x10^4	1.1x10^4	1.00
	Hazard range (m)	160	113	1.42
4	Area (m^2)	2.5x10^4	3.3x10^4	0.76
	Hazard range (m)	339	270	1.26
5	Area (m^2)	*4.8x10^4	5.2x10^4	0.92
	Hazard range (m)	*450	353	1.27
6	Area (m^2)	3.3x10^4	4.6x10^4	0.72
	Hazard range (m)	399	330	1.21
7	Area (m^2)	>7.3x10^4	1.5x10^5	>0.49
	Hazard range (m)	>450	559	>0.81

* 300ppm contour extended just beyond edge of CFD domain

Table 4 Comparison of CFD and WEDGE/HEGADAS results for 300ppm concentration criterion

Run		CFD	W/H	Ratio
1	Area (m^2)	9.9x10^3	1.9x10^4	0.52
	Hazard range (m)	140	134	1.05
2	Area (m^2)	4.0x10^4	6.2x10^4	0.65
	Hazard range (m)	289	309	0.94
3	Area (m^2)	5.0x10^3	3.9x10^3	1.28
	Hazard range (m)	88	33	2.67
4	Area (m^2)	1.2x10^4	1.8x10^4	0.67
	Hazard range (m)	171	159	1.08
5	Area (m^2)	3.2x10^4	3.6x10^4	0.89
	Hazard range (m)	430	250	1.72
6	Area (m^2)	2.0x10^4	3.0x10^4	0.67
	Hazard range (m)	263	220	1.20
7	Area (m^2)	4.3x10^4	6.3x10^4	0.68
	Hazard range (m)	331	310	1.07

Table 5 Comparison of CFD and WEDGE/HEGADAS results for 500ppm concentration criterion

It can be seen from Table 4 that the results for the 300ppm contour are in reasonable quantitative agreement, with a tendency for the CFD to predict slightly smaller areas and longer hazard ranges. The agreement is slightly less good for the 500ppm contour, with the biggest ratio for the rather short hazard ranges predicted for Run 3. A measure of the goodness of fit can be determined by averaging the ratios in the final columns of these tables. From Table 4, Run 7 is omitted in the averaging because contour extended significantly beyond the CFD domain. From Table 5, Run 3 is omitted because the rather short hazard ranges give rather large ratios. The results of averaging the remaining ratios are presented in Table 6.

	300ppm	500ppm
Area ratio	0.79	0.68
Hazard range ratio	1.19	1.18

Table 6 Average of ratios [CFD/(WEDGE/HEGADAS)] from Tables 4 and 5

The CFD results can also be compared with the HEGADAS results with no building effects (Run 8). Hence, this comparison would represent the effects of ignoring the buildings in the simple modelling. The results are given in Table 7.

	300ppm		500ppm	
	Run 1	Run 8	Run 1	Run 8
Area ratio	0.63	0.50	0.52	0.31
Hazard range ratio	1.00	0.55	1.05	0.44

Table 7 Comparison of CFD against HEGADAS results with (Run 1) or without (Run 8) WEDGE, for 3.6kg/s in D2.4 wind conditions

It is clear from these comparisons that the modelling of building effects using WEDGE gives a significantly better fit of simple modelling to the CFD results than using HEGADAS alone. It is also clear from the other comparisons in Tables 4-6 that there is generally good agreement between CFD and the current implementation of WEDGE/HEGADAS modelling.

Summary of Comparisons Although simple modelling will never predict the detailed features of the dispersion as predicted by CFD, the above comparisons have shown that simple wake modelling can be used effectively to provide a good indication of near field concentrations and plume dimensions, which should be sufficient for typical risk assessment applications. The greatest difficulties arise when channelling between/along buildings is significant, as this will probably require some user expertise in the choice of which buildings should be used to define the wake in WEDGE.

The very near field, inside the wake and between buildings, is also not well predicted by simple modelling, and so the importance of such situations may need to be considered as a separate issue in the risk assessment process. For example, it would probably be appropriate to assume that any location within the building complex would experience a dangerous toxic load for any wind direction for any of the releases considered above.

A revised risk assessment was undertaken, incorporating wake effects for all wind speeds, and for a SW direction. The results are given in Figure 8, which shows that the inclusion of wake effects has surprisingly little influence on the overall risks results for distances in the near field from 150m to about 700m. There are several reasons for this, namely:

- Even in the base case QRA, where building wake effects are not included, dense gas clouds of chlorine tend to spread laterally quite rapidly. In this particular study, it appears that, at distances of around 150m from the release point, this lateral spread is often broadly comparable to the spread that is induced by the building wake, and consequently there is little difference in the associated risks.

- The methodology used to calculate the risks (ie. the typical HSE approach using a Dangerous Toxic Load with different probabilities for escape depending on the concentration) is not particularly sensitive to the precise cloud concentrations in the near field (ie. the cloud has relatively sharp edges and a person is either in the cloud or is outside the cloud). Different approaches, such as using the AIChE probit to calculate the risk, could lead to greater differences.

- Examination of the results for individual scenarios reveals that the inclusion of building wake effects can lead to either greater or lesser risks in the near field, depending on the precise combination of weather conditions, release rate, release duration, etc. When the results for all scenarios are combined, many of these differences cancel out, giving relatively little change in the overall risk. If the choice of events or their frequency were different, then this cancelling out effect may not occur.

- Building wake calculations have only been undertaken for the six most significant scenarios (see Table 1). If these wake conditions were also undertaken for the other 30 continuous releases scenarios in the risk assessment (ie. a total of 4 x 30 = another 120 WEDGE runs), then it is likely that greater differences would be observed in the near field.

Examination of Figure 8 shows that the greatest differences in risk occur in the medium to far field (beyond 700m). For example, at 1000m the inclusion of wake effects increases the risk from 6.8×10^{-7} to 1.0×10^{-6}/yr (a factor of 1.5 increase), whereas at 1500m the inclusion of wake effects reduces the risk from 2.7×10^{-7} to 1.0×10^{-7}/yr (a factor of 2.7 decrease). At most distances, the differences in risk are less than a factor of two, but it should be emphasised that this could still shift the 3×10^{-7}/yr contour by up to several hundred metres, which could have implications for the size of the consultation zone around this type of major hazard site.

CONCLUSIONS

CFD modelling

- The lessons learned from previous studies applying CFD to external atmospheric flows enabled robust and efficient CFD modelling of wake effects to be undertaken.

- Scoping studies using an unobstructed terrain model are useful to set the size of the computational domain.

- CFD modelling enabled certain features of the flow to be identified which could not be predicted with the simple models. These included near source effects such as upwind spreading and overall cloud width, both of which are dependent on source size and momentum.

- The results from the CFD modelling also demonstrated the importance of wind direction, and of including all the nearby buildings on site, at least for the range of release rates and wind speeds considered.

Simple modelling

- Simple wake models can give a substantial improvement in risk estimates, although single building wake models may have limited use on complex sites.

- As a result of the comparison with CFD, improvements were made to the application of WEDGE.

- the wake width has been modified (increased) for moderately dense releases
- modifications have been made to allow for different effects of release location (upwind/downwind/side)
- the model has been applied by treating the group of buildings as a single effective building

- With the improvements outlined above, good correlation was obtained between CFD and simple modelling results, within the range 300-500ppm. Beyond 100ppm, direct comparison was unreliable, since the CFD predicted contours which extended beyond the computational domain. Similarly, for concentrations in excess of 500ppm, near wake effects become important, and again the comparison is unreliable.
- The building wake concentration should be assumed to apply over the whole region of the building complex, which may have the effect of increasing the on-site calculated risk.

Risk calculation

- Wake effects can increase or decrease the risk calculated downwind of a group of buildings, although the overall changes identified for a typical chlorine installation are all within a factor of around 2. While the effects are generally greatest in the near field, they can also extend to the far field (ie. beyond about 1km in this case).
- The effects on the individual contribution to risk will vary between scenarios, and may exceed the factor of 2 identified above at certain distances from the source.
- The significance of wake effects on risk depends on release rate and wind speed, being generally greatest for low wind speed and small to moderate release rate. Large releases tend to engulf buildings, whose presence then has only a minor effect on the dispersion.

REFERENCES

1. Lines I G, Hall R C, Gallagher P and Deaves D M 1994. 'Dispersion of releases of hazardous materials in the vicinity of buildings - Phase I'. HSE report WSA/RSU8000/006.
2. Hall R C 1996. 'Dispersion of releases of hazardous materials in the vicinity of buildings'. Phase II CFD modelling. HSE report WSA/RSU8000/013.
3. Lines I G and Deaves D M 1996. 'Dispersion near buildings - application of simple modelling'. HSE Contract Research Report WSA/RSU8000/013/2.
4. Carter, D.A., Deaves, D.M., & Porter, S.R., 1995 'Reducing the risks from major toxic gas hazards', Loss Prevention and Safety Promotion in the Process Industries, 1: 679-689, Edited by J.J. Mewis, H.J. Pasman and E.E. De Rademaeker, Elsevier Science B.V.
5. Post L 1994. 'HGSYSTEM 3.0 Users Manual'. Shell Research Report TNER.94.058.
6. Fackrell J.E '1984 'Parameters characterising dispersion in the near wake of buildings', J. of Wind Eng. & Ind. Aerodynamics 16:97-118.
7. Brighton P.W.M 1986 'Heavy gas dispersion from sources inside buildings or in their wakes', IChemE North Western Branch Symposium 'Refinement of Estimates of the Consequences of Heavy Toxic Vapour Releases', UMIST, Manchester.

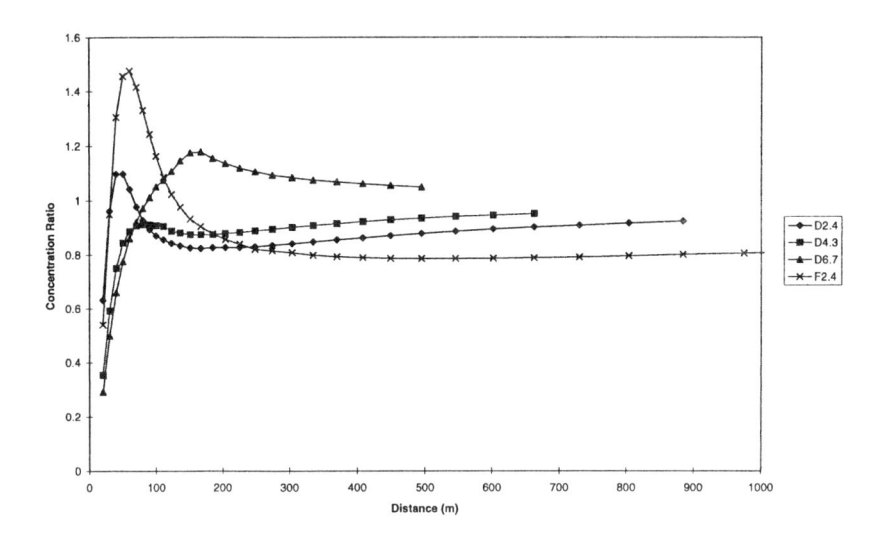

Figure 1. Effect of varying wind speed on the concentration ratio for a 3.6kg/s release

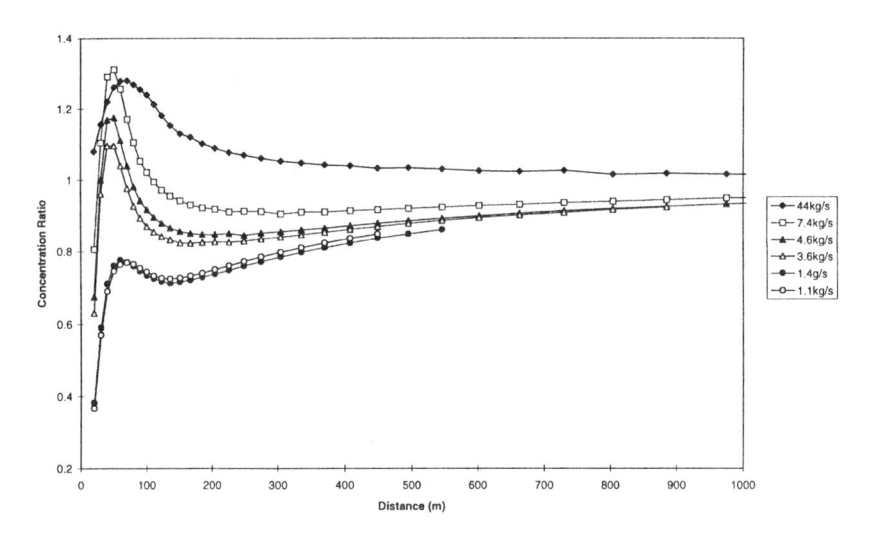

Figure 2. Effect of varying the release rate on the concentration ratio at a constant wind speed of 2.4m/s

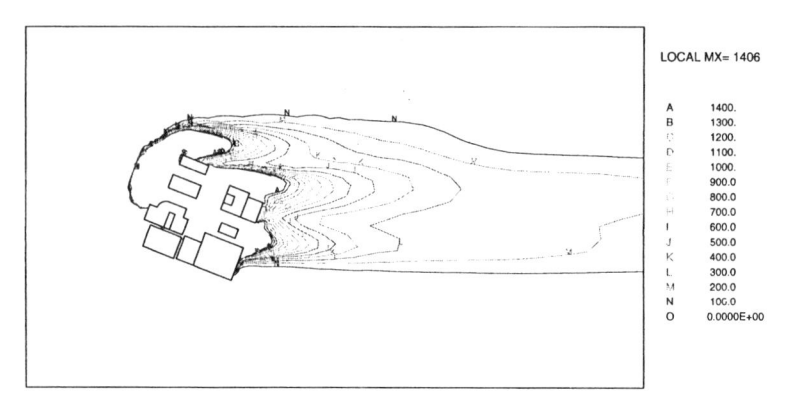

Figure 3. Ground level concentration contours for Run 1
(3.6kg/s, 2.4m/s, D stability, SW wind)

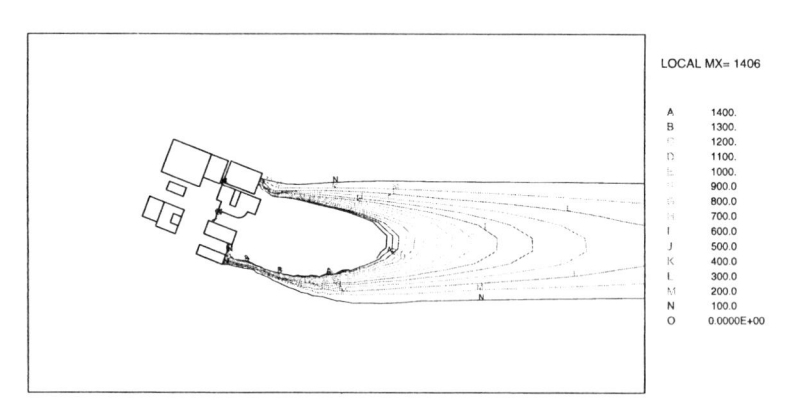

Figure 4. Ground level concentration contours for Run 5
(3.6kg/s, 2.4m/s, D stability, NE wind)

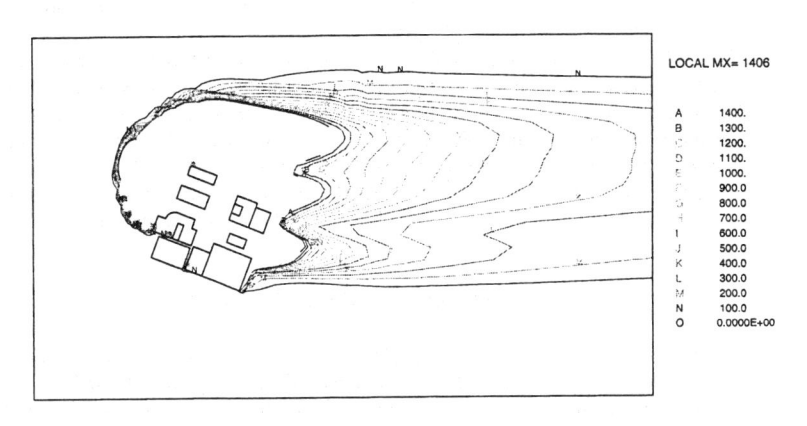

Figure 5. Ground level concentration contours for Run 7
(3.6kg/s, 2.4m/s, F stability, SW wind)

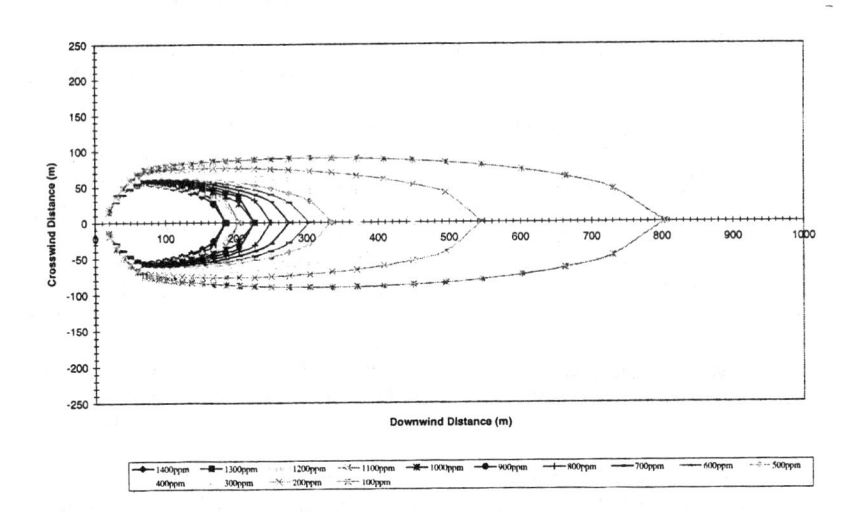

Figure 6. WEGDE/HEGADAS results from Run 1, single building modelling

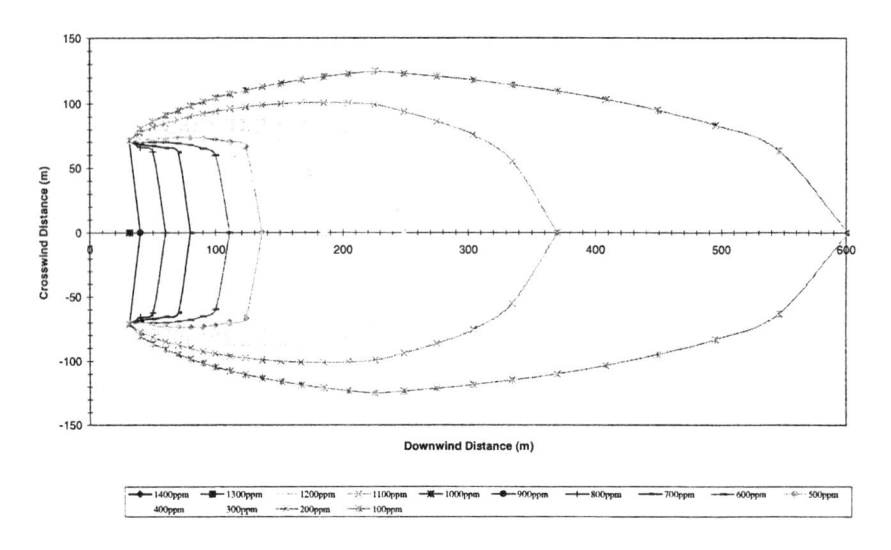

Figure 7. WEGDE/HEGADAS results from Run 1, complete site modelling

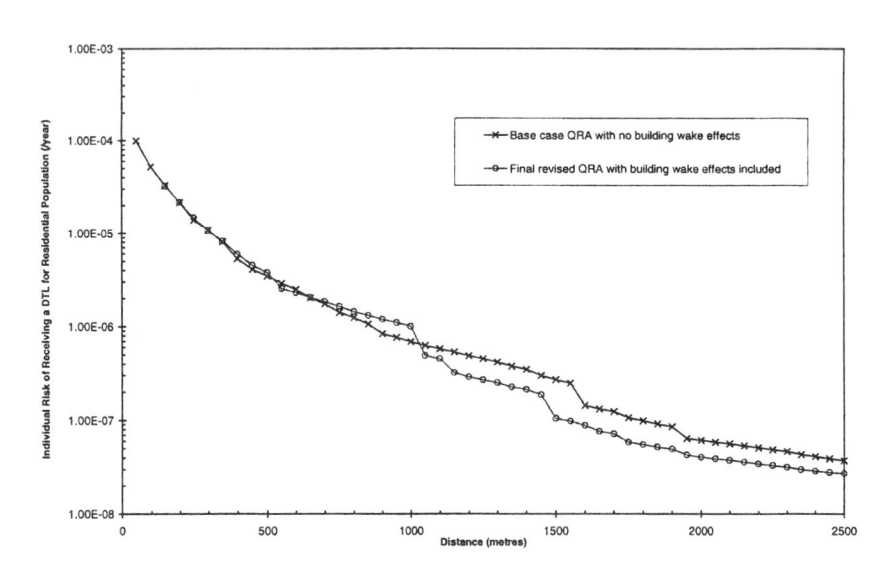

Figure 8. Comparison of base case and revised QRAs for a typical chlorine bulk storage site, showing the effect of including building wake effects

THE USE OF RISK-BASED ASSESSMENT TECHNIQUES TO OPTIMISE INSPECTION REGIMES

G R Bennett, M L Middleton, P Topalis
Det Norske Veritas, Stockport Technical Consultancy, Highbank House, Exchange Street, Stockport, SK3 0ET.

Regulators and insurance companies now recognise the acceptability of a risk based approach to the optimisation of inspection and maintenance intervals. Qualitative approaches to Risk Based Inspection (RBI) have been developed for general use and more detailed quantitative methods exist for activities with major loss potential or large preventive expenditure requirements. DNV pioneered the use of RBI in the chemical process industry in 1992 and has produced a resource document for the API (API 581). DNV subsequently further developed Quantified RBI software. A number of case studies and client projects have been conducted and the RBI techniques and software are becoming valuable tools for the oil and chemical industry.

Keywords: Risk Based Inspection, Inspection Planning, Consequence, Likelihood, Corrosion

WHY CONSIDER RISK IN INSPECTION PLANNING ?

"The first duty of business is to survive and the guiding principle of business economics is not the maximisation of profit, it is the avoidance of loss"

(Peter Drucker)

All industrial organisations use people, equipment and property to add value to a commodity and thus generate a return on their investments. If the returns are in excess of the costs, the company is in profit, if they are less, a loss occurs and if the losses continue to be sustained, the organisation will ultimately fail.

People and equipment are not infallible. People do make mistakes and equipment does fail. All mistakes and failures have consequences, some small and others large. The magnitude of those consequences influences the ability of the organisation to achieve a sustained profit and thus survive. It is therefore good business practice to consider the consequences of failures and how likely they are to occur, i.e. to systematically consider the risks arising from failures.

All forward thinking organisations practice some form of loss avoidance policy. Equipment maintenance, inspection and testing, and the training of personnel are all undertaken to avoid failures and reduce or mitigate their associated losses

How many organisations however, know in detail how effective their investment in loss prevention activity is? Does the extent and frequency of inspection reflect the magnitude of the consequences of an undesired failure? Is the inspection activity likely to identify the degradation mechanisms that exist? Do the planned maintenance routines actually influence the chance of the equipment breaking down? Is money being spent on inspecting and maintaining equipment which, if it fails, has very little effect on the organisation?

An organisation's ability to consider these questions has traditionally been driven by legislative rather than business need. For example, inspection of pressure retaining equipment or structures has been primarily calendar based, driven purely by legislative requirements. This no longer needs to be the case. The new goal setting legislative environment provides the opportunity to plan loss avoidance activities by linking any increase in the level of activity, to the reduction in risk achieved by that increased level of activity.

Both regulators and insurance companies now recognise the acceptability of a risk based approach to the optimisation of maintenance.

RISK BASED INSPECTION METHODS

Risk Based Inspection (RBI) techniques have been developed along two complementary routes. Qualitative approaches to RBI have been developed for general use and detailed quantitative methods have been developed and are being refined for activities with major loss potential or large preventive expenditure requirements. This paper will concentrate on recent developments in the quantitative assessment approach.

QUALITATIVE RISK BASED INSPECTION

Qualitative RBI is based on answering a series of questions regarding likelihood of failure and the consequences of failure and assigning notional levels (High / Medium / Low) to the answers to place the item on a risk matrix, see Figure 1.

The closer an item is to the top right corner of the matrix, the more critical the item is, and the greater the inspection activity warranted. It should be noted however, that changes to inspection regimes can only effect the frequency of an event, and not its consequences. Therefore if an item is in the high risk category primarily due to it's consequences, then no amount of additional inspection will improve it, and design changes to the system may be required instead. This statement is equally true for quantitative assessment techniques.

Software packages are now available which allow a qualitative estimate of inspection frequency to be made. They vary in complexity, and in the degree of "engineering judgement" to be applied, but they can be used as an effective screening tool in order to determine which equipment should be subjected to a more detailed quantified assessment. An example screen shot from one of the DNV software packages is shown in Figure 2.

The software is driven simply by the selection of various options from drop down menus, which prompt the user to select the most appropriate category for equipment type, location, fluid type and inventory, business interruption consequence, and specific details relating to the material of construction of the equipment, and its susceptibility to various failure mechanisms. Based upon these inputs, the software then uses a set of pre-defined rule sets to calculate a risk ranking and a corresponding recommendation for the next inspection interval.

THE QUANTIFIED RISK BASED INSPECTION METHOD

The Quantified Risk Based Inspection philosophy and method has been endorsed and is being promoted by the American Petroleum Institute Committee on Refining Equipment. The group is comprised of representatives from the following companies:

Amoco	Dow	Pennzoil
Aramco	DNO Heather	Petro Canada
Arco	DSM	Phillips
Ashland	Exxon	Shell
BP	Fina	Sun
Chevron	Koch	Texaco
Citgo	Marathon	Unocal
Conoco	Mobil	

DNV pioneered the use of RBI in the chemical process industry in 1992. In 1993 the API approached DNV with the request to jointly fund a larger development effort, aimed at producing a resource document for how to establish risk based inspection in the petroleum industry. DNV obliged, and in 1994 it produced the Base Resource Document on Risk-Based Inspection. This document is now being promoted by API as a standard referred to as API 581, it will be followed shortly by API 580 which will become an API Recommended Practice.

In 1995, the API commissioned DNV to develop software for its members to allow them to automate some of the processes required by API 581. This software based system has now been used successfully on a number of studies conducted both by DNV, and the sponsor group members. DNV decided to further develop the Quantified Risk Based Inspection Software in 1996. The new software, called ORBIT was initially released in March 1998 and is available to the industry as part of an integrated package of Risk Based Inspection services. ORBIT provides a fast interpolation approach for consequence analysis.

The quantified RBI approach provides a methodology for determining the optimum combination of inspection methods and frequencies. Each available inspection method can be analysed and its relative effectiveness in reducing failure frequency can be estimated. Given this information and the cost of each procedure, an optimisation program can be developed. The key to developing such a procedure is the ability to quantify the risk associated with each item of equipment and then to determine the most appropriate inspection techniques for that piece of equipment.

SYSTEMATIC APPROACH TO REDUCING RISKS

A fully integrated Risk Based Inspection system should contain the steps shown in Figure 3. The system includes inspection activities, inspection data collection and updating, and continuous quality improvement of the system. Risk analysis is "state of knowledge" specific and, since processes and systems are changing with time, any risk study can only reflect the situation at the time the data were collected. Although any system, when first established, may lack some needed data, the risk based inspection program can be established based on the available information, using conservative assumptions for unknowns. As knowledge is gained from inspection and testing programs and the database improves, uncertainty in the analysis will be reduced. This results in reduced uncertainty in the calculated risks.

The combination of elements required as inputs to a quantitative RBI analysis are shown in Figure 4.

The two major elements of a quantitative RBI analysis, as with any risk based study are an assessment of the probability (or likelihood) of an event occurring, and its consequences should it occur.

LIKELIHOOD ANALYSIS

When considering the likelihood of a failure occurring, the RBI process utilises a series of technical modules, in order to establish a damage rate for the equipment. The calculated damage rate depends upon the item's material of construction, the process fluids it is exposed to, its external environment and the process conditions (pressure, temperature etc.). Details of the current technical modules are shown in Figure 5.

It is also necessary to consider the current inspection regime, and to identify when the equipment was last inspected, how it was inspected, and what the results of those inspections were.

For example, if an equipment is subject to damage by corrosion or erosion, and if no inspections have been performed, then the likelihood of failure may be high. If however, many inspections of sufficient quality have been performed (and the equipment still meets its design intent) then the likelihood of failure will be quite low, even if there has been significant corrosion, as the rate of corrosion will be well understood.

Inspection activity is not however guaranteed to provide precise details of actual corrosion rates, each inspection has an error band associated with the particular technique. These error bands can be established by trials and review of historical surveys. Inspection activity does not change a corrosion rate, it reduces the error band and increases our confidence that we know the actual corrosion rate.

Statistical methods can be used to evaluate the likelihood that damage severe enough to cause a failure could exist given the amount of appropriate inspection activity that has been performed. As the damage rate is time based, future inspection techniques and intervals can be planned based upon the amount of damage expected to be seen at some point in the future. A proper balance must be established between advancing damage and increased knowledge of the amount of damage, to ensure safe and economic operation.

An example of a screen shot of the equipment details from the equipment specification module of the software is shown in Figure 6. A screen shot of the likelihood module is shown in Figure 7.

CONSEQUENCE ANALYSIS

The consequence analysis conducted within the RBI software is based upon look-up tables calculated using the DNV software package ***PHAST***. Consequences are calculated in terms of the area of equipment damage, and the areas within which personnel will be adversely affected by flames, explosions, or the toxic effects of the product concerned.

Using the input data on process pressure and temperature, material properties, and inventories, the system determines the release rate for a range of representative hole sizes, and also determines the release type. After determining whether a release is continuous or instantaneous (as in a vessel rupture), the software calculates the final phase in the environment (liquid or gas) and then determines the toxic or flammable consequences. In evaluating the consequences, the software also allows modeling of account mitigating features such as isolation and shutdown systems.

An example of the Consequence data module can be seen in Figure 8.

RISK ASSESSMENT

Having evaluated both the likelihood of an event, and its consequences, the system then combines this data to produce the overall risk evaluation for each piece of equipment. This allows the assessment of the overall risk levels of the plant, and the identification of where the high risk items on the plant are, so that inspection effort can be focused initially on the high risk items. Various reports can be automatically generated to produce a wide range of analyses. These reports include:

- Action damage/mechanism summary reports.
- Financial risks.
- Inspection planning.
- Risk ranking.

Graphs can also be produced for specific items of equipment showing the optimised number of inspections (cost of risk per number of inspections versus years of inspection) and the percentage of equipment versus the percentage of risk. An example plot is illustrated in Figure 10. The curve demonstrates the general principle that 80-90% of the risk is contributed by only 10-20% of a plants fixed equipment.

PLANT EXPERIENCE

It should be clear from the forgoing that regardless of whether a qualitative or quantitative approach is followed, it cannot be implemented without the active involvement of personnel familiar with the plant and its operation. Corrosion engineering experience is required to determine the damage mechanisms possible. Inspection management personnel are required to extract the knowledge of past inspection, and operations personnel are required to assist with establishing the safety and production implications of failures. It must be a team effort, in order to be effective.

CASE STUDIES

DNV has now completed a number of both pilot and full scale studies, in order to validate the data in the model, and to evaluate the usage of the technique. Results demonstrate that the RBI techniques and software are becoming valuable tools for the oil and chemical industry.

On one site, an RBI analysis of nearly 2,000 piping sections in an ethylene plant showed that less than 10% fell into the high risk category. Failure of those items constituted a business interruption and asset damage risk of $11.5 million per year. The application of improved inspection techniques reduced this risk to $4.1 million per year, a saving of $7.4 million per year. There was of course a cost associated with this risk reduction, and the improved inspection technology utlised was estimated to cost $250,000 per year. In this case the benefits clearly outweighed the costs. On the other hand, a review of the bottom 10% of risk items showed that the application of the same inspection techniques could still result in a risk reduction from $12,000 per year, to $4,300 per year a saving of $7,700 per year. However the costs of the improved inspection of those items would still cost $250,000, and this was clearly not cost effective.

On another site, 10 vessels were removed from the annual inspection plan, at an annual cost saving of $25,000 per vessel, and some pipework materials were upgraded at a 20% increased cost of materials, but avoiding a possible $3 million loss in business interruption.

SUMMARY

The RBI methodology and its supporting computer program has already gone a long way toward an integrated risk management program.

The technology is still being developed by DNV, and the author wishes to take this opportunity to thank the RBI sponsor group for their advice and support during the development phase. The RBI philosophy and database has drawn upon the experience of both DNV and the project sponsors in order to evaluate failure mechanisms, corrosion rates, consequences etc.

The author would also like to acknowledge the contribution of his colleagues Mark Middleton (DNV Stockport), Panos Topalis (DNV Software Products), Angus Lyon (DNV Aberdeen) and Gert Koppen (DNV Rotterdam) to the production of this paper.

Figure 1 Example Risk Matrix

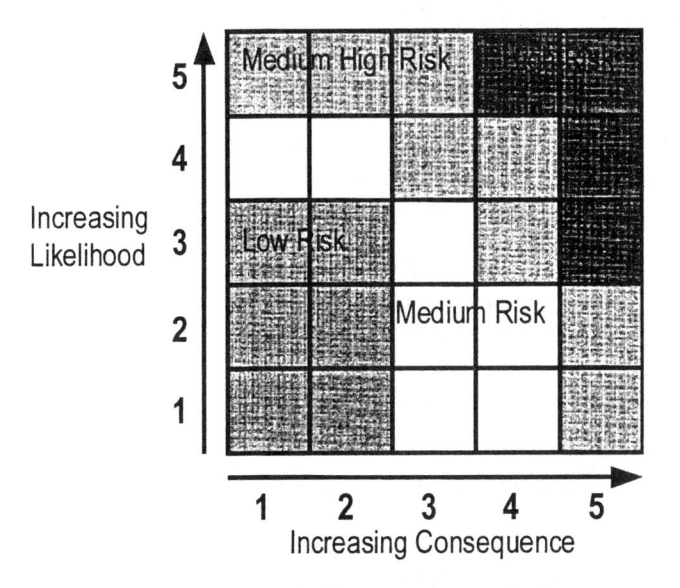

Figure 2 Example of Qualitative RBI Software

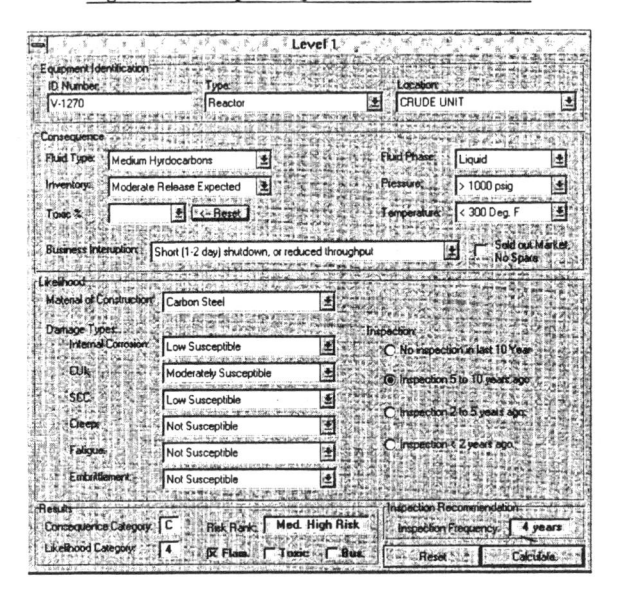

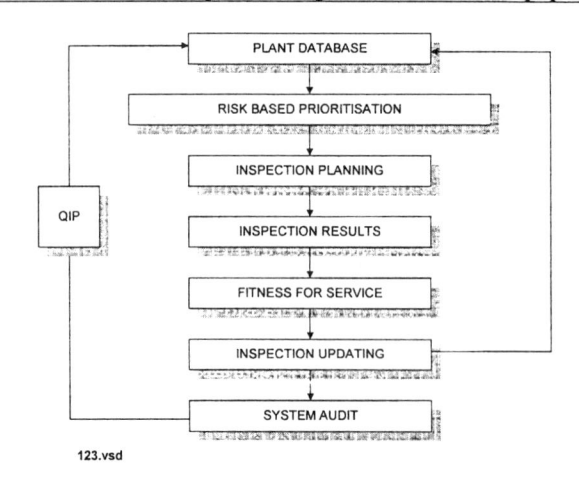

Figure 3 Risk Based Inspection Program for In-Service Equipment

Figure 4 Overview of Risk Based Inspection

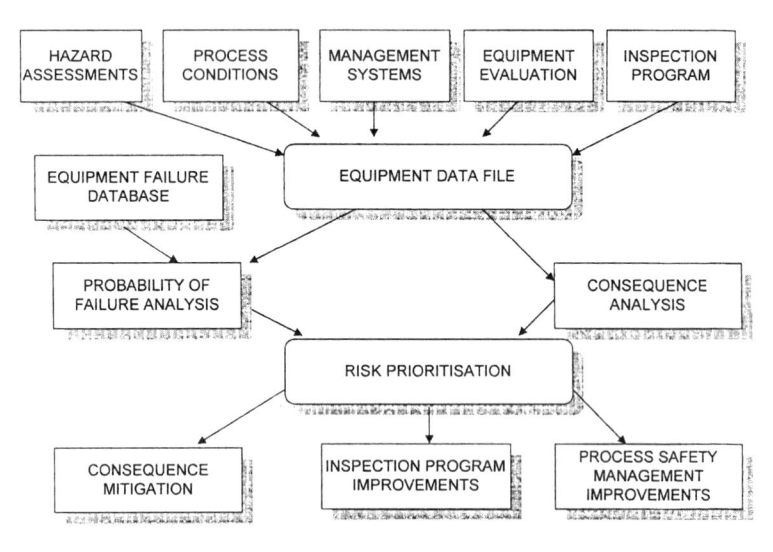

Figure 5 RBI Technical Modules

```
                        Technical Modules
                            Set-up
```

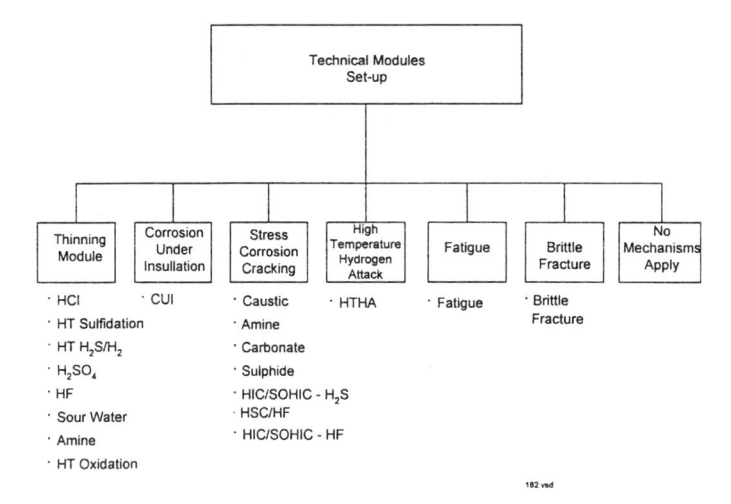

Thinning Module	Corrosion Under Insullation	Stress Corrosion Cracking	High Temperature Hydrogen Attack	Fatigue	Brittle Fracture	No Mechanisms Apply
· HCl	· CUI	· Caustic	· HTHA	· Fatigue	· Brittle Fracture	
· HT Sulfidation		· Amine				
· HT H$_2$S/H$_2$		· Carbonate				
· H$_2$SO$_4$		· Sulphide				
· HF		· HIC/SOHIC - H$_2$S				
· Sour Water		· HSC/HF				
· Amine		· HIC/SOHIC - HF				
· HT Oxidation						

Figure 6 General Equipment Module

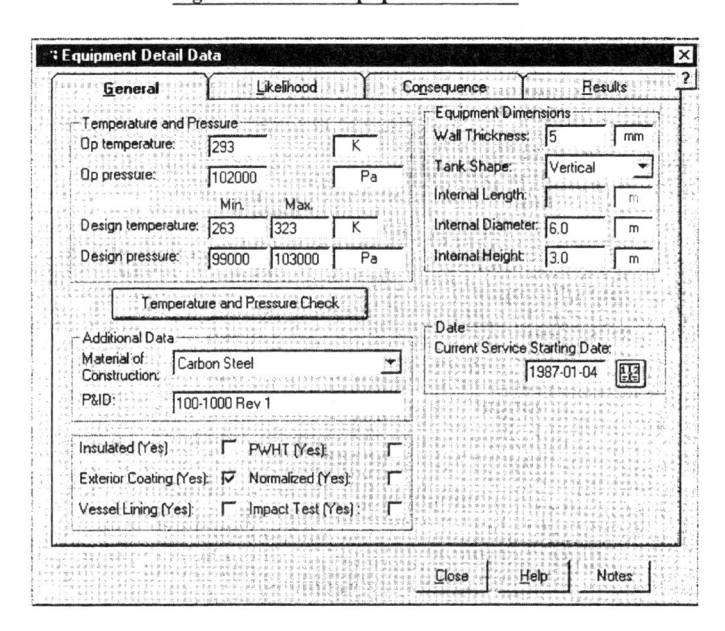

Equipment Detail Data

General | Likelihood | Consequence | Results

Temperature and Pressure
- Op temperature: 293 K
- Op pressure: 102000 Pa
- Design temperature: Min. 263 Max. 323 K
- Design pressure: 99000 103000 Pa

[Temperature and Pressure Check]

Additional Data
- Material of Construction: Carbon Steel
- P&ID: 100-1000 Rev 1

- Insulated (Yes): ☐   PWHT (Yes): ☐
- Exterior Coating (Yes): ☑   Normalized (Yes): ☐
- Vessel Lining (Yes): ☐   Impact Test (Yes): ☐

Equipment Dimensions
- Wall Thickness: 5 mm
- Tank Shape: Vertical
- Internal Length: m
- Internal Diameter: 6.0 m
- Internal Height: 3.0 m

Date
- Current Service Starting Date: 1987-01-04

[Close] [Help] [Notes]

511

Figure 7 Likelihood Analysis Module

Figure 8 Consequence Analysis Module

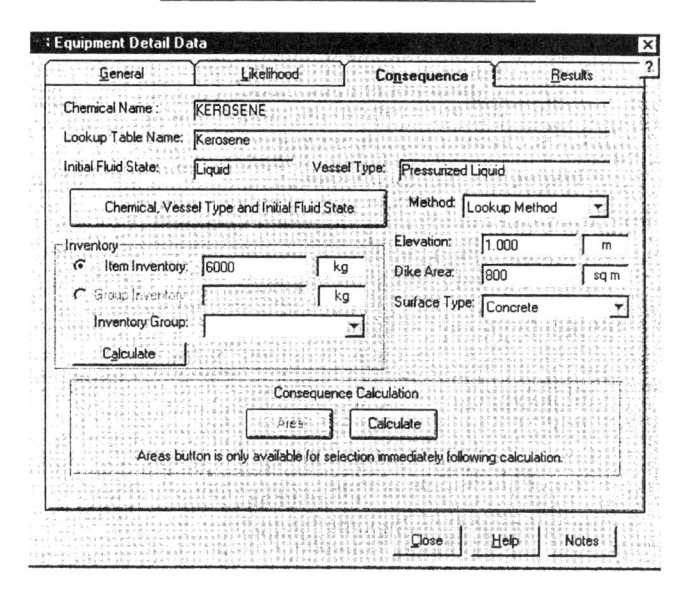

Figure 9 Results Module

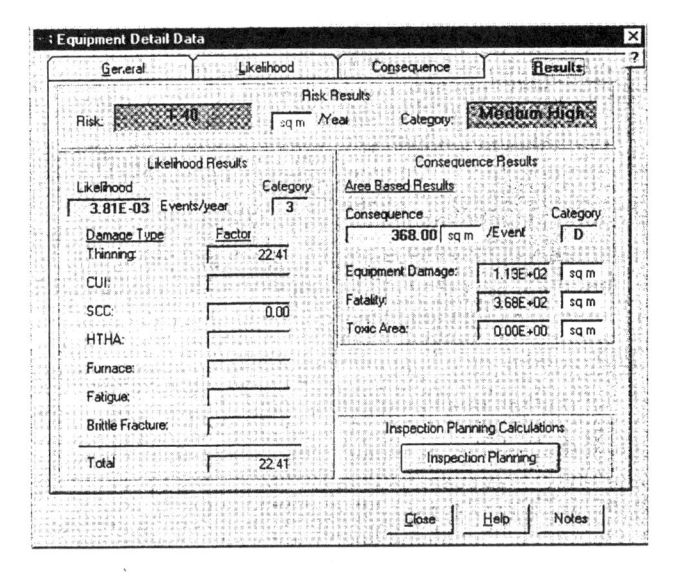

Figure 10 Risk Results

Percentage of Equipment vs. Percentage of Risk

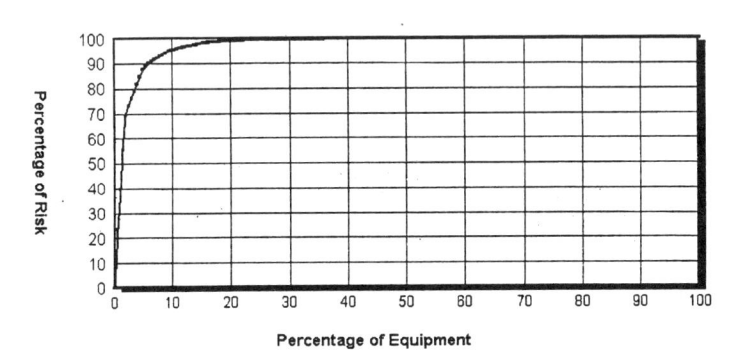

CATEGORIZATION OF RISK

GOOD, BAD OR INDIFFERENT, SITE RISK GRADING

Paul Clarke
Allianz Cornhill International Division, 32 Cornhill, London EC3V 3LJ

How does your site stack up against others? Are you more hazardous, or more or less likely to have a loss? If so, can you minimise the impact of the incident? Finally, where can you best spend your Capital to protect the site or the Group? Insurance Engineers regularly have to evaluate the quality of sites of the clients we do (and don't) insure. Most use some assessment methods to provide quality analysis. For Allianz Cornhill, we are able to do this by use of a Risk Grading System, which differentiates between General/Manufacturing industries, and Chemical/Petrochemical Risks. Because a client will often insure both Property, and Business Interruption, this is also one area we analyse. This paper focuses on what goes into a Risk Grading system like ours for Oil and Petrochemical/Chemical business, and how we feed back the information. The comparative nature of the system means that similar facilities can also be looked at. This can also compare the relative premium that a good performing site should pay as against a poor performer.

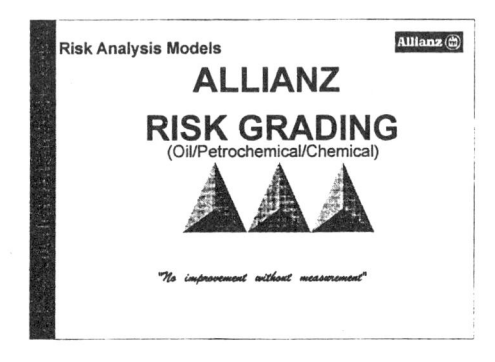

Risk Analysis Models

ALLIANZ RISK GRADING
(Oil/Petrochemical/Chemical)

"No improvement without measurement"

GOOD, BAD, OR INDIFFERENT?

Do you feel your site comes into any or all of these categories?. However, on what basis do you make your assessment, and do others use that as well?. How do you compare yourselves to other Group locations, and how do you rate?.

The facilities in which you work, will be subject to a series of inspections over the years. These can take the form of internal and external audits, self assessments and can include studies carried out by consultants on your behalf. As your Insurers, we also want to survey the premises, to then advise the Underwriters on the quality of the risk. To carry out Risk Assessments, we ask for data, talk to site personnel and walk through the site.

The aim of this paper is to give an insight on how we as Insurers assess the site from a property-damage and business interruption standpoint. Other Insurers and brokers have their own methods, but this is based on the Allianz Risk Grading methodology.

Grading of Risk

The standard of any grading system is that it allows for comparisons between plants of the same or similar occupancies. This is to highlight the strength of the protection, "the defence in depth", or weaknesses that need to be addressed. At the same time, it can be recognised here that some processes have a greater inherent hazard, should control and safety systems fail to allow safe shutdown.

Once we know the occupancy, we can start to look at the expected protections and layout. At the same time, the age of the site, and its development over the years can have an important impact on the quality of the risk as fire divisions are knocked down, or breached, a spare capacity is utilised, and as obsolescence of equipment creeps in.

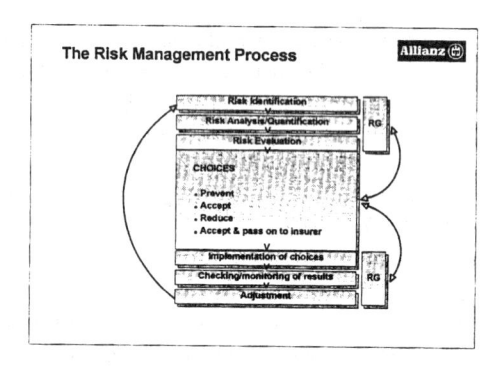

Consistent questions

As a Group, the occupancies may vary considerably, potentially involving refineries and petrochemical works, or chemical works with mixtures of hazardous and non-hazardous.

Our assessment method covers a full range of the site operations, under the general headings of Property Damage and Business Interruption :-

- Occupancy- Inherent Hazard Factor
- Process Modifying Hazards/ Safety Features
- Spacing/Layout
- Construction
- Utilities
- Management
- Fire Protection
- Security
- Additional Perils
- Business Interruption
- Loss History
- Risk Information

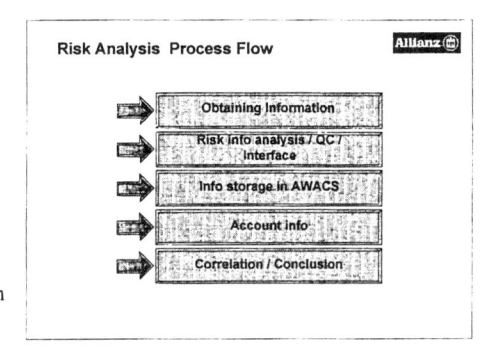

Risk Analysis Process Flow — Allianz

- Obtaining Information
- Risk Info analysis / QC / Interface
- Info storage in AWACS
- Account info
- Correlation / Conclusion

The important thing is to ask questions consistently, and therefore assess from a level playing field. The reason for this can be plainly seen if you compare a facility in an earthquake zone, compared with one outside such an area.

Once this information has been obtained, it is plugged into the Grading model, and produces a grading for the site based on the site as seen at the time of the survey. This is the "As is" position.

At the same time, sites are often constantly changing, and as insurers, we also make recommendations to improve the risk areas(which are themselves changes). On the basis of how these proposed changes could affect the way we view the facility, we then come up with a "To be " scenario. It is the difference between the "As Is " and "To be " gradings, that show how much improvement can be made at any one site.

Such differences are normally positive, as you would expect a site to improve over the years. In this way, the "To be " position would be better than the "As is". In fact, this system can also reflect negative changes, which are caused by the changes in operations, site layout, or some other deterioration. An example of this could be the location of a new unit in amongst other process equipment might reduce previous good spacing, and potentially expose an existing control room.

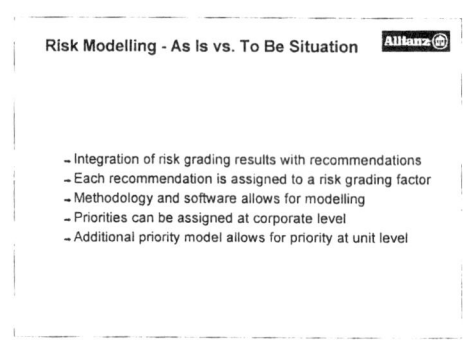

Risk Modelling - As Is vs. To Be Situation Allianz

- Integration of risk grading results with recommendations
- Each recommendation is assigned to a risk grading factor
- Methodology and software allows for modelling
- Priorities can be assigned at corporate level
- Additional priority model allows for priority at unit level

What does the Client gain?

An Insurance survey, whether it is carried out by the Insurer or a Broker, is for information gathering, so that an assessment of the quality of the site risks can be carried out. Once complete, you get feedback, often in the form of Recommendations, but these relate just to one site.

As part of a group of companies, your Risk or Insurance Manager (or Finance Director) will be insuring a range of locations, and your site(s) will be a part of this programme of surveys. The premium paid is a pot of money, allocated centrally or on a site by site basis. Insurance is, in the end, about taking risk, and protecting against it. By having a grading system applied to the sites in the programme, the better quality sites can actually benefit, by paying a lower proportion of the overall pot. As the average quality of the sites in the programme improves, this also helps your insurance buyer (The Risk Manager) get as good a deal as possible from the Insurance Underwriters.

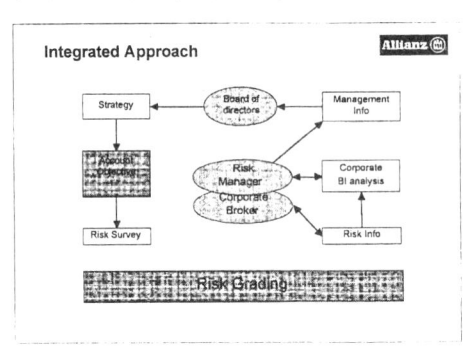

Integrated Approach   Allianz

Where to spend your effort/money

Where safety issues need to be addressed at a site, there is an in-built priority associated with them. With Risk Improvement Proposals, these also go into the loss of Property and the interruption of Business, which might follow a major incident. There is no question that the safety of lives is paramount, but it is very worthwhile for you to have a job to come back to. However, where is the best place to put your effort, or spend money?.

When improvements are proposed, the relative importance, the impact of it being accomplished, and the benefit to the overall programme of assessment can also be evaluated.

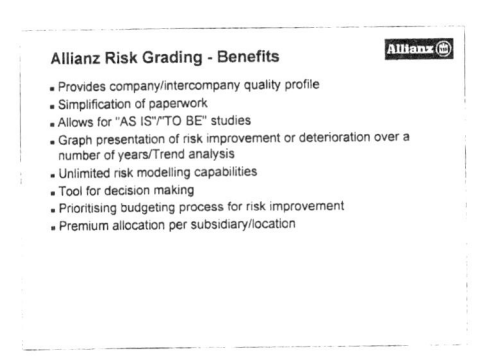

Allianz Risk Grading - Benefits   **Allianz**

- Provides company/intercompany quality profile
- Simplification of paperwork
- Allows for "AS IS"/"TO BE" studies
- Graph presentation of risk improvement or deterioration over a number of years/Trend analysis
- Unlimited risk modelling capabilities
- Tool for decision making
- Prioritising budgeting process for risk improvement
- Premium allocation per subsidiary/location

In what order of priority should we:-

- fireproof the reactor supports in Scunthorpe?
- repair the cracked/spalling fireproofing at Milford Haven, Scunthorpe & Hamburg
- put subsurface foam injection in the Storage tanks at Milford Haven?
- upgrade the control room and/or sprinkler protect the warehouse at East Kilbride?
- reprint and review the hot work procedures and permit system? (and are the US and Spanish sites using the same concept)
- develop a Group-wide Business Recovery plan?

Which will help the Group as a whole, and which is very site specific? If the E. Kilbride control room is for a bottlenecking plant that affects the whole Business unit, and the warehouse is the sole distribution centre for the UK, some answers are less easy than others.

This method is to be used as an aid to decision making, for a client Risk Manager and the Management of Industrial sites. When used across the range of sites belonging to the company, it can strengthen the decision process, and is capable of being used as a client model, as well as a tool for insurers' purposes only.

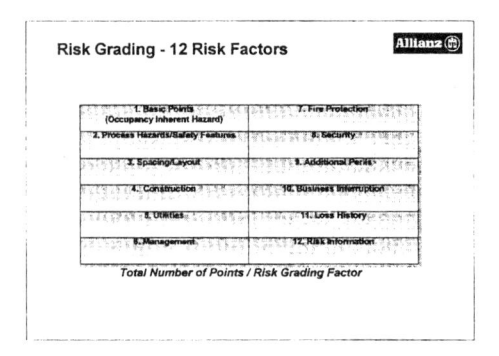

Risk Grading - 12 Risk Factors	Allianz
1. Basic Points (Occupancy Inherent Hazard)	7. Fire Protection
2. Process Hazards/Safety Features	8. Security
3. Spacing/Layout	9. Additional Perils
4. Construction	10. Business Interruption
5. Utilities	11. Loss History
6. Management	12. Risk Information

Total Number of Points / Risk Grading Factor

HOW DO WE WORK WITH THE INFORMATION?

This involves the conversion of the information into a numerical format. But first we break down the site, it's occupancy, and the elements of exposure and protections into bite-size chunks. From there we can more easily work with the data.

There are a series of headings under which we assess the available facts.

Occupancy-Inherent Hazard

This relates to the overall occupancy of the facility, and therefore by extension the potential risk. It looks at whether the site is a complex oil refinery, a primary pharmaceuticals manufacturer, or a chemical plant. Different levels of severity of operating conditions provide different relative hazards, and can be recognized as such.

Process Modifying Hazards/Safety Features

Even within a "standard" facility, there are features which make the potential likelihood of an incident more or less. Construction of a new unit, or the mothballing of an old unit can possibly change the risks.

Process Safety Features include relief systems, flares, MOVs, and how they apply to the processes being carried out. At the same time, for older units, and equipment, the designs and safety features themselves have changed over the years. In this case, the systems must be reviewed against latest thinking, compared with what was appropriate at the time of construction.

LAYOUT

In the end, any facility must fit into the plot space provided for it. Having said that, the plot space should be consistent with what is being asked to be fitted into it. Where the site has become "fully developed" there is often little room for manoeuvre and certainly no room for error.

Where possible, separation between different occupancies should be by physical distance. Where this can't be achieved, fire break walls with protected access ways are important.

CONSTRUCTION

Once we have sorted out the layout, then the construction elements come into play. For the enclosed areas, we will be looking at the company design guidelines addressing standard building code issues such as wind, resistance and snow loading, then naturally, the ability of the building construction to not only not support a fire, but even resist it. We will be looking for fire proofing of open or enclosed structures acting as supports to reactor vessels in flammable liquid/gas service, and in principle, fire separation between the manufacturing/ operating area and the warehouse, or between the office area and the warehouse area/manufacturing area, with appropriate subdivisions for other areas. As control systems become more and more automated, any control room should be adequately separated from the process area, including protection and detection systems, as necessary.

Drainage systems should allow for adequate removal of both rain water, and if this is a worse case, fire water. The drainage system should be to an appropriate impounding basin to allow fire water run-off not to pollute water courses or rivers.

UTILITIES

Power, steam, gas etc. How resilient is the design, and how loaded up is it?

MANAGEMENT

This section looks at the areas of Risk Management, Operations, Maintenance, Inspection, etc.

FIRE PROTECTION

From sprinklers and deluge systems, through fire and gas detection, to water supplies, and the fire teams to use them (and the planning that goes into it)

SECURITY

Active and passive protection to make the site secure.

ADDITIONAL PERILS

These cover other possible causes that could bring about a loss at a site, and include:-

<u>Natural</u>

    Storm
    Subsidence
    etc.

<u>Social</u>

    Vandalism
    Riot
    etc.

<u>Third Party</u>

    Collision
    Adjacent facilities
    etc.

<u>Consequential Aspects</u>

    Contamination
    Debris removal
    etc.

BUSINESS INTERRUPTION

How much does a loss hurt, and for how long?

How much have you, can you (or can others) do on your behalf, to mitigate the circumstances of a loss.

LOSS HISTORY

What has the loss history been like in the past?. We use a five year period as a selected time frame for this model.

We look at the issues of repeated losses, or one big loss at a particular site.

RISK INFORMATION

Where does the information come from, how much does it answer the questions, and how old/ potentially out of date is the data?

USING THE RESULTS

When the data has been put in, the results come out for individual sites. The detailed issues of areas of strength, and opportunities for improvement within each site are then available. Summaries of the data can then be presented for overview and trending purposes. This can be based over:-

-Years

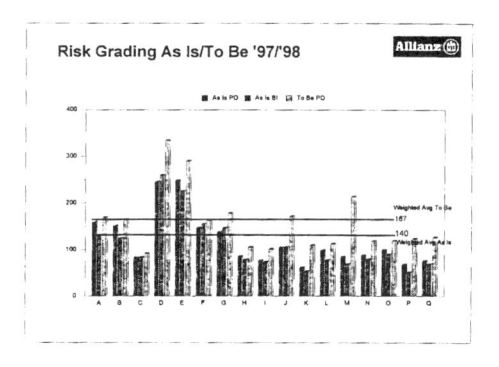

-Group locations

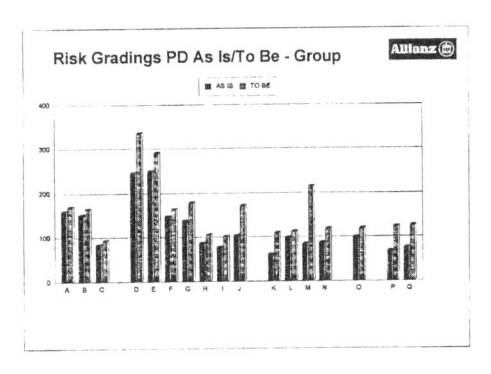

-Regions

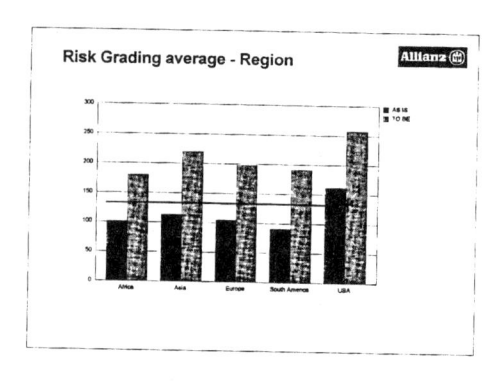

Risk Grading average - Region

-Loss estimates

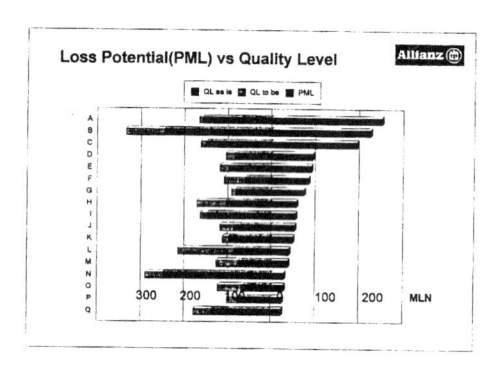

Loss Potential(PML) vs Quality Level

-Specific types of recommendation, or area for improvement

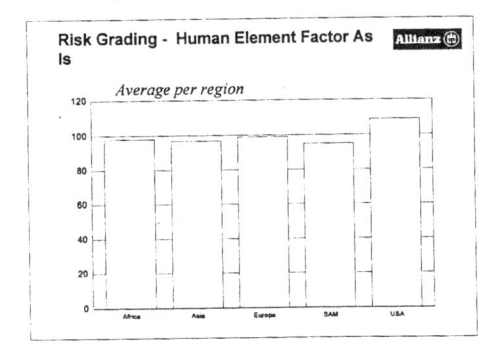

Risk Grading - Human Element Factor As Is
Average per region

Because the analysis looks at both the Business Interruption as well as the Property Damage aspects, we can present the results in a number of different ways.

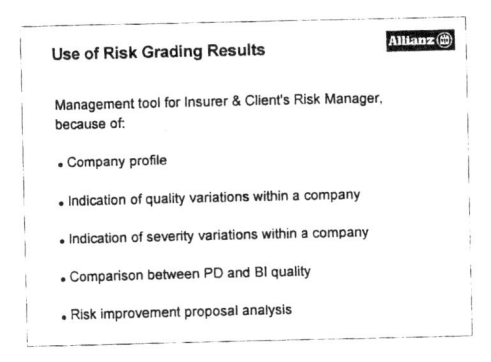

Use of Risk Grading Results

Management tool for Insurer & Client's Risk Manager, because of:

- Company profile
- Indication of quality variations within a company
- Indication of severity variations within a company
- Comparison between PD and BI quality
- Risk improvement proposal analysis

Once the trends are able to be analysed, then this can be integrated into the decision-making process for Companies/Corporations, to best advise them on where to spend their effort/ capital investment.

Not only that we have also developed client application modules which can provide more option-analysis for a client's Group Risk Management, for their own benefit.

If you have a way of comparing individual sites within a Group in a consistent way, then the comparison itself can be used to help allocate premium to those individual sites, as a proportion of the overall programme premium. Higher quality graded sites would then pay a lower proportion of the overall premium. Obviously, as the years pass, an overall improvement in the quality of the sites gives the Group Risk Manager the best ammunition to get the best price for the insurance of the programme.

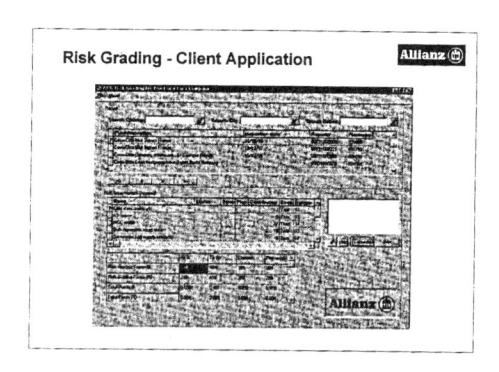

Risk Grading - Client Application

CONCLUSION

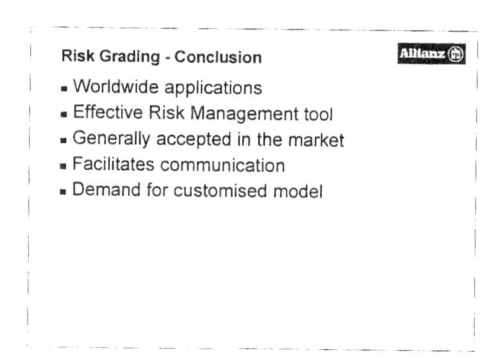

Risk Grading - Conclusion
- Worldwide applications
- Effective Risk Management tool
- Generally accepted in the market
- Facilitates communication
- Demand for customised model

Adopting any form of grading system is better than doing nothing. We believe that this method can help us as insurers and you as clients to assess more clearly the quality of the risks you deal with on a day to day basis.

By adopting a proactive approach, you may not be good or bad, but you certainly won't be indifferent.

Remember:-

What you can't measure, you can't improve

THIRTY YEARS OF QUANTIFYING HAZARDS

J T Illidge, M L Preston, A G King*†
ICI Technology, PO Box 8, The Heath, Runcorn, Cheshire, WA7 4QD
* ICI Technology, PO Box 90, Wilton, Middlesbrough, Cleveland TS90 8JE
† Author to whom correspondence should be addressed

© 1998 Copyright Imperial Chemical Industries PLC

>This paper reviews the development of hazard quantification tools, techniques and methodologies in ICI since the 1960's. From early application to Instrumented Protective Systems and toxic gas releases, the techniques have evolved to cover a wide range including safer process design, environmental spills and full Quantitative Risk Assessment (QRA).
>
>Experience from a number of applications is briefly described and difficult areas discussed. Finally issues and needs for training for effective hazard assessment are outlined.
>
>Keywords: Hazard assessment, Hazard analysis, HAZAN

INTRODUCTION

A **Hazard** has been defined (Ref. 1) as 'a physical situation with a potential for human injury, damage to property, damage to the environment or some combination of these'. As so defined, hazards in the process industry can be associated with physical, chemical or biological effects in the storage, transport or use of chemicals and in the operation of other assets. As long as the hazard remains only a potential to cause harm there is no problem. If, however, the potential is realised, an event causing actual harm occurs and this has been defined as a **hazardous event**.

Where hazardous events can occur, their likelihood and consequences may be quantified.

Hazard Assessment, Hazard Analysis or **HAZAN**, is a process for identifying undesired events that lead to the materialisation of a hazard, the analysis of the mechanisms by which these undesirable events could occur and usually the estimation of the extent, magnitude and likelihood of any harmful effects. The relationship of HAZAN to HAZOP is described in the by Trevor Kletz (3) and the development of ICI's HAZOP process over the same period as this paper is detailed in Ref 4.

WHAT IS HAZARD ASSESSMENT USED FOR?

There is little point in quantifying hazards if there is no objective to be met. In many cases, criteria need to be set to allow the acceptability of assessed risks to be judged. In other cases there may be alternative designs or ways of carrying out some operation and a quantification of the hazards may allow the 'safer' alternative to be identified. Hazard assessment may also be used to assess the cost-effectiveness of a proposal. So, within ICI, hazard assessment has been accepted primarily as a decision-making aid rather than as a means of 'proving' that a process is 'safe'. Used in this way, the benefits have been in improved safety of our operations.

HISTORY OF HAZARD ASSESSMENT IN ICI

Within ICI, the first experience of quantifying hazards was in 1966. A serious explosion occurred, which killed two employees, and this was caused by failure of an automatic instrumented protective system (trip system). Advice was sought from the United Kingdom Atomic Energy Authority, and they were able to provide the methodology to re-design the trip system to meet a very high standard of reliability (high integrity). No effort was made to assess either the frequency of demands (events requiring the trip system to operate to prevent hazardous events) or the tolerable frequency of explosions. Procedures and Guidance were developed within ICI for the design and operation of Instrumented Protective Systems to meet a range of reliability targets - the sophisticated and very expensive design of a high integrity system would clearly be inappropriate for a minor hazard.

In 1970, an incident occurred which released chlorine to atmosphere and affected people outside the Works. In response, a major study, using hazard assessment techniques, was started. This study looked at all the operations on the Works and quantified the risks to the public and where these were unacceptably high, appropriate improvements to reduce the likelihood or consequences were identified. The initial stage of the study involved setting quantitative criteria to be met for different potential effects on members of the public outside the Works. The study covered 26 plants and took some nine man-years of effort to complete. Quite a commitment for the first full hazard assessment in ICI.

From 1970 onwards, quantification of hazards on projects to meet appropriate ICI criteria became an established approach to ensuring that process designs were safe enough. The hazard assessment allowed the selection of an appropriate reliability of protective systems, but was also used to assess the acceptability of many diverse aspects of design, including the acceptability of standard pipe joints, bellows, etc., the provision of spare equipment, optimising power supply arrangements, etc. Techniques were developed for assessing flammable hazards and development of improved methods for modelling the effects of gaseous emissions continued (and still continues).

In 1988, the techniques used to assess toxic gas emissions were extended into the assessment of the hazards of spills to the aqueous environment and a number of these studies have been carried out.

A developing technology

When the first significant hazard assessments were being started in the early 1970s, hazard scenarios were developed in words only. This could result in difficulties of understanding with more complicated scenarios and made communications difficult. As experience was gained, the presentation was changed to the graphical form as fault trees, using logic symbols to represent 'AND' and 'OR' logic gates. These fault trees were initially drawn by hand and from the earliest stages it was felt in ICI that fault trees using the international standard symbols for 'AND' and 'OR' gates were difficult to draw, inflexible and not easily

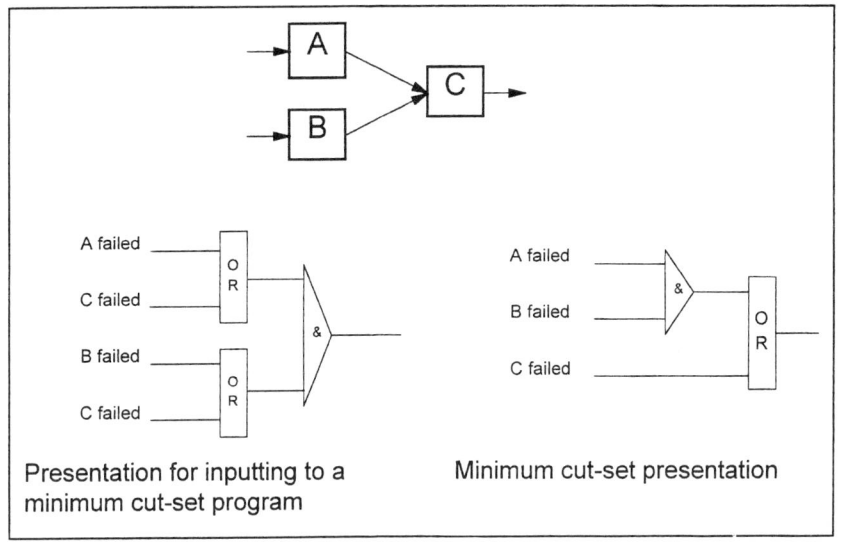

Figure 1

Presentation for inputting to a minimum cut-set program

Minimum cut-set presentation

understood by those designing and operating processes. It was decided that the use of simple shapes - a rectangle for an 'OR' gate and a triangle for an 'AND' gate was much simpler to produce and understand. Furthermore, the Western world reads from left to right, so it was decided that fault trees should also read from left to right. The decision to use a non-standard notation has been shown in practice to be very well received by the most important people - those managing and operating the processes.

As time went by, computer programs were developed to allow the drawing of logic diagrams to be carried out. At an early stage, these graphics packages simply drew out what the hazard analyst would previously have drawn, and carried out no automatic layout or calculations. Later programs allowed the automatic calculation of the fault tree and also laid out the fault tree automatically to fill the available paper area.

Within ICI we have felt that it was vital that the end user - those managing and operating the processes - should understand both the logic and the calculations in the fault tree and we have therefore avoided the use of sophisticated 'black box' minimum cut-set calculating programs, which require the user to trust what cannot be seen. We have developed the approach to hazard assessment to develop the fault tree directly as a minimum cut-set representation - see figure 1.

Experience over the years is that for hazard assessment, data shortage has been a serious problem only in particular areas. Hazard assessment generally involves assessing the frequency of events with a significant probability of happening in the life of a plant. It entails synthesizing the frequency from data on likely equipment failures, such as pump, compressor and instrumentation failures. Since these are usually relatively common events and there is a large population of similar systems, there is usually good quality data for such events and the hazard assessment of such scenarios is therefore quite accurate, particularly where the final event of concern can arise from several different scenarios. However, there are areas where hazard assessment is particularly difficult. These include:

- Assessment of ignition probability. Even where there is no obvious source of ignition, there remains some probability that flammable mixtures will be ignited by electrical static or some unforeseen event. Estimating the likelihood of ignition in these situations is essentially an experienced guessing game. The value used is particularly problematical where low values of ignition probability have to be predicted. In many cases the ignition probability is a multiplier for all the scenarios in the analysis and so can be very critical.

- Assessment of human error likelihood. Two areas are relevant here - the assessment of how often someone makes a mistake and the prediction of the likelihood of failing to respond correctly to an alarm or other stimulus, particularly in situations which may cause high stress. It is not easy to predict the probability of such events, particularly where the operator may on occasions be subject to many demands at the same time.

- Dependent failures. Dependent failures cause problems particularly where duplication of identical equipment is used to improve reliability. However, even where diversity is used (using different means of achieving the same function) there still remains the potential for dependent failures. Assessing the likelihood of such failures is difficult to do with confidence. Even more difficulties exist in the area of dependency between demands and protective system failures, which is an area where soundly-based methods based on real data seem to be virtually nonexistent.

Following prediction of the frequency by hazard assessment, consequence assessment may be very simple - for example, it will be either a big incident or a minor incident - through more graduated consequences to a full-blown Quantified Risk Assessment (QRA).

CRITERIA FOR HAZARD ASSESSMENT

The first criteria developed within ICI dealt with two areas. The first was the need to ensure that we were good neighbours to those living round our chemical plants as far as the accidental emissions of toxic gases were concerned. Criteria were set according to the severity of potential effect outside the Works (Ref. 2). The second area was in the establishment of criteria for potential fatalities among ICI employees from accidents relating to the process operations. Based on historical data, criteria were set to ensure that as time went by, the risk at work would be no greater than from accidents at home.

Following on from the major incident at Flixborough, further criteria were needed to deal with incidents with the potential for causing multiple fatalities. With the development of QRA methodology, criteria have been required for risk of fatality to a member of the public and for multiple fatalities. Criteria have also been developed for use in assessment of accidental spills to the environment.

In the use of all these criteria, the purpose of hazard quantification has been to ensure that effective and cost-effective decisions about process improvements are made.

WHAT'S IN A NAME?

There has been confusion between the terms Hazard Assessment, Hazard Analysis or Hazan and Quantified Risk Assessment (QRA).

Although QRA can in theory be an assessment of any type of risk, it has come by popular usage in the process safety area to refer to the assessment of risk of fatality from an accident and, in particular, affecting members of the public as far as land-based installations are concerned. Such risks are usually only significant for major loss of containment accidents and these are predominantly caused by 'generic' major failures of pipes and vessels, which are rare events. Since these generic failures are rare events, there is very little data on such failures, and usually any data applies to different designs or situations or operating conditions from that which is being assessed. The assessment of the risk in QRA uses very sophisticated consequence models to predict the likelihood of fatality to an individual or the frequency of killing a number of people. The overall accuracy of a QRA is severely limited by the reliability of the data and the consequence assessment.

EXPERIENCE WITH HAZARD ASSESSMENT

Hazard assessment in the process industry has been used in a wide range of applications. Some examples of these are:

- Assessment of the risks of accidental toxic gas emissions affecting people, to help make decisions about the need for improvements and the effectiveness and cost-effectiveness of modifications
- Assessment of the risks of accidental spills to the environment and the benefits of different improvements
- Assessment of the risks of explosions within the process affecting the safety of people and determining suitable improvements
- Assessment of risks from chronic toxic releases (long-term health risks) and deciding appropriate control measures
- Assessment of the risks from fires, causing serious damage to plant, buildings or equipment and resulting in major business interruption
- Assessment of the risks of major plant down-time affecting business
- Assessment of the suitability of a plant location
- Assessment of transport risks and selection of optimal transport routing
- Assessment of the most effective plant configuration
- Balancing the risk and benefit of a proposed modification
- Assessment of the required capacity of vent and flare systems handling multiple streams
- Identification of the failure modes of equipment to help improve the design, operation or maintenance strategies

Some areas of quantification have consistently proven difficult for the non-expert hazard analyst. A particular area of difficulty has been the calculation of Fatal Accident Rate (FAR). This is a measure of individual risk of death, dimensioned as deaths per 100 million hours exposed to a hazard. It is essentially a simple concept, but as experience within ICI and in the published literature has shown over the years, it is a calculation very prone to error. A more serious error has been where a dominant cause of hazard has been missed, or when a failure has been analysed as being independent of a protective system, when in fact it is functionally dependent on parts of the protective system. The latter type of error can and has caused errors of a factor of 1000 in the overall assessment.

TRAINING FOR EFFECTIVE HAZARD ASSESSMENT

It is very difficult for a non-expert to check a hazard assessment carried out by someone else. It is therefore likely that errors in hazard assessments will not be detected by checking. Errors may well not be detected in operating experience, particularly where hazard assessment is being carried out on relatively rare events. If an assessment is made that the frequency is 1 in 100,000 years, even an error of under-estimating by a factor of 1,000 in the analysis would still only make the event frequency 1 in 100 years and so it is unlikely that the event will occur in the life of the plant. As mentioned above, errors of this size have occasionally been found. In most cases large errors are caused by a mistake in the logic. To ensure that trustworthy hazard assessment is carried out, there need to be

- Effective training. This needs to deal with many potential pitfalls in the essentially simple methodology.

- Effective early practice and mentoring. It is vital that those who have been trained in hazard assessment start practising the techniques on significant hazard analyses very shortly after training. An experienced hazard analyst should ensure that early hazard assessments are monitored and that the learning has been absorbed effectively and help to build up the competence of the novice hazard analyst by mentoring.

- Refresher training or team working. Working as part of a team with experienced hazard analysts is an ideal way of ensuring an increasing level of competence. However, for those who only spend a very small proportion of their time carrying out hazard assessment or who may not have carried out an assessment for many months, refresher training may be the best way to restore competence.

- HELP. A problem facing the newly-trained hazard analyst, and also the more experienced hazard analyst facing a problem beyond his/her experience, is where to find help. Expertise in hazard assessment is relatively thinly spread and many of the people with expert knowledge are specialists in a particular limited area, such as QRA or protective systems or the mathematics of hazard assessment. The number of people with extensive experience of applying hazard assessment in the process industry and who are acknowledged as 'expert' hazard analysts is small. Within ICI there is a new initiative to use the experts' experience to raise the level of expertise in hazard analysis throughout the company.

TOOLS TO HELP WITH HAZARD ASSESSMENT

It is common experience that in the development of fault trees many different trees are drawn out before the final logic is agreed. A computer program to allow the development of the logic and its neat presentation for inclusion in reports is virtually essential. Within ICI, the original graphics based packages were replaced in the 1980s by a user-friendly DOS based package able to draw fault trees in the ICI preferred format on standard printers. The program featured routines to ensure that the fault trees were logically valid and to lay out the print-out automatically. This original program has recently been pensioned off and a full Windows 95 based 'LOGIDRAW' has replaced it. This is a much more powerful package with a graphical interface and it allows printing on any standard Windows printer and offers flexibility to develop many enhanced features.

A further development has been aimed at helping to streamline the assessment of environmental spills. A program RASP (Rapid Assessment of Spill Potential) was originally developed for assessing the risk of liquid spills from stock tank installations. A generic hazard assessment is built into the program. The user can select various different configurations and see the effect on the risks to the aqueous environment. Where the risk is unacceptably high, different improvements can be tried very quickly to see what will be most effective and cost-effective. RASP is being developed to allow hazard analysis of further installations.

From the earliest assessments in 1970, gas dispersion calculations have been essential, particularly for assessing the consequences of toxic gas emissions. ICI has continued development of a gas dispersion program to allow the calculations of concentration to be combined with wind and weather data to assess risks.

DOES HAZARD ASSESSMENT HELP IN OPERATING PROCESSES SAFELY?

A survey of opinions of people within ICI in 1980 (11) showed that a large majority believed that hazard analysis had made process operations safer. This was borne out by the statistics on Fatal Accident Risk caused by process hazards, which had shown a reduction of about a factor of 4 between 1960-69 and 1970-79 and which has continued to reduce through the period since 1979. This improvement has resulted in avoiding about 50 fatalities in the company since 1970. Hazard assessment used in other areas of concern, such as toxic emissions and spills to the aqueous environment, has been shown to provide reliable advice on controlling risks to specified levels.

CONCLUSIONS

Hazard assessment has proven to be a valuable and effective means of making decisions to help improve the safety of processes effectively and cost-effectively. Despite the concerns about the accuracy of QRA, it is helping to ensure that efforts to improve processes are concentrated in those areas where the risk is greatest.

Resources for reducing the risks of processes will always be scarce, so the use of reliable hazard assessment and QRA to ensure that money is spent where it will most improve safety has much to commend it.

REFERENCES

1. David Jones, 1985. Nomenclature for Hazard and Risk Assessment in the Process Industries Second Edition. . Published by the Institution of Chemical Engineers

2. J G Sellers, 1976, Quantification of Toxic Gas Emission Hazards. I Chem E Symposium Series No. 47.

3. T A Kletz, 1992, HAZOP and HAZAN. Notes on the identification and assessment of hazards Loss Prevention Hazard Workshop Modules

4. C D Swann and M L Preston, 1995, 25 Years of HAZOPs, J. Loss Prev. Process Ind. Vol 8, No.6. 349-353

5. T A Kletz, 1971, Hazard Analysis - A Quantitative Approach to Safety, I Chem E Symposium Series No. 34: 75-81

6. R M Stewart, 1971, High Integrity Protective Systems, I Chem E Symposium Series No. 34: 99-104

7. Bulloch B C, April 1974, The development and Application of Quantitative Risk Criteria for Chemical Processes, Fifth Chemical Process Hazard Symposium, Institute of Chemical Engineers, Manchester, England

8. Gibson S B, Design of new chemical plants using hazard analysis. I Chem E Symposium Series, No. 47.

9. Gibson S B, July 1977, Major Hazards - should they be prevented at all costs?, XXIII Annual Meeting of the Institute of Management Sciences

10. Illidge J T and Wolstenholme J, 1978, Hazards of Oxyhydrochlorination, Loss Prevention, Vol 12: 111-117

11. Illidge J T, October 1983, Hazard Studies and Hazard Analysis 10 years on, Loss Prevention Bulletin, issue 53, 1-6

Index

A

accident database	133, 181
Accident database: Capturing corporate memory, The, Powell-Price, M., Bond, J. and Mellin, B.	133
Ali, M.W. (see Wilday, A.J.)	475
Anderson, M., European state-of-the-art research: Integrating technical and management/organisational factors in major hazard risk assessment	209
Andrews, G.E. (see Gardner, C.L.)	279
Ansell, J.C., Mullins, J. and Voke, R., The impact of the new Seveso II (COMAH) regulations on industry	123
Application of case-based reasoning to safety evaluation of process configuration, Heikkilä, A-M., Koiranen, T.K. and Hurme, M.	461
Ashurst, J.A.S. (see Facer, R.I.)	447
assessment	11, 41, 515
Assessment of COMAH safety reports: Emergency response criteria, McDonald, K.K.	41
Assessment of the predictive aspects of COMAH safety reports, Welsh, S.	65
Assessment of technical aspects of COMAH safety reports, The, Evans, R.F.	27
Assessment of the thermal and toxic effects of chemical and pesticide pool fires based on experimental data obtained using the Tewarson apparatus, Costa, C. Treand, G. and Gustin, J-L.	433
ATEX 100a	221
Atkinson, C.J., A qualitative approach to criticality in the allocation of maintenance priorities to manufacturing plant	195
audit	209

B

Beale, C.J., A methodology for assessing and minimising the risks associated with firewater run-off on older manufacturing plants	167
Bennett, G.R., Middleton, M.L. and Topalis, P., The use of risk-based assessment techniques to optimise inspection regimes	503
Blackmore, E., Meeting the demands of the regulator and litigator: An international approach	101
Bond, J. (see Powell-Price, M.)	133

Bours, R. (see De Vries, S.)	221
Brindley, J., Griffiths, J.F., Hafiz, N., McIntosh, A.C. and Zhang, J., Criteria for autoignition of combustible fluids in insulation materials	417
Britton, T.J., The regulators approach to assessing COMAH safety reports	11
Broadbent, J.K. and O'Donoghue, T., Operational safety reviews	143
building wakes	489
buildings	293
bursting discs	305
business risk	101

C

CARMAN: A systematic approach to risk reduction by improving compliance to procedures, Embrey, D.	153
case-based reasoning	133, 461
Categorization of risk: Good, bad or indifferent, site risk grading, Clarke, P.	515
CBA	447
CFD	293, 345, 489
chemical reaction hazards	373
chemical yield	433
Chung, P.W.H. (see Iliffe, R.E.)	181
Clarke, P., Categorization of risk: Good, bad or indifferent, site risk grading	515
COMAH regulations	95, 123
COMAH safety reports	65
combustion	345
combustion data	433
combustion efficiency	433
compact heat exchanger	405
competency	257
complex terrain	293
compliance programme	249
computer-based training	257
computer integration	181
consequence	503
continuous process	405
Control of major accident hazards regulations 1999 (COMAH), The, MacDonald, G. And Varney, L.	1
Cooke, P. (see Cooper, S.)	359
Cooper, S., Explosion venting: The predicted effects of inertia	305
Cooper, S. and Cooke, P., Suppression of high violence dust explosions using non-pressurised systems	359
corporate memory	133

corrosion	503
Costa, C., Treand, G. and Gustin, J-L., Assessment of the thermal and toxic effects of chemical and pesticide pool fires based on experimental data obtained using the Tewarson apparatus	433
criteria	95
Criteria for autoignition of combustible fluids in insulation materials, Brindley, J., Griffiths, J.F., Hafiz, N., McIntosh, A.C. and Zhang, J.	417
criticality	195

D

Dangers of grating floors: Dispersion and explosion, The, Holdo, A.E., Munday, G. and Spalding, D.B.	345
De Vries, S., Van den Schoor, M. and Bours, R., Design for safety applying IEC 6-1508 "from the manufacturer's point of view"	221
Deaves, D.M. (see Lines, I.G.)	489
deflagration	321
Design for safety applying IEC 6-1508 "from the manufacturer's point of view", De Vries, S., Van den Schoor, M. and Bours, R.	221
detonation	321
dispersion	293, 345
distance learning	257
domain models	133
doors	305

E

E/E/PES	221
embedded chips	249
Embrey, D., CARMAN: A systematic approach to risk reduction by improving compliance to procedures	153
emergency planning	123
emergency response	41
environment	475
European state-of-the-art research: Integrating technical and management/organisational factors in major hazard risk assessment, Anderson, M.	209
Evaluation of CFD modelling of gas dispersion near buildings and complex terrain, Hall, R.C.	293
Evans, R.F., The assessment of technical aspects of COMAH safety reports	27
explosion	279, 305, 359
Explosion venting: The predicted effects of inertia, Cooper, S.	305

F

Facer, R.I., Ashurst, J.A.S. and Lee, K.A., Top level risk study - A cost effective quantified risk assessment	447
Fernando, D., Information technology and training in safety	257
fire	167, 321
firewater runoff	167
flame quenching	279
flammable gas clouds	265
flash fires	265
flow	345

G

Gardner, C.L., Phylaktou, H. and Andrews, G.E., Turbulent Reynolds number and turbulent-flame-quenching influences on explosion severity with implications for explosion scaling	279
gas dispersion	489
gas generator	359
gratings	345
Green, A. (see Wood, M.)	405
Griffiths, J.F. (see Brindley, J.)	417
guidelines	305
Gustin, J-L. and Laganier, F., Understanding vinyl acetate polymerization accidents	387
Gustin, J-L. (see Costa, C.)	433

H

Hafiz, N. (see Brindley, J.)	417
Hall, R.C., Evaluation of CFD modelling of gas dispersion near buildings and complex terrain	293
Hall, R.C. (see Lines, I.G.)	489
Hamilton, I., Industry experience from the pilot exercise	95
HAZAN	527
hazard analysis	527
hazard assessment	527
Heikkilä, A-M., Koiranen, T.K. and Hurme, M., Application of case-based reasoning to safety evaluation of process configuration	461
high violence	359
Holdo, A.E., Munday, G. and Spalding, D.B., The dangers of grating floors: Dispersion and explosion	345

HSE management	235
Hunt, P.J., VOC abatement and vent collection systems	321
Hurme, M. (see Heikkila, A-M.)	461

I

ignition criteria	417
Iliffe, R.E., Chung, P.W.H. and Kletz, T.A., More effective permit- to-work systems	181
Illidge, J.T., Preston, M.L. and King, A.G., Thirty years of quantifying hazards	527
Impact of the new Seveso II (COMAH) regulations on industry, The, Ansell, J.C., Mullins, J. and Voke, R.	123
Incorporation of building wake effects into quantified risk assessments, Lines, I.G., Deaves, D.M. and Hall, R.C.	489
Index method for cost-effective assessment of risk to the environment from accidental releases, Wilday, A.J., Ali, M.W. and Wu, Y.	475
Industry experience from the pilot exercise, Hamilton, I.	95
inertia	305
Information technology and training in safety, Fernando, D.	257
inherent safety	461
inspection planning	503
intranet	257

K

King, A.G. (see Illidge, J.T.)	527
Kletz, T.A. (see Iliffe, R.E.)	181
Koiranen, T.K. (see Heikkila, A-M.)	461

L

Laganier, F. (see Gustin, J-L.)	387
lagging fires	417
Lambert, P.G., Managing hazards and risks in fine chemical and peroxygen operations	373
Lardner, R., Safety implications of self-managed teams	109
Lee, K.A. (see Facer, R.I.)	447
legislation	1, 27, 123
lessons learned	133
Lidstone, A., THESIS: The health environment and safety information system keeping the management system "live" and reaching the workforce	235
life cycle	221

likelihood	503
Lines, I.G., Deaves, D.M. and Hall, R.C., Incorporation of building wake effects into quantified risk assessments	489
litigator	101
loss control	373

M

MacDonald, G., The control of major accident hazards regulations 1999 (COMAH)	1
McDonald, K.K., Assessment of COMAH safety reports: Emergency response criteria	41
McIntosh, A.C. (see Brindley, J.)	417
Maddison, T. (see Rew, P.J.)	265
maintenance	195
major hazards	1, 11, 27
	41, 53, 475
management	209
Managing hazards and risks in fine chemical and peroxygen operations, Lambert, P.G.	373
MAPP	53
mass	305
Meeting the demands of the regulator and litigator: An international approach, Blackmore, E.	101
Mellin, B. (see Powell-Price, M.)	133
Methodological approach to process intensification, A, Wood, M. and Green, A.	405
Methodology for assessing and minimising the risks associated with firewater run-off on older manufacturing plants, A, Beale, C.J.	167
Middleton, M.L. (see Bennett, G.R.)	503
millennium bug	249
More effective permit-to-work systems, Iliffe, R.E., Chung, P.W.H. and Kletz, T.A.	181
Mullins, J. (see Ansell, J.C.)	123
multimedia	257
Munday, G. (see Holdo, A.E.)	345

N

non-pressurised	359

O

O'Donoghue, T. (see Broadbent, J.K.)	143
onshore/offshore project	109

Operational safety reviews, Broadbent, J.K. and O'Donoghue, T.	143
option studies	447
organisational	153, 209

P

participation	153
permit-to-work	181
Phylaktou, H. (see Gardner, C.L.)	279
pilot exercise	95
polymerization accidents	387
pool fires	433
Powell-Price, M., Bond, J. and Mellin, B., The accident database: Capturing corporate memory	133
predictive criteria	65
Preston, M.L. (see Illidge, J.T.)	527
prioritisation	143
priority	195
procedures	153
process configuration	461
process intensification	405
public image	101

Q

QRA	209, 489
Qualitative approach to criticality in the allocation of maintenance priorities to manufacturing plant, A, Atkinson, C.J.	195
quantified risk assessment	265

R

reactivity	321
regulator	101
Regulators approach to assessing COMAH safety reports, The, Britton, T.J.	11
relational databases	235
Rew, P.J., Spencer, H. and Maddison, T., The sensitivity of risk assessment of flash fire events to modelling assumptions	265
risk	235, 321, 475, 515
risk assessment	167, 209, 447
risk-based inspection	503
risk index	475

risk perception	101
risk ranking	143
risk reduction	447

S

safety	321, 345
Safety implications of self-managed teams, Lardner, R.	109
safety integrity	221
Safety issues and the year 2000, Storey, R.	249
safety management	447
Safety management system assessment criteria, Snowball, D.	53
safety reports	1, 11, 27, 41, 53, 95
scaling	279
self-managing teams	109
Sensitivity of risk assessment of flash fire events to modelling assumptions, The, Rew, P.J., Spencer, H. and Maddison, T.	265
Seveso II	1, 123
simpler	195
site risk grading	515
Snowball, D., Safety management system assessment criteria	53
software design	221
Spalding, D.B. (see Holdo, A.E.)	345
Spencer, H. (see Rew, P.J.)	265
spontaneous ignition	417
standards	101
static mixer	405
Storey, R., Safety issues and the year 2000	249
strategic	195
suppression	359
Suppression of high violence dust explosions using non-pressurised systems, Cooper, S. and Cooke, P.	359

T

task analysis	153
technical	209
Tewarson apparatus	433
THESIS: The health environment and safety information system — keeping the management system "live" and reaching the workforce, Lidstone, A.	235
Thirty years of quantifying hazards, Illidge, J.T., Preston, M.L. and King, A.G.	527

Top level risk study — A cost effective quantified risk assessment, Facer, R.I., Ashurst, J.A.S. and Lee, K.A.	447
Topalis, P. (see Bennett, G.R.)	503
training	153
Treand, G. (see Costa, C.)	433
Turbulent Reynolds number and turbulent-flame-quenching influences on explosion severity with implications for explosion scaling, Gardner, C.L., Phylaktou, H. and Andrews, G.E.	279

U

uncertainty	293
Understanding vinyl acetate polymerization accidents, Gustin, J-L. and Laganier, F.	387
Use of risk-based assessment techniques to optimise inspection regimes, The, Bennett, G.R., Middleton, M.L. and Topalis, P.	503

V

Van den Schoor, M. (see De Vries, S.)	221
vapour phase explosions	373
Varney, L. (see MacDonald, G.)	1
vent collection	321
venting	305
vinyl acetate	387
VOC abatement and vent collection systems, Hunt, P.J.	321
Voke, R. (see Ansell, J.C.)	123

W

Welsh, S., Assessment of the predictive aspects of COMAH safety reports	65
Wilday, A.J., Ali, M.W. and Wu, Y., Index method for cost-effective assessment of risk to the environment from accidental releases	475
Wood, M. and Green, A., A methodological approach to process intensification	405
Wu, Y. (see Wilday, A.J.)	475

Y

Y2K	249

Z

Zhang, J. (see Brindley, J.)	417